LAKE CHAD

MONOGRAPHIAE BIOLOGICAE

VOLUME 53

Series editor

H. J. DUMONT

1983 **Dr W. JUNK PUBLISHERS**
a member of the KLUWER ACADEMIC PUBLISHERS GROUP
THE HAGUE / BOSTON / LANCASTER

LAKE CHAD

Ecology and Productivity of a Shallow Tropical Ecosystem

Edited by

J.-P. CARMOUZE, J.-R. DURAND and C. LÉVÊQUE

1983 **Dr W. JUNK PUBLISHERS**
a member of the KLUWER ACADEMIC PUBLISHERS GROUP
THE HAGUE / BOSTON / LANCASTER

Distributors

for the United States and Canada: Kluwer Boston, Inc., 190 Old Derby Street, Hingham, MA 02043, USA
for all other countries: Kluwer Academic Publishers Group, Distribution Center, P.O.Box 322, 3300 AH Dordrecht, The Netherlands

Library of Congress Cataloging in Publication Data

Main entry under title:

Lake Chad.

   (Monographiae biologicae ; v. 53)
   Includes bibliographies and index.
   1. Lake ecology--Chad, Lake. 2. Biological
productivity--Chad, Lake. 3. Chad, Lake. I. Carmouze,
Jean-Pierre. II. Durand, Jean René. III. Lévêque, C.
IV. Series.
QP1.P37 vol.53 574s [574.5'26322'096743] 83-4288
[QH195.C46]
ISBN-13: 978-94-009-7268-1   e-ISBN-13: 978-94-009-7266-7
DOI: 10.1007/978-94-009-7266-7

Cover design: Max Velthuijs

Contents

Preface

In 1963 the director general of the 'Office de la Recherche Scientifique et Technique Outre-Mer' (ORSTOM) asked me to form a well structured team of hydrobiologists based at Fort-Lamy (now N'Djamena) where a laboratory could be built as a branch of the already existing ORSTOM Center. The International Biological Program, IBP, had at this time recommended an integrated study of ecosystems selected in proportion to their representation of a particular zone or ecological conditions. I took this opportunity to propose Lake Chad as the model of a tropical lake in a semi-arid climate as part of the activity of the section Productivity of freshwater communities (PF). With the efficient help of Dr B. Dussart a research program was established involving the establishment of a permanent team of ten researchers in Chad.

Apart from myself, ORSTOM only had three hydrobiologists: Mr R. Gras, Mr G. Loubens and Mr A. Iltis. These valuable researchers who had been working for the 'Centre Technique et Forestier Tropical' and had acquired considerable experience in African freshwaters, unhesitatingly accepted to join me at Fort-Lamy. It was necessary to enlarge this team with some young and enthusiastic collaborators, namely, Mr C. Lévêque, Mr C. Dejoux, Miss S. Duwat, Mr L. Lauzanne and Mr. J. R. Durand, later joined by Mr J. P. Carmouze and Mr J. Lemoalle.

At first we stayed briefly at the laboratory of the 'Institut de Médecine Vétérinaire et Tropicale' (IMVT) at Farcha until the construction of the hydrobiology laboratory at Fort-Lamy was completed. It is with pleasure that I pay respect not only to the scientific value of this first team which established itself by producing seven Ph.D's but especially to the good atmosphere and friendship established between the researchers and technicians at Fort-Lamy as well as in the field. Due to this exceptional atmosphere, there were only scientific problems to solve and the two years that I spent at Chad have left me with unforgettable memories.

With the installation of the team in its laboratory, opened in 1965 and fairly well equipped, the research program was well on its way and I returned to Paris. Here, with the help of Dr B. Dussart and until 1975, I supervised the scientific

Photo 1 The ORSTOM Hydrobiological team in 1966.

coordination of research and the recruitment of new scientists and technicians to ensure the maintenance or reinforcements of the first team.

This experience proved to be extremely beneficial for the training of men who have become highly qualified specialists as well as for the importance of their scientific results. The director general of ORSTOM took a certain risk by regrouping all his hydrobiologists at the same time on the same program. What appeared at first to be a challenge soon became a total success as the reader will find out for himself when reading this book.

It should, however, be recognized that the interest and the scientific scope of the study presented here partly depends on a happy conjuncture: the presence of a team of scientists having had the time to study the state of the lake called 'Normal Chad' before and during the time when years of drought led to the state of 'Lesser Chad'. This indeed appeared to be a cyclic phenomenon that had already occurred at the beginning of the century as shown by Tilho's observations, and which would recur again after a new period of high waters. However, a researcher would only have the chance to observe this once during his career. This complete team of hydrobiologists then, was able to follow in detail the lowering of the lake level. If this event had occurred ten years earlier, there

VIII

would have been no one to study it; if it had occurred ten years later, it would have passed unobserved as research in Chad would have been impossible.

When I first became interested in Lake Chad, I was struck by two factors in particular. In spite of the approximately 20 000 km² of area then known, Hutchinson had not considered it worthy to be included in the list of African lakes considered in his classical treatise of Limnology. Furthermore as far as the fish fauna was known, there were no true lacustrine species present. *Alestes dageti*, described as endemic to the lake by Blache and Miton, only appeared to me as an ecophenotypic form of *Alestes nurse* characterized by its dwarfism, a poorly known deterministic phenomenon, although it is relatively frequent in some fish populations. In 1967, I wrote that 'Ecologically Lake Chad is closer to the flood zones of large tropical rivers than true lakes ... it thus appears probable that a slight worsening of climate could involve a complete drying of the lake'. The exceptional drought of the years 1972–73 which was dramatic for the Sahelian populations showed that it did not behave as a simple system, as drought first affected the north, i.e. the deepest part. It will be interesting to read how the major types of population changed over the course of the drying phase.

Finally, it is necessary to recall that the fisheries in Lake Chad were developing around the sixties in relation to the introduction of nylon gill nets and especially the development of transport facilities to carry the dried and smoked fish towards Nigeria and the southern territories. Now, paradoxically, the driest years turned the production rocketing to 220 000 tons in 1974, causing a huge concentration of fishermen in the lake. In addition the productivity of these tropical waters and the resilience of their species was seen to be much higher than provisionally estimated by comparison with other environments.

Following several notes and papers published in specialized journals, the summary of 15 years of research presented here in an easily accessible form appeared welcome to me. It is not premature as the field-work had to be interrupted in 1979 nor is it too late because the problems of management and the improvement of the Chad basin are not yet solved.

I agreed and accepted the invitation to preface this scientific work with real pleasure for it is a factual document. I would like to address my personal thanks to the authors for their collaboration as well as to the editors who agreed to publish it and distribute it to the international scientific community. I am sure that it will obtain from the latter all the success that it deserves.

Jacques Daget
Professor, National Museum
of Natural History, Paris

Foreword

Lake Chad is a large, endorheic, shallow basin south of the Sahel zone. Although it is subject to important annual variations of area and volume, it is one of the largest lakes of the African continent, following those in the Rift valley (Victoria, Tanganiyka and Malawi).

The biotopes of Lake Chad are similar to those found in lakes in the central Niger delta, southern Sudan, Lake Chilwa, Okavango delta and a number of other less important aquatic ecosystems. They are characterized primarily by the instability of hydrological conditions and the abundance of macrophytes (*Cyperus, Typha* and *Phragmites*, in particular). Thus the interest of studies conducted on Lake Chad extends beyond regional limits because many problems are common to all these regions even if each has its own specificity.

Since the beginning, the research program on Lake Chad was part of the French contribution to the International Biological Program, entitled 'Study of a Tropical Lake in a Semi-Arid Climate'. Its principal objective was to describe the lake hydroclimate and evaluate the biomass and production of the main groups of organisms in order to establish a balance of matter and energy and of their cycles on the different trophic levels. In addition to the fundamental value of this research for biological phenomena in a climatic zone that was fairly unknown until now, this study could also have economic consequences for the development projects for Chad basin, and particularly for the fisheries. Indeed Lake Chad is a very productive environment due to its morphology. Its average fisheries catch, estimated at more than 100 000 tons per year, provides an important economic activity for the nearby communities.

Between 1964 and 1978, a team of several limnologists worked at the ORSTOM Center at N'Djamena, where a modern laboratory of 220 m² was built in 1965. The first studies were limited to the east zone of the lake where it was possible to make preliminary tests for sampling methods and analysis with the Nausicaa, a 10 m long boat specially adapted to these shallow environments.

It was soon apparent that a general view of the lake was essential for the proper interpretation of several observed phenomena. Thus, a study of large

Photo 2 The ORSTOM Hydrobiological Laboratory in N'Djamena.

ecological zones of Lake Chad was designed providing the first synthesis of the physico-chemical characteristics of the lake and the qualitative and quantitative distribution of diverse plant and animal populations (Carmouze et al., 1972). These studies pointed out the heterogeneity of the ecosystem with considerable ecological and demographic differences existing between the north and south basins, as well as in the types of natural zones (open water or archipelago). Consequently, it was necessary to have a fair distribution of the sampling stations in the major biotopes in order to obtain a general view of the ecosystem dynamics. It was not possible to investigate all areas due to the size of the environment (20 000 km^2) and the distances to be covered. The laboratory boat 'Jacques Daget' was placed at our disposal in 1970 and partially solved this problem by allowing fairly safe navigation in the open water.

The research was also complicated by the rapid fall of the lake level which started in 1972 and involved important modifications in the biotopes and the populations, as well as the drying up of a large part of the lake (the north basin) in 1974. Priority was then given to the study of the effects of the drying up on the development of the environment and the populations. This brief history largely explains the chronology of the studies as well as the orientation and choices that could have been made in the course of this study.

In the various chapters of this book, a distinction between the 'Normal Chad'

Photo 3 The research boat 'J. DAGET'.

period from 1964 to 1971 and the period of drying up that began in 1972 will be made. The former corresponds to a relatively steady state, although the lowering of the lake's level had already become noticeable. During the latter period the environment and populations were greatly modified. It should also be noted that a large part of this work, on the various topics, was carried out between 1968 and 1971. It was only during this period that an attempt could be made to establish a balance of the functioning of the ecosystem in relatively stable conditions.

The hydrobiological research carried out by the team at N'Djamena was made possible by several scientists who stayed there for a number of years. We should mention J. P. Carmouze, J. M. Chantraine and J. Lemoalle (chemistry and primary production), A. Iltis (phytoplankton), R. Gras and L. Saint-Jean (zooplankton), C. Dejoux, L. Lauzanne and C. Lévêque (benthos and trophic relations), V. Bénech, J. R. Durand, J. Franc, G. Loubens, J. Quensière, A. Stauch and G. Vidy (fish and fisheries), without forgetting J. J. Troubat, who was responsible for navigation requirements. To this team must be added the participation of the young French scientists and foreigners who stayed in Chad for a training period or as volunteers for the National Service: Mr Cambrony, Mr and Mrs Lek, Mr Matelet, Mr Mok, Mr Srinn (ichthyology and trophic relations), J. Vieillard (ornithology), Mr Bacque and Mr Masson (inverte-

XIII

Photo 4 The research boat 'Nausicaa'.

brates). Finally, the French and foreign scientists who participated in the systematic study of the lake organisms were: Mr Compère (algae), Mr Dragesco (protozoa), Mr Léonard (macrophytes), Mr Pourriot (rotifers) and Mrs Rey (cladocera).

Beside the hydrobiological research, a number of researchers belonging to other disciplines made an equally large contribution to the present and past knowledge of this ecosystem. In particular we cite, B. Dupont, J. Y. Gac, M. Servant and S. Servant (geology), B. Billon, A. Bouchardeau, P. Carre, A. Chouret, R. Lefèvre, M. Roche (hydrology), G. Maglione (hydrogeology), C. Riou (climatology), P. Couty (economics), C. Cheverry (pedology).

It should also be noted that a small team of researchers, including A. J. and J. Hopson, was established at the same time at Malamfatori station on the Nigerian side, and worked primarily on the ichthyology of the north basin. We were able to establish profitable contacts with them and exchange information.

The Lake Chad study represents a large undertaking, in consideration of the number and diversity of scientists who worked and the importance of the resource input. To this day, these studies constitute one of the rare examples of a multidisciplinary approach to a continental aquatic environment in tropical Africa.

In particular, we would like to thank Mr Daget and Mr Dussart who initiated the Hydrobiological section of ORSTOM and the research program at Lake Chad. Their experience, competence and advice provided the impetus for research at a time when the team was mainly composed of young and inexperienced researchers. We also thank the General Director of ORSTOM who financed the entire project and provided us with excellent conditions throughout the project.

Finally, we should not forget the Chadian technical team, who participated largely in the field and laboratory work and were very competent in dealing with the difficult material and climatic conditions. Without them it would not have been possible to carry out this comprehensive study in such a short time.

The Scientific Editors
ORSTOM
24, rue Bayard
75008 Paris, France

Historical background

Jean-Pierre Carmouze

The first available data about the hydrography of Central Africa date back to the 14th century. At that time, a Moor born at Granada in 1483 and known by the name of Leo Africanus, first spoke of a great lake situated in a desert region and fed by underground water coming from an arm of the Nile. He suggested that the River Niger originated from this lake.

In 1632, a map drawn by a Dutch geographer, Guillaume Blavew, from the description made by Leo Africanus and new data, pointed out a *Borno lacus*, situated approximately at the site of present Lake Chad and fed by a river from the southern hemisphere, north of Lake Zaire. The Niger has its source in the western part of this lake and moves along an east–west line. In this document, the Shari, Yoobé and Niger rivers form a part of the *Niger Fluvius*.

At the beginning of the 19th century, Ali Bey, the son of the Prince of the Abassides argued that a large inland sea must exist in the centre of Africa. He thought that the Niger must disappear into this 'African Caspian' and he gathered numerous facts to support this hypothesis. For his part, the Englishman, Jackson, also noticed the observations made by travellers which corroborated the existence of a connection between the Nile from Egypt and the Nile from the Sudan (Niger). This connection took place in a region called Bahr Kulla. This name is comparable with that of Koulou 'great mass of water' by which the Kanembous describe present Lake Chad. The two authors came to the same conclusion concerning the situation of this inland sea: it would be situated at a 15-day walk east of Tombouctou.

At the same time, a German, Frederic Hornemann, appears to be the first European who heard the name of Zad during his travels in Fezzan and Egypt (1797–1800). It was not a lake but a very wide river (at an 8-hour journey). He also learnt about the existence of the Boudoumas, considered as primitive people and always found in the middle of this river. According to Hornemann, the Zad was nothing more than the name given to the Niger by the population of Bornou, and flowed into the Nile south of Darfou.

So, at the beginning of the 19th century, the existence of a great lake in

J.-P. Carmouze et al. (eds.) Lake Chad
© *1983, Dr W. Junk Publishers, The Hague/Boston/Lancaster*
ISBN 978-94-009-7268-1

Central Africa was still not certain. On the other hand, there was the prevailing notion of the Zad as a river connecting the Niger with the Nile.

But the idea of an 'African Caspian' aroused imagination and stirred up much enthusiasm.

Ritchie from England, and his colleague Lyon left Tripoli for Central Africa in March 1819, but Ritchie died at Mourzouk and Lyon did not go much further. His notes inform us that the Zad described by Hornemann had to be a lake and not a river.

The nature of Lake Chad became more clear for geographers, but the first Europeans to reach Lake Chad from Tripoli in 1822, by following the western side to Kouka, were three English travellers: Denham, Clapperton, and Oudney. Denham passed around the southern part of the lake, crossing the Shari downstream at Showy (Fig. 1), passed the foot of Hadjer el Hamis and reached Tangalia in the dried up groove of Bahr el Ghazal. He considered Lake Chad to be an overflow of the Nile. Then he went down the Shari to its mouth, and entered the lake on February 3rd, 1824. Upset by strong waves which filled up the boats, he sought refuge in the delta after covering a distance of about two miles, without even being able to see the nearest islands inhabited

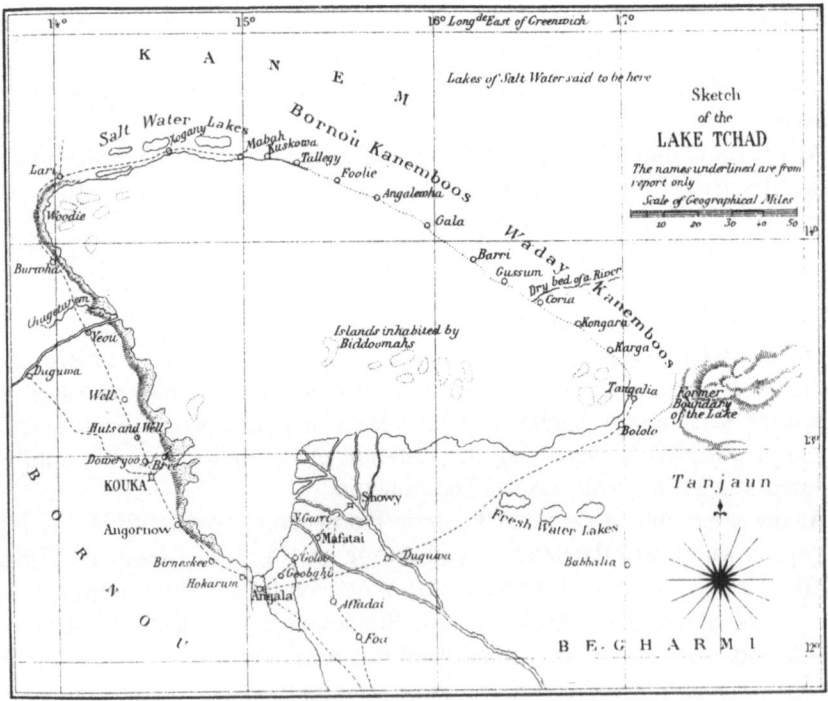

Fig. 1   Map of Lake Chad according to Denham (1824).

2

by the Boudoumas, whose existence was known to him. After resting at Kouka, he again set out in May, with the aim of going around the lake, with a Bornouan expedition army fighting against the Assalé Arabs in the region of Kanem. He was stopped at Tangalia near the mouth of Bahr el Ghazal after Bornouan troops suffered a defeat and he had to give up the pursuit of his journey northwards. He persisted in his determination to survey the periphery of the lake and he again set out northwards on August 16th, 1824. At Woodie, he again met the Bornouan troops whom he had accompanied some months earlier, now on their way back from Kanem, tired out and defeated. Yet he pursued his journey, but he had to turn back at Kiskaoua (Kuskowa) because of lack of security in this area. Finally, he reached Tripoli on January 26th, 1825. Despite a 19-month trip around Lake Chad, Denham could not draw a map of the whole contour. He left an unfinished map (Fig. 1). No offshore surveying was done and islands were not seen even in the distance.

So, on March 30th, 1850, when the Englishman Richardson left Tripoli with the Germans Barth and Overweg, he took a boat called 'The Lord Palmerston' which was specially intended to sail on Lake Chad. This boat was divided into four parts in order to be carried on a camel. Richardson, exhausted by the climate, died in Bornou before seeing the lake. Barth took over and arrived at Kouka on April 2nd, 1851. When he set out for Adamaoua, Overweg undertook to survey the lake aboard the 'Lord Palmerston'. He visited several islands after a two-month journey (Fig. 2). Then he went with Barth, who wanted to survey the regions surrounding Lake Chad, up to Bahr el Ghazal in order to resume Denham's project. But they also encountered the hostility of the population and returned to Kouka. Then, Overweg undertook to study the Komadougou Yobé and he specified that it was a tributary and not an effluent. Completely exhausted he returned to Kouka, and was taken to Madouari on the shore of the lake where he died on September 27th, 1852 without being able to give a detailed report about his travels through the lake. However, 'he can boast of having been the first European who sailed and visited part of the Archipelago' (Tilho 1910).

Barth continued to survey the western regions of Bornou alone. When he returned to the shores of the lake two years later, he noted that the shore had suffered profound changes from Baroua to Nguigmi as a result of the flood of the lake. He returned to London on September 6th, 1855 after travelling through the Black continent for five and a half years, two of which were spent in the regions near the lake without going around it. As Overweg's notes could not be completed, several aspects of Lake Chad remained unknown.

Vogel, who was sent to meet Barth, reached Kouka in early 1854 with the sole task of exploring the lake. His astronomical observations made through the sextant were the only basic point that were used by the cartographers of these regions for fourteen years. He was murdered during a journey through the Ouadai in February 1856.

Gerhard Rohlfs, also a German, went on a long journey through Africa from

Fig. 2   Map of Lake Chad according to Barth (1851).

1865 to 1867. He reached Nguigmi on July 14th, 1866. His observations relate only to a small number of points on the northern and western sides. According to him, at that time the lake had reached a level higher than those which were previously observed. He thought that part of the water which did not evaporate flowed underground through the Bahr el Ghazal.

Nachtigal, a German doctor, undertook a journey through Central Africa in 1869. He arrived at Nguigmi on June 28th, 1870 where he caught his first glimpse of the lake. He stayed at Kouka for one year, then visited the northern side of the lake on his way to Kanem and Borkou. On a trip through Baguirmi,

4

he passed along the southwestern side. Four years later, he returned to Europe through the Ouadai and the Nile, and so had the opportunity of going through the Bahr el Ghazal. He could not sail on the lake or visit its archipelago, but he brought back considerable information about the geography of Chad, its islands and its inhabitants (Fig. 3).

On April 10th, 1891, Monteil, a French lieutenant colonel, reached Kouka (or Koukaoua) by following the road towards the west (Saint-Louis–Karo) for the first time. Meanwhile, Gentil succeeded in reaching the lake from the south aboard a small steamer called the 'Léon Blot' which sailed from the basin of the Oubangui to that of the Shari. But, there was a lack of wood! Moreover, Rabah, the leader of the Lower Shari, forced Gentil to go back to the Shari. Seventy-five years after Denham's first attempts, it was once more impossible to sail far from the lake and to visit the archipelago. But this period of dangerous raids by the European explorers was coming to an end. France decided to occupy the Chad basin and sent three missions there, led respectively by Gentil, Joalland-Meynier and Fourreau-Lamy. On April 22nd, 1900, the powerful Rabah, the 'Emir of the believers' was overthrown at Kousseri and the exploration of the environment resumed.

Fig. 3   Map of Lake Chad according to Nachtigal (1870–1873).

At that time, Lieutenant Colonel Destanave ordered scientific explorations to be made in the whole archipelago, from the mouth of the Shari to the delta of Bahr el Ghazal and on the southwestern, western and northwestern sides of the lake which were still not well known. They showed the existence of an archipelago including several hundred islands, occupied by about 45 000 inhabitants and many droves of oxen. Consequently, a fairly complete map of the archipelago was drawn for the first time.

Commander Largeau carried on similar explorations. A series of cross routes provided information about the area and the depth of the open waters of the lake. The increasing dryness of this period and the resulting development of vegetation led them to consider Lake Chad as more of an immense marshland than a lake.

At the end of 1903, Delevoye came from the mouth of the Shari sailing alone through the open waters aboard a big steel barge called the 'Benoît Garnier', starting from the basin of the Oubangui (as was the case for the 'Léon Blot' twelve years earlier).

Finally, the Franco English Commission led by Captain Tilho (mission Moll) arrived at Koukaoua in January 1904. They were in charge of delimiting the borders between Niger and Chad and established the first precise map of the whole region. The circumstances did not allow him to study the southern shore. However, he did identify the zone of shallows spreading from the archipelago to the open waters and that separating the southern and the northern basins (Fig. 4).

The Anglo-German Commission which was in charge of delimiting Yola-Lake Chad studied the still poorly known southwestern part of the lake. Therefore the southern side of the lake which had been positioned 15 kilometers too far south by Vogel was modified in its position.

Although the geographic situation of the lake was finally defined, the lake continued to astonish the travellers and scientists who ventured into the area, because of the constant changes in its appearance and zones which occurred in relation to the variations in water level.

Explorations continued, and in 1904 and 1905, Boyd Alexander, leading an English expedition visited the mouth of the Komadougou Yobé and the reed islands of the northern archipelago. He gave up attempting to reach the island inhabited permanently because the lowering of water made navigation very difficult. He then visited the archipelago of the southern basin after dismounting and carrying his boats up to Seyorom and ended his journey after reaching the mouth of the Shari. He could only make very limited explorations since he arrived during the period of very low water.

Meanwhile, the French officers from the military territory of Chad continued

Fig. 4 Map of Lake Chad and its archipelago according to Tilho (1904).

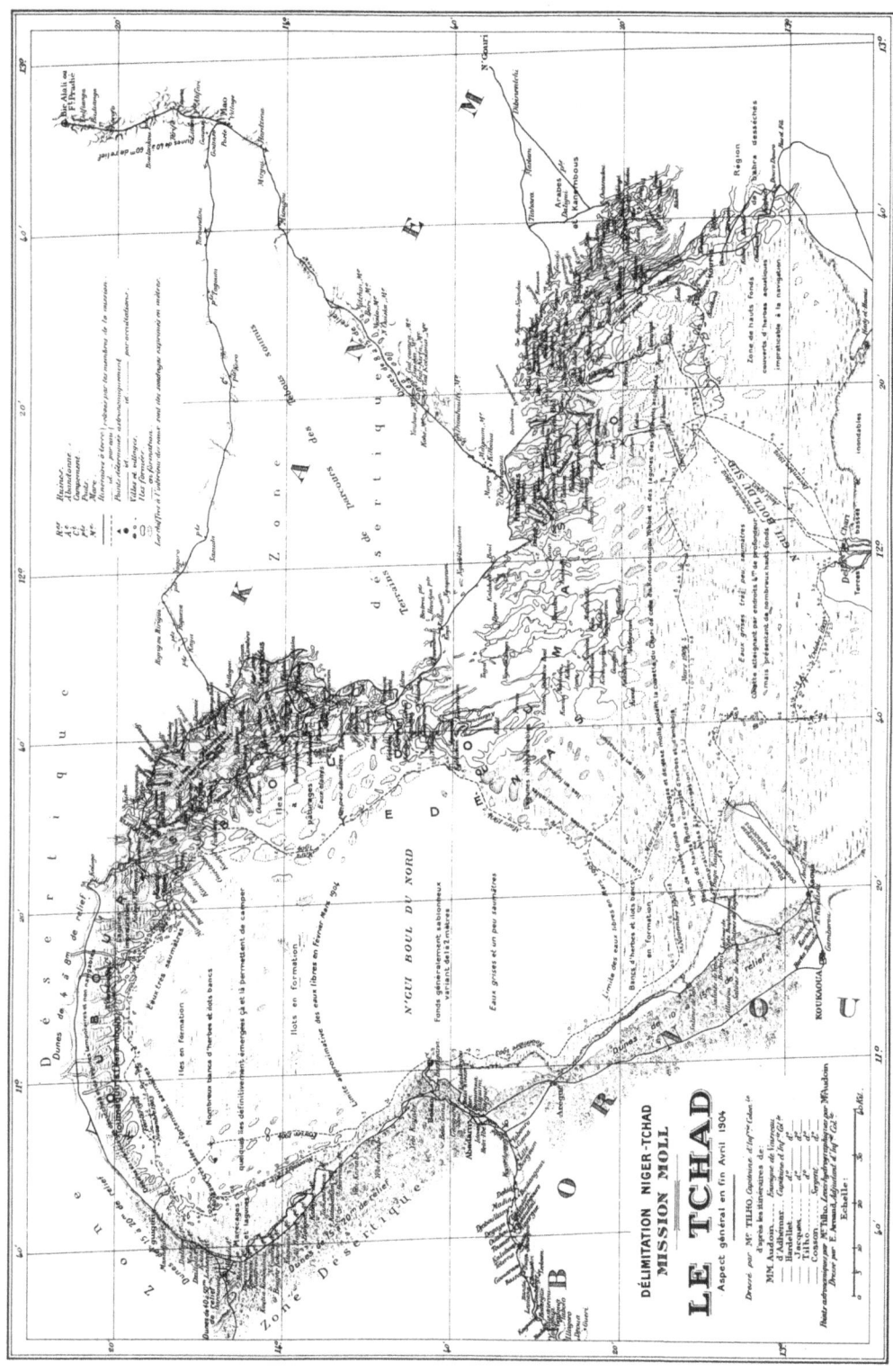

DÉLIMITATION NIGER-TCHAD
MISSION MOLL
LE TCHAD
Aspect général en fin Avril 1904

to study the lake. In particular, Captain Freydenberg had the opportunity of visiting the whole northern basin of the lake, in 1905, when the lowering of the lake level was considerable. He made very important observations about the flora, the fauna and the inhabitants of the whole lake and its islands and presented them in a thesis submitted to the Faculté des Sciences in Paris in 1908.

Lake Chad which had remained mysterious for a long time was rather well known when on August 26th, 1906, the second mission led by Tilho arrived with the aim of delimiting on land the frontier between French and English territories in Central Africa. This delimitation was made in consultation with a British mission led by Commander in Chief R.P.O. Shee. The scientific documents from Tilho's mission, which were gathered over two years were decisive elements in the knowledge of the lake and its entire basin. (Tilho 1910). The works included not only the survey of the lake and its eastern regions that were still poorly known but also studies on the variations in levels of the lake and neighbouring depressions, the hydrology of the main tributaries of the lake, the climatology and the meteorology of the region, the magnetism, the history and the ethnography of the populations. Numerous samples were gathered in all the scientific fields, especially malacology and ichthyology. The malaco-logical fauna was then studied by L. Germain (Tilho 1910).

During this period of exploration, the explorers were generally puzzled by two problems: the role of Bahr el Ghazal in the hydrology of the lake and the low salinity of the water in an environment situated in the sub-arid zone.

Some, such as Barth, thought that the Bahr et Ghazal was an effluent of Lake Chad, while others, such as Nachtigal, thought that it was a tributary of the lake. During his second mission (1906–1909), Tilho confirmed that it was an effluent after making surveys in the region under consideration.

As far as the low salinity of waters is concerned, the following two phenomena are often quoted:

— a freshening of lake waters in outlying ponds containing natron (Huart *in* Destenave 1903).
— a seepage of water into the sands of Kanem and then an underground run-off in the old arm of Bahr el Ghazal (Truffert *in* Destenave 1903).

Lahache and Marne (*in* Tilho 1910) confirmed the low salinity of water which was revealed previously through the 'taste test'. But they were in favour of the second explanation. We shall see in Chapter 5 that these two phenomena play an important part in maintaining the low salinity of the water.

In fact, the main concern at the beginning of this century was elsewhere. Was the lake not disappearing? As a matter of fact, explorers such as Denham, Barth, Rolfs and Nachtigal had observed a 'great lake' with swampy zones situated on a coastal margin of 5 to 10 km since the discovery of the lake in 1823 and throughout the 19th century. In 1854, Vogel saw a considerable flood invading bordering towns such as Ngornou with 30 000 inhabitants. On the contrary, from 1903 to 1915, the lake turned gradually into a marshland, unfit for navigation. In 1908, the water surface was reduced by a half as compared with that of 1903. For this reason those who forecast the drying up of the lake were then numerous.

It seemed that such changes could be accounted for only by a final or temporary modification in the climate and the water supplies. In order to evaluate the importance of the variations in the water level and to reveal their possible frequency, during his third journey to Chad, Tilho used the measurements from his previous journey to Bol, on the eastern coast. Apart from measurements of the water level, they included measurements of temperatures, pressures, winds and rainfalls taken from November 1912 to September 1919. During this period, with a rise in the water level, the idea of a lake drying up for a short time was replaced by that of periodicity with large variations in the water level. At the end of this study, Tilho (1928) defined three main hydrological situations which characterized the lake: a 'Greater Chad' of 25 000 km^2 (in the 19th century), a 'Chad normal' of 18 000 km^2 (1916–1919) and a 'Lesser Chad' (from 1904 to 1915). He also tried to compare the fluctuations of the lake with those of the glaciers of Bossons and Grindewald. He inferred that the fluctuations in lake levels would follow those of the alpine glaciers within one or two years and while referring to the periodicity of the Bossons floods, he thought that the next 'Greater Chad' would occur about 1952–57 and 1993–97, the next 'Lesser Chad' at about 1940–50 and 1975–85. He was not far from the mark as far as the levels are verifiable today. In any case, he showed that there was no fear of a future drying up of Lake Chad from the climatological point of view. On the contrary, he revealed another danger which could threaten the existence of the lake itself, as a result of modification in the regime of its main tributaries. Tilho thought that a geological danger would originate from the capture of the Logone and afterwards the Shari by the Benoué which is an affluent of the Niger (Tilho 1926). If the number of his publications on this subject since that time are considered, his fear at such a capture can be evaluated, and he felt it was urgent to consider arrangements suggested by him in order to avoid this (Tilho 1926, 1927, 1928, 1932, 1934, 1935, 1936 and 1937).

Tilho's mission ended a period of twenty years that provided valuable information about Lake Chad. Then nearly half a century elapsed before scientific works resumed. In fact, the scientific commission about the Logone and the Chad was created at Fort-Lamy (now N'Djamena) in 1954. A base including a meteorological station manned by scientists from ORSTOM. (Lefevre, Bouchardeau and Guichard) was set up at Bol in 1955. At the same time, hydrological studies were undertaken in the region of Bol where polders were created by scientists from INRA (Pirard and Schneider).

Finally, in 1964, a hydrobiological team was formed under the leadership of J. Daget and B. Dussart.

References

Destenave, L., 1903. Le lac Tchad. 1° partie: le lac, les affluents, les archipels. Rev. gén. Sci., Paris: 649–662.

Destenave, L., 1903. Le lac Tchad. 2° partie: les habitants, la faune, la flore. Rev. gén. Sci, Paris: 717–727.

Freydenberg, L., 1908. Le Tchad et le bassin du Chari. Thèse, Paris, 187 pp.

Lenfant, E. A., 1904. De l'Atlantique au lac Tchad par le Niger et la Bénoué. Géographie 9: 321.

Tilho, J., 1910. Documents scientifiques de la Mission Tilho (1906–1908). Imp. Nat., Paris, 3 vols., 1 atlas.

Tilho, J., 1926. Sur une aggravation du danger de capture par le Niger des principaux affluents du Tchad. C. r. Ac. Sci. Paris 182: 1063–1065.

Tilho, J., 1927. Devons-nous sauver le lac Tchad? Rev. sci. Paris 15: 454–463, 4 figs.

Tilho, J., 1928. Variations et disparition possible du Tchad. Ann. Géogr. Paris 37: 238–260, 5 figs.

Tilho, J., 1932. Le problème du désssèchement du Tchad. Dépêche col. Paris, 1er juill.

Tilho, J., 1934. Sur l'éventualité de la capture du Logone, affluent du lac Tchad, par le Niger, C. r. Ac. Sci. Paris 199: 752–755.

Tilho, J., 1935. Le Logone quittera-t-il le bassin du Tchad? Rev. gén. Sci., Paris 23: 652.

Tilho, J., 1936. Sur l'état actuel de la zone de capture du Logone par la Bénoué. C. Sci. Paris 202: 2109–2113.

Tilho, J., 1937. La capture du Logone par la Bénoué. Bull. Ass. Géogr. fr. Paris 000: 49–53.

I. The lacustrine environment and its evolution

1. Paleolimnology of an upper quaternary endorheic lake in Chad basin

M. Servant and S. Servant

A research program on the quaternary lakes of Chad basin, initially focused on stratigraphic, geomorphologic and sedimentological studies, (Servant 1973), but was completed by diatom (Servant-Vildary 1978) and pollen analysis (Maley 1980) from lacustrine deposits. The main purpose was to reconstruct the paleoenvironments of the quaternary lakes in order to determine climatic variations. The results are incorporated into a more comprehensive program on the paleogeography and paleolimnology of southern Sahara regions, from the Atlantic to the Red Sea.

From the end of the tertiary, the bottom of Chad basin was occupied by very large lakes which deposited diatomaceous clayey sediments up to several hundred meters in thickness. In the upper Pliocene or at the beginning of the Pleistocene, the first climatic change led to the drying up of the basin.

The whole late quaternary history of the basin is characterized by complex climatic fluctuations that resulted in wind modifications during dry periods and in the resurgence of lakes during rainy periods. This situation has been studied in detail for the last forty millennia and will be the object of our study (Fig. 1).

1.1 Variations in lacustrine levels

The first observations during the nineteenth century, suggest that the water level of Lake Chad undergoes significant variations of about 50% in its depth and extent (Tilho 1947) that are directly related to changes in the annual average rainfall in the whole drainage basin (Servant 1974) (Fig. 2). These variations were observed over ten years and reflect changes of greater magnitude which occurred over a century or a millennium scale. For instance, it is known that Lake Chad was as large as the Caspian during a recent geological time of about 6000 years BP (Tilho 1925; Schneider 1967). It is also known that this lake dried up temporarily during the quaternary, evidence for this being the existence of wind dunes which are almost completely submerged by the present lake. The lacustrine oscillations which have been obvious since the first observation are supported by

J.-P. Carmouze et al. (eds.) Lake Chad
© *1983, Dr W. Junk Publishers, The Hague/Boston/Lancaster*
ISBN 978-94-009-7268-1

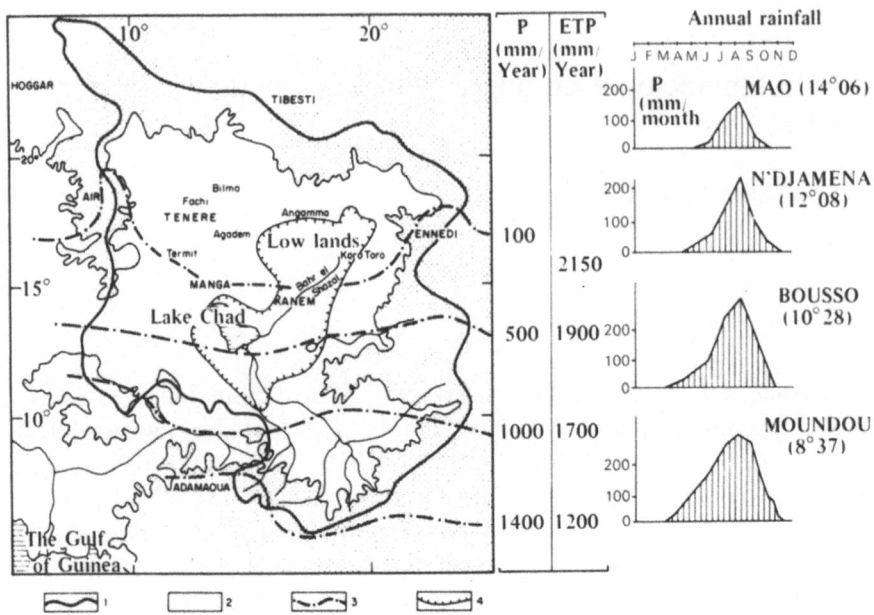

Fig. 1 The Chad basin and its main climatic features. 1: limits of the basin; 2: altitudes above 500 m; 3: isohyets; 4: limit of Lake Chad in about 6000 years BP (Schneider 1967); P: yearly average rainfall; ETP: potential yearly average evaporation.

numerous ^{14}C datings (Servant 1973), carried out for the last 40 000 years. From these three main geological periods can be singled out.

— From about 40 000 to 20 000 or 18 000 years BP, the Chad basin was occupied by small lakes, much more numerous than at present, which generally deposited calcareous sediments with ostracods and diatoms. These deposits are often stratified with eolian sandy layers or desiccation cracks which indicate that the lakes went through temporary drying up periods.

— From 20 000 or 18 000 to 13 000 years BP, these lakes completely disappeared and the bottom of Chad basin was reworked by wind as now happens in the Sahara. Therefore the southern limit of the desert belt was displaced at least 500 km towards the Equator.

— From 13 000 years BP, numerous lakes again appeared in the basin. However, eolian sand layers, reworked layers, desiccation cracks or sometimes, evaporation deposits in lacustrine series, show that these lakes went through several drying up periods.

Figure 3 shows the methodology used to analyze the lacustrine oscillations of these three periods in various parts of Chad basin (Servant 1973; Servant-Vildary 1978; Maley 1981). It deals with lacustrine series deposited in the interdune depression of an erg currently fixed in the northeast of Lake Chad.

12

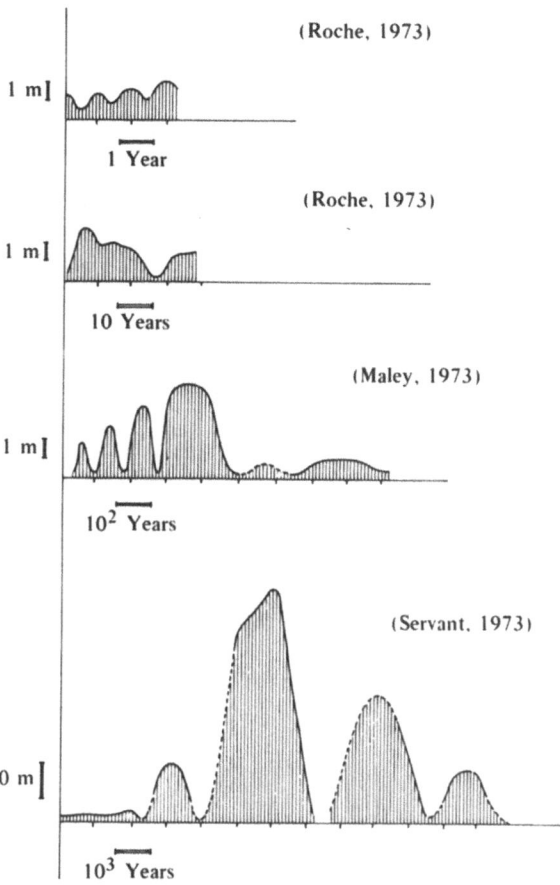

Fig. 2 Oscillations of the level of Lake Chad over different time scales (Servant and Servant-Vildary 1980).

The lithological development of this series makes it possible to specify the main periods of decreasing water levels or drying up periods, which were characterized by desiccation cracks (for instance at about 7.80 m depth). The altitude of the highest lacustrine deposits on the edges of the depression determines the water-level position (Martin, unpublished). The evolution of the diatom flora is another source of detailed information. Diatom assemblages have been divided into six different groups according to the extent and depth of the ancient lake:

Group I is mainly composed of periphytic and aerophilic species (*Hantzschia amphioxys, Pinnularia borealis, Navicula mutica* ...) Numerous phytolitaria are associated with these diatoms.

Group II is composed of benthic species typical of very shallow ponds

13

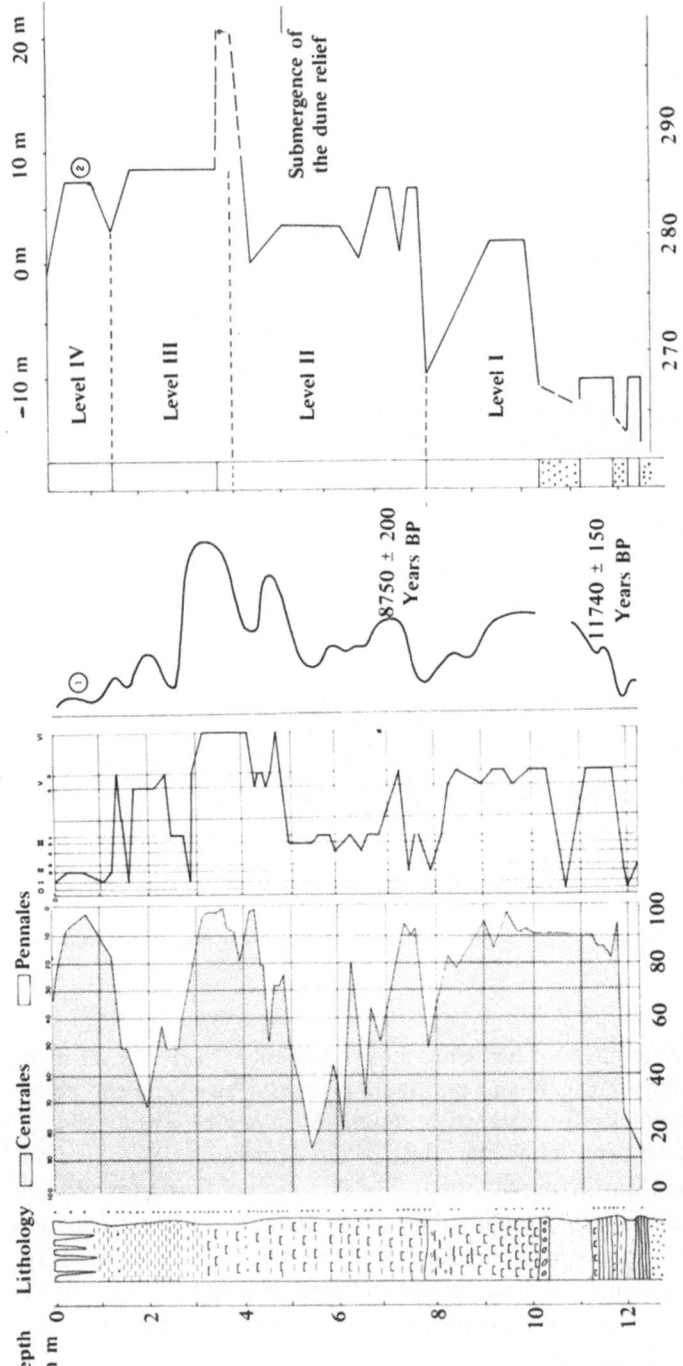

Fig. 3 Example of the determination of oscillations in a quaternary lake (section of the interdune depression in Tjeri) (Servant-Vildary 1978). 1st column: lithological column; 2nd column: relative variations in Centrales and Pennales diatoms; 3rd column: distribution along the sedimentary column of the associations of diatoms typical of the depth or the extent of the lake free water; curve 1: relative variations in the water level compared with all the available data; curve 2: absolute variations according to the altitude of the highest lacustrine deposits (Martin 1974). Palynological information which are not reproduced here, have also been used (Maley 1977).

14

(*Navicula, Anomoeoneis, Campylodiscus*), or littoral species (*Fragilaria, Nitzschia, Coscinodiscus...*).

Group III is related to a lake whose level was probably rather unstable (*Gomphonema, Fragilaria* were present when the level was low, *Navicula* and *Cyclotella* when the rise in the water level led to an open water surface etc....).

Group IV is dominated by the genus *Cyclotella* with *Melosira* as subdominant or epiphytic species depending on the bottom topography.

Groups V and VI are characterized by the prevalence of euplanktonic species (*Melosira, Stephanodiscus*) mixed with small quantities of periphytic diatoms (V) or not (VI).

The distribution of these different groups along the sedimentary columns and comparative results from the different methods make it possible to estimate the relative variations in water level.

For the lacustrine period from about 40 000 to 20 000 years BP, it was not possible to date accurately the water-level fluctuations which clearly appeared in the sedimentary series. It must be noted that the ^{14}C datings for this period involve considerable margins of error (from ± 1000 to ± 2000 years), with other possible causes for error, related to the type of deposits analyzed. Datings were performed on lacustrine limestones whose isotopic composition may have been modified by recrystallization of calcite. Therefore it would be futile to try to date with high accuracy the main stages in the changes in water level in spite of all the precautions taken in the analysis and interpretations of ^{14}C datings (Servant 1973).

For more recent periods, datings are much more accurate (from ± 100 to ± 300 years) and have been made on various materials (plant waste, shells, lacustrine limestones). Therefore, it is possible to make a detailed study of the lacustrine oscillations over the millennium. The comparative study, based on numerous lacustrine depressions (the majority of which are now dried up) is represented in diagrams of the absolute or relative variations in ancient water level.

The topographical, hydrological and hydrogeological context of each lake must be taken into account in order to interpret these diagrams. The bottom of Chad basin was never occupied by a single lake but by numerous lakes which could be completely isolated from one another. These lakes can be divided into three groups:

— Interdune lakes were situated in the depressions of a fossil erg which appeared during the arid period from 20 000 to 13 000 years BP. Completely isolated from the hydrographic system, they were fed by rainfall and especially by underground water. Numerous examples of this group now exist in the north of Lake Chad.

— Piedmont lakes were located at the foot of low mountains that gave a broken appearance to the bottom of the Chad basin. Water originated from these mountains but their small drainage basins excluded a remote water origin.

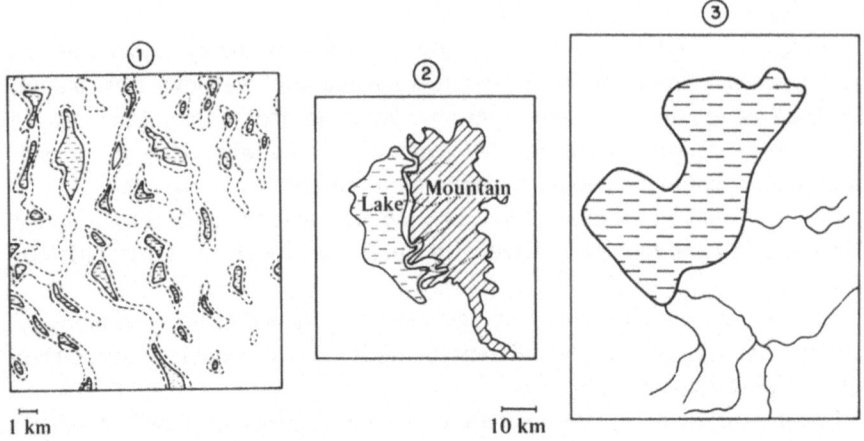

Fig. 4 Classification of the lacustrine depressions (Servant 1978). 1: interdune lakes related to the outcrop of the groundwater level; 2: piedmont lake fed by a small drainage basin; 3: lake fed by a very large drainage basin (here, the Holocene lake in about 6000 years BP, Schneider 1967).

— Hydrographic lakes were fed by a large river system and were thus similar to the present Lake Chad (Fig. 4).

During the last 13 000 years BP, the evolution of these different lakes was parallel and synchronous. It may be noted for instance, that from 8000–6000 years BP floods were observed in all the depressions, even where drainage basins occupied small areas, as the lacustrine floods were independent of local topography. Therefore, the remote supplies were not the main reason for the rise in water levels. Local climatic factors, such as an increase in rainfall and/or a decrease in evaporation must have played a decisive role. On the contrary, at about 3000 years BP, a very different paleogeographical situation is observed. A substantial rise in lacustrine levels only occurred in depressions (such as the 'Pays-Bas') which were fed by river systems, while the endorheic basins of the Niger were occupied by very shallow lakes.

The comparison of two different types of lakes, at the same time, and at the same latitude, suggests that at 3000 years BP, there was no significant increase in the rainfall to evaporation ratio (R/E), contrary to what had happened in about 6000–8000 years BP. These examples provide the major principles upon which the interpretation of lake oscillations was based (Servant 1973). Only a comparative analysis allows us to determine the relative variations in the rainfall and evaporation ratio over time (Fig. 5).

Many other data likely to be meaningful from the paleoclimatic point of view can be added profitably to the paleohydrological interpretation. For example, the palynological analysis shows that with time, the relative ratios of pollens

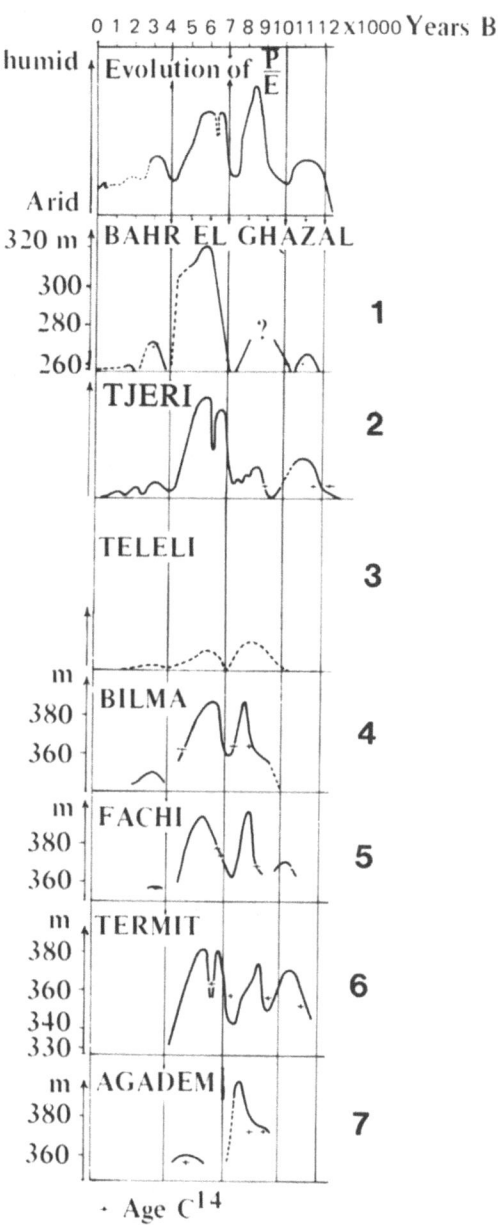

Fig. 5 Variations in the level of some lakes in the Chad basin and paleoclimatic interpretation (Servant 1973, 1974; Servant and Servant-Vildary 1980). 1: oscillations of a lake fed by the river system; 2: oscillations of a lake fed by the groundwater level with river supplies during some periods (above all in about 6000 years BP); 3: oscillation of an interdune lake always isolated from the river system; 4, 5, 6 and 7: oscillations of piedmont lakes in Eastern Niger fed by small-sized drainage basins (see Fig. 4, classification of the lacustrine depressions).

17

from Soudano-Guinean, Sudanese or Sahelian plants have varied considerably. These variations are related to changes in the vegetation cover near the ancient lakes or in the drainage basins as a whole (Maley 1977). Prehistoric sites, near the ancient lakes, help to determine the paleoenvironments of prehistoric man (Maley et al. 1971).

1.2 Lacustrine paleoenvironments — paleosalinity

1.2.1 Regions related to the change from a dry to a wet period

At about 12 000 years, after a very dry period, numerous small lakes appeared in the bottom of the Chad basin, especially in the interdune depression located in the north and northeast of the present Lake Chad. The benthic and epithytic flora, characteristic of shallow water ponds, shows great geographical variability in the chemical composition of the water. (They were derived from the outcrop of groundwater in topographic hollows.)

— high pH value, sodium carbonate-rich water with *Anomoeoneis costata, Rhopalodia gibberula, R. musculus.*

— sodium chlorosulphate-rich water with *Campylodiscus clypeus*, particularly numerous in the eastern Niger.

— calcium carbonate-rich water with *Navicula oblonga, Nitzschia denticula, Cymbella microcephala.*

In fact at about 12 000 years BP, lake regions were very similar to those observed now near Lake Chad (Maglione 1974; Iltis 1974). Under similar climatic conditions, we observe the juxtaposition of environments with different hydrochemical features, which can be explained through the local hydrological and hydrogeological conditions. The lakes could be completely closed and under those circumstances, evaporation led to a rapid increase in salinity. The differences in hydrochemistry result mainly from the initial composition of water with regard to drainage basins (chlorides are not very abundant in Chad and more abundant in Niger). Some of these ponds could have had underground outlets which drove some of the dissolved salts out of the lacustrine environment, thus lessening the increase in salinity through evaporation.

1.2.2 Regions related to the maximum of a wet period

We shall take the example of a lacustrine extension which occurred in about 6000 years BP, when a large lake occupied a large part of the bottom of Chad basin. Towards the northwest there were isolated oligotrophic lakes in endorheic depressions of eastern Niger. In the west, the large lake was bordered by regions very similar to those located on the northern edge of present Lake

Chad: they corresponded to an ancient dune system, more or less completely submerged, where the diatom flora was characterized by the prevalence of the genus *Fragilaria*. The existence of fully grown herbs is shown by the great number of Phytolitaria in lacustrine sediments. The center of the basin was occupied by a large deep lake where planktonic diatoms developed well (*Stephanodiscus astrea, Melosira granulata*, etc.). The main characteristics of this lacustrine period were generally low paleosalinity in the basin, sedimentation of smectites and diatomites and the absence of carbonate precipitation.

1.2.3 *Regions related to the change from a wet to a dry period*

It is difficult to study the evolution of lacustrine environments under climatic conditions which moved towards increasing drought, because their deposits often disappeared through erosion after the total dryness which may have occurred at the end of this development.

The best preserved sedimentary series were observed in the bottom of interdune depressions in the northeast of Lake Chad (Bahr El Ghazal). The lowering of lake levels resulted in a decreased percentage of diatoms, an increase in detrital sandy fractions from reworked neighbouring dunes and frequent carbonated accumulations in diffuse form or in slabs and limestone nodules. Phytolitaria are numerous and associated with aerophilic diatoms such as *Navicula mutica, Hantzschia amphioxys, Pinnularia borealis*. A slight increase in paleosalinity (*Cyclotella meneghiniana, Coscinodiscus* sp.) was followed by a greater increase which did not appear in the diatom flora. Evaporites were deposited when lake levels were very near the soil surface and sodium layers were preserved at the top of recent Holocene lacustrine deposits in eastern Niger (Bilma) and Chad (Largeau).

Diatom study shows that the paleosalinity was low (less than $1 \mathrm{g}\,l^{-1}$ throughout the basin, and dissolved salts reached a high value only during short periods of time and in localized areas.

It is surprising to find low paleosalinities in an endorheic basin where salts, brought into solution by surface water, should accumulate in lacustrine basins with time. But for the Quaternary, it should be emphasized that those basins which are endorheic from the hydrological point of view are not so from the sedimentological point of view. During drying up periods, lacustrine sediments and in particular evaporites derived from this drying up are affected by wind erosion. Thus a considerable part of the deposits is driven out of the basin by the wind as is shown by the presence of continental diatoms in the quaternary deposits of the western Atlantic.

A similar phenomenon occurred on the shores of Lake Chad (Roche 1970) on another time scale. Saline alluviums appeared on the edge of this lake when there was a lowering of the water level and they may have been reworked by

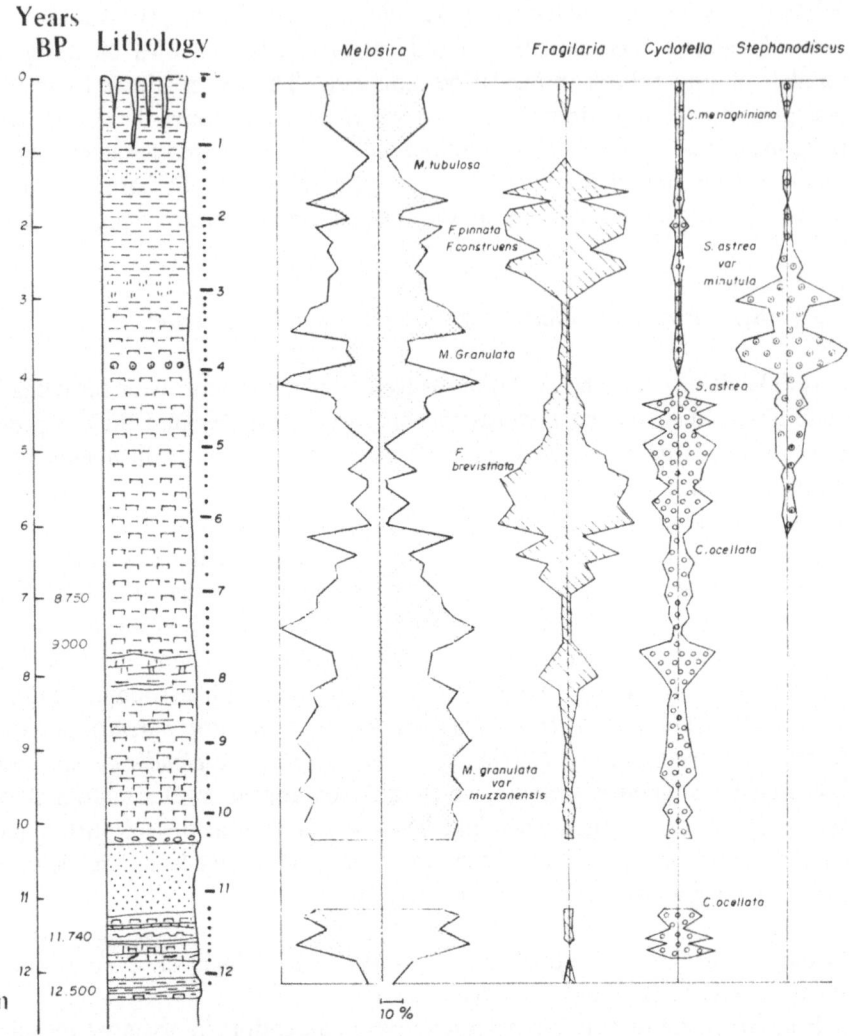

Fig. 6 Relative variations in the main genera of diatoms in a lacustrine series (section of Tjeri, cf. Fig. 3) (Servant-Vildary 1978).

wind erosion. A subsequent advance of the lake recovered only part of the saline losses from the lake waters thus limiting an increase in salinity.

The present salinity regulation in Lake Chad is accounted for by other mechanisms which can be translated into geological terms. Water infiltration, which may have been particularly active in large lakes transgressing permeable lands, helps dissolved substances to move towards underground water, and

20

clayey neo-formations (smectites) lead to immobilization of some dissolved elements as shown geologically by Dupont (1970) and geochemically by Carmouze (1976). Organisms fix numerous substances, mainly calcium and silica which are at least partly incorporated into sediments. Higher aquatic plants can also play a decisive role during recession periods. Very numerous on the edge of the lake, they are left on the surface when the water level drops. So a considerable amount of matter can be removed from the aquatic environment, even when some of the plants appear again through a new extension of the lake (Carmouze et al., 1978).

The limited paleosalinities can be accounted for by the composition of the water in Chad basin which always has a very low chloride content (Roche 1973; Carmouze 1976). Moreover, chlorides which are not fixed by sedimentation can reach high concentrations only in endorheic environments with high evaporation.

Finally, it should be noted that the mechanisms which limit paleosalinities are compounded by water-level fluctuations. Decreasing water levels deposit surface saline alluviums and a large amount of organic matter which leads to an elimination of dissolved substances. The transgression of a lake onto permeable sandy land leads to leakages by infiltrations etc. Therefore, the main reason for the low percentage of dissolved salts is probably the constant variation in lake levels. In other tropical regions, it is possible that the quaternary water coverage in endorheic basins did not suffer such sudden and distinct variations.

1.3 Trophic conditions

At 7000 years BP, interesting changes in the diatom flora were observed in ten 'interdune' lakes, located near the Bahr el Ghazal river. *Stephanodiscus astrea, Melosira granulata* var. *tubulosa, Fragilaria construens* replaced *Cyclotella ocellata, Melosira granulata* var. *muzzanensis, Fragilaria brevistriata* (Fig. 6). This modification of diatom assemblages was connected with a change in the trophic conditions. The evolution from oligotrophic to eutrophic conditions can be explained by an internal change but is better related to a major paleogeographical event. Interdune depressions were fed before 7000 years BP by subterranean waters and rainfall, but after 7000 years BP, they were fed by the flooding of river water. The comparison between many topographically and hydrologically different lakes supports this hypothesis.

At 8000–9000 years BP, interdune lakes were oligotrophic but at the same time, hydrographic lakes (such as the lakes of the 'Pays-Bas du Tchad', the lower part of the basin) were characterized by the presence of 94% of the eutrophic species, *Stephanodiscus astrea*. Yet these lakes were fed by southern and northern rivers. In the Nigerian part of the basin, the lakes were fed only by runoff, never by rivers, and the complete absence of *Stephanodiscus astrea* was

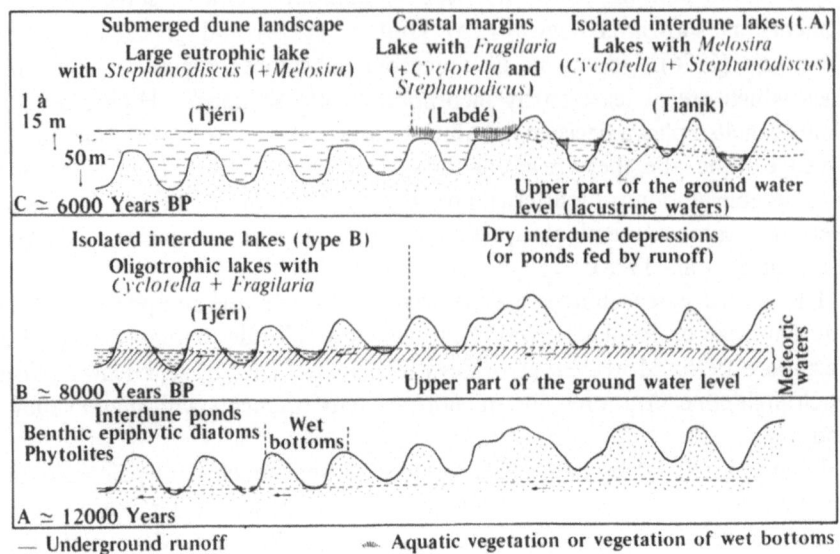

Submerged dune landscape
Large eutrophic lake
with *Stephanodiscus* (+ *Melosira*)

Coastal margins
Lake with *Fragilaria*
(+ *Cyclotella* and
Stephanodicus)

Isolated interdune lakes (t.A)
Lakes with *Melosira*
(*Cyclotella* + *Stephanodiscus*)

1 à
15 m

(Tjéri)

(Labdé)

(Tianik)

50 m

Upper part of the ground water
level (lacustrine waters)

C ≃ 6000 Years BP

Isolated interdune lakes (type B)
Oligotrophic lakes with
Cyclotella + *Fragilaria*

(Tjéri)

Dry interdune depressions
(or ponds fed by runoff)

B ≃ 8000 Years BP

Upper part of the ground water level

Meteoric waters

Interdune ponds
Benthic epiphytic diatoms
Phytolites

Wet
bottoms

A ≃ 12000 Years

— Underground runoff   ⸺ Aquatic vegetation or vegetation of wet bottoms

Fig. 7 Cross section of some quaternary lacustrine landscapes in the northeast of Lake Chad. Here lakes develop in a dune landscape resulting from the previous periods (Servant-Vildary 1978).

observed, even at the maximum of the wet period. An example of the development of the regions is given in Fig. 7.

Trophic conditions were different from one lake to another at the same time in the Chad basin.

1.4   Paleotemperatures

It is difficult to find methods for estimating changes in temperature in lacustrine deposits; however, some information can be provided by diatoms. In Chad sediments, we found some psychrophilic diatoms (diatoms which preferably live in high- and mid-latitude regions, or in high tropical mountains). The only explanation for their presence in the ancient lakes is that they had found suitable conditions for life. At 11 500 years BP, they developed in oligotrophic conditions and then disappeared although the oligotrophic conditions remained, as a result of temperature variations.

As far back as the ancient Quaternary, associations of psychrophilic diatoms appeared in Chad, whose climatic significance is particularly interesting:

— The species which developed between 25 000 and 20 000 years are: *Cyclotella ocellata* (20–80%), *Cyclotella comta* var. *radiosa* (5–20%), *Diploneis domblittensis* (1–5%), *Fragilaria lapponica* (0–1%), *Melosira italica* var. *subarc-*

22

tica (5–20%). They completely disappeared at 18 000 years BP, when cosmopolitan and tropical species developed.

— Some rather thin layers inserted in lacustrine sediments (one dated back 11 500 years) gave rise to another psychrophilic flora: *Melosira italica subarctica* (0–1%), *Cymbella augustata* (1–5%), *Cymbella hybrida* (0–1%), *Cymbella naviculoïdes* (1–5%), *Cyclotella ocellata* (20–40%).

— At about 9000 years (9610±155 years BP Kamala): *Cyclotella ocellata* (5–20%), *Cymatopleura elliptica* and var. *hibernica* (0–1%). *Navicula scutelloïdes* (1–20%), *Cymbella augustata* (0–1%) almost completely disappeared after 7000 years BP in favour of tropical diatoms such as *Nitzschia lancettula* (5–20%), *Melosira granulata* var. *valida* (1–5%), var. *tubulosa* (1–5%), *Navicula confervacea* (5–40% Tjeri), etc....

1.5 Discussion: paleoclimatic interpretations of paleolimnological data

The most conspicuous characteristics of the history of the lake in this region are the instability of lacustrine levels and water volume over all time scales (10, 10^2, 10^3 years).

This phenomenon is inevitably related to climatic changes on the southern edges of the Sahara. These changes which were analyzed over the last decade resulted mainly from a decrease or an increase in tropical rainfall and a greater or lesser movement of the Intertropical Convergence Zone inside the African continent during the recent Quaternary; for instance, at about 6000 years BP, when large areas were submerged, paleogeographical studies show that at that time, hydrographic lakes were fed as today by southern rivers. Moreover, diatoms and pollen emphasize the tropicality of the lacustrine environments and paleoenvironments. Therefore, a northern movement of the isohyets may have occurred in relation to their current position.

But some paleohydrological and paleolimnological data suggest that other phenomena may have occurred. For instance, it is known that from the beginning of the Holocene (8000–9000 years BP) lakes received river supplies simultaneously from the north (Tibesti) and the south, and the latitudinal climatic zonation which is now very distinct was not clear. At that time, streams in central Saharian mountains deposited fine alluvial and stratified sediments whose presence cannot be explained under tropical climatic conditions which are characterized by seasonal and stormy rainfall. It is also known that the diatom flora of ancient lakes, at least temporarily, included species which are now unknown in tropical plains.

This biogeographic information appears still more clearly for some more ancient periods: for example, from about 25 000 to 20 000 years BP, tropical diatoms disappeared as a whole in favour of cosmopolitan or paleoarctic species.

All these characteristics required particular paleoclimatic conditions. Servant (1973) has suggested that advection of cold polar air may have played a decisive role in tropical latitudes. This would provide an appropriate explanation for paleoclimatic anomalies such as untropical rainfall, reduction of latitudinal climatic zonation, development of psychrophilic diatoms, etc....

Finally, in the Chad basin the evolution of lakes is closely related to instability or modifications in the climates of the South Saharian tropical zone.

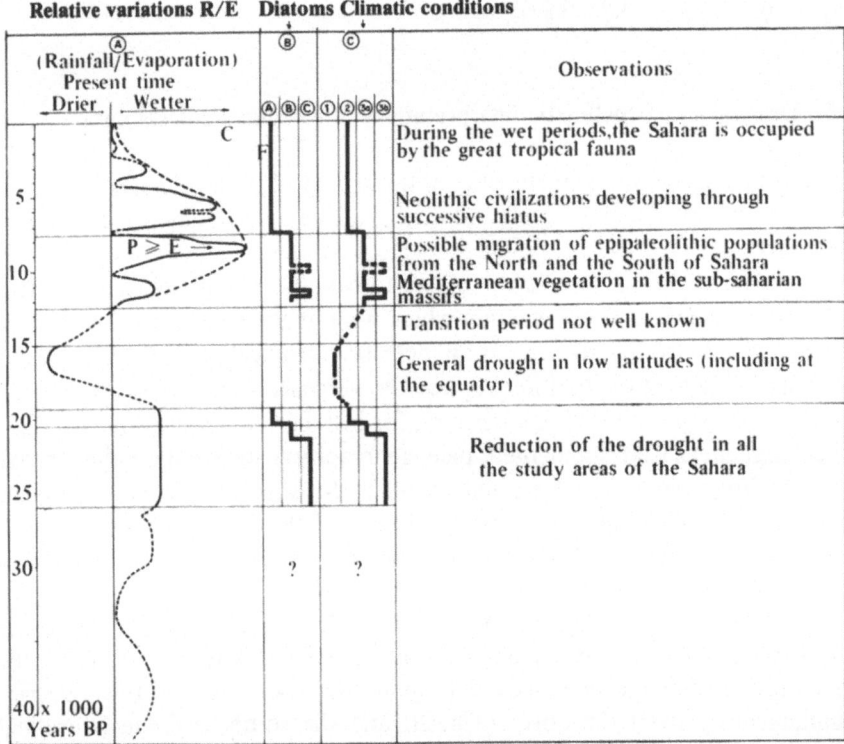

Fig. 8 Evolution of Chad basin over the last forty millenniums. A: relative variations in the rainfall to evaporation ratio according to paleolimnological interpretations (towards latitude 14° N (Servant 1973, 1974); B: distribution of the associations of diatoms (Servant-Vildary 1978); (BA) tropical species; (BB) preferential species in cold water associated with rare species from high and middle latitudes with very rare tropical species; (BC) species from high and middle latitudes, no tropical species; C: hypotheses about the modifications in atmospheric circulations; (C1) hyper-arid period, eastern circulation, subtropical anticyclone deviate towards the equator; (C2) exceptional advections of polar air or rare advections in low latitudes; tropical climates with seasonal and thunder rainfall in Chad; (C3a and C3b) frequent advections of polar air in the latitudes of Chad; climates without current equivalents in tropical regions of the African continent.

24

This instability and these modifications are part of global atmospheric circulation. Therefore, the paleolimnology of the tropical waters extends far beyond the regional framework where it is applied (Fig. 8).

References

Carmouze, J. P., 1976. La régulation hydrogéochimique du lac Tchad. Contribution à l'analyse biogéodynamique d'un système lacustre endoreique en milieu continental cristallin. Trav. Doc. ORSTOM No. 58 Paris, 418 pp.

Carmouze, J. P., Fotius, G., and Lévêque, C., 1978. Influence qualitative des macrophytes sur la régulation hydrochimique du lac Tchad. Cah. ORSTOM Sér. Hydrobiol., 12: 65–69.

Dupont, B., 1970. Distribution et nature des fonds du lac Tchad. Cah. ORSTOM Sér. Géol. 2: 9–42.

Faure, H., 1969. Lacs quaternaires du Sahara. Verh. int. Ver. Limnol. 17: 131–146.

Gasse, F., 1975. L'évolution des lacs de l'Afar Central (Ethiopie et TFAI) du Plio-Pléistocène à l'Actuel. Thèse d'Etat Univ. Pierre et Marie Curie Paris, 406 pp.

Hebrard, L., 1972. Contribution à l'étude géologique du Quaternaire du littoral mauritanien entre Nouakchott et Nouadhibou, 18–21° latitude nord. Publ. Lab. Géol. Fac. Sci. Univ. Dakar, 2 vols., mimeo.

Iltis, A., 1974. Le phytoplancton des eaux natronées du Kanem (Tchad). Thèse d'Etat Univ. Pierre et Marie Curie Paris, 269 pp.

Maglione, G., 1974. Géochimie des évaporites et silicates néoformés en milieu continental confiné. Thèse d'Etat Univ. Pierre et Marie Curie Paris, 331 pp.

Maley, J., 1973. Les variations climatiques dans le bassin du Tchad durant le dernier millénaire, essai d'interprétation climatique de l'Holocène. C. Acad. Sci. Paris 276: 1673–1675.

Maley, J., 1977. Analyses polliniques et paléoclimatologie des douze derniers millénaires du Bassin du Tchad (Afrique centrale), Rech. Fr. sur le Quater. Bull. AFEQ 50: 187–197.

Maley, J., 1981. Etudes palynologiques dans le bassin du Tchad et paléoclimatologie de l'Afrique nord-tropicale de 30.000 ans à l'époque actuelle. Trav. Doc. ORSTOM No. 129, Paris, 586 pp.

Maley, J., Roset, J. P. and Servant, M., 1971. Nouveaux gisements préhistoriques au Niger Oriental, localisation stratigraphique. Bull. Asequa. 31–32: 9–17.

Martin, R., 1974. Stratigraphie du quaternaire récent dans le Kanem, région de Moussoro. Rapport ORSTOM Fort-Lamy, 6 pp.

Roche, M. A., 1970. Evaluation des pertes du lac Tchad par abandon superficiel et infiltrations marginales. Cah. ORSTOM Sér. Géol. 2: 67–80.

Roche, M. A., 1973. Traçage naturel salin et isotopique des eaux du système hydrologique du lac Tchad; Thèse d'Etat Univ: Pierre et Marie Curie Paris, 397 pp.

Schneider, J. L., 1967. Evolution du dernier lacustre et peuplements préhistoriques au Pays-Bas du Tchad. Bull. Asequa. 14–15: 18–23.

Servant, M., 1973. Séquences continentales et variations climatiques: évolution du Bassin du Tchad au Cénozoïque supérieur. Thèse d'Etat Univ. Pierre et Marie Curie Paris, 368 pp.

Servant, M., 1974. Les variations climatiques des régions intertropicales du continent africain. Soc. Hydrotechn. France. XIIIᵉ Journées Hydrauliques, Paris, 10 pp.

Servant, M. and Servant-Vildary, S., 1980. L'environnement quaternaire du bassin du Tchad. In William, M. A. J. and Faure, H. (eds.), The Sahara and the Nile, pp. 133–162. Balkema, Rotterdam.

Servant-Vildary, S., 1978. Etude des Diatomées et Paléolimnologie du Bassin du Tchad au Cénozoïque supérieur. Trav. Doc. ORSTOM No. 84, Part 1, 329 pp.

Servant-Vildary, S., 1978. Les Diatomées des dépôts lacustres quaternaires de l'Altiplano bolivien. Cah. ORSTOM Sér. Géol. 10: 25–36.

Tilho, J., 1911. Documents scientifiques de la Mission Tilho (1906–1908). Imp. Nat., Paris, 3 vols.

Tilho, J., 1925. Sur l'aire probable d'extension maxima de la mer paléotchadienne. C. Acad. Sci. Paris 181: 643–646.

Tilho, J., 1947. Le Tchad et la capture du Logone par le Niger. Gauthier-Villars, Paris, 202 pp, 15 plates.

2. The lacustrine environment

Jean-Pierre Carmouze and Jacques Lemoalle

2.1 Geography of the lake

The lake is situated between 12° and 14°20 latitude north and between 13° and 15°20 longitude east. It occupies the entire part of a 25 000 km² closed basin which has an outlet towards the Chad Bahr-el-Ghazal above the altitude of 283 m (Fig. 1). Lake waters rise above this point only after several consecutive high river floods and then flow towards the lower regions in the northern part of the basin. Around this level, the lake looks like a small inland sea, as was the case during the second half of the 19th century and more recently in 1963–64. On the other hand, it can be reduced to a number of residual ponds of a few hundred square kilometers which are covered with vegetation, as in the beginning of the century (Tilho 1910) or in 1973–76.

These changes depend upon the lake water regulation, its morphology and the growth of higher aquatic plants. Water regulation, associated with the morphology of the lake basin, is such that the mean depth of the lake is 4 m (or 281.8 m a.s.l.) and the seasonal and between years variations in water level are respectively about 0.5 and 5 m. However, it should be noted that the variations in depth are in the range of about 4–5 m excluding the extreme values. Therefore, large areas are likely to be emerged or immersed with small variations in the water level. Moreover, the bottom of the basin is characterized locally by a dune system which determines three types of landscapes according to the altitude of the dune crests compared with that of the water level:

— a region of islands when the dune crests are emerged
— a region of vegetation island called reed islands when the crests are immersed in 0.50–1.50 m of water and occupied by semi-aquatic phanerogams (mainly *Cyperus papyrus* and *Phragmites*)
— a region of open waters when the crests are inundated and without emergent vegetation, generally the case when they are situated at depths greater than 1.5 m.

A fourth zone can appear in very shallow and protected regions. It is

J.-P. Carmouze et al. (eds.) Lake Chad
© *1983, Dr W. Junk Publishers, The Hague/Boston/Lancaster*
ISBN 978-94-009-7268-1

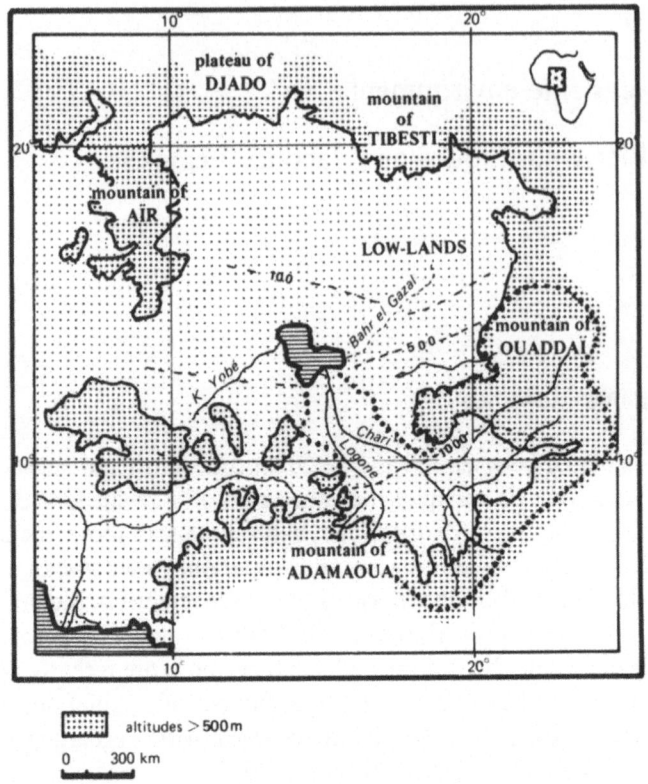

Fig. 1 The Chad basin.

composed of reed islands and swampy meadows with *Potamogeton, Vallisneria*
and *Ceratophyllum demersum.*

In fact, aquatic and semi-aquatic phanerogams which create the landscapes
of reed islands and swamps do not always settle in such a simple way, as they
also depend largely upon the recent past history of the water level (Fotius and
Lemoalle 1976).

On the whole, the distribution and extent of the lake regions depend
upon:
— the bottom topography and dune relief
— the height of the water level
— the plant cover which is itself dependent upon the seasonal and annual
changes in water level.

We shall first describe the morphology of the basin, defining the various
lacustrine environments in relation to the water level during the period of our
study.

28

2.1.1 *Morphology of the lake basin*

The basic maps used for the Niger, Cameroon and Chad parts of the lake are those made by the Institut Géographique National during 1950 to 1954. However, the maps used for the Nigerian part were made by the Federal Survey Department in 1957. They were set up with lake levels ranging from 280.6 to 282.2 m, corresponding to the most usual aspect of the lake.

Carmouze and Dupont in 1970 and Roche in 1971 gave respectively, a bathymetric map for the level of 282 m and a map of the bottom altitude. Aerial observations conducted during the partial drying up of the lake in 1973 were used to make a new map which showed the bottom altitude and served as an extension of the two preceding ones (Carmouze 1976). The contour of this map corresponds to the altitude of 281.9 m and the relief is described by curves showing equal bottom altitudes which were drawn from a thousand measurements taken regularly from the whole lake (Fig. 2). In fact, this kind of representation makes it possible to define properly the easily flooded zones for a given height of the water level but it does not show the bottom dune relief which largely determines the distribution of the zones. So, after the description of the bottom topography, we shall determine the dune relief.

2.1.1.1 *Bottom topography*. In its northern part, from Baga-Kawa to Daboua (Fig. 2), the lacustrine basin is bounded by a continuous dune chain whose altitude is always above 284 m; in the northeast and east, from Daboua to Kouloudia, by a fixed erg which gives a very meandering contour; finally in the south, from Kouloudia to Baga-Kawa, by a low flat plain with slight contour lines (282–283 m). Therefore, the southern limits are likely to be greatly displaced during fluctuations in the lake level.

The lake is divided into a south and a north basin by a line from Baga-Kawa to Baga-Sola because of a narrowing coastal periphery at this level, as well as a bottom shift (the bottoms of the south basin range usually from 280 to 278.5 m and those of the north basin from 277.5 to 275.5 m (Fig. 2).

The lowest region of the lake bottom, with an average altitude of 275.5 m lies in the north basin. Its area is 4000 km² and its position is slightly north of the geographical center of the basin. The slope is smoother towards the south than towards the northeast and west. In the south basin where the relief is less sharp there are three regions of higher altitudes (278.5 m on an average). The one which is near the Shari delta has an area of 1100 km², that in the north, of 1000 km² and, finally the other one in the south is 550 km².

Morphometric curves showing the graphical determination of the surface occupied by the lake in the north and south basins as well as the corresponding volume are given in Fig. 3 (Carmouze et al. 1972; Carmouze 1976; Lemoalle 1978). Furthermore, data from Lansat satellites in 1972–73 and 1975–76 during

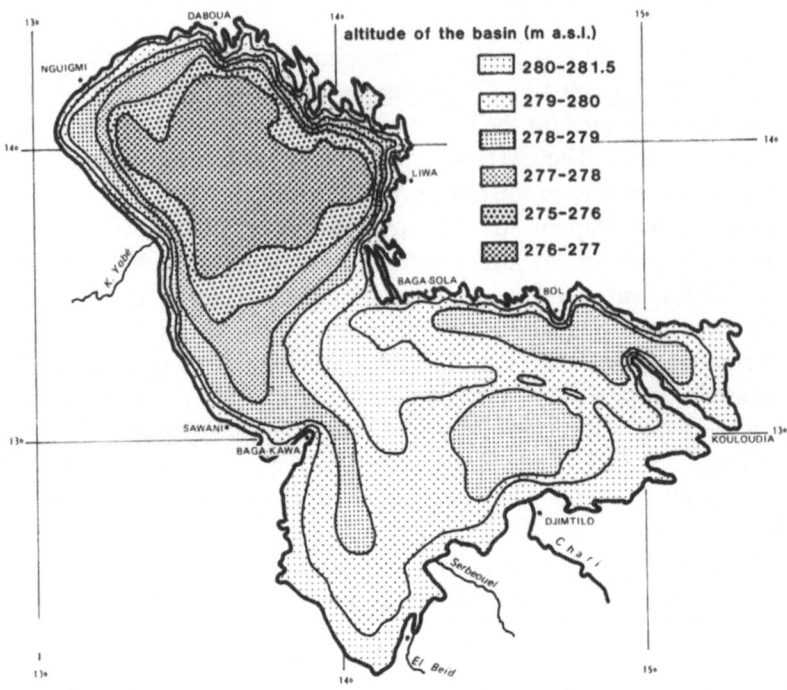

Fig. 2 The altitudes of the lacustrine basin.

the successive emergence of different regions of the lake make it possible to define precisely the relief of the major part of the lake bottom.

2.1.1.2 *Dune relief.* The erg bordering the lake in the north and the east provides the bottom of the lake basin with a very irregular relief. The orientation of the dunes is southeast–northwest. Their height as well as the altitude of their crests decreases from the shoreline towards the inner part of the lake. Therefore, areas of lake bottom furthest from the northern and eastern limits of the lake are much less influenced by the dune system.

The coastal dunes of the south basin are longer (5 to 10 km against 3 to 6 km) and higher (10 to 15 m against 7 to 9 m) than those of the north basin. On the contrary, beyond the coast, the dune relief diminishes more rapidly in the south basin than in the north basin. The wave amplitude is usually below 1.5 m in the southern part of the south basin but it is higher in threequarters of the north basin. The dune crests lower than 285 m in altitude are usually eroded, probably as a result of the erosive action of the lake floods during the periods of lacustrine transgressions.

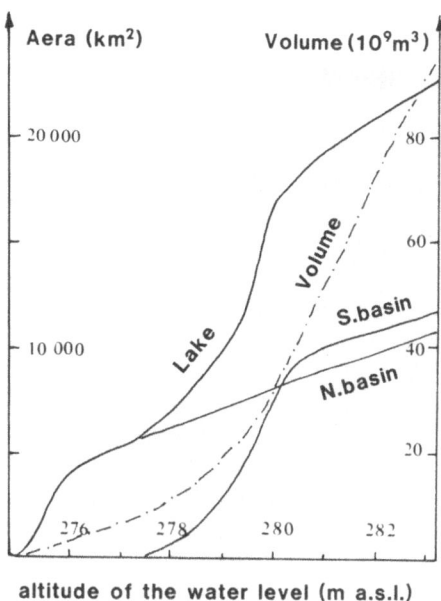

Fig. 3 The morphometric curves of the lake.

Finally, though it is shaped by a sharper dune relief, the bottom of the south basin is flatter and higher than that of the north basin. This may result from lake sedimentation, which is more important in the south than in the north (Bouchardeau and Lefèvre 1957).

2.1.2 *Changes and development of the lake regions*

From 1967 to 1975, the water level continued to decrease each year from a mean of 281.9 m to 279–279.5 m and was followed by irregular changes in the lacustrine environment. Thus, until the level of 280.5 m, there was only a slight change in the different regions from 1967 to 1971. On the contrary, profound changes have been recorded below this level from 1972.

In the first case, we are dealing with a lake under average circumstances or 'Normal Chad' and in the second case, with a lake undergoing reduction or 'Lesser Chad' (Tilho 1910).

2.1.2.1 *The 'average' lake or 'Normal Chad'* (water level 281.9 m a.s.l.). Lake waters in the north and south basins occupy, respectively, 10 000 km² and 11 000 km² (Fig. 4).

In the north basin, dune crests above 282 m which are situated along the northern and eastern coasts create an archipelago landscape called the North-

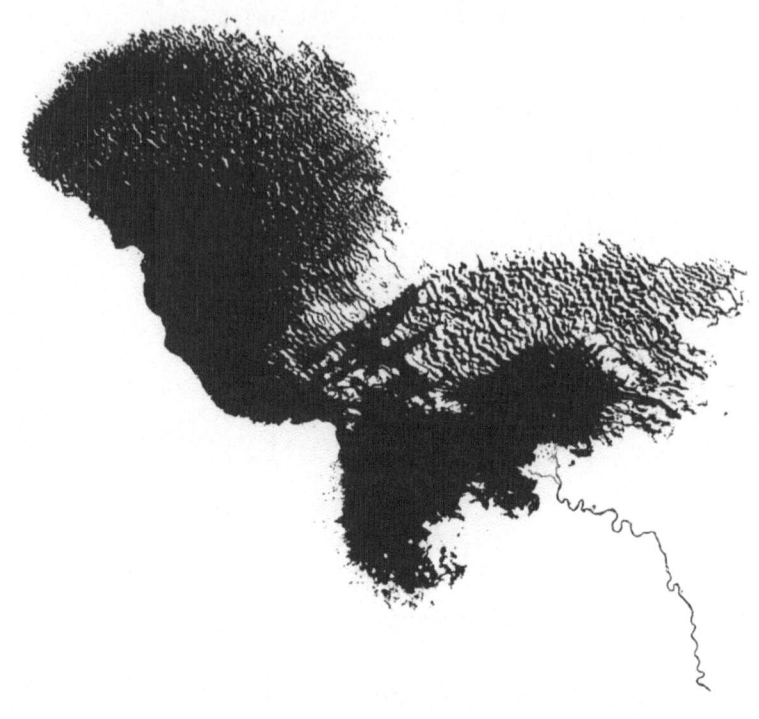

Photo 5 Water surface of Lake Chad (January 1973) from a satellite photograph.

eastern Archipelago, along a stretch of about 25 km wide (Fig. 4). About 500 flat islands occupy 40% of the total area amounting to 2400 km^2, and are covered with herbaceous vegetation. They are often surrounded by a narrow stretch, a few meters wide, which is occupied by macrophytes, mainly *Phragmites* and sometimes extended by *Ceratophyllum demersum* and *Potamogeton demersum*. Depths range from 2.5 to 6.5 m.

Northwest of the archipelago, most of the dune crests whose altitudes range from 279 to 281.5 m are immersed and covered by macrophytes, mainly *Phragmites*. The archipelago zone is followed by a region of reed islands extending to the Nigerian coast. It has a total area of 3600 km^2, of which 17% is occupied by reed islands with an average area of 2.5 km^2. In some cases this region is divided into northern and northeastern reed islands (Fig. 4). To the east of Nguigmi, there is a small area with few reed islands forming a zone of open waters, with depths ranging from 4 to 7 m.

The southwest part of the basin with dune crests under 1.5 to 3 m of water is less influenced by the dune system. It also has a large zone of open water, the northern open waters, which is bordered to the west and the southwest by sandy beaches of a dune chain. Their area amounts to 4200 km², and depth increases regularly from the south to the north, ranging from 2 to 6.5 m.

The northern open waters are bordered in the southeast by a region of low dune undulations of low interdune depth that constitute a mixture of archipelago and reed islands. This shallow region with numerous sandy islands and vegetation extends to the south basin and interferes with water transfers between the two basins; it is therefore called the Great Barrier (Fig. 4).

In the south basin, a fixed erg borders the northern and eastern coasts as in the north basin, thus creating a 20 to 35 km wide archipelago. The islands can reach 10 m in height and become lower and flatter with increasing distance from the coasts. Some doom-palms (*Hyphaene tebaica*) and acacias grow here and there, while the lowest islands are covered with herbaceous vegetation. Some of them are partially lined with a narrow stretch occupied by *Phragmites, Cyperys papyrus* and *Typha*.

A dune chain from Kouloudia to the northwest divides this archipelago into Southeastern Archipelago and Eastern Archipelago, which occupy respectively areas of 1400 km² and 1800 km². Islands cover 40% of the total area in the first region and 45% in the second one and depths usually range from 2 to 4 m

Fig. 4 The main natural regions of the lake.

The rather low and flat regions situated immediately to the east and north of the delta have a large open water region of 3900 km². This is divided into Southern Open Waters and Southeastern Open Waters which are located, respectively, west and east of the shallow Great Barrier which advances towards the mouth of the El Beïd (Fig. 4).*

The regions around this zone of open waters are higher. Dune crests are immersed under 0.5 to 1 m of water, thus contributing to the establishment of macrophytes (*Cyperus papyrus*, *Typha* and *Phragmites*). To the east of Baga-Kawa up to Kouloudia, there is a 5 to 80 km wide coastal stretch with reed islands, covering an area of 1500 km². 20% of the area is occupied by the reed islands themselves (Southern Reed Islands) (Fig. 4). To the east of the delta where the bottoms are flat and high (280 to 281 m), there is a swampy zone with meadows occupied by *Potamogeton* and *Vallisneria*; the Southerneastern Reed Islands (1700 km²) are developing between the islands and the open waters. They continue to the southerm border of the Southeastern Archipelago and join the Great Barrier which has a region of reed islands only in its eastern part.

On the whole, the open waters, archipelagos and reed islands comprise respectively 38, 23 and 39% of the total water area which is about 21 000 km². Lake Chad belongs to the category of shallow lakes, being a unique example because of its great extent.

Table 1 shows the main hydrological features of the large natural regions of the lake at the 281.9 m level.

2.1.2.2 *The change towards 'Lesser Chad' in 1973*. Since 1973, the appearance of the lake has been profoundly modified following a very small flood in 1972–73 (17.3×10^9 m³ against 40×10^9 m³ annual average). As it was insufficient to initiate any seasonal rise of the level, and barely succeeded in compensating for the evaporation losses over two to three months. Then, a marked lowering occurred, resulting in the emergence of the shallow zones of the lake such as the Great Barrier, the Eastern Archipelago and the Southern Reed Islands. From April 1973, the lake became split into three unequal water masses (Fig. 5). The north basin retained its water, however the shoreline was displaced towards the center of the basin from between a few hundred meters to a few kilometers according to the region. About half the south basin was exposed and only the Southeastern Open Waters, the South Open Waters which looked like an appendix to the north basin, as well as the central part of the Southeastern Archipelago retained their waters (Fig. 5).

During July 1973, in the Northern Open Waters, previously submerged dune crests appeared; the Southern Open Waters dried up and the Southeastern

---

* The differences between the North and Northeastern Reed Islands and between the South and Southeastern Open Waters appear *a priori* arbitrary. However, they are very useful in studying some hydrological, hydrochemical and sedimentological aspects.

Table 1 Hydrological characteristics of the main natural regions for an average water level of 281.9 m (above sea level).

	Z (m)	S (km²)	V (10⁹ m³)		Z (m)	S (km²)	V (10⁹ m³)
Southeastern	3.60	2000	7.2	Southeastern	3.30	875	2.9
Open Waters	3.20	1975	6.3	Archipelago	2.95	850	2.5
	2.75	1925	5.3		2.55	825	2.1
Southern	2.75	2000	5.5	Eastern	3.10	1075	3.3
Open Waters	2.35	1975	4.6	Archipelago	2.70	1050	2.8
	1.95	1925	3.6		2.40	1025	2.5
Southern	2.05	1525	3.1	Great Barrier	2.10	1700	3.5
Reed Islands	1.55	1475	2.3	(southern part)	1.80	1675	3.0
	1.10	1425	1.6		1.55	1610	2.5
Southeastern	2.55	1800	4.6	Southern	2.75	10 975	30.1
Reed Islands	2.15	1750	3.7	Basin	2.35	10 750	25.3
	1.65	1700	2.8		1.95	10 435	20.4
Great Barrier	3.40	1550	5.3	Northeastern	6.00	975	5.8
(northern part)	2.95	1525	4.5	Reed Islands	5.65	950	5.3
	2.55	1455	3.7		5.15	925	4.7
Northern	4.90	3750	18.3	Northern	5.35	1475	7.9
Open Waters	4.40	3700	16.3	Archipelago	5.00	1450	7.2
	3.90	3660	14.2		4.70	1400	6.7
Northern	5.35	2700	14.5	Northern	5.00	10 325	51.8
Reed Islands	5.00	2650	13.2	Basin	4.60	10 145	46.7
	4.60	2550	11.7		4.15	9865	41.1
	3.85	21 300	82.0				
	3.45	20 900	72.0				
Lake	3.05	20 300	61.5				

Z = depth; S = area; V = volume.
For a given region and parameter, three values should be noted: the upper number corresponds to the flood level (282.4 m), the lower number to the lower water level (281.5 m) and the intermediate figure to the average level.

Archipelago was composed only of some residual ponds (Fig. 6). From August 1972 to August 1973, the water level decreased from the average level of 280.1 to 278.4 m, i.e. by 1.6 m. This resulted in the occupation by the macrophytes of some newly emerged zones, especially those of the Great Barrier, the South-eastern Reed Islands and the Southern Open Waters. While the immersed or half-immersed vegetation around the islands or along the coastal zones had

35

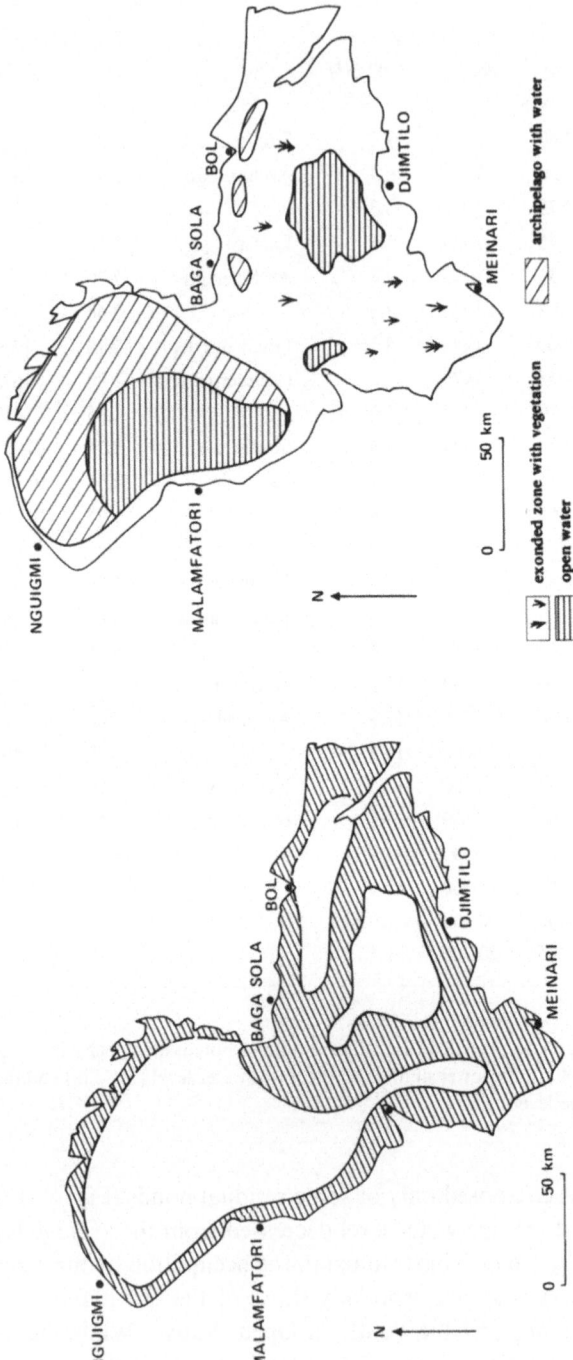

Fig. 5 The 'Lesser Chad' in April 1973.

Fig. 6 The 'Lesser Chad' in July 1973.

archipelago with water

exonded zone with vegetation

open water

50 km

N

36

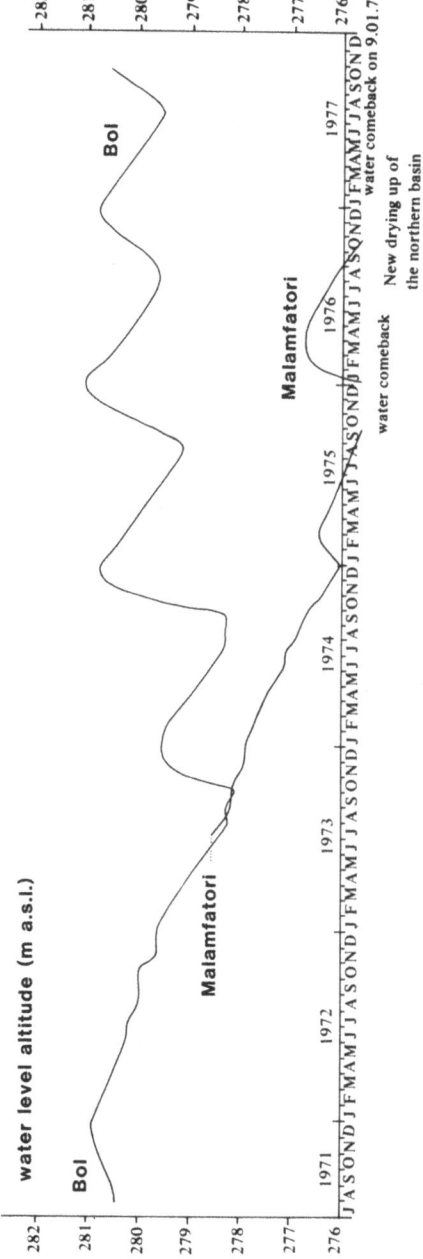

Fig. 7 Water levels of the lake in the southeastern archipelago and the northern basin from 1971 to 1977.

37

virtually disappeared during the 'Normal Chad' period in 1972 as a result of the rapid contraction of the lake, the exposure of the above-mentioned regions made it possible for seeds in the surface sediments to germinate (Fotius and Lemoalle 1976). This vegetation which was mainly composed of *Cyperus papyrus*, *Typha vossia*, *Ipomea* and *Ludwigia* largely persisted until the water level increased again. The 'Ambadjs' (*Aeschynomene* ssp., *Sesbania* spp.) were equally prominent. In the south basin, the areas thus occupied were equal to about 5000 km^2.

In conclusion, apart from the reduction and the fragmentation of the water area, the second important phenomenon which occurred in 1973, was the development of a thick vegetation over a large part of the emergent areas in the southern basin.

2.1.2.3 *The development of the lake during the period of 'Lesser Chad' (1973–77)*. (a) *1973–74*: partial drying up of the south basin during the period of low water and the beginning of drying up in the north basin.

In 1973–74, the Shari flood was scarcely greater than that of 1972–73 (18.4×10^9 m^3). Until mid-September, the Shari low waters flowed into the restricted Southeastern Open Waters that grew larger when the Shari flood occurred. From the beginning of October, the Southern Open Waters and a part of the Southeastern Archipelago, which were covered with abundant vegetation at that time, were filled up by water (Fig. 8). Although macrophytes were not entirely immersed, the rise of the water level was large and rapid (Fig. 7) amounting to about 1.30 m between early October and the end of November for the Bol region and to 1.00 m for the Baga-Kawa region.

The south basin reached its maximum water level in December (Fig. 9) and open waters of this basin occupied nearly the same area. In the Southeastern Archipelago, open water areas appeared, resulting in the total immersion of macrophytes and swampy zones were displaced southwards and northwards. At that time, water flowed through the thick vegetation of the Great Barrier towards the north basin, in a diffuse and minor way. As it received practically no water supplies, except those from the Komadougou Yobé and slight rainfall, the north basin underwent profound changes. The water level decreased by 85 cm in the central zone, dividing the open waters into two remnant large ponds separated by an archipelago of sandy islands without vegetation.

During 1974, the flood in the south basin was similar to that of the previous year with the decrease in water less marked during the first months (Fig. 7). Dune crests and shallows gradually reappeared but they were covered with vegetation before being exposed. Ambadjs developed to such an extent that in places they looked like forests. In July, the water level was nearly the same as that of the previous year at the same time, but with more abundant vegetation, especially at the level of the Great Barrier.

In the north basin, which had not been supplied with water since the

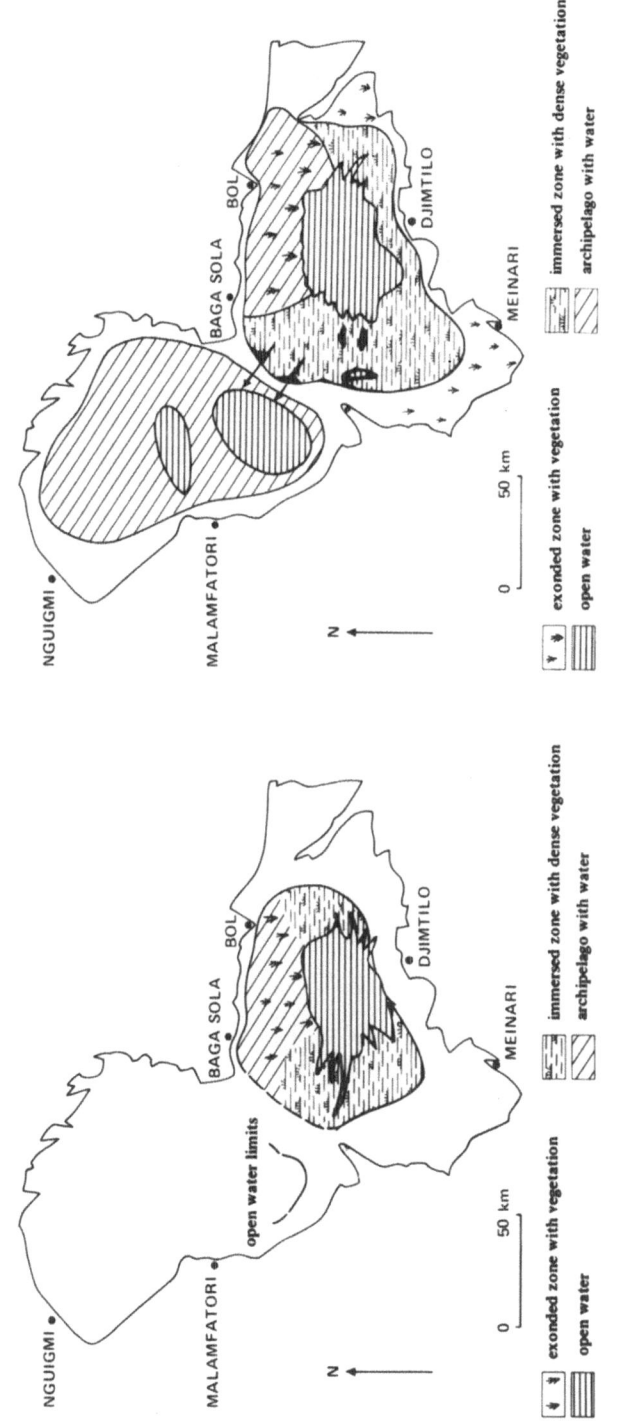

NGUIGMI •

MALAMFATORI •

open water limits

BAGA SOLA •
• BOL

• DJIMTILO

MEINARI •

N ←

0   50 km

☐ exonded zone with vegetation   ☐ immersed zone with dense vegetation

▨ open water   ▨ archipelago with water

Fig. 8   The 'Lesser Chad' in October 1973.

NGUIGMI •

MALAMFATORI •

BAGA SOLA •
• BOL

• DJIMTILO

MEINARI •

N ←

0   50 km

☐ exonded zone with vegetation   ☐ immersed zone with dense vegetation

▨ open water   ▨ archipelago with water

Fig. 9   The 'Lesser Chad' in December 1973.

39

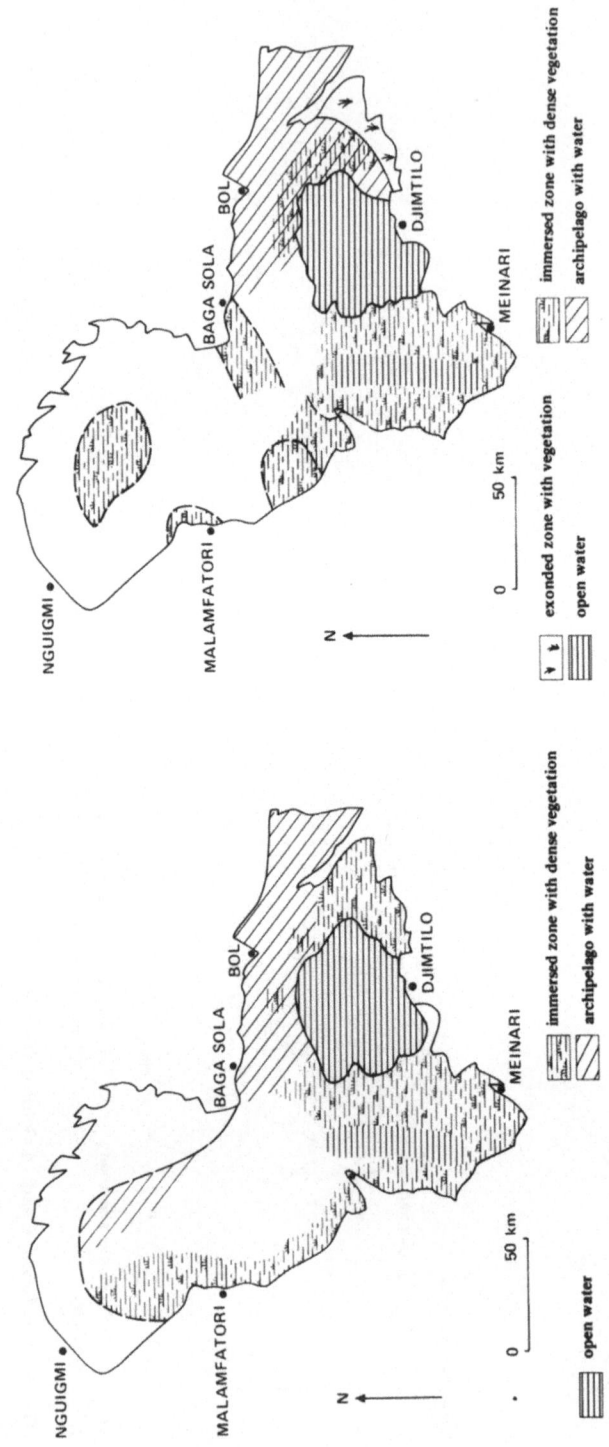

immersed zone with dense vegetation

archipelago with water

open water

Fig. 10 The 'Lesser Chad' in January 1975.

exonded zone with vegetation

open water

immersed zone with dense vegetation

archipelago with water

Fig. 11 The 'Lesser Chad' in July 1975.

beginning of the year, the water level decreased by 1.90 m from July 1973 to July 1974 (Fig. 7), and it turned into a large archipelago.

(b) *1974–77*: return to a nearly normal situation in the south basin and almost total drying up of the north basin during the period of low water.

In 1974–75, the Shari flood was clearly that of the two previous years, while being still below the average flood (30.5×10^9 m^3). There was an abrupt rise in the water level in the south basin in October–November (Fig. 7). At the end of 1974, the water level at Bol was similar to that recorded from the same station at the end of 1971.

The vegetative cover did not prevent waters from moving into the south basin, but it restrained their movement towards the north basin which was supplied with an insufficient amount of water to be able to compensate for the annual evaporation of 2.20 to 2.50 m. From May 1975, the water area was thus gradually reduced to a few isolated ponds around Kindjeria which were completely dried up by October 1975. The floods of the Komadougou Yobé had a limited influence since a vegetation zone existed only around its mouth.

During the period of low waters examined above, a hydrological discontinuity between the Southeastern Archipelago and the Southeastern Open Waters existed only during 1973 and 1974. The south basin later returned to a normal state, except that seasonal variations in the water level were accentuated.

During 1975–76, the Shari flood was similar to the average flood (36.6×10^9 m^3) as in 1974–75. In the south basin, the water level variations were close to those of 'Normal Chad' but with a slight increase in the subsidence period. Around the pocket of the Eastern Open Waters, numerous ambadjis disappeared because of an increase in the water level. On the other hand, those from the northern side of the Great Barrier remained along with abundant macrophytes, and acted as a buffer to water penetration. Events occurred as if the communication between the two basins has been raised higher, since in 1971–72, for the same level of the south basin, the water penetration into the north basin was normal. So, the north basin was in the same situation as in the previous year, although the Shari flood was higher. In zones of temporary water, vegetation became thicker and thicker year by year.

The Shari flood of 1976–77 was again small (28.7×10^9 m^3). However, the south basin was only slightly affected and the average water level decreased by 10–20 cm compared with the water level of the previous year. Figures 10 and 11 give an idea of the environmental modifications since 1974, respectively, during the flood and the subsidence periods.

2.1.3 *Sedimentological characteristics of the lake bottom*

Dupont (1967, 1970) was the first to study the surface sediment facies and their geographical distribution. The classification of sediments according to their form, colour, degree of coherence, texture and organic matter content was reduced to four main types: mud, clay, sand and pseudo-sand, which are subdivided into varieties. In some zones of the lake, these materials are encrusted with limestone.

From 1970 to 1973, Dupont and Carmouze continued this study. A new map with surface sediments was drawn from 1500 observations that were uniformly distributed over the entire lake (Carmouze 1976) (Fig. 12).

2.1.3.1 *Materials with muddy clay facies.* Mud is a structureless material which usually appears in the form of a greyish-black, fine, homogeneous suspension and sometimes as big brownish flakes. Its water phase is always very important since it represents 250 to 500% of the dry weight. Its mineral portion, representing 80% of the average dry weight, is distributed approximately equally among three types of metric granules: clay up to 0.002 mm, silt from 0.002 to 0.050 mm and sand from 0.05 to 2 mm. Its organic fraction which is mainly composed of decaying macrophyte debris is relatively high usually ranging from 10% to 16% of the dry weight but reaching more than 30% when it is genuine peat. The average percentages of carbon and nitrogen are equal to 90% and 8% respectively of the dry weight; the C/N ratio is almost 11, indicating the leading part played by the higher plants in the accumulation of this organic matter.

The most abundant material, mud, is found particularly in the zones of reed islands and archipelagos (Fig. 12). Thus, the lake bottoms of the Southeastern Archipelago, the Southeastern Reed Islands, the Southern Reed Islands, the majority of the Great Barrier and finally the eastern and southern edges of the Northern Reed Islands and the Northern Archipelago are covered with flaky brown mud (flakes can reach 1 cm in diameter). The eastern half of the Eastern Archipelago is also covered with it. Peat like mud is found along the northern and eastern coasts of the north basin, as well as in some parts of the Southeastern and Eastern Archipelago arms.

Clay is a material with a variable consistency but always firmer than mud. It appears either in a homogeneous soft and structureless form or in a more or less structured form (cleaving into angular polyhedrons, a few centimeters in diameter and into smaller aggregates) or finally in the granular form (composed of thick particles resistant to manual weathering). Whatever the variety may be, clay is characterized by a mineral portion containing more than 50% of components less than 2 mm in diameter. It is usually composed of 10 to 40% slime and 10 to 20% sand. The water phase represents 120 to 130% of the dry weight when it is soft clay and 40 to 120% of the dry weight when it is structured clay. The organic portion is smaller in clays than in muds and does not exceed 5% of the dry weight. The average percentages of carbon and nitrogen in organic matter amount respectively to 25% and 2.5% and the C/N ratio of 10 is similar to that of muds.

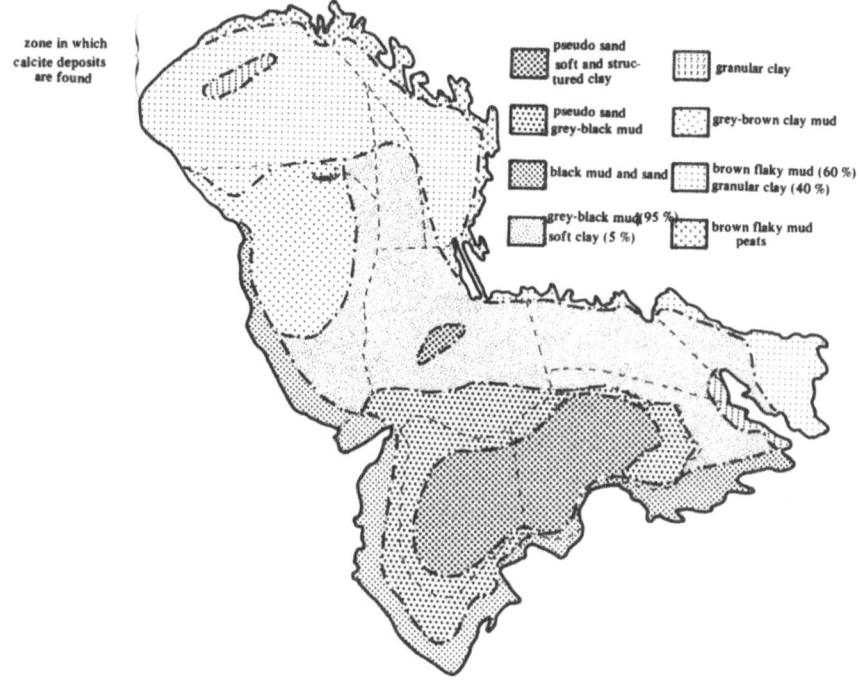

Fig. 12 The different types of sediments.

Clay appears in large amounts in open water zones such as the Southeastern, Southern and Northern Open Waters. In the first two regions, it appears in the form of soft clay and covers the depressions, while the shallows are covered with pseudo-sand as is shown further on. In the second region, it forms a slightly flaky material between soft clay and greyish-brown mud. The Northern Reed Islands and the Northern Archipelago are usually covered with 30 to 40% clay, either soft or granular (Fig. 12). Either type of clay is still found in the bottoms of the Southeastern and Eastern Archipelagos, the Great Barrier and the Southeastern Reed Islands which are swept over by currents, representing 5 to 15% of the area of these regions.

This schematic distribution of muddy and clayey materials over the entire lake whose limits are not always easy to define, leads us to make several observations.

Mud is especially found in the zones of reed islands and archipelagos, due to the large amount of macrophytes whose debris constitute the main part of this sediment. Thus, the mud layer reaches 30 to 80 cm in thickness in regions with a high macrophyte production. However, as it is a fine suspension, it is not found in places swept over by currents. Finally, the flaky mud is located

43

in regions where highly saline waters cause flocculation, when the pH is higher than 8.8.

Unlike mud, clay is found in large amounts only in regions with few macrophytes such as the Southeastern Open Waters, the South Open Waters and the Northern Open Waters. From a distribution study of the different varieties a general interpretation (Dupont 1970) could be suggested. This author noticed that soft clay is found in the deepest zones where vegetation could generally not settle during lake recession; on the contrary, structured clay is found in shallower zones where macrophytes could occasionally grow during temporary drying up periods. Finally, granular clay is related to emergent regions either as a result of their shallowness or because of their isolation from the Shari delta, which exposes them to temporary drying up during large decreases in the lake. Dupont inferred that soft clay would have been the original material and could undergo a structural change resulting from the development of the root system as a result of macrophyte settlement. This would result in the isolation of polyhedrons and a decrease in water content. When sediments are exposed, this process would be accentuated leading to the formation of clay granules. Desiccation which isolates polyhedrons could be increased in certain cases by fires by farmers.

2.1.3.2 *Materials with sandy facies.* Fine and well-sorted sands originate from two different stocks. The first one corresponds to the eastern and northern edges of the submerged erg and is composed of aeolian quartz sand whose average diameter is about 0.250 mm. The second one derived from river sources is mainly situated in the coastal zones of the Southern and Southeastern Reed Islands. It is micaceous and the average diameter of its grains is about 0.16 mm. Both varieties are often mixed. Their organic matter content remains low, equal to or less than 1% of the dry weight (Dupont 1970).

It should be noted that sands are more abundant than is shown in Fig. 12, for, they actually cover the submerged periphery of the islands as well as most of the lake shallows where the reed islands are.

2.1.3.3 *Materials with granular facies (psuedo-sand).* These materials are composed of small granules which are variable in size and were first described by Guichard (1957) as 'marc de café'. Their unimodal frequency curves reach a peak around 0.250 mm (Dupont 1970) and in the samples under study, the median range is from 0.205 to 0.283 mm. Moreover, cumulative curves are very straight, thus showing that it is a well-sorted material (Fig. 13). There are also pseudo-sand samples smaller in average size in which the measurement* of granules cemented into clay or mud is difficult. They appear in different forms,

---

* The phylogeny of these granules is studied in Chapter 4.

44

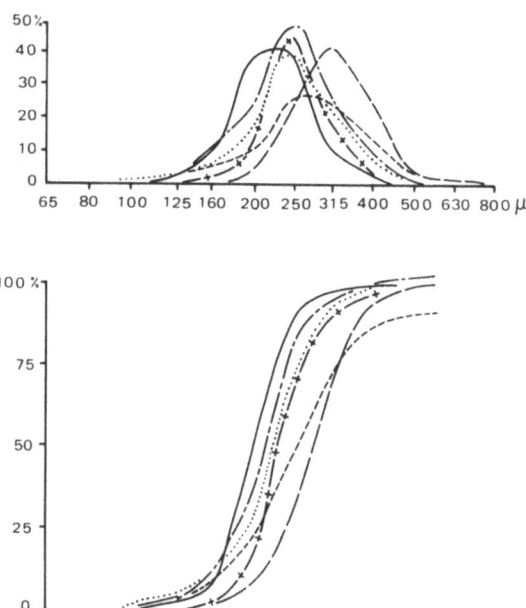

Fig. 13 The granulometry of the pseudo-sands.

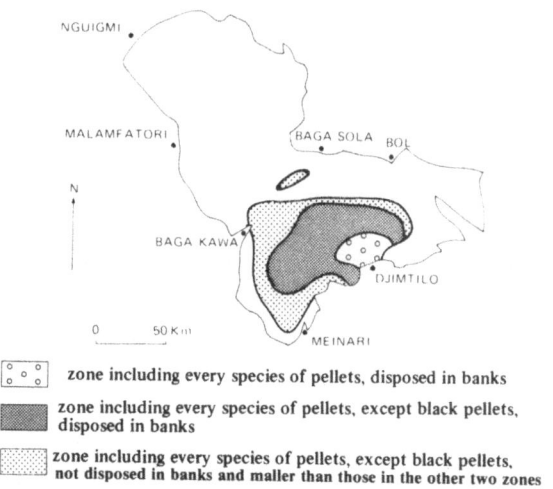

○ ○
○ ○ zone including every species of pellets, disposed in banks

▓▓ zone including every species of pellets, except black pellets,
      disposed in banks

▒▒ zone including every species of pellets, except black pellets,
      not disposed in banks and maller than those in the other two zones

Fig. 14 The localization of the different types of pseudo-sands.

45

colours and hardness, ranging from the spherical, unstriped black and manually non-friable form, to the ovoid form with big superficial yellow-greenish irregularities which are easy to crumble by hand.

These pseudo-sand sediments are mainly situated in the Southeastern and Southern Open Waters (Fig. 14). They are scattered on banks and lie almost always on clay, either of a soft or, often, structured type. Sometimes, they are covered with a layer of soft clay or mud, 5 to 20 cm in thickness (up to 50 cm in the Great Barrier). The pseudo-sand layer is usually 5 to 15 cm thick and reaches 50 cm in certain places. In the depressions, it is not very abundant and is largely related to clay. Outside the Southeastern and Eastern Open Waters, there are pseudo-sands in the South and Southeastern Reed Islands and the Great Barrier. However, in these regions they are not very abundant and are closely mixed with mud, except in the open water zone situated southeast of the Great Barrier where they appear in greater amounts.

2.1.3.4 *Materials with calcareous facies.* Materials situated along the northern and eastern coasts of the north basin can consist of up to 10% carbonates. In this zone, numerous clay blocks and granules are encrusted with calcium carbonate which is no more than calcite. In the wadis and the former arms of the lake which are now dried up, these encrustations are much more numerous* (Dupont 1970; Roche 1973; Carmouze 1976).

2.2 The lake climate

The Chad basin is situated between the anticyclones of Libya and Saint Helena which result, respectively, in the continental tropical masses of dry air and maritime equatorial masses of humid air. These two air masses converge in the Intertropical Convergence Zone (ITCZ) which moves along a south–north axis over the year**. A humid climate prevails south of the ITCZ and dry climate north of it. Continental winds, or harmattan, blow over the whole basin during winter and lead to drought, while marine winds or monsoons are associated with rainfall.

So, the position of the ITCZ in relation to the lake determines the climatic features of the latter (winds, air temperatures, rainfall, humidity, insolation and evaporation); and its movements (an example of which is given in Fig 15) determine the seasonal variations.

---

* The formation of these calcareous deposits is specified in Chapter 4.

** This zone covers the whole Sahel; its movement is dependent upon the general circulation whose interannual irregularities can result in droughts that were catastrophic for agriculture and stock farming in 1972–73.

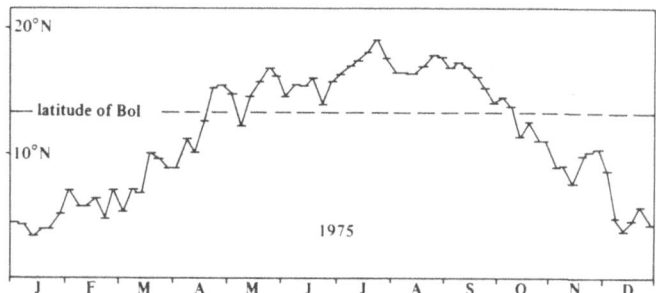

Fig. 15 The displacements of the intertropical convergence zone.

2.2.1 *Winds*

The harmattan is a sry northeast wind that blows from October to April, while the monsoon is a humid southwest wind that blows from May to September. These rather strong winds blow mainly in the morning from 0600 to 1200 hours. The daily average at Bol (Southeastern Archipelago) was measured over the period 1965 to 1970. The frequency curve for velocity which is shown in Fig. 16 (after Billon et al. 1968) indicates that the average wind velocity is above 5 m/s for 6 hours a day. The two main winds are associated with local winds that result from anabatic winds and therefore blow from the land towards the sea in the night and in the other direction during the day.

Figure 17 summarizes the directions and average velocities of winds over ten days at Bol in 1969 and 1970 (after Roche 1973).

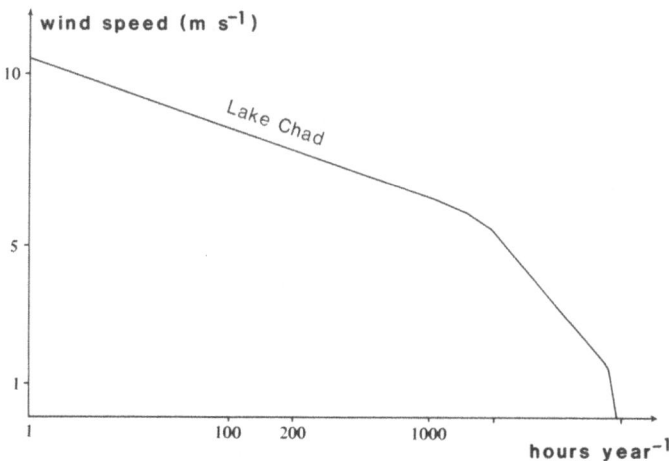

Fig. 16 Wind speed frequencies at the Bol station.

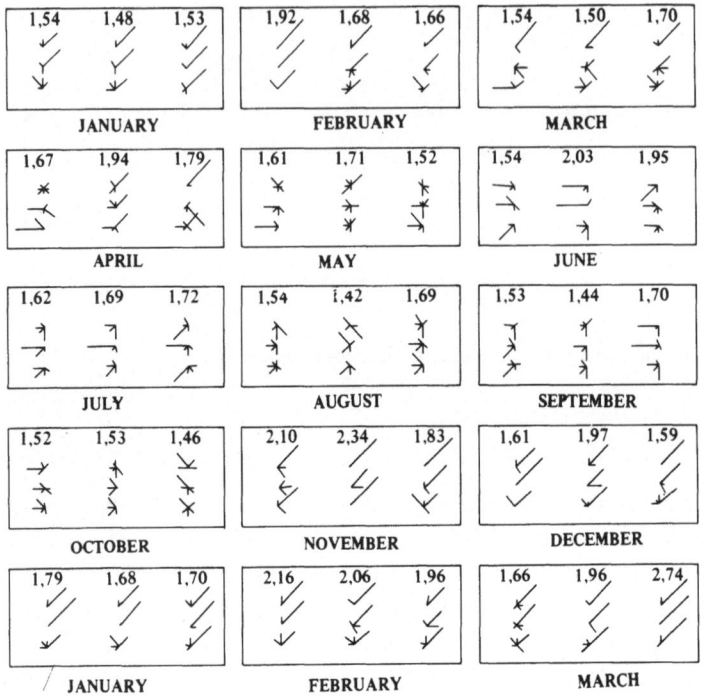

Fig. 17 The winds: mean speeds and directions at Bol (calculated upon 10 years).

2.2.2 *Rainfall*

The rainy season begins in May–June and ends in October with maximum rainfall in August when the lake receives half the rainfall. The annual rainfall decreases considerably from the south to the north. The lake, situated between the isohyets 550 and 240 mm (Fig. 18), had an average rainfall of 315 mm at Bol from 1954 to 1972 with large between years variations (the extremes recorded were 125 and 565 mm)*.

2.2.3 *Air temperature*

The average annual air temperature is about 28°C. From 1936 to 1970, it was 28.7°C at N'Djamena and 28°9 at Bol from 1957 to 1970 (Toucheboeuf de Lussigny 1969; ORSTOM 1974a, 1974b). During the warm season, the average monthly temperatures vary from 29 to 32°C from March to October with a slight decrease in August (27.5°C). During the cool season, they range from 22

* Brunet-Moret (1968) undertook a detailed study on the frequency of rainfall intensity.

48

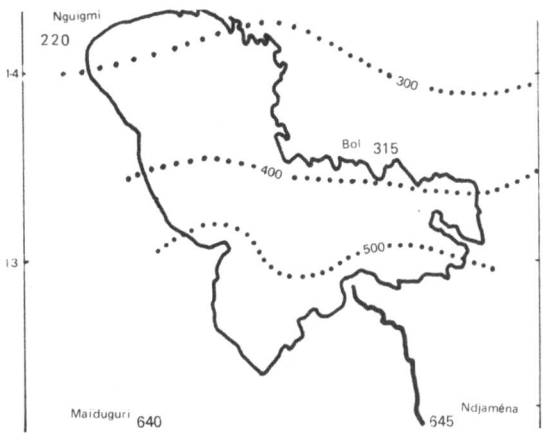

Fig. 18 The rainfall repartition on the lake (isohyets in mm year^{-1}).

to 24°C from December to February with a minimum of 21 to 23°C in January when diurnal variations are in the order of 16 to 17°C compared with 7 to 9°C in August* (Fig. 19).

2.2.4 *Air humidity*

The air humidity recorded at the peripheral stations of the lake (N'Djamena-Bol-Nguigmi) is maximum at night and minimum at about 1200 hr. The seasonal maximum, ranging from 72 to 81% occurs in August and the minimum, ranging from 23 to 31% occurs in February–March. In fact these values are not the exact representation of those calculated for the lake. Roche (1973) showed the influence of the water mass on this parameter through a comparison of measurements made in March–April 1967 over the whole lake, with those recorded at the same time at the Bol station. He pointed out that air masses moving from the northeast at that time become saturated with water when passing over the lake. A gradient is established according to the size of the water areas that were previously swept through and this represents an average absolute variation of 50% between the northern and southern regions. Therefore, during the dry season of the lake, the average humidity is higher than that measured at the Bol station. The opposite phenomenon must occur during the humid season.

---

* This temperature range was thoroughly analyzed according to the global radiation when entering the atmosphere and according to rainfall (Riou 1972).

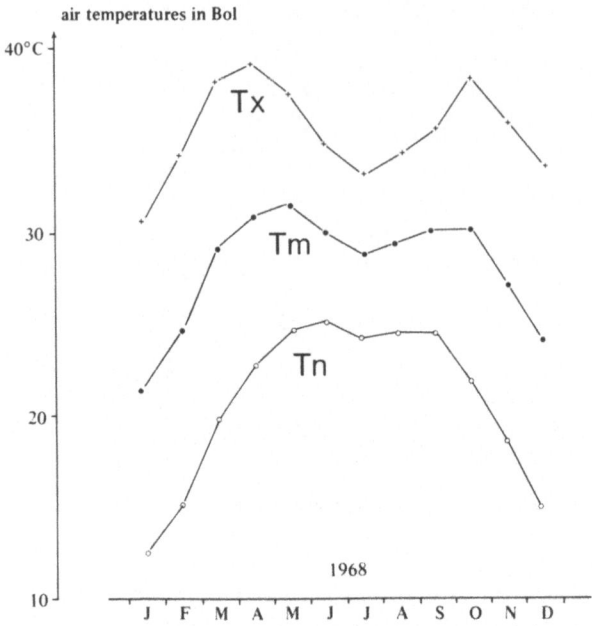

air temperatures in Bol

Fig. 19 Air temperatures at Bol; Tx, Tm and Tn are respectively maximum, mean and minimum temperatures.

2.2.5 *The total incident radiation*

The annual average, the daily total incident radiation at the ground amounted to 2308×10^4 Jm^{-2d-1} at N'Djamena from 1968 to 1973. It varies only slightly from one year to the other, and its seasonal variations are low in amplitude with a minimum of 2140×10^4 Jm^{-2d-1} in January and a maximum of 2580×10^4 Jm^{-2d-1} in March. Therefore, Lake Chad receives a relatively high irradiance with little variation in comparison with temperature zones (Fig. 20).

2.2.6 *Evaporation*

The different climatic parameters just examined lead to considerable evaporation through their characteristics of high wind frequency and intensity, high temperatures, low humidities and high insolation.

Evaporation, one of the main constituents of the hydrological balance was evaluated in numerous ways, according to different approaches. Annual estimations ranging from 2.10 to 2.20 m were made by Riquier (1963), Turc (1968) and Riou (1972) from theoretical formulae. Calculations from direct

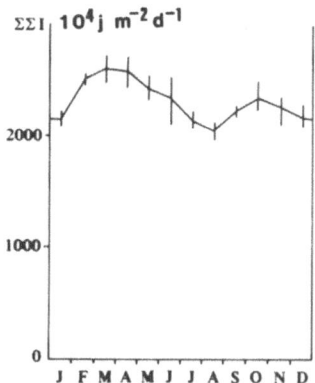

Fig. 20 Mean monthly irradiance on the lake surface.

measurements on an evaporation tank are based on data collected by the ORSTOM hydroclimatologists since 1964 along the Southeastern Archipelago at Matafo station situated near Bol. Of these data, the measurements of evapotranspiration give the best values of potential evaporation.

2.3 General circulation of waters

Large water movements in the lake result mainly from the regimens of river supplies and winds. Their direction and amplitude depend upon the morphology of the basin.

The Shari is related to a transition tropical regime which is characterized by a period of rising water level from the end of July to September, a period of high waters from October to November, a period of subsidence from December to early February and a period of low waters from February to July. Therefore, the seasonal distribution of water supplies is such that the lake is necessarily fed irregularly, receiving nearly 50% of the annual river supplies from October to November and 75% from September to December. Therefore, the deltaic zone is under a very high hydraulic gradient over this period with a maximum at the end of October–early November. The considerable kinetic energy from flood waters also leads to large water movements starting in September which are repeated and dampen in the whole lake until January.

From mid-October to March, the harmattan rises regularly in the middle of the night and dies down at the end of the morning. It pushes waters towards the western and southern coasts. From May to mid-October, the phenomenon is reversed and the monsoon winds tend to accumulate water on the eastern and northern coasts. It should be noted that the action of wind is reduced in the archipelagos and reed islands since they blow at right angles to the islands.

51

Furthermore, the winds never lead to mixing of waters between the south and north basins.

The shape of the lacustrine basin, the arrangement of islands and shallows facilitate some movements and prevent others. Thus, the Great Barrier which is a shallow region covered with islands and reed islands acts against exchanges between the Southeastern and Eastern Open Waters and the Northern Open Waters. Similarly, in all the zones of archipelagos and reed islands, the islands and shallows which are generally aligned southeastwards and northwestwards act as a brake on water movements to the northeast.

Finally, the distribution of forces exerted on the major water movements (wind regime and river supplies) is practically the same from one year to another: the period of wind reversal occurs at the same time with a difference of 15–20 days and the period of the maximum Shari flood with a difference of 10 days. The occurrence of large water movements in the lake does not vary much from year to year but the sizes of the movements are variable and depend upon the strength of the Shari flood and the lake volume*.

These movements were shown by the study of distribution in space and time of sodium concentration in waters (Carmouze 1971, 1976) as well as that of the water conductivity (Roche 1973).

We will describe these movements over an annual cycle by taking as a starting point the period of June–July which is just before the abrupt occurrence of the Shari flood waters (Fig. 21).

2.3.1 *Water movements from mid-June to mid-August*

River supplies which are very low during this period (8% of the annual supplies) cannot cause water movements of great amplitude but their influence does become considerable at the end of July or at the beginning of August. The Shari waters enter the Southeastern Open Waters, however, without resulting in a major water renewal in the regions close to the delta.

However, water moves from the Great Barrier towards the Northeastern Reed Islands and Northeastern Archipelago. This water is pushed back by the water of the Southern Reed Islands and the Southern Open Waters which is driven by the southwest monsoon winds.

The other regions of the lake are not affected by large-scale water movements. In the Southeastern Archipelago, water penetration into the Southeastern Open Waters remains low. In the Northern Open Waters and the Northern Reed Islands, water movements are low in amplitude.

---

* During the period of 'Lesser Chad', the water circulation is totally modified as a result of the exudation of the shallows.

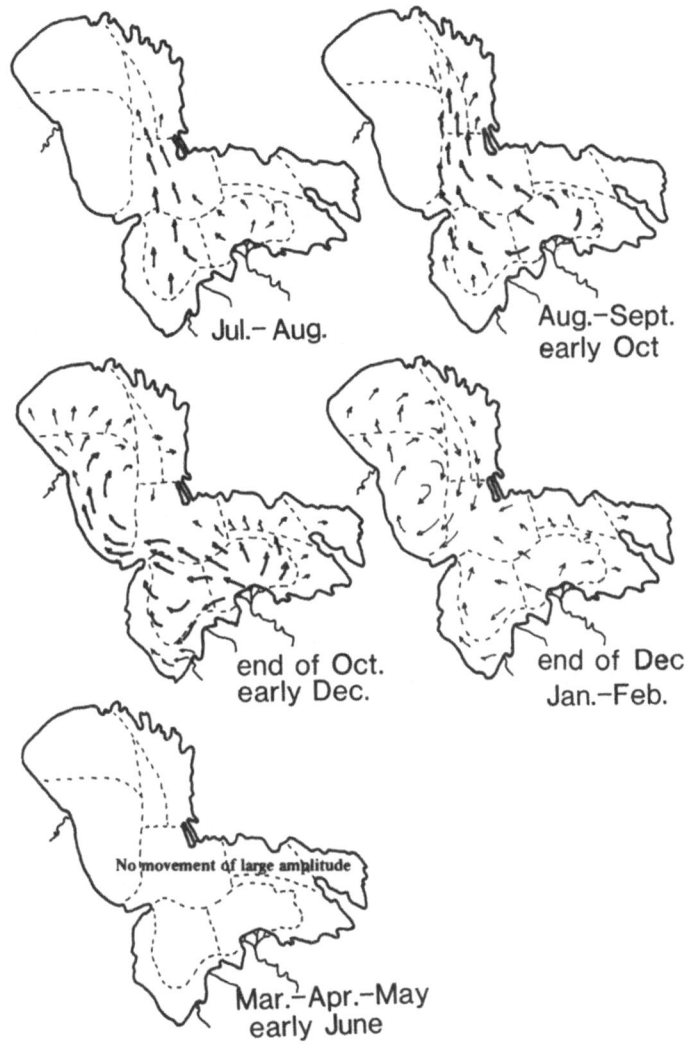

Fig. 21 Water movements in the lake.

2.3.2 *Water movements from mid-August to early October*

During this period, the Shari water supplies, representing 15 to 20% of the annual river supplies, initiate a rise in the lake water level. The waters from the Southeastern Open Water pushed back towards the periphery without mixing new water with old one.

53

The main flow always moves towards the Northeastern Archipelago and the Northeastern Reed Islands. However, during this period now, it results not only from the monsoon winds but also from the great upsurge of the Shari flood waters. The extent of this water movement is such that water from the Northeastern Reed Islands and the Northeastern Archipelago is also moved towards the Northern Reed Islands.

Water which occupied the Southeastern Open Waters in August begins to move to the Southeastern Archipelago and the Southern Open Waters. In the Northern Open Waters and the Northern Reed Islands, a few large water movements can be noted.

2.3.3 *Water movements from mid-October to December*

During October–November, the lake receives 50% of the river supplies raising the lake level by 25 to 30 cm. The upsurge of flood waters is then maximal and this water abruptly stems the flow from the Southeastern Open Waters with no mixing turning the region into a genuine extension of the Shari. Water no longer reaches the Northeastern Archipelago but now flows directly into the Northern Open Waters. This change results from the reversal in the wind regime, for the northeast harmattan blows from mid-October.

Residual water from the Northern Open Waters is pushed back towards the Northern Reed Islands and the Northeastern Archipelago. Nevertheless, a large part is quickly mixed with flood waters. In November the center of the Northern Open Waters is occupied by mixing waters, an important water supply that results in a general water movement northwards. Therefore, the water from the Northern Reed Islands is held back.

In the Northeastern Archipelago, water is also restrained and pushed back towards the coast under the upsurge of water from the Northern Open Waters along with partial mixing of water in this region as well as in the region of the Northeastern Reed Islands. Finally, the flood waters move further into the Southern Open Waters and the Eastern Archipelago.

2.3.4 *Water movements from mid-December to February*

Over this period, the lake receives 8 to 10% of the annual supplies and its level begins to decrease. Water no longer flows into the north basin but on the contrary, new water moves towards the Southern Open Waters and the Southern Reed Islands. In the Southeastern Archipelago, there is still a considerable water influx.

In the Northern Open Waters, incoming water mixes with residual water and in the heart of this region mixing is complete. On the other hand, the northern

Photo 6 General view of a saline lake in Kanem (northeast of Lake Chad).

water mass from the Northern Open Waters is pushed back towards the Northern Reed Islands, while water from the Northeastern Reed Islands mixes with newer water pushed the same way to the south of the Northern Open Waters.

So, in the Northern Open Waters, new waters take a circular course, while mixing with residual waters. This is favoured by environmental topography of the Island Barrier and Reed Islands of the Northern and Northeastern Reed Islands and the Great Barrier and by the influence of the harmattan.

2.3.5 *Water movements from March to early June*

At this time, water supplies become very low, representing 4 to 5% of the annual supplies, water movements are very reduced and the wind regime is unstable. This is a transitional period since the harmattan is gradually replaced by the monsoon.

Partial mixtures of waters from contiguous regions are caused by the Northern Reed Islands waters mixing with those from the Northern Open Waters, the Northeastern Reed Islands waters with those of the Northeastern Archipelago, the Southeastern Open Waters with those from the Southeastern

55

Archipelago and finally the Southern Reed Islands waters with those from the Southern Open Waters.

In summary, during June, at the end of the low water period, the monsoon winds contribute to the general water movement from the south to the north. This small water discharge into the north basin is followed at the end of July by the upsurge of the first flood waters of the Shari which flow in the same direction. It continues up to the beginning of October and contributes to the movement of the residual waters from the Northeastern Reed Islands and the Northern Archipelago further towards the north. At this time, the monsoon which favours water flow in the northeastern part of the Great Barrier is replaced by the harmattan which, on the contrary, contributes to a movement of the water discharge zone towards the south of the Great Barrier. The maximum Shari flood waters then penetrate directly into the Northern Open Waters during October, November and December. Initially they go along the Nigerian coast, then they take a circular course in the Northern Open Waters, while pushing back part of the residual waters towards the Northeastern Archipelago. In the midst of the south basin, they penetrate gradually into the Southern Open Waters and to a lesser extent, the Southern Reed Islands and the Southeastern Archipelago.

At the end of January, no water discharge is observed in the north basin. In the Northern Open Waters, the flood waters that took a curved course in November and December is followed by waters from north of the Great Barrier which is pushed back towards the western coast. Then, from February to March, the small water supply from the Shari spread into the Southeastern and Eastern Open Waters and the Eastern Archipelago, thus resulting in a general water movement towards the periphery of the basin. Finally, from April to June, there is no longer any marked movement. On the other hand, it should be noted that the irregular supply of the lake raises the water level from the end of August to the end of December and lowers the water level from the beginning of January to the end of July.

2.4 The river system and its tributaries

Billon et al. (1968) and Durand (1978) described the river system in detail, and we will deal with the essential part of it.

2.4.1 *The Shari*

The Shari whose lower course is 25 km has a very shallow gradient of 5 to 7 m km^{-1}, and a meandering course obstructed downstream by sandbanks. In the most intricate meanders, there are troughs up to 27 m in depth downstream from N'Djamena.

The Shari receives the Bahr Erguig on its right and its main tributary, the Logone, on its left. Several distributaries flow away from the latter and include the Loumia which flows into the Shari and the Logone during the period of high waters and then downstream from N'Djamena. Two others, the Sebbewel and the Taf-taf flow into the south bank of Lake Chad only during the period of high waters (Fig. 22).

The hydrological regime of the Shari is related to the tropical type, and characterized by annual flood and minimum flow. However, the variations in flow are less abrupt than in the purely tropical type as shown in Fig. 23.

The average annual flow of the lower Shari amounted to 1270 m^3 s^{-1} from 1953 to 1976. The between year variations were considerable since the values ranged from 537 m^3 s^{-1} in 1972–73 to 1720 m^3 s^{-1} in 1955–56.

The average annual flow of the Shari upstream from the Logone amounted to 815 m^3 s^{-1} from 1953 to 1976. In fact, this value covers a very high variability since during this period, the annual flow varied from 306 m^3 s^{-1} to 1290 m^3 s^{-1}, that is, they were quadrupled (Durand 1978).

Although the Shari floods are large, the river waters do not overflow the high water bed because the banks are high. Therefore, there is no flooding zone *sensu stricto* but a high water bed approaching 6 km in width.

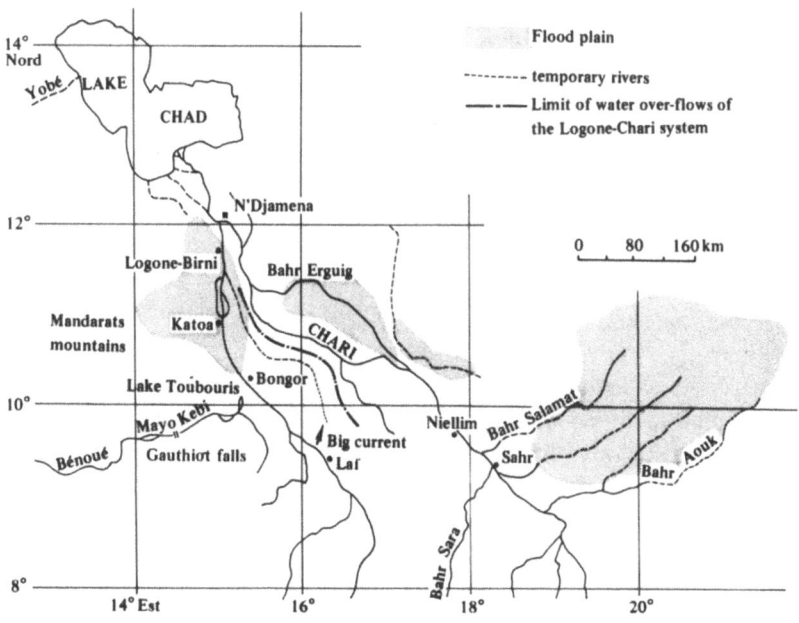

Fig. 22 The hydrographic system of the Shari.

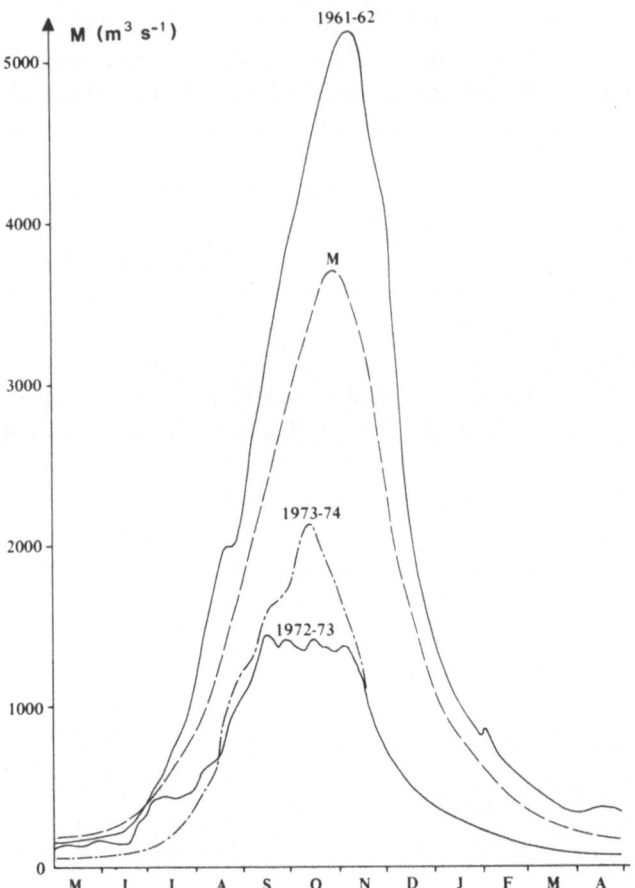

Fig. 23 The Shari floods in N'Djamena.

2.4.2 *The Logone*

The hydrological regime of the Logone is different from that of the Shari, for the degradation of its bed in its lower course allows side discharges and distributary emissions on both banks during the flood period. On the right bank, there is a considerable discharge only upstream from Bongor, contributing to inundation of the plain between the Logone and the Shari mainly through the 'Grand Courant' which has its source downstream of Laï. Part of the water returns towards the high water bed slightly upstream of Logone Gana. During the period of high water, the Logone and the Shari are connected through the Loumia.

On the left bank, there are considerable discharges: the Logomatia brings back only a small part of the discharged water to the Logone forty kilometers further.

Thus, profound changes occur in the regime of the Logone because of the large losses along its course, resulting in a decrease in flow from upstream to downstream and a corresponding decrease in the peak of the flood is recorded (Fig. 24). The average flow of 578 $m^3 s^{-1}$ at Bongor decreases by 125 $m^3 s^{-1}$ at Kaboa and by 103 $m^3 s^{-1}$ from Kaboa to Logone Birni in an average hydrological year. It is obvious that in the case of low floods, these differences are less while they are increased in the case of high floods.

The result is an extraordinary stability at the junction of N'Djamena where the Logone flow varied from 47 to 360 $m^3 s^{-1}$ between 1953 and 1971 with an average of 387 $m^3 s^{-1}$, half that of the Shari.

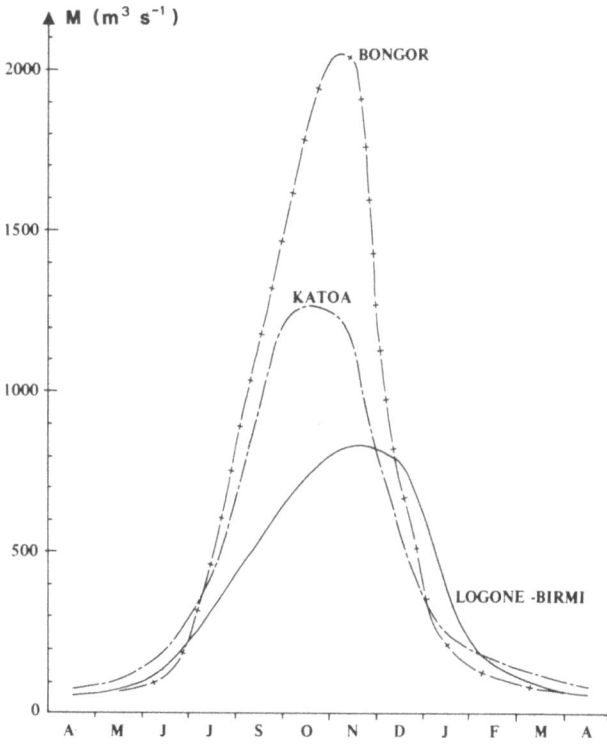

Fig. 24 The Logone floods in Bongor, Katoa and Logone-Birni.

2.4.3 Temporary environments

(a) *The great flood plain in the North Cameroons.* A very large flooding zone or 'Yaéré' extends over about 5000 km² from the Logone and the Shari on the east to the Mandaras Mountains in the west and south and runs into Lake Chad on the north (Fig. 22). This plain is very vast, covered in places with hillocks that are usually artificial and occupied by Kotokos villages where the vegetation is mainly herbaceous. During the period of the river floods, it is immersed under 0.70 to 1 m of water. By mid-July on the average, the flooding regime begins with rainfall which, a month later, increases the water level by about 30 cm. Generally, the Logone inundation appears only in September. In December, waters subside and are collected by the El Beïd which flows into the southern-most region of Lake Chad.

(b) *The El Beïd and Yobé rivers.* Although the water supplies of these two intermittent rivers to Lake Chad represent only 5% of the total, they are important, especially the El Beïd, for the migrations of numerous species (Durand 1970, 1971, 1978; J. Hopson 1969, 1972) (Fig. 22).

The El Beïd flows only five to eight months in a year and is supplied by water from the plain in the North Cameroons. Its bed is 40 to 60 m in width and is composed of only a string of muddy ponds during the period of low waters from April to July. The first runoff from the Yaéré occurs during August and September and the flood reaches its maximum in December, when the El Beïd overflows its bed. It then subsides until the end of March. From 1953 to 1976, the average flow was $46 \, m^3 \, s^{-1}$. The annual variations of $87.5 \, m^3 \, s^{-1}$ in 1954–55 and 19.5 in 1955–56 and 1965–66 were great.

Throughout the periods of flood and subsidence, the transparency and the conductivity vary to a great extent according to the mixture ratio of two water masses of different origin. On the one hand, water from the first flood corresponds to the rainfall in July and August. This water is highly mineralized, resulting mainly from the dissolution of salts accumulated during the dry season. On the other hand, the Logone water has a lower conductivity of about $60 \, \mu S \, cm^{-1}$ (Roche 1973).

The first ones prevail at the beginning of the humid season, the second ones at the end of the humid season. Transparency measured with Sacchi disc is never high and ranges from 65 cm at the beginning of the flood to 10 cm in February and 3 cm in May. This abrupt change corresponds to the arrival in the El Beïd of the last drainage water from the flood plain which is very rich in clayey materials.

The Yobé river whose flood occurs generally at the end of July and ends in April—May (Hopson 1972) also has an intermittent regime. The annual flow, very variable, is of about $0.6 \times 10^9 \, m^3$ (Toucheboeuf de Lussigny 1969).

2.5 Conclusion

On consideration of its morphological, climatic and hydrological features, Lake Chad appears very unstable in time and very heterogeneous in space.

The lacustrine basin occupies temporally a large region of 25 000 km² with few depressions. The hydraulic regulation plus the morphology of this basin is such that the mean depth of the lake amounts to 4 m. This depth is modified by: 1) seasonal variations reaching amplitudes of ±0.5 m, resulting from the irregular feeding of the lake by the Shari which has a regime related to the tropical type (the lake receives 50% of its water supplies in October and November) and from the lack of surface effluent likely to allow an adequate regulation of the seasonal and irregular supplies. On the contrary, evaporation is responsible for 20% of the losses and infiltration for 10%, their distribution is fairly regular throughout the year; 2) considerable annual variations reaching about 5 m. These result from the fluctuations of river supplies which can be trebled from one year to the other. They become more pronounced when there is a series of 'humid' or 'dry' years.

Consequently, if a comparison between the amplitude of these variations in the water level and the mean depth of the environment is made, it is obvious that *large areas are likely to emerge or be immersed from one period to the other.* So in 1964, the lake occupied an area of 25 000 km², while ten years later, it was reduced to an area of 9000 km² as a result of the drying up of the northern part of the basin. Only the region bordering the Shari delta is almost permanently occupied by water.

These variations in the water area are accompanied by profound changes in the natural lake regions. The lake is composed of three predominant zones which are related to the presence of an erg bordering the lake on the north and the east. These are: a region of islands when the crests are emerged, a region of islands occupied by macrophytes (*Cyperus Papyrus* and especially *Phragmites*) or reed islands when crests are immersed under 0.5–1.5 m of water, a favourable condition for the development of these plants; and a region of open water when crests are inundated and without emergent vegetation, generally the case when they are situated at depths above 1.50 m.

The fluctuations of the water level are such that, even if the type of colonization or decolonization of the regions is not as simple as mentioned previously, *profound changes in lacustrine zones must be expected both in time and space.* So, after a lake subsidence, a region of open water can be turned into a region of reed islands, a region of archipelago and a dry region within a few years. During 1973–75, numerous shallows appeared in the northern open waters and became an archipelago before drying up in 1975. During the period of flood, the opposite situation can occur.

The hydrological regime determines not only the water area and the distribution of lake zones the renewal of water in the different regions. The general water circulation is such that the renewal of water is increased by proximity to the Shari delta and by the river going through a period of flood from September to January. The first flood waters move towards the archipelago under the influence of the southwest winds up to mid-October. At that time, the harmattan, a northeast wind, takes over and changes the movement of the Shari flood waters towards the zones of open water. This change in direction allows a more thorough mixing of 'new' and 'residual' waters.

According to the rate of renewal of waters, it can be said that *the northern basin has a proper lacustrine appearance while the southern basin or at least its central zone has a more riverine appearance*. From this point of view the northern basin is more stable than the southern basin as long as the lake is not divided. This is not true during a severe contraction of the lake when the Shari no longer supplies the northern basin. It can then dry up either partially or totally over a short period of time, while the region connected to the Shari delta is always occupied by water.

The nature of the lake bottom varies considerably from one zone of the lake to the other. It is dependent mainly upon the distribution of solid suspensions from the Shari, which is itself dependent upon the general water circulation as well as the distribution among the macrophytic vegetation. Generally, the deposited materials are a clay facies in the zones of open water, a muddy facies in the zones of reed islands, and a clay-muddy facies in the archipelago zones. These three main types are subdivided into varieties. *Therefore, the nature of sediments is far from being regular and it leads to a strengthening of the spatial heterogeneity of the environment.*

In short, before dealing with the physical as well as chemical features of the waters, it is observed that Lake Chad is composed of a mosaic of biotopes, each of them likely to undergo profound modifications in time, even disappearing and reappearing according to the lake floods and recessions. Only the open waters of the southern basin do not disappear during the drying up periods.

All these environments are subject to the influence of a tropical climate. Rainfall, winds, air temperature, air humidity and evaporation are determined by the Intertropical Convergence Zone (ITCZ) which moves along a south–north line throughout the year. South of the ITCZ, a humid climate prevails and north of it, a dry climate. Continental winds such as the northeast harmattan blow over the entire lake in winter leading to the occurrence of drought from October to February–March, while the marine winds such as the southwest monsoons blow from June to September and are associated with rainfall. The lake is situated between the isohyets 550 mm in the south and 240 mm in the north and receives 50% of its rainfall in August. At that time, evaporation falls to a minimum of 16–17 cm per month, while in October–November, it increases up to 21–23 cm per month. The mean monthly air temperatures range from 29 to 32°C from March to October (with a minimum of 27°C in August) and from 22 to 24°C from December to February. Diurnal variations are considerable in winter, up to 16 to 17°C compared with 7 to 9°C in August. Air humidity reaches maximum values in August (70–80%) with minimum values in February–March (23–31%). The incident radiation amounts to 2310×10^4 J m^{-2} d^{-1} on average with low seasonal variations. These seasonal fluctuations are more pronounced than in equatorial regions, but less so than in temperate ones.

References

Billon, B. et al. 1968. Monographie hydrologique du Chari. ORSTOM Paris, 5 parts in 6 vols., 610 pp., mimeo.

Bouchardeau, A. and Lefèvre, R., 1957. Monographie du Lac Tchad. C.I.E.H. Paris, 114 pp., mimeo.

Carmouze, J. P., 1971. Circulation générale des eaux dans de lac Tchad. Cah. ORSTOM Sér. Hydrobiol. 5: 191–212.

Carmouze, J. P., Dejoux, C., Durand, J. R., Gras, R., Iltis, A., Lauzanne, L., Lemoalle, J., Lévéque, C., Loubens, G. and Saint-Jean, L., 1972. Grandes zones écologiques du lac Tchad. Cah. ORSTOM Sér. Hydrobiol. 6: 103–169.

Carmouze, J. P., 1976. La régulation hydrogéochimique du lac Tchad. Contribution à l'analyse biogéodynamique d'un système lacustre endoréïque en milieu continental cristallin. Trav. Doc. ORSTOM No. 58 Paris, 418 pp.

Dupont, B., 1967. Nature des fonds dans la zone est du lac Tchad. ORSTOM Fort-Lamy, 8 pp., mimeo.

Dupont, B., 1970. Distribution et nature des fonds du lac Tchad. Cah. ORSTOM Sér. Géol. 2: 9–42.

Durand, J. R., 1970. Les peuplements ichtyologiques de l'El Beïd. Première note. Présentation du milieu et résultats généraux. Cah. ORSTOM Sér. Hydrobiol. 4: 3–26.

Durand, J. R., 1971. Les peuplements ichtyologiques de l'El Beïd. 2 ème note. Variations inter et intraspécifiques. Cah. ORSTOM Sér. Hydrobiol. 5: 147–159.

Durand, J. R., 1978. Biologie et dynamisme des populations d'*Alestes baremoze* (Pisces, characidae) du bassin Tchadien. Trav. Doc. ORSTOM, No. 98 Paris, 332 pp.

Fotius, G. and Lemoalle, J., 1976. Reconnaissance de l'évolution de la végétation du lac Tchad entre janvier 1974 et juin 1976. ORSTOM N'Djaména, 13 pp., mimeo.

Guichard, E., 1957. Eaux du lac Tchad et mares permanentes au nord d'Ira. ORSTOM Fort-Lamy, 26 pp., mimeo.

Hopson, A. J., 1968. The gill net fisheries of Lake Chad. Federal Fisheries Service Maïduguri, 64 pp.

Hopson, J., 1972. Breeding and growth in two populations of *Alestes baremoze* (Joannis) (Pisces, Characidae) from the northern basin of Lake Chad. Overseas Res. Publ. 20, 50 pp.

Lemoalle, J., 1978. Application des images Landsat à la courbe bathymétrique du lac Tchad. Cah. ORSTOM Sér. Hydrobiol. 12: 83–87.

Riou, Ch., 1972. Etude de l'évaporation en Afrique Centrale (Tchad, R.C.A., Congo). Contribution à la connaissance des climats. Thèse Univ. Paris VII.

Riquier, J., 1963. Formules d'évapotranspiration. Cah. ORSTOM Sér. Pédol. 1: 33–50.

Roche, M. A., 1973. Traçage naturel salin et isotopique des eaux du système hydrologique du lac Tchad. Thèse d'Etat Univ. Paris VI, ORSTOM Paris, 398 pp., mimeo.

Tilho, J., 1910. Documents scientifiques de la Mission Tilho (1906–1908). Imp. Nat., Paris, vol. 1, 412 pp., vol. 2, 568 pp.

Toucheboeuf de Lussigny, P., 1969. Monographie hydrologique du lac Tchad. ORSTOM Paris, 169 pp., mimeo.

3. Physical and chemical characteristics of the waters

Jean-Pierre Carmouze, Jean Marie Chantraine and Jacques Lemoalle

Lake Chad shows several aspects of continuous change and therefore, permanent changes in the physical and chemical features of the lake must be expected both spatially and temporally. It is thus very difficult to describe the environment. We have subdivided the lake into a certain number of regions and considered parameters such as water turbulence, temperature, oxygenation and transparency as well as the chemical characteristics of the water. They were examined during 'Normal Chad' when the lake was not divided, with the north and south basin communicating, and when the lake became 'Lesser Chad' as a result of a lowering of water level, when the two basins were separated.

3.1 Physical characteristics of the water

The main physical features such as turbulence, temperature, oxygenation and transparency were largely determined by a combination of morphological, sedimentological and climatic features of the environment (cf. Chapter 2).

3.1.1 *Wind disturbance*

Winds blow from the northeast during winter and from the southwest in summer (cf. para. 2.1, Chapter 2). They blow mainly in the morning with an average velocity above $5 \mathrm{m\,s^{-1}}$ for 6 hours a day and above $6.4 \mathrm{~m~s^{-1}}$ for 2.7 hours a day. They generate both vertical circular movements of the water which decrease with depth and water movement along the direction of the waves which are equivalent to a current. An eddying current, whether it is synchronous or not, is associated with this water movement.

These two kinds of movement depend upon the wave height which itself depends upon the range of wind action over the water surface, the fetch. The wave height, H, increases with the fetch and, for a given fetch, H is proportional to the average wind velocity over the fetch insofar as the wavelength remains

J.-P. Carmouze et al. (eds.) Lake Chad
© *1983, Dr W. Junk Publishers, The Hague/Boston/Lancaster*
ISBN 978-94-009-7268-1

below four times the depth. Generally, this was not the case in the open water areas during the period of 'Normal Chad'. Let us take an example: for a fetch of 20 km in length and a wind of 4.5 m s⁻¹ velocity (in these regions, these values are often higher), the theoretical wave height is 0.4 m and the wavelength λ is 12 m. In the regions concerned, where depths 6–8 m, the whole water column is submitted to a high turbulence for its depth amounts to $\lambda/2$. (Smith and Sinclair 1972).

Moreover, considering the direction of the prevailing winds, waves move either towards the low shores of the western or southern coast during the dry period or towards the reed islands which correspond to shallow bottoms during the monsoon period. Therefore, the coastal zone and the reed islands experience maximum turbulence respectively in December–January and July– August.

Water transport by waves results in diurnal oscillations of the water level, whose amplitude generally reached 10 cm during the period of 'Normal Chad'. The same phenomenon was observed in the Southeastern Archipelago at Bol, in the center of the north basin at the island of Kindjéria, and on the west coast of the Northern Open Water at Malamfatori. These variations in level were associated with currents in the whole lake and especially in the channels between the islands of the archipelago.

During the period of 'Lesser Chad', the various open water areas were reduced and the fetch was diminished by the presence of new islands or reed islands. However, the principal obstruction to the fetch in this zone was the orientation of the islands, perpendicular to the direction of the main winds. Wave propagation was therefore limited in these environments. Moreover, the vegetation considerably dampened water movements, especially in the Great Barrier and in the Archipelago, resulting in a disappearance of the diurnal oscillations and their corresponding currents (Fig. 1).

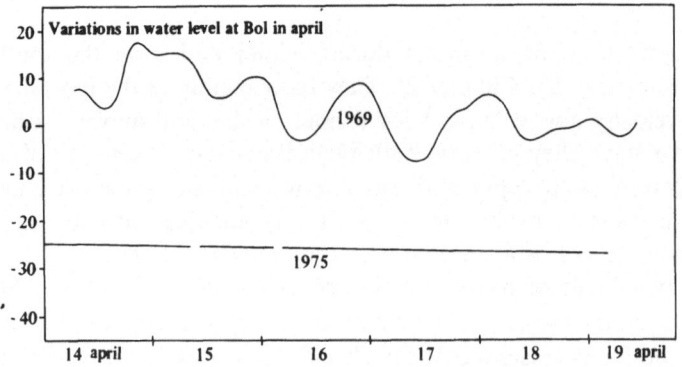

Fig. 1 Variations of the lake water level at Bol during April 1969 and 1975.

3.1.2 *Water temperature*

The surface water temperatures in the lake were studied mainly in the Southeastern Archipelago at Bol (Roche 1973; Lemoalle 1979) and in the Northern Open Water at Malamfatori (Robinson 1968; Tobor, personal communication).

3.1.2.1 *Annual variations.* The annual variations in water temperature were low as shown by the data in Table 1. This table gives the average values of temperature measured twice a day at Bol, in the Southeastern Archipelago during different periods ranging from 2 to 3 years. These temperatures are applicable to the whole lake, because the annual average values of the two basins are quite similar (Lemoalle 1979).

3.1.2.2 *Monthly averages: influence of total radiation.* Generally, rainfall results in a slight decrease in surface water temperature during July–August, which is barely noticeable when the rainy season is slight as during the recent period from 1972 to 1975. Riou (1972) showed that the decrease in the average air temperature during the rainy season resulted mainly from a drop in maximum temperatures. Minimum air temperatures are much less influenced by rainfall (Fig. 19, Chapter 2) and follow after a slight delay the variations in total radiation, G_0. The same is true for the average water temperature (Fig. 2). At first glance solar radiation, G_0, is the main factor controlling the surface temperature of the lake.

In other words, over the year, variations in lake temperature depend upon a factor which is directly related to its geographical location. Thus Lake Chad is similar to several other African lakes in which a relationship between the latitude and the amplitude of temperature variations over the year has been shown (Talling 1969).

Table 1 Annual averages of the surface water temperature in the Southeastern Archipelago, at Bol.

Period	Temperature in °C	Authors
1956–60	27.4	Billon et al., 1963
1964–66	26.6	ORSTOM hydrology
1968–70	25.7	Roche 1973
1974–75	25.6	Lemoalle 1979
1908[a]	26	Tilho 1910

[a]Insufficient data for adequate calculation of the average temperature at Bol.

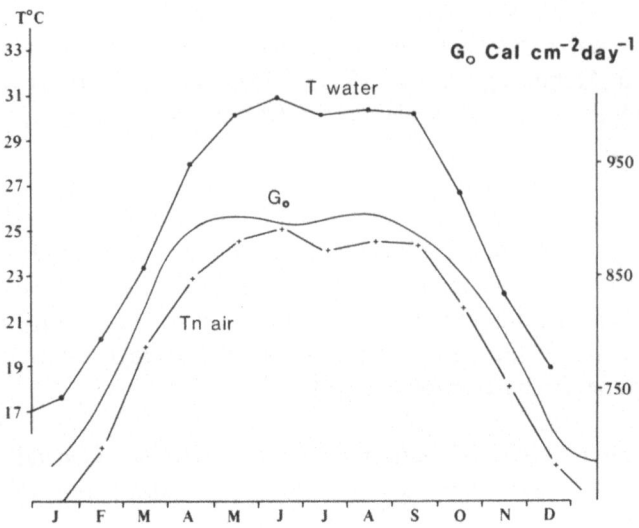

Fig. 2 Air and water temperatures and global irradiance during the year.

3.1.2.3 *Short-term seasonal variations; influence of the Intertropical Convergence Zone (ITCZ).* We have just seen that the general pattern of water temperature change over the year is mainly determined by the variations in solar radiation G_0. However, it depends upon numerous minor climatic factors such as air temperature and humidity, wind velocity and total radiation at the ground, which modulate heat transfers between the lake and the atmosphere. None of these parameters, when taken individually, can account for the variations in water temperature. So we considered the oscillations in latitude of the Intertropical Convergence Zone (ITCZ)* which include these different parameters. These oscillations, averaged over 5 days (pentads) for the year 1975 are illustrated in Fig. 15, Chapter 2. South of the ITCZ the humid winds from the southwest, the monsoon, prevails, and the dry winds from the northeast, the harmattan, prevails north of this zone. When moving over the lake, the ITCZ modifies not only the conditions for insolation and evaporation, but also the wind regime which, during the period of 'Normal Chad' changes the slope of the water surface (Toucheboeuf et al., 1969). Thermal transfers between the lake and the atmosphere are facilitated when the ITCZ moves from the north to the south of the lake.

Generally, when moving from the north to the south in September–October, the ITCZ results in an abrupt drop in temperature, but when moving from the

---

* The ITCZ is defined in para. 2, Chapter 2.

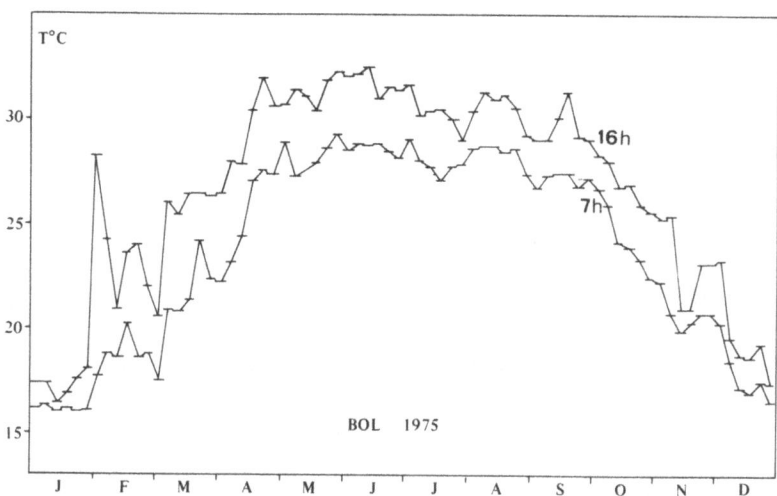

Fig. 3 Surface temperatures at 7.00 and 16.00 hr at Bol. Mean values calculated for five consecutive days.

south to the north in April–May, it prevents the water from warming, as is clearly shown in Figs. 15 (Chapter 2) and 3.

In the middle of the cool season, the oscillations of the ITCZ are felt in the far north of the convergence zone. A northern movement of the ITCZ induces a temporary decrease in the pressure gradient over the lake which results in calmer weather and a considerable diurnal increase in temperature.

During the periods of drought which sweep down the Sahel, the ITCZ is situated a few degrees further south than during the rainy periods (Dorize 1974). The dry season then appears earlier in the year and the lake is affected by northern air masses sooner, giving rise to an early water cooling and lower annual average temperatures than during normal periods.

3.1.2.4 *Diurnal variations; temperature profiles at Bol.* The diurnal variations in surface temperatures represent the heat accumulated at the surface in the daytime. Therefore, they are more important when the water is shallow and when the turbulent transfers with the underlying water layers are limited. The annual average of the diurnal variations measured by Roche (1972) for the period 1956–70 at Bol was 2.14°C, while the variations between 1967 and 1970 were lower amounting to 1.87°C. From 1972, the variations have become much greater with a maximum in 1973 (Table 2) due to a lowering of the water level and to the changed hydrological conditions.

3.1.2.5 *Vertical distribution of temperature.* Temperature profiles were measured in relation to depth at different times of the day in the Southeastern

69

Table 2 Annual averages of diurnal variations.

Years	From 1956 to 70	1972	1973	1974	1975
Differences in °C	2.1	2.7	4.6	3.7	2.8

Archipelago over two distinct periods: during 'Normal Chad' at a station east of Bol (A) in 1968–69; and during 'Lesser Chad' at a station south of Bol (C) (Fig. 4).

At station A, the water–sediment interface was not well defined since mud was very fluid at that time. So the deepest measurements sometimes indicate the temperature of the surface sediment which may be slightly different from the temperature of the bottom water. In 1974 the station C corresponded to a trough compared with the whole Southeastern Archipelago (the point A then being in the middle of an ambadj forest). This station was not representative of the whole region which nearly became a swamp at that time. However it could be considered as showing existing conditions in the open water areas.

Temperatures at Bol during 1968–1969. At 7 a.m. the water column was homogeneous from the surface to the bottom during all periods considered, except one which was characterized by a current when the measurements were

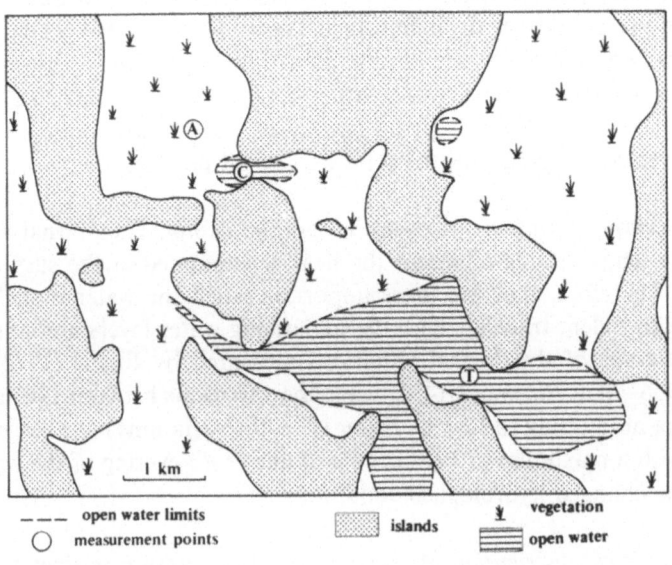

Fig. 4 Localization of temperature measurements in the Southeastern Archipelago.

70

taken (on January 15th, 1969). Therefore, no stratification exceeding 24 hours was observed over this period. Stratification itself was thus an unusual and short-lived phenomenon which appeared in the middle of the afternoon when surface waters warmed up and disappeared at the end of the day. The difference in temperature between the surface and bottom averaged 0.7°C at 18.00 hours.

Isothermy was likewise infrequent over a period of 24 hours and sometimes appeared during the lake cooling at the beginning of the dry season or during the rainy season when the weather is particularly overcast. The most frequent case was thus the occurrence of a temperature gradient at the beginning of the afternoon between the water surface and the bottom which disappeared at the end of the day. So, turbulence was insufficient for fast and complete propagation of the surface heat but it was sufficient to destroy stratification at the end of the day. The same was true for the Northern Open Water where larger waves allowed heat to be distributed in a similar way in deeper zones (Robinson 1968; Hopson, personal communication). Therefore, during the period of 'Normal Chad', Lake Chad as a whole could be considered as a tropical polymictic lake.

Temperature profiles at Bol during 1975 (Fig. 5). At Bol, in 1975, there was only a large pond of open water where the measurements were taken (Fig. 4). At the beginning of warming up of the water in February–March, a stable stratification began to develop and remained constant until the flood of early September, in spite of a considerable lowering of the water level. Throughout the flood, there was homothermy but in February 1976, stratification appeared again when the water began to warm up. Under such circumstances, therefore, the lake arm at Bol behaved as a warm monomictic lake.

A comparison of the two periods reveals that an annual cycle of stratification which did not exist in 1969 appeared at Bol in 1975. Since the average water level was almost the same for these two periods, this change was mainly explained by the presence of vegetation in 1975 which divided the open water areas and provided them with better shelter from the wind. Therefore, with similar winds, waves were lower and turbulence was reduced. As the wind action was reduced, the daily oscillations in the water level became insignificant (Fig. 1) and the corresponding currents were obstructed and even suppressed by the vegetation.

Finally, during the contraction from 'Normal Chad' to 'Lesser Chad' there were two important changes:

— there was a drop in the annual average temperature during the lake major reduction. In fact, the disturbances of the annual ITCZ oscillation accounted for this cooling (Lemoalle 1979).

— during the normal period, the entire lake could be considered as a tropical polymictic lake. During the period of 'Lesser Chad', a warm monomictic regime appeared in the zones of the archipelago, while the open water areas retained their polymictic character.

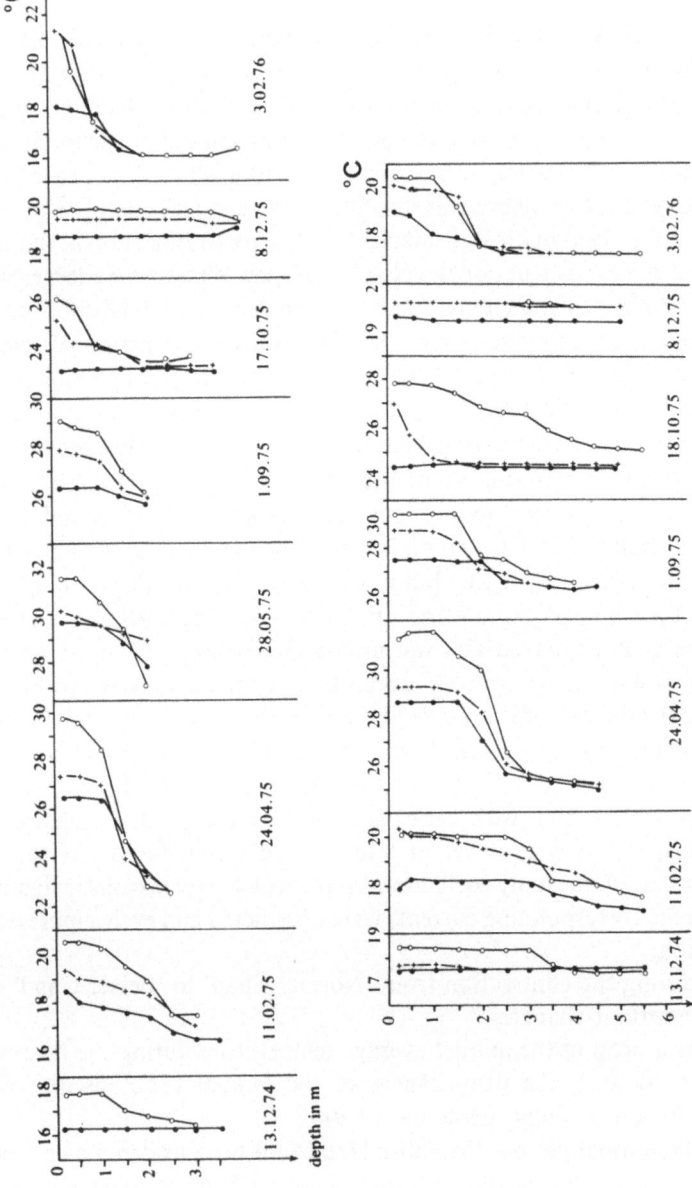

Fig. 5 Temperature profiles at Bol and Berim.

3.1.3 *Dissolved oxygen content of the water*

The dissolved oxygen content of the water was studied mainly by Lemoalle (1979) in the Southeastern Archipelago from 1968 to 1975 and in the north Basin in 1974 and 1975, as well as by Robinson (1968) in the Northern Open Waters. The most detailed study was undertaken at Bol along with the above-mentioned study of temperatures. The main conclusions were drawn from this and allowed a more precise definition of the conditions for oxygenation in the Southeastern Archipelago and in the whole lake, for which we have very limited measurements.

3.1.3.1 *Oxygen profiles at Bol.* The oxygen profiles were determined by a galvanic probe (Mackereth type, 1964). In the same profile, sensitivity was 1% of the saturation value; and the reproducibility between the different profiles was 3%. The results, shown by the probe in % saturation have the advantage over those expressed in concentration (mg O_2 1^{-1}) of giving a better representation of the deviation in relation to the equilibrium when temperature is not mentioned.

a) *Oxygen profiles at Bol in 1968.* The vertical distribution of oxygen concentrations was homogeneous at the beginning of the day: the values found at 7.00 hr ranged from 80 to 100% saturation. During the day, the oxygen pressure increased from the surface towards more or less important depth. During the warm season, a diurnal stratification appeared which did not exist from October to December (Fig. 6).

This temporary stratification could result from a combination of several phenomena: warming of the surface water which increased the oxygen partial pressure even if there was no change in its concentrations; photosynthetic activity of the phytoplankton which generated oxygen and finally, respirations of organisms causing a decrease in oxygen content in the water especially in the bottom layers.

Without turbulence, a saturation gradient appeared, as clearly seen in the example from May 4th, 1968 (Fig. 6). However, it was an exception because generally the water turbulence resulted in an homogenization of the upper layer down to a depth that varied with weather. Generally, this turbulence extended to the whole column during the morning while a stratification appeared in the afternoon.

b) *Oxygen profiles at Bol in 1975.* After the lowering of the water level in 1973, the water area was reduced to a pond which was surrounded by macrophytes through which water flowed, especially during floods (Fig. 4).

The vertical homogeneity appeared in the morning only during the cool season (from October to December 1974 and 1975) and unlike 'Normal Chad', the oxygen pressures were close to 0 (Fig. 6). In the course of the day, oxygen in the upper layers reached 12% saturation. At that time, water flowed through

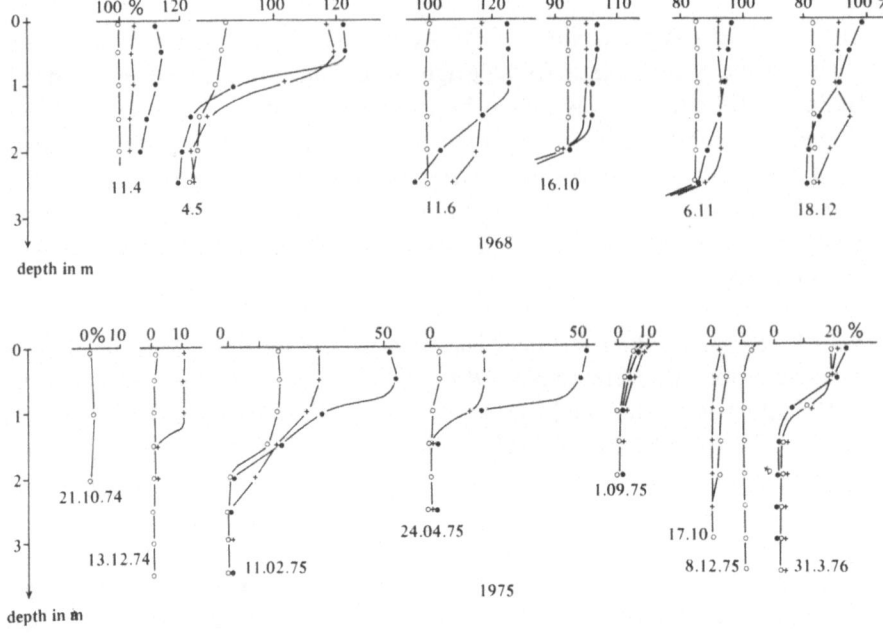

depth in m

depth in m

Fig. 6   Oxygen profiles at Bol.

the macrophytes which retained the oxygen-generating phytoplankton while releasing dissolved oxygen consuming organic matter.

During the warm season, the water circulation and mixing were less, so that warming of the water gave rise to a stable stratification. The oxygen pressure in the epilimnion changed considerably from 5 to 60% during the afternoon, while the hypolimnion remained anoxic. Some tornadoes such as that which occurred on August 22nd, 1978 can destroy this stratification, which reappeared quickly. Phytoplankton concentrations (1-9-75) could also be reduced by a horizontal small scale circulation of water.

Generally, the small open water zones in the archipelago were more influenced by water circulation than the larger ones which were more markedly of lacustrine type. The decrease in oxygen due to horizontal circulation was less in these larger areas.

3.1.3.2   *Oxygenation in the Southeastern Archipelago*. During the period of 'Normal Chad', the station at Bol, in the middle of the open water had no special feature compared with the rest of the Archipelago. Turbulence may have been slightly higher here than closer to the shore and lower than in the channels situated between two islands. But the difference was undoubtedly insignificant and the same phenomena were likely to occur everywhere.

74

During the period of 'Lesser Chad', three different environments must be distinguished:

— the archipelago open waters which were stratified in relation to their area and depth during the warm season;

— the zones of emergent vegetation where major decomposition of organic matter occurred. The oxygen content was very low except on occasions in surface water when there was a supply of oxygen from the surface open water or even from the photosynthesis of the periphyton (from 5 to 10% of the saturation);

— the coastal zones close to the shore which were often devoid of dense vegetation. Benthic algae grew there in water 5 to 30 cm deep. During the day, photosynthesis produced considerable supersaturation, while during the night, the high surface/volume ratio prevented complete deoxygenation. Such environments were occupied particularly by very young *Tilapia*.

In the Southeastern Archipelago as a whole, the period 'Lesser Chad' was characterized by a general oxygen deficit in comparison with the period of 'Normal Chad' with important consequences for the fish populations as will be shown in Chapter 10.

3.1.3.3 *Oxygenation in the whole lake.* During the period of 'Normal Chad', measurements were taken in the Northern Open Water (1967 and 1968) in the middle of the day. Surface supersaturation often occurred, especially during the warm season when a diurnal thermal stratification occurred. These oxygen data of the Northern Open Water were similar to those of the station at Bol in 1969, whose results can thus be considered as representing the whole lake at that time.

In the north basin, water became more and more turbid during the drying up period due to the increase in bottom turbulence and phytoplankton concentration. Various combinations of these two factors resulted in a high variability in oxygen content temporally and spatially even on a small scale. For instance, in the Northern Reed Islands, in June 1974, the influence of high phytoplankton concentration was observed during the day and resulted in variations of 20 mg O_2 l^{-1} in 12 hours occurring at 10 cm below the surface. Since the environment consumed 3.9 mg O_2 $l^{-1}h^{-1}$, anoxia occurred during part of the night. At that time, on overcast sky during one day only could result in a decrease in photosynthesis, creating even more drastic conditions.

Finally, during the period of 'Normal Chad', the distribution was always homogeneous over the entire water column in the whole lake at the beginning of the day (75 to 100% saturation). In the course of the day, there was an increase in surface oxygen; when the weather was calm a diurnal stratification appeared, especially during the warm season.

During the period of 'Lesser Chad', the conditions for oxygenation were very variable from one region to the other. In the Southeastern Open Water, conditions remained similar to those of 'Normal Chad' with some modifications appearing near the shore due to vegetation. The Southern Archipelago was characterized by very low oxygenation which varied according to the period: during the cool season, oxygen in the surface layers reached 12% saturation in the course of the day compared with 5 to 60% during the warm season. The hypolimnion generally remained anoxic. In the north basin, during the drying up period in 1974 and 1975, no stable stratification was revealed but there were variations of great amplitude which were marked by periods of frequent anoxia. In 1976, when the water level of this basin increased again, water was enriched by organic matter and the conditions for oxygenation remained very variable.

3.1.4 *Water transparency*

3.1.4.1 *Local characteristics of transparency*. Since 1964, water transparency has been measured by a Secchi disc (Gras et al. 1967; Roche 1973; Lemoalle 1979). This allowed three main environments to be distinguished: the open water of the south basin, the archipelagos of the south basin and the north basin, each characterized by seasonal and multi-annual variations.

(a) In the open water of the south basin, the seasonal change was characterized by maximum transparency in December–January during the high water period and minimum transparency in June–July during the low water period. This was true during the period of 'Normal Chad' as well as during the period of 'Lesser Chad' (Fig. 7a,b). Over the years, the lowest value recorded was related to the lowest levels observed during 1973 (Fig. 7c). Generally, the predominant transparency factor was the variation in water level. The clayey suspensions carried by the Shari at the beginning of the flood reached the lake only in July–August. So, although they contributed to the seasonal minimum transparency, they were not its main cause.

(b) In the archipelago of the south basin, water transparency increased towards the north. In other words, transparency increases with the movement of water between islands. Seasonal variations were less defined than in open water. At the station of Bol, since 1964, the change over the years has been characterized by a decrease in water transparency along with a gradual decrease of the water level up to 1973, and from that time, by an increase in transparency. This increase was related to the development of macrophytes. The considerable decrease in transparency that was continuous from 1964 to 1973 largely concealed possible seasonal changes. On the contrary, from September 1973, seasonal variations in transparency became abrupt and marked when flood water penetrated into the large vegetation banks bordering the archipelago. The transparency measured with the Secchi disc, SD, increased from 5 to 40 cm over 10 days at the end of 1973, from 15 to 35 cm in December

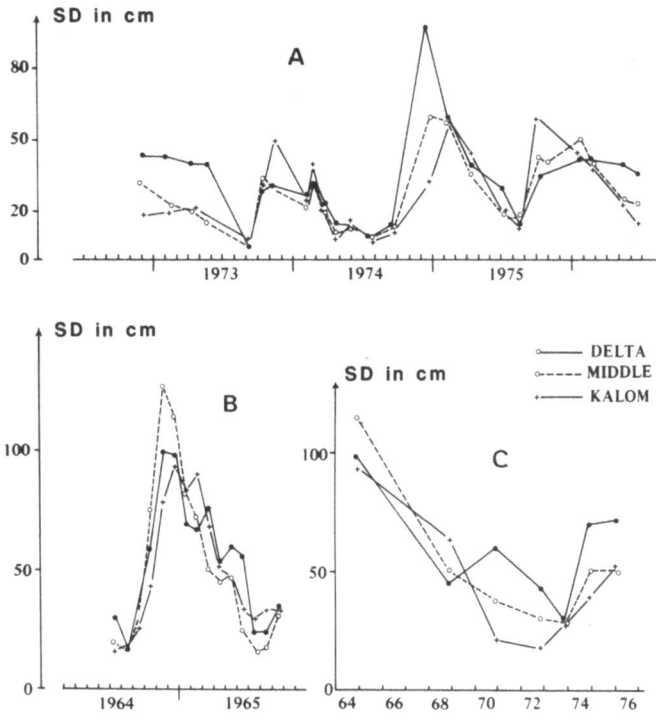

Fig. 7 Changes in transparency in the Southern Basin.

1974 and from 40 to 85 cm in January 1975. The highest values were observed at the beginning of the year (70 to 90 cm) and the lowest ones during the period of low water (Fig. 8). Vegetation acts as a filter for waters penetrating the archipelago. These waters deposit their clayey suspensions that are largely

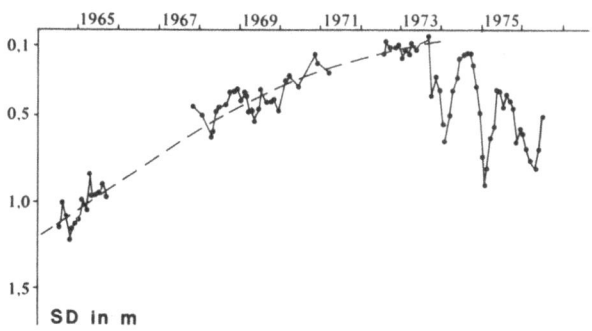

Fig. 8 Long-term changes in transparency at Bol.

responsible for their turbidity and in return they take up dissolved matter. After flowing through the macrophytes, the water is therefore relatively clear but, given the conditions of agitation and the concentration of dissolved iron, a rustcoloured precipitation may form, further reducing the transparency.

(c) Throughout the north basin, the change in transparency over the years was studied from April 1968 to April 1976. This period corresponded to the gradual drying up of the basin before its water level increased again at regular intervals after 1976 (Fig. 9).

In April 1968, transparency measured with a Secchi disc, SD, was 50 cm in the open water and the region of the reed islands and 80 cm in the archipelago itself, the values being higher closer to the shore. In April 1971, SD ranged from 28 to 41 cm in the center of the basin. The difference between the open water and the archipelago, which was fairly well marked for the rest of the year was not clear at that time.

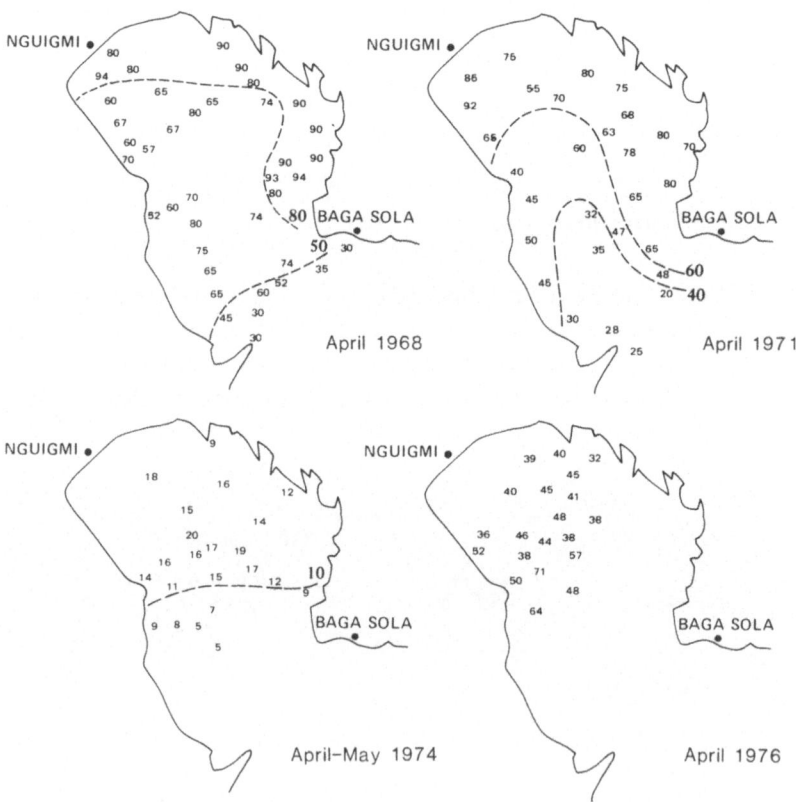

Fig. 9 Water transparencies in the Northern Basin (cm).

During April–May 1974, the north basin was isolated from the rest of the lake because of the drying up of the Great Barrier. The south basin and some parts of the archipelago with limited depth had a particularly low transparency of about 10 cm. The high phytoplankton biomass, as well as a resuspension of sediments by wave action caused short-term fluctuations of SD usually ranging from 10 to 20 cm in the center of the zone still under water.

In February 1975, SD ranged from 10 to 20 cm around Kindjéria. The water was very clayey and transparency was more reduced in the rest of the basin whose mean depth was 50 cm. Finally, in April 1976, after the complete drying up of the north basin in October 1975, water masses that percolated through the 'Great Barrier' were clearer with SD ranging from 32 to 71 cm, without any distinct zonation (Fig. 10).

3.1.4.2 *Spectral characteristics of the water.* Two series of measurements were undertaken: the first one in 1970–1971 covered the whole lake during the period of 'Normal Chad', and the second one covered only the south basin during the period of 'Lesser Chad'. For all stations, the vertical attenuation coefficient of the decreasing illumination, K_{min}, or extinction coefficient, was maximum in the blue region. However, according to the least absorbed wavelength, two extreme shapes and an intermediate one for attenuation spectra could be distinguished (Fig. 11):

— a spectrum with minimum absorption around 560 nm corresponding to grey waters, loaded with clay particles. This type of water is called clayey water;
— a spectrum with a considerable decrease from the blue to the red colour was caused by dissolved organic matter and almost no suspended material. This type of water is called organic water;
— finally, an intermediate spectrum reaching a plateau between 600 and 640 nm, was found for water rich in plankton, when water concentrated during the period of 'Lesser Chad'. This type of water is called concentrated water.

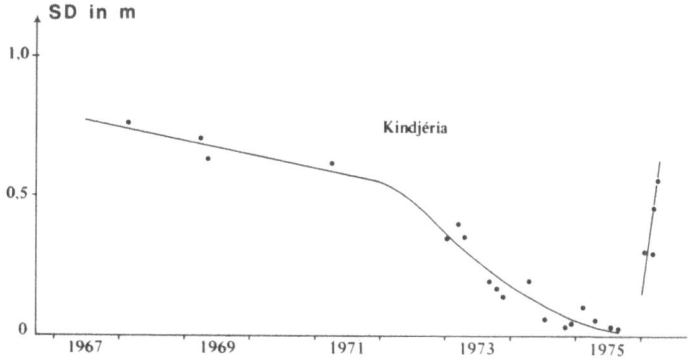

Fig. 10 Long-term changes in transparency in the Northern Basin.

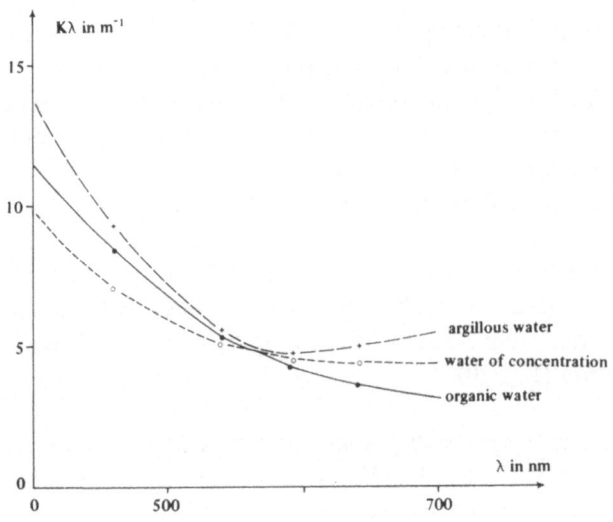

Fig. 11 Three spectral types of the attenuation coefficients Kλ of light.

Thus, the shape of the spectra as well as the values of the extinction coefficients extended over a wide range, emphasizing the heterogeneity of the lake water both in time and space. The differences in water types agreed with the results obtained in other lakes. Dokulil (1973) showed the difference between shore water occupied by *Phragmites* stands with dissolved organic matter but little suspended matter, whose K_{min} was around 700 nm, and water in the center of Neusiedlersee with significant mineral turbidity whose K_{min} was about 550 nm. For Lake George which is very rich in phytoplankton and has a marked organic colour, Ganf (1974) gives an attenuation spectrum with a plateau between 520 and 680 nm. The role played by phytoplankton in the evolution of K_{min} towards 590 nm was emphasized by Bindloss (1976) and it has also been observed elsewhere (Vollenweider 1961; Sauberer 1962). Talling et al. (1973) gave results ranging from the pure organic type Lake Aranguadi to the eutrophic organic type, Lake Kilotes (in July) with coefficients generally much higher than in Lake Chad.

As most of the transparency measurements were made with a Secchi disc, Lemoalle (1979) related the transparency values obtained with the Secchi disc to those of the attenuation coefficients from the two main types of water, the 'organic' water and clayey water. These results are shown in Table 3.

3.1.4.3 *Transparency factors.* Three main absorption factors, which are not mutually exclusive, help to account for the observed phenomena and for the three types of water.

During the period of 'Normal Chad', mineral particles were resuspended

Table 3 Relation between the different optical parameters of the two main types of water in Lake Chad.

Relation	Clayey water	Organic water
Energy E_{SD} at depth SD	$E_{ED} = 0.17 \, E_0'$	$E_{SD} = 0.09 \, E_0'$
Depth of the euphotic zone Z_{eu} (1% E_0')	$Z_{eu} = 2.79$ (SD)	$Z_{eu} = 1.88$ (SD)
Relation between K and SD (on Z_{eu})	$1/K = 0.61$ (SD)	$1/K = 0.41$ (SD)
Relation $K_{min} - SD$	K_{min} (SD) $= 1.39$	K_{min} (SD) $= 1.95$

E_{SD} = irradiance measured at the depth of Secchi disc;
E_0' = subsurface irradiance;
Z_{eu} = depth of 1% of E_0';
K = vertical attenuation coefficient related to the entire visible spectrum;
K_{min} = vertical attenuation coefficient related to the least attenuated wavelength.

from the sediment. Their concentration was dependent upon the fetch, the depth and the type of sediment, and they were responsible for light attenuation. With a decrease in water when the different basins were isolated, a period of concentration occurred during which the phytoplankton became very abundant and were responsible for most of the attenuation despite the considerable amount of suspended sediment. Finally, during the period of 'Lesser Chad', two types of water mass were observed: that which flowed through the macrophyte barriers, and that of the open water of the south basin, originating mainly from the Shari. In the latter, the water quality was the same as that of the period of 'Normal Chad' with, however, a fetch that was limited by the decrease in open water areas and the emergence of numerous reed islands. Elsewhere, higher plants filtered the water masses and dampened turbulence. The particles originating from the sediments then had a very low concentration. The dissolved organic matter collected by the water when flowing through the macrophytes contributed to the colour of 'organic water' and to the attenuation of light, while the intervention of the phytoplankton had a variable role in changes in transparency.

The influence of depth which thus appeared as a decisive factor in the determination of transparency, is shown in Fig. 12.

3.2 Chemical characteristics

Since the purpose of this chapter is to give basic data characterizing the different environments, we will describe the chemical composition of water in the different regions during the period of 'Normal Chad' and 'Lesser Chad' without, however, dealing with controlling factors which are covered in Chapter 4.

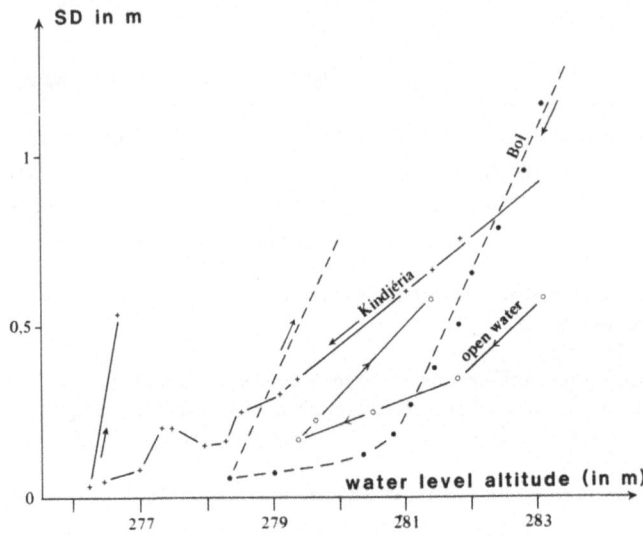

Fig. 12 Relation between transparency and lake water level at Bol and Kindjéria.

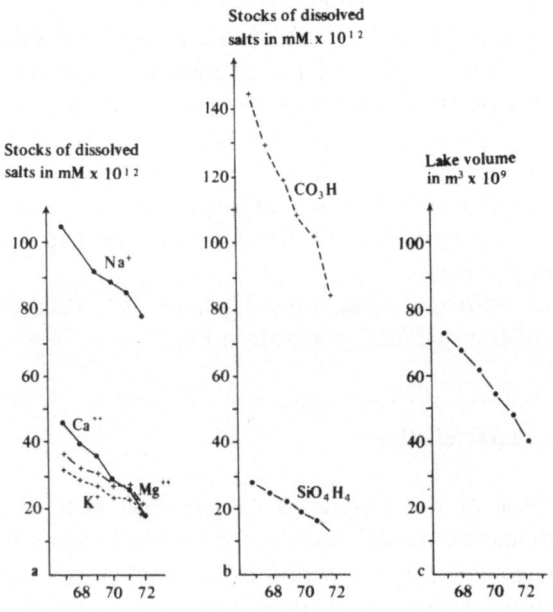

Fig. 13 Changes in the dissolved salt stocks in the lake during lake recession.

82

The hydrochemistry was characterized by profound changes throughout the environment and by strong seasonal and annual variations at each point, and it is thus difficult to describe. However, from 1966 to 1972, the annual variations can be ignored. In fact, changes in the stock of dissolved salts were roughly similar to those of lake volume during this period of slow subsidence (Fig. 13), suggesting that there were few modifications in the dissolved salt concentrations. Moreover, spatial modifications were reduced by dividing the lake into eleven main regions (Carmouze 1976). The mean, maximum and minimum values of the main hydrochemical features were determined seasonally for eleven regions (Table 4).

Due to variations in conductivity, the water salinity increased with increasing distance from the Shari delta. Thus, water from the South-eastern Open Water near the delta was barely 1.2 times as concentrated as that of the Shari (cf. Chapter 4), while water of the Northern Reed Islands and the North-eastern Archipelago was respectively 11.8 and 11.7 times as high. More generally, the water salinities in the north basin were on average four times as high as those of the south basin which were twice as high as those of the river water (60 mg l^{-1}).

Salinity was composed of sodium, carbonates and bicarbonates, calcium, potassium and magnesium ions and dissolved silica. Its relative composition changed considerably from one region to the other*. The relative HCO$_3^-$–CO$_3^-$ content (represented by (Alc) in Table 4) was higher in the north basin than in the south basin (91.1% compared with 75.6%). It was the reverse with dissolved silica (8.9% to 24.3%). The relative sodium content and, to a lesser extent, potassium content was higher in the north basin (35.8 to 29.9% for sodium; and 10.4 to 9.4% for potassium). It was the reverse with calcium and magnesium content (29.5 to 35.0% for calcium and 24.3 to 25.6% for magnesium). Actually, all these modifications were gradual, so that lake water was closer to the bicarbonate-sodium pole with increasing distance from the Shari delta.

The local seasonal variations gain importance as the regions concerned receive flood waters directly from the Shari. Such was the case for the Southeastern and Eastern Open Waters, the Southeastern and Southern Reed Islands, the Great Barrier, the Northern Open Water and the Northeastern Reed Islands. In this region, variations in salinity ranged from 30 to 50% with a maximum of 80% observed in the Southern Reed Islands. As far as this last feature is concerned, the archipelago regions were more stable.

Generally, the pH of water in the south basin did not exceed 8; and therefore, this water was low in alkalinity. On the contrary, in the north basin, the pH

---

* The relative chemical composition of water was defined as follows: the absolute concentrations (i) are expressed in mé l^{-1} or mM l^{-1}, and the relative concentrations (i)$_r$ in %.

Table 4 Hydrochemical features of the main natural regions of 'Normal Chad'. $[1]_m$ $[i]_M$ and $[i]$ represent respectively the minimum, maximum and mean concentrations of the element, i, expressed in mé l^{-1} (except SiO_4H_4 which is expressed in mM l^{-1}). Conductivity is expressed in $\mu S\,cm^{-1}$.

$[Na]_r = [Na]/[\Sigma \text{ cations}]$; $[Ca]_r = [Ca]/[\Sigma \text{ Cations}]$; $[Mg]_r = [Mg]/[\Sigma \text{ cations}]$; $[K]/[\Sigma \text{ cations}]$; $[Alc]_r = [Alc]/[Alc] + [SiO_4H_4]$; $[SiO_4H_4]_r = [SiO_4H_4]/[Alc] + [SiO_4H_4]$. $[Alc] = \text{Alkalinity} = [HCO_3] + 2[CO_3]$

	South Eastern Open Waters				Southern Open Water			
	$[i]_m$ (Sept.)	$[i]_M$ (June)	$[i]$	$[i]_r$	$[i]_m$ (Sept.)	$[i]_M$ (June)	$[i]$	$[i]_r$
Na	0.115	0.20	0.18	24.9	0.21	0.44	0.28	30.7
Ca	0.17	0.30	0.24	37.5	0.30	0.42	0.31	33.9
Mg	0.14	0.22	0.18	28.5	0.20	0.36	0.23	25.2
K	0.045	0.07	0.055	8.7	0.08	0.13	0.09	9.8
Alkalinity	0.47	0.75	0.62	71.0	0.80	1.30	0.89	74.0
SiO_4H_4	0.20	0.25	0.25	29.0	0.34	0.30	0.31	26.0
pH	7.0	7.4	7.2	—	7.0	7.5	7.2	—
Conductivity	45	70	58	—	75	124	83	—

	Southern Reed Islands				Southeastern Reed Islands			
	$[i]_m$ (Dec.)	$[i]_M$ (June)	$[i]$	$[i]_r$	$[i]_m$ (Dec.)	$[i]_M$ (June–July)	$[i]$	$[i]_r$
Na	0.30	0.09	0.52	35.6	0.20	0.32	0.24	25.3
Ca	0.40	0.80	0.43	29.4	0.30	0.48	0.36	37.9
Mg	0.30	0.62	0.36	24.6	0.23	0.37	0.27	28.4
K	0.11	0.25	0.15	10.2	0.065	0.10	0.08	8.4
Alkalinity	1.08	2.50	1.42	78.9	0.77	1.25	0.93	73.8
SiO_4H_4	0.26	0.44	0.38	21.1	0.28	0.43	0.33	26.2
pH	7.3	7.7	7.5	—	7.2	7.7	7.4	—
Conductivity	101	234	134	—	72	118	87	—

	Southeastern Archipelago				Eastern Archipelago			
	$[i]_m$ (Dec.)	$[i]_M$ (July)	$[i]$	$[i]_r$	$[i]_m$ (Feb.)	$[i]_M$ (August)	$[i]$	$[i]_r$
Na	0.23	0.38	0.32	27.1	0.75	0.92	0.84	32.5
Ca	0.36	0.52	0.44	37.3	0.80	1.00	0.91	35.3
Mg	0.27	0.38	0.32	27.1	0.53	0.65	0.59	22.8
K	0.08	0.13	0.10	8.5	0.22	0.27	0.24	9.3
Alkalinity	0.92	1.40	1.15	74.2	2.25	2.77	2.53	78.8
SiO_4H_4	0.36	0.46	0.40	25.8	0.60	0.75	0.68	21.2
pH	7.2	7.8	7.4	—	7.5	8.0	7.7	—
Conductivity	86	133	105	—	211	259	237	—

Table 4 (Continued).

	Great Barrier				Northeastern Reed Islands			
	[i]$_m$ (Oct.–Sept.)	[i]$_M$ (May–June)	[i]	[i]$_r$	[i]$_m$ (Sept.)	[i]$_M$ (May)	[i]	[i]$_r$
Na	0.30	1.28	0.68	25.8	1.40	1.85	1.62	33.6
Ca	0.37	1.06	0.66	31.9	1.20	1.68	1.48	30.7
Mg	0.35	0.90	0.52	30.2	0.94	1.40	1.23	25.5
K	0.14	0.35	0.21	12.0	0.41	0.61	0.49	10.1
Alkalinity	1.18	3.98	2.03	81.8	3.85	5.62	4.73	89.7
SiO$_4$H$_4$	0.43	0.56	0.45	18.2	0.50	0.57	0.54	10.2
pH	7.3	8.5	8.0	—	8.0	8.8	8.5	—
Conductivity	111	373	190	—	360	526	443	—

	Northeastern Archipelago				Northern Open Water			
	[i]$_m$ (Sept.)	[i]$_M$ (July)	[i]	[i]$_r$	[i]$_m$ (Nov.)	[i]$_M$ (Sept.)	[i]	[i]$_r$
Na	2.58	2.95	2.82	36.7	1.40	1.97	1.59	35.5
Ca	1.90	2.18	2.00	27.4	1.20	1.54	1.32	29.5
Mg	1.79	2.00	1.89	24.6	0.99	1.35	1.10	24.5
K	0.78	0.92	0.86	11.2	0.38	0.52	0.47	10.5
Alkalinity	7.05	8.05	7.34	91.8	3.60	4.90	4.35	90.6
SiO$_4$H$_4$	0.59	0.69	0.65	8.1	0.43	0.53	0.45	9.4
pH	8.3	8.9	8.7	—	7.5	8.5	8.1	—
Conductivity	660	754	687	—	337	458	407	—

	Northern Reed Islands				Southern Basin		Northern Basin		Lake
	[i]$_m$ (Dec.)	[i]$_M$ (Oct.)	[i]	[i]$_r$	[i]	[i]$_r$	[i]	[i]$_r$	[i]
Na	2.67	2.95	2.82	35.3	0.35	29.9	2.06	35.8	1.46
Ca	2.20	2.49	2.36	28.6	0.41	35.0	1.70	29.5	1.25
Mg	1.94	2.20	1.98	24.8	0.30	25.6	1.40	24.3	1.05
K	0.72	0.89	0.82	10.3	0.11	9.4	0.60	10.4	0.43
Alkalinity	7.15	8.30	7.74	92.4	1.15	75.6	5.63	91.1	4.06
SiO$_4$H$_4$	0.55	0.69	0.64	7.6	0.37	24.3	0.55	8.9	0.49
pH	8.3	8.9	8.7	—	7.5	—	8.4	—	8.15
Conductivity	669	777	725	—	108	—	525	—	380

Table 5 Mean hydrochemical characteristics of the Southeastern Open Water between 1973 and 1977. Na, Ca, Mg, K, and alkalinity are expressed in me l^{-1}, SiO$_4$H$_4$ in nM l^{-1} and conductivity in μS cm^{-1}.

	1973	1974										1975		
	7/10	5/2	25/2	21/3	20/4	21/5	12/6	16/7	6/9	12/12	5/2	3/4	2/6
Na	0.16	0.17	0.19	0.20	0.23	0.25	0.28	0.20	0.10	0.11	0.14	0.12	—
Ca	0.20	0.30	0.35	0.41	0.45	0.42	0.50	0.30	0.17	0.28	0.31	0.32	0.40
Mg	0.18	0.24	0.29	0.28	0.32	0.30	0.33	0.26	0.15	0.19	0.25	0.30	0.41
K	0.07	0.06	0.07	0.07	0.09	0.09	0.10	0.08	0.05	0.05	0.06	0.07	—
Alkalinity	0.47	0.71	0.83	0.89	0.98	0.98	1.09	0.77	0.41	0.60	0.72	0.77	0.85
SiO$_4$H$_4$	0.20	0.12	0.14	0.13	0.13	0.15	0.16	0.11	0.16	0.17	0.12	0.16	0.17
pH	7.10	7.45	7.50	7.65	7.15	6.85	7.65	8.00	7.65	7.15	6.90	7.15	8.15
Conductivity	51	70	84	86	99	103	110	80	43	58	71	73	81

	1975	1976										1977
	1/7	14/8	10/10	31/10	26/1	27/2	5/5	17/6	28/8	21/10	11/12	31/3
Na	—	0.18	0.11	0.10	0.13	0.15	0.18	0.21	0.12	0.12	0.14	0.21
Ca	0.37	0.25	0.22	0.25	0.32	0.36	0.40	0.38	0.20	0.25	0.25	0.31
Mg	0.41	0.27	0.29	0.25	0.20	0.16	0.18	0.27	0.16	0.11	0.13	0.17
K	—	0.05	0.06	0.05	0.05	0.07	0.08	0.09	0.07	0.06	0.04	0.06
Alkalinity	0.85	0.57	0.50	0.53	0.66	0.69	0.83	0.86	0.46	0.50	0.55	0.67
SiO$_4$H$_4$	0.16	0.14	0.18	0.19	0.17	0.13	0.17	0.21	0.16	0.21	0.19	0.18
pH	7.30	7.90	7.25	7.25	7.20	7.15	7.50	7.50	7.70	7.70	7.90	6.65
Conductivity	81	61	50	52	63	70	89	97	50	53	58	73

often reached values of 8.5–8.8, especially in the northern part where water was quite alkaline.

3.2.2 The period of 'Lesser Chad'

Unlike the previous period when there were few variations in the chemical characteristics of the water despite the gradual reduction in water level, this new period was characterized by profound hydrochemical modifications due to considerable hydrological disturbances and vegetation development.

The appearance of the lake began to change in 1973. At that time it was divided into three distinct zones: they were the Southeastern Open Water with a connection to the river system; the Southeastern Archipelago composed of a limited number of relatively small ponds and fed seasonally by the flood of the Shari; the north basin which was nearly always isolated from the rest of the lake by the drying of the Great Barrier. The hydro-chemical characteristics of these three zones between 1973 and 1977 are shown in Tables 5, 6 and 7 (Chantraine 1978).

Table 6 Mean hydrochemical characteristics of the Southeastern Archipelago. Na, Ca, Mg, K and alkalinity are expressed in mé l^{-1}, SiO_4H_4 in mM l^{-1} and conductivity in μS cm^{-1}.

	1973		1974				1975	
	17/9	20/12	28/2	14/5	8/10	17/12	13/2	25/4
Na	3.70	0.33	0.42	0.61	0.30	0.19	0.24	0.47
Ca	1.02	0.78	0.91	1.69	1.20	0.53	0.59	0.91
Mg	0.95	0.57	0.66	1.13	1.18	0.42	0.43	0.70
K	0.59	0.08	0.07	0.05	0.25	0.09	0.08	0.09
Alkalinity	5.40	1.43	1.73	2.89	2.40	1.08	1.23	1.63
SiO_4H_4	0.19	0.44	0.39	0.29	0.50	0.29	0.31	0.32
pH	8.90	6.90	7.25	7.55	8.00	7.25	6.90	8.15
Conductivity	524	164	193	279	254	105	119	169

	1975		1976				1977	
	3/9	9/12	4/2	April–May	8/8	1/10	6/1	7/3
Na	0.31	0.15	0.19	0.25	0.30	0.30	0.19	0.26
Ca	0.83	0.34	0.43	0.56	0.68	0.58	0.36	0.47
Mg	0.84	0.37	0.24	0.33	0.45	0.42	0.23	0.29
K	0.05	0.08	0.08	0.07	0.07	0.09	0.07	0.08
Alkalinity	1.70	0.77	0.88	1.13	1.37	1.28	0.82	1.02
SiO_4H_4	0.25	0.21	0.22	0.25	0.32	0.31	0.7	0.32
pH	7.45	6.75	6.96	7.35	7.30	7.20	7.10	7.40
Conductivity	184	75	84	109	141	124	83	105

Table 7 Hydrochemical characteristics of the north basin from 1973 to 1976. Na, Ca, Mg, K and alkalinity are expressed in mé l^{-1}, SiO$_4$H$_4$ in mM l^{-1} and conductivity in μS cm^{-1}.

	1973		1974				1975				1976	
	Sept.	Nov.	May	July	Sept.	Dec.	Feb.	April	28/8	17/12	April	July
Na	4.10	5.59	8.71	13.23	17.49	34.25	3.20	3.33	22.90	0.71	1.04	1.87
Ca	1.39	1.33	1.04	0.70	0.57	0.55	1.47	0.94	0.71	2.17	2.39	2.22
Mg	1.42	1.59	2.32	1.48	1.61	1.72	1.55	1.55	1.51	1.60	1.52	1.81
K	0.75	1.06	1.55	1.93	2.46	4.10	0.86	1.33	3.48	0.75	0.57	0.76
Alkalinity	7.30	8.65	12.18	16.85	19.55	88.96	6.66	6.67	28.69	4.51	5.15	6.27
SiO$_4$H$_4$	0.32	0.35	0.06	0.18	0.20	0.25	0.17	0.28	0.63	0.56	0.55	0.42
pH	8.60	8.60	8.70	8.90	9.20	9.20	7.30	7.80	8.60	7.30	8.15	8.70
Conductivity	649	791	1120	1489	1889	3530	696	672	2568	467	508	633

3.2.2.1 *The Southeastern Open Water*. This region was less affected by lake contraction as it was permanently fed by the Shari water. Its hydrochemical characteristics were similar to those of the period of 'Normal Chad' (Table 5).

There was only a slight change in the mean water salinity while the average conductivity was 70 μS cm^{-1} from 1974 to 1976 compared with 60 μS cm^{-1} during the period of the average lake. On the contrary, seasonal variations were more pronounced: in June, maximum conductivity was 110 μS cm^{-1} compared with 75 μS cm^{-1} and in September, minimum conductivity was 43 μS cm^{-1} compared with 47 μS cm^{-1}.

3.2.2.2 *The Southeastern Archipelago*. Unlike the Southeastern Open Water, the Southeastern Archipelago was profoundly modified in appearance by 1973 when considerable vegetation development occurred. Only the zone situated southeast of Bol was studied.

The mean salinity of the water increased considerably during maximum recession of the lake in 1973 and 1974. Water conductivity reached high seasonal maximum values of 530 μS cm^{-1} in September 1973 and October 1974 respectively, compared with 135 μS cm^{-1} during the period of 'Normal Chad'. In 1975, these values decreased to 170 μS cm^{-1} and in 1976 to 140 μS cm^{-1} following an almost normal water supply of the region (Table 6).

The relative chemical composition was also profoundly modified at certain times. The relative potassium content decreased considerably from the end of 1973 following the development of vegetation (1.4% in June 1974 compared with 8.5% during the period of 'Normal Chad'). There was a sharp increase in the relative sodium content during the major drying up in 1973 (59.1% in September 1973 compared with 27.1% during the period of 'Normal Chad') and a decrease in the relative dissolved silica content (5.4% in September 1973 compared with 35.8% during the period of 'Normal Chad'). In October 1974 when the water supply in the region returned to normal, the relative chemical composition of the water approached that of the 'Normal Chad'.

The pH of the water ranged from 6.9 to 8.20.

3.2.2.3 *The North basin*. Measurements were taken on a few 10 km radials having the center of symmetry at Kindjéria (situated in the middle of the northern basin (Table 7).

Lake recession had considerable effects by the beginning of 1974, During that year, water salinity tripled and conductivity increased from 1200 to 3700 μS cm^{-1}. In 1975, salinity returned to a 'normal' value as a result of river supplies and increased again to high values in August with conductivities of 2800 μS cm^{-1}. During the first half of 1976, conductivity decreased again to about 500–600 μS cm^{-1}.

The relative chemical composition of the water changed profoundly from one year to another, each characterized by strong seasonal variations. When salinity was high, there was always a considerable increase in the relative sodium content (for example, it was 79% in September 1974 compared with 35.5% during the period of 'Normal Chad'). It was the reverse with dissolved silica (1.6% in September 1974 compared with 11% during the period of 'Normal Chad').

During this period of considerable recession of the water pH remained high, ranging from 8.6 to 9.2. However, in December 1975, and January 1976 when residual water mixed with flood water, it decreased to about 7.3–7.8.

3.3 Conclusions

A. Regular northeast winds blow over Lake Chad from October to April and southwestern winds from May to September. They generally blow from 6 to 12 a.m. and therefore, the environment is subjected to turbulence nearly every morning. Given the low depths, turbulence generally affects the entire water column and it is very unequal from one region of the lake to another. Obviously, it is much higher in the open water (where the fetch can reach 50 km in length) than in an archipelago (where the fetch reaches only 2 to 5 km). During the period of 'Lesser Chad', it was very reduced as a result of the decrease in open water areas and the development of vegetation.

Moreover, winds help to move the water masses that tend to accumulate on the southern and western coasts in winter and on the northern and eastern coasts in summer. This causes diurnal oscillations of the water level whose amplitude reaches 10 cm throughout the lake during the period of 'Normal Chad'. They can disappear during the period of 'Lesser Chad' when macrophytes develop.

B. Water transparency is generally low (1.20 m). This is not surprising in a turbulent and shallow environment where clayey mud and mud sediments are easily resuspended and the plankton population is relatively large. It varies considerably from one region to the other in relation to the height of the water level.

In the open water of the southern basin, water transparency measured with Secchi disc, decreased from 20–100 cm in 1964 during the period of 'Greater Chad' to 25–30 cm in 1973 during the period of 'Lesser Chad'. A seasonal minimum was recorded in July–August when the first flood waters of the Shari loaded with solid materials entered the lake. In the Southeastern Archipelago at Bol, the decrease was still more spectacular, ranging from 120 cm to 10–15 cm during the same period. But, after 1974 water transparency again increased from 25 to 100 cm due to the development of vegetation, and seasonal fluctuations in transparency remained low.

In the northern basin, water transparency ranged from 45 to 65 cm in the open water and from 70 to 90 cm in the archipelago during the period of 'Normal Chad'. The minimum value recorded in the open water corresponded to the arrival of the Shari flood water in February. During the period of 'Lesser Chad', water transparency became reduced to 10 cm before the drying up of the basin. It rapidly increased again to 30–70 cm when water returned, bearing with it a solid loading, through the Great Barrier which acted as a filter.

C. Water temperature is closely related to the annual, seasonal and diurnal variation in air temperature, since Lake Chad like all the flat lakes has a low thermal inertia.

The mean water temperature ranged from 25.5 to 27.5°C and varied little from one year to another. This minimum value was recorded during the period of 'Lesser Chad'.

Seasonal variations depended upon the movement of the Intertropical Convergence Zone. When it moved from the north to the south, in September–October, the temperature dropped abruptly from 27–29°C to 19–21°C; and when it moved from the south to the north, in April–May, it halted the process of warming which started in February. Over this period the temperature increased from 18–20°C to 29–31°C.

Diurnal variations were 2°C on average, and they reached 3.5°C during the period of 'Lesser Chad'.

During the period of 'Normal Chad', there was no seasonal thermal stratification. A temperature gradient of 0.5–1°C could appear in the mid afternoon and disappear through the night. The lake then appears as a warm polymictic lake. The same situation occurred during 'Lesser Chad' except in some zones of the archipelago where a thermal stratification developed in February–March and remained constant up to September. Flood waters then appeared and isothermy remained until February. The lake turned into a monomictic environment only in the zones of the archipelago and during considerable recession. Generally speaking, Lake Chad has a very low thermal inertia due to its shallow depth and it has large variations from one year to the other depending on the floods and the recession. Also it poorly absorbs the influence of a tropical climate with well-defined seasons. Therefore, it is an environment with low thermal stability.

D. At saturation, the oxygen content of the water ranges from 9.32 mg l^{-1} at 18°C to 7.09 mg l^{-1} at 32°C, depending on the altitude of the lake (280 m).

During the period of 'Normal Chad', and a fortiori, during the period of 'Greater Chad', the oxygen content of the water was close to saturation. A vertical stratification could develop along with diurnal thermal stratification in the different zones of the lake early in the afternoon and disappear at night: at the end of the day, saturation reached 120% in surface layers, while it was seldom less than 80% in deep layers.

During the period of 'Lesser Chad', conditions for oxygenation in the waters of the southern basin were similar to those recorded during the period of 'Normal Chad'. In the zones of the archipelago, they underwent large variations from one arm to another, and from one period to another. Oxygenation of the water was low at Bol with oxygen reaching 12% saturation in surface layers during the warm season. The hypolimnion generally remained anoxic. Finally, in the northern basin, the oxygen content of the water varied considerably and the environment was often anoxic.

On the whole, Lake Chad is well oxygenated in the deep layers as well as in the surface layers during the periods of high and mean water levels. It can have a temporary or permanent anoxia in any region, except in the open waters bordering the delta of the Shari during the period of low water.

E. The hydrochemistry of the environment is not simple. The chemical composition of the water has profound qualitative and quantitative changes from one region of the lake to another and considerable annual and seasonal variations inside each region.

The Shari waters which are the main source of dissolved salts have a low salinity of 40–50 mg l^{-1}. They include major elements such as Na, Ca, K, Mg HCO_3 and SO_4H_4 and minor elements such as Cl and SO_4. In the lake, the salinity becomes higher with increasing distance from the delta.

During the period of 'Normal Chad' the open water of the southern basin was 1.2 to twice as saline as that of the Shari (60 to 120 mg l^{-1}), while in the zone of the archipelago, water was two to three times as saline (100 to 150 mg l^{-1}). In the arm bottoms, salinities generally reached higher values (300–400 mg l^{-1}).

Water in the northern basin was on average four times as saline as that of the southern basin. In the open water, the salinity ranged from 250 to 400 mg l^{-1}, and from 400 to 800 mg l^{-1} in the archipelagoes, the maximum values being related to the inshore waters.

Together with this increase in the salinity of the water, there was a progression of the water towards a sodium bicarbonate facies with increasing distance from the delta, resulting from chemical sedimentation in the environment which mainly affected SiO_4H_4, Ca and Mg. The northern water of the lake was composed of 35.8% Na, 91.1% HCO_3, 29.5% Ca, 24.3% Mg, 10.4% K and 8.9% SiO_4H_4 compared with 29.9% Na, 75.6% HCO_3, 35% Ca, 25.6% Mg, 9.4% K and 24.3% SiO_4H_4 in the Shari waters.

The seasonal variations could be considerable in the regions that directly received the flood water of the Shari (open water from both basins and their meeting point, the Great Barrier where the differences in salinities were about 50%). On the contrary, they were low in the zones of the archipelago.

During the period of 'Lesser Chad', profound modifications appear which were dependent upon the change in water level and the development of the vegetation. The region of open water in the southern basin which was directly fed by the Shari was the least affected by the contraction of the lake. The mean salinity of the water remained close to 50–70 mg l^{-1}.

In the archipelago of the southern basin, the water of the Bol region reached 500 mg l^{-1} during severe lake recession (1973). However, this region could be fed 'almost normally', while the northern basin dried up, resulting in salinities similar to those of a normal period.

In the northern basin when the environment dried up, the salinity of the water could reach several grams per liter with Na and HCO_3/CO_3 dominating and when normal water supply was established, salinity fell to 500 mg l^{-1}. Generally, the northern basin was chemically more stable than the southern basin during the period of 'Normal Chad' in relation to seasonal fluctuations. However it was much more unstable over several years since it could undergo a partial or even a total drying up during the period of 'Lesser Chad'. In the southern basin, the open water which had large seasonal fluctuations (salinity could double from 40 to 80 mg l^{-1}) was not subject to major chemical changes during the severe floods and contractions.

References

Billon, B. et al., 1973. Monographie hydrologique du Chari. ORSTOM Paris, 5 parts in 6 Vols., 650 pp., mimeo.
Bindloss, M. E., 1976. The light climate of Loch Leven, a shallow Scottish lake, in relation to primary production. Freshwat. Biol. 6: 501–518.
Carmouze, J. P., 1976. La régulation hydrogéochimique du lac Tchad. Contribution à l'analyse biogéodynamique d'un système lacustre endoréïque en milieu continental. Trav. Doc. ORSTOM, No 58, 418 pp.
Chantraine, J. M., 1978. Evolution hydrochimique du lac Tchad de Septembre 1973 à Septembre 1975 au cours d'une phase de décrue. Cah. ORSTOM Sér Hydrobiol. 12: 3–17.
Dokulil, M., 1973. Planktonic primary production within the *Phragmites* community of lake Neusiedlersee (Austria). Pol. Arch. Hydrobiol. 20, 175–180.
Dorize, L., 1974. L'oscillation pluviométrique récente sur le bassin du lac Tchad et la circulation atmosphérique générale. Rev. Géogr. Phys., Geol. Dyna. 16(4): 393–420.
Ganf, G. G., 1974. Incident solar irradiance and under water light penetration as factors controlling the chlorophyll *a* content of a shallow equatorial lake (Lake George, Uganda). J. Ecol. 62: 593–604.
Gras, R., Iltis, A. and Leveque-Duwat, S., 1967. Le plancton du Bas-Chari de la partie est du lac Tchad. Cah. ORSTOM Sér. Hydrobiol. 1: 25–100.
Hopson, A. J., 1968. Seasonal changes in the pattern of salinity distribution in the northern basin of lake Chad. Lake Chad. Res. Stat. Malamfatori, Ann. Rep. 1966–67, pp. 13–26 Lagos.
Lemoalle, J., 1969. Premières données sur la production primaire dans la région de Bol (avril-octobre 1968) (lac Tchad). Cah. ORSTOM Sér. Hydrobiol. 3: 107–120.
Lemoalle, J., 1979. Biomasse et production phytoplanctonique du lac Tchad (1968–1976). Relations avec les conditions du milieu. Thèse d'Etat, Univ. Paris VI, ORSTOM, Paris, 311 pp.
Riou, C., 1972. Etude de l'évaporation en Afrique Centrale (Tchad, RCA, Congo) contribution à la connaissance des climats. Thèse Univ. Paris VII, 205 pp., mimeo.

Robinson, A., 1968. Notes on diurnal and seasonal changes in temperatures and oxygen regimes in lake Chad. In: Annual report 1966–67, pp. 26–34. Fed. Fish. Services Lake Chad Res. St. Malamfatori, Nigeria.

Roche, M. A., 1972. Eléments sur la température des eaux dans l'ensemble du lac Tchad. Rapp. ORSTOM N'Djamena, 20 pp., mimeo.

Roche, M. A., 1973. Taçage naturel salin et isotopique des eaux du système hydrologique du lac Tchad. Thèse d'Etat. Univ. Paris, ORSTOM, Paris, 398 pp., mimeo.

Sauberer, 1962. Empfehlungen für die Durchfuhrung von Strahlungsmessungen an und in Gewässern. Mitt. int. Ver. Linnol., II: 1–240.

Smith, J. R. and Sinclair, I. J., 1972. Deep water waves in Lakes. Freshwat. Biol. 2: 387–399.

Talling, J. F., 1969. The incidence of vertical mixing and some biological and chemical consequences in tropical African Lakes. Verh. int. Ver. Limnol. 17: 998–1012.

Talling, J. F., Wood, R. B., Prosser, M. V. and Baxter, R. M., 1973. The upper limit of photosynthetic productivity by phytoplankton: evidence from Ethiopian Soda Lakes. Freshwat. Biol. 3: 53–76.

Tilho, J., 1910. Documents scientifiques de la mission Tilho 1906–1909. Imp. Nat., Paris, Vol. 1, 412 pp. Vol. 2, 598 pp.

Toucheboeuf, de Lussigny P. 1969. Monographie Hydrologique du lac Tchad. ORSTOM Paris, 169 pp. mimeo.

Vollenweider, R. A., 1961. Relations existing in the spectral extinction of light in water. Memorie Ist. Ital. Idrobiol. 13: 87–113.

4. Hydrochemical regulation of the lake

Jean-Pierre Carmouze

4.1 Introduction

Chad basin, situated between 5° and 25° latitude north (Fig. 1, Chapter 2) has a sub-arid climate, mainly characterized by strong variations in rainfall. During the course of this study, the rainfall was about 1500 mm year^{-1} in the south and less than 200 mm in the north. This rainfall gradient was associated with a topographic gradient; tropical waters flowed from south to north towards arid zones. Collected by the river Shari, these waters spread into a large, shallow and closed basin located between 12°5 and 14°5 latitude north to form Lake Chad (Fig. 1).

This water carries dissolved materials which have been picked up into the lake. They are mainly Ca^{++}, Mg^{++}, Na^+, K^+, HCO_3^- and SiO_4H_4 arising from monosialitic alteration of crystalline rocks (Carmouze 1976).

Due to its endorheic nature and its climate, the lake appears to be a basin that concentrates salts. However the water salinity is low and relatively unchangeable in time. Our purpose is to explain these apparent contradictions by analyzing the different factors that control the concentration of dissolved constituents in the lake waters.

Although Lake Chad is geographically closed, it is, like every aquatic environment, an open system in which material transfers take place. Therefore, the problem is to determine the different flows of water and dissolved elements which control the various lake stocks of dissolved salts, and to determine the volume of water at all times and in each chemically homogeneous region of the lake, since the salt concentration is the ratio of these two parameters.

Generally, flows vary in time and space. Therefore, especially in Lake Chad, local heterogeneities and temporal fluctuations in the environmental hydrochemistry occur.

For this reason, two simplifications were introduced:
— in order to reduce spatial variations, the lake was divided into two less heterogeneous parts representing the south and north basins of the lake;
— over a long period of time the hydrochemical state of a lake oscillates around

J.-P. Carmouze et al. (eds.) Lake Chad
© *1983, Dr W. Junk Publishers, The Hague/Boston/Lancaster*
ISBN 978-94-009-7268-1

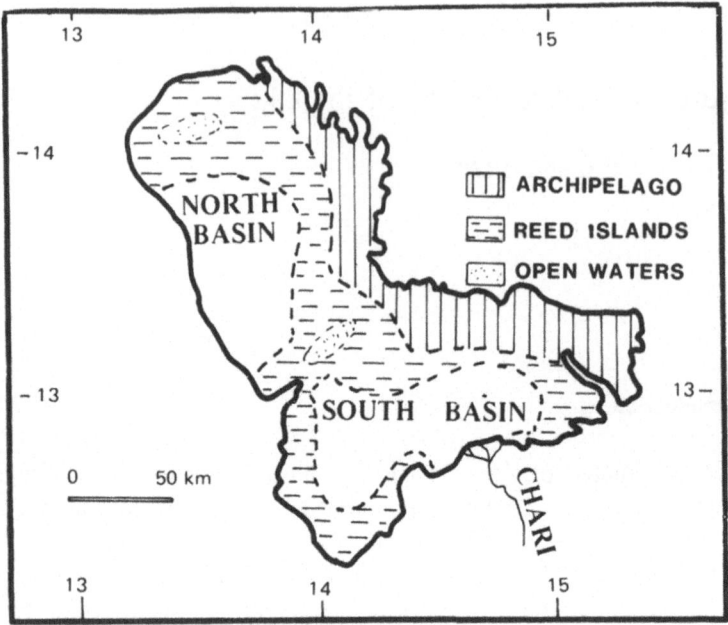

Fig. 1 Lake Chad and its different ecological zones.

an average condition because material input and output balance. If one ignores these fluctuations, the environment can be considered as a steady-state system. So the chemical characteristics of water become independent of time. To compare Lake Chad to such a system, a period of 18 years was examined (1954–1972).

Depending upon their nature the material flows may be classified in two groups: those associated with hydrological currents and those associated with physical, chemical, geochemical or biochemical reactions (Carmouze and Pedro 1977). The water currents primarily control the 'climato-geographical' dissolved salt currents (climato-geographical regulation). Depending upon this pre-regulation, biogeochemical dissolved salt currents occur (biogeochemical regulation) that control the total.

Choosing this approach, the lake will first be considered as a steady-state system. The average water and dissolved salt currents and the corresponding dynamic equilibrium for the period 1954–72 will be examined successively for the whole lake and its north and south basins. The contribution of the two hydrochemical reactions representing the main aspect of the biogeochemical reactions, i.e. mineralogical changes inside the lake and the influence of molluscs and macrophytes will be also examined.

Then, the hydrochemical regulation of the lake will be analyzed during flood and recession periods.

4.2 The hydrology of the lake and its contribution to hydrochemical regulation during a permanent state

Lake levels have been estimated from 1895 to 1954 (Toucheboeuf et al. 1969) and measured since 1954. Complete data about climatology and hydrology are also available after this date. Therefore the period 1954–1972 has been chosen to describe the average hydrological state of the lake. This state was similar to that of the longer period 1895–1972, according to the lake level fluctuations (Fig. 2).

4.2.1 *The total hydrology: mean hydrological balance of the lake*

4.2.1.1 *Input regime. (a) River supplies*: almost all of the river supplies come from the Shari and El Beïd. The annual Shari supplies are between 19.5×10^9 and $54.5 \times 10^9 m^3$, the average being $40 \times 10^9 m^3$. The seasonal distribution is similar to a transition tropical regime with the lake receiving 73% of the average annual supply between September and December.

The river El Beïd flows from October to March. The annual supply varies from 0.6×10^9 to $2.7 \times 10^9 m^3$ with a mean value of $1.35 \times 10^9 m^3$. Over the period under consideration, the lake received $41.5 \times 10^9 m^3$ water from rivers.

(b) *Rainfall*: the rainy season lasts from May to October. In fact the lake receives 50% of the annual supply in August with an average of 300–310 mm year^{-1}. It progressively decreases from south (600 mm) to north (250 mm). The annual volume ranges between 2.7×10^9 and $8.7 \times 10^9 m^3$, with a mean value, $(\bar{v})_M$ of $6.3 \times 10^9 m^3$.

Overall, the annual supply of water varies from $23 \times 10^9 m^3$ to $61.4 \times 10^9 m^3$ with an average of $48 \times 10^9 m^3$. 87% of this is from the rivers and 13% from rainfall.

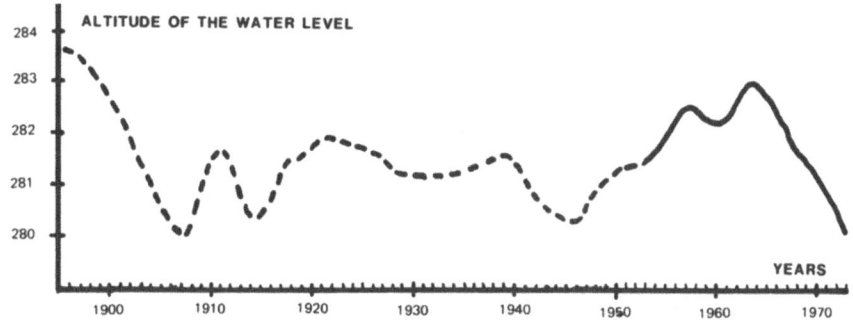

Fig. 2 Annual variations in the water level of Lake Chad from 1895 to 1973 (----- estimated; ——— measured).

4.2.1.2 *Output regime.* (a) *Evaporation losses*: with an annual evaporation rate of 2.11 m, the volume of water evaporated varies between approximately 47 and $38 \times 10^9 m^3$ year^{-1} with a mean value of $44 \times 10^9 m^3$.

(b) *Infiltration losses*: the lake is directly connected with a 50–70 m thick groundwater bed. According to the piezometric gradients the lake is above. This is a situation favouring infiltrations. Moreover they are facilitated because of the sandy nature of the shores and a part of the lake bottom (10 to 15%) (Dieleman and De Ridder 1963; Pirard 1963; Schneider 1967, Fontes et al. 1969; Roche 1973; Maglione 1974).

The infiltrations have been indirectly estimated, using the Na budget and assuming that this element is only eliminated by infiltration. During the period 1954–1972, the Na output of $5.6 \times 10^{12} mM$ year^{-1} has balanced the Na input. This quantity $\overline{(Na)}_I$ is eliminated in a volume of infiltration $(\bar{v})_I$ equal to $\overline{(Na)}_I / [\overline{Na}]_I$, where $[\overline{Na}]_I$ is the mean Na concentration in the infiltrated water. Assuming that the infiltrations occur throughout the lake, $[\overline{Na}]_L$, whose value is 1.46mM 1^{-1} (Carmouze 1976). Thus the infiltration volume can be calculated as:

$$(\bar{v})_I = 5.6 \times 10^{12} / 1.46 \times 10^3 = 3.85 \times 10^9 m^3 \text{ year}^{-1}$$

The annual evaporation and infiltration losses representing a water height of 2.29 m have been approximately calculated for each year of the period 1954–1972, by allowing the total losses to be roughly proportional to the lake surface. The total losses varied from 41.3×10^9 to $51.8 \times 10^9 m^3$, of which evaporation represented 92% and infiltration 8%.

4.2.1.3 *Hydrological characteristics of the lake.* The volume, area and depth are controlled by the basin morphology, and the input/output regimes of water previously described. These parameters were calculated from morphometric curves of the lake (Carmouze 1976; Lemoalle 1978).

The lake levels registered on the northern shore of the south basin from 1954 to 1972 changed from 281.5 to 280.5 m with a maximum of 283.3 m; and was always modified by seasonal changes of 0.85 to 0.80 m.

The lake volume varies between $42.5 \times 10^9 m^3$ and $91 \times 10^9 m^3$, with an average value, $(\bar{v})_L$, of $72 \times 10^9 m^3$. The considerable volume fluctuations are caused by very variable inputs: from 1954 to 1972 the water supply varied from 23 to $61.4 \times 10^9 m^3$ year^{-1} while the annual output variations were less important, the losses ranging from 41.3 to $51.8 \times 10^9 m^3$ year^{-1}. The small volume of the lake poorly matches the output–input inegality: on average 66% of the lake water is renewed each year.

The seasonal fluctuations in volume represent about 15% of the total volume and result from 80% of the water supplies being received over 5 months while the losses are distributed more evenly through the year.

The lake area varies between 22 600 km² (1962) and 1 800 km² (1972) with a mean of 20 900 km². The seasonal fluctuation in area during an average hydrological regime of the lake is ±3%.

4.2.1.4 *The mean hydrological regulation and its contribution to hydrochemical regulation.*

From the data just presented, it is possible to evaluate the annual mean hydrological balance of the lake (Fig. 3) and point out the following features:
— the input mainly comes from the river (86.5% against 13.5% from rainfall) while the losses due to evaporation are greater than those due to infiltration (92% against 8%);
— the input and output regimes and the shape of the basin are such that the lake volume remains small. Two thirds of the lake volume is renewed each year with a turnover time of about 1.5 years.

Because of the first characteristic, the lake is a salt concentration basin whose concentration factor is easy to calculate. Let us consider a dissolved element, i, whose circulation is only controlled by water flows. This means that the input of i is annually balanced by infiltration.

$$[i]_F \cdot (\bar{v})_F = [i]_I \cdot (\bar{v})_I = [i]_I \times [(\bar{v})_F + (\bar{v})_M - (\bar{v})_E] \ (1);$$

$[i]_F$ and $[i]_I$ = mean value of respectively, river and infiltrated water. $(\bar{v})_F$, $(\bar{v})_I$, $(\bar{v})_M$ and $(\bar{v})_E$ = annual mean volume respectively, of river water, infiltrations, rain and evaporation.

Let us assume that the infiltration occurs approximately uniformly all over the lake. Then, $[i]_I = [i]_L$ = mean lake concentration of i. According to (1)

$$\frac{[i]_L}{[i]_F} = \frac{(\bar{v})_F}{(\bar{v})_I} = \frac{(\bar{v})_F}{(\bar{v})_F + (\bar{v})_M - (\bar{v})_E} = 10.8$$

This dissolved salt concentration factor controlled by the 'climato-geographical' regulation is close to 11. It is not very high for a closed lake because of the relative importance of infiltrations.

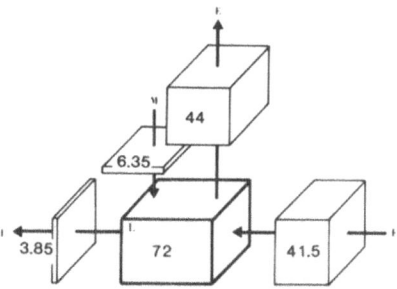

Fig. 3 Mean annual hydrological balance of the whole lake. F = fluvial input; E = evaporation; M = meteoric input; I = infiltration. All the flows and the lake volume are expressed in m³ × 10⁹.

The second characteristic of the hydrological regulation denotes the lake instability in regard to the input irregularities. The effects on the hydrochemistry are examined in para. 5.1.

For example, an annual excess of the river input of about 33% produces a 15% decrease in the lake concentration of i. On the contrary, for a deficit of 33% the increase in concentration of i is 21.5%. So, when only geographical regulation takes place, the variation in saline concentration corresponding to the input–output imbalances (frequently having these extents), is relatively important.

4.2.2 *Regional hydrology: mean hydrological balance of the south and north basins and its influence on hydrochemistry*

The south and the north basins are two distinct geographical entities with different characteristics. So, to obtain a better hydrological knowledge of the lake the mean annual hydrological balance of the two basins has been determined in a way similar to that for the whole lake (Carmouze 1976).

From the results shown in Fig. 4 the following features can be pointed out:
— the north basin contains twice as much water as the south basin, while having roughly the same area;
— the losses due to evaporation are almost equal in both environments because the evaporation rates are similar;
— the infiltration losses are twice as important in the north basin as in the south (65% of the total infiltration against 32%);

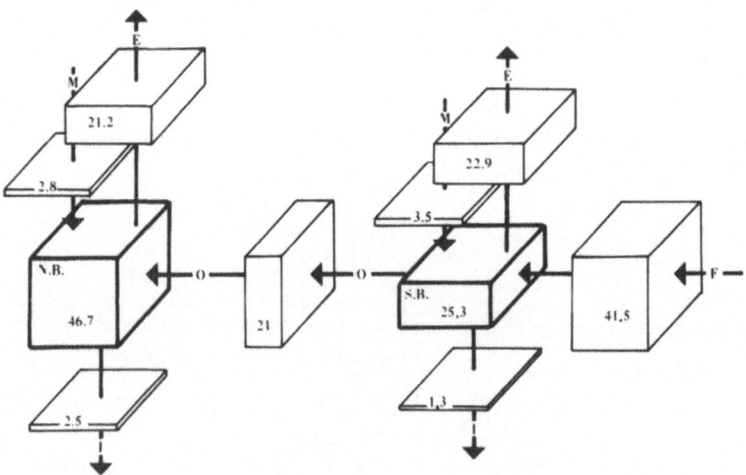

Fig. 4 Mean annual hydrological balance of the south and the north basin. F = fluvial input; E = evaporation; M = meteoric input; O = north basin input coming from south basin. All the flows and volumes of the north (N.B.) and south (S.B.) basins are expressed in m^3 × 10^9.

100

— the turnover rate of the water in the south basin is very high and the residence time of the water is 6 months. 50% of the river input flows into the north basin where the average residence time of the water is about 2 years.

These hydrological regulation differences cause some hydrochemical modifications in the two basins by determining the rate of increase of the different salinities.

For an element i, whose regulation is only climato-geographical (a conservative element) 8.2% of its annual input remains in the south basin, while 91.8% flows into the north (Carmouze 1976). A rough calculation indicates that the rate of increase of the concentration of i, due to the geographical regulation is about 2.6 in the south basin and 8.5 in the north basin. Consequently, the climato-geographical increase of salinity is 3.25 times higher in the latter basin (Carmouze 1976).

4.3 Contribution of biogeochemical sedimentation to the regulation of steady-state hydrochemistry of the lake

The mean annual dynamic equilibrium of each constitutent of the water salts corresponding to 1954–1973 can be established for the lake and its north and south basins. The constituents are: Na^+, Ca^{++}, Mg^{++}, K^+, HCO_3^- and SiO_4H_4. To achieve this, some approximations must be made because hydrochemical data are only available for part of this period.

The river salinity data are only available from 1967 (Carmouze 1969, 1972, 1976; Roche 1969, 1973), but they point out that, for a given season, the salt concentrations do not change from one year to another, whatever the water flow may be. So, knowing the mean monthly concentrations, it is possible to calculate the different salt inputs if the monthly water flows are available.

The mean hydrochemical characteristics of the lake can be approximately obtained, using the fact that lake salt stocks depend mainly on lake volume. In 1967, the lake had reached a balanced situation and so, for this year the average stock values can be extrapolated from the data of 1968 to 1972.

The total mean annual losses can be estimated as being equal to the inputs for the entire period 1954–1972. Moreover, supposing that Na is only eliminated by infiltration, it becomes possible to evaluate the fraction of the other constituents that are biogeochemically eliminated.

4.3.1 *The hydrochemistry: mean annual balances of the dissolved salts*

4.3.1.1 *River input of dissolved salts.* (a) *The lower Shari supplies*: The ionized salinity of the Shari varies between 0.38 and 0.78 mé 1^{-1} with a minimum at the beginning of the flood (August–September) of 0.38–0.42 mé 1^{-1} and a maximum at the time of low water (May–June) of 0.68–0.78mé 1^{-1}.

The mean monthly values of the dissolved components vary between 0.41 and 0.72mM 1^{-1} for HCO_3^-, 0.071 and 0.137mM 1^{-1} for Ca^{++}, 0.067 and 0.115mM 1^{-1} for Mg^{++}, 0.105 and 0.155mM 1^{-1} for Na^+, 0.04 and 0.06mM 1^{-1} for K^+ and 0.300 and 0.425mM 1^{-1} for SiO_4H_4. The salinity of these waters is half that of water in continental rivers (Davies and De Viest 1966). However the dissolved silica concentration of the Shari river is 1.5 times higher than the average.

The salt inputs of each constituent have been calculated on a monthly basis for the period 1954–1972 using monthly concentration values and the corresponding hydrological data (Billon et al. 1969). The annual inputs generally vary between 75 and 125% of the mean values which are:

$$(\overline{Ca^{++}})_{CH} = 4.01 \times 10^{12} mM; \quad (\overline{Mg^{++}})_{CH} = 3.1 \times 10^{12} mM; \quad (\overline{Na^+})_{CH}$$
$$= 5.25 \times 10^{12} mM;$$
$$(\overline{K^+})_{CH} = 1.9 \times 10^{12} mM; \quad (\overline{HCO_3})^- = 20.8 \times 10^{12} mM; \quad (\overline{SiO_4H_4})_{CH}$$
$$= 15 \times 10^{12} mM$$

Generally speaking, the supply regimes for dissolved salts parallel those for water supplies.

(b) *The El Beïd supplies*: the mean chemical composition of the El Beïd water is:

$$(\overline{Ca^{++}})_{EB} = 0.215 mM\ 1^{-1}; \quad (\overline{Mg^{++}})_{EB} = 0.15 mM\ 1^{-1}; \quad (\overline{Na^+})_{EB}$$
$$= 0.35 mM\ 1^{-1};$$
$$(\overline{K^+})_{EB} = 0.10 mM\ 1^{-1}; \quad (\overline{HCO_3^-})_{EB} = 1.15 mM\ 1^{-1}; \quad (\overline{SiO_4H_4})_{EB}$$
$$= 0.71 mM\ 1^{-1}\ (Roche\ 1973).$$

From these chemical data and the hydrological data given by Billon et al. (1974) the annual inputs for 1954–1972 have been calculated:

$$(\overline{Ca^{++}})_{EB} = 0.3 \times 10^{12} mM; \quad (\overline{Mg^{++}})_{EB} = 0.2 \times 10^{12} mM; \quad (\overline{Na^+})_{EB}$$
$$= 0.5 \times 10^{12} mM;$$
$$(\overline{K^+})_{EB} = 0.15 \times 10^{12} mM; \quad (\overline{HCO_3^-})_{EB} = 1.6 \times 10^{12} mM; \quad (\overline{SiO_4H_4})_{EB}$$
$$= 1.1 \times 10^{12} mM.$$

Hence, the mean annual values of the total dissolved salt inputs which include the Shari and El Beïd supplies are:

$$(\overline{Ca^{++}})_F = 4.35 \times 10^{12} mM; \quad (\overline{Mg^{++}})_F = 3.3 \times 10^{12} mM; \quad (\overline{Na^+})_F$$
$$= 5.6 \times 10^{12} mM;$$
$$(\overline{K^+})_F = 2.05 \times 10^{12} mM; \quad (\overline{HCO_3^-})_F = 22.4 \times 10^{12} mM; \quad (\overline{SiO_4H_4})_F$$
$$= 16.1 \times 10^{12} mM.$$

4.3.1.2 *Output rates of dissolved salts.* The dissolved salt losses arise from infiltration, biogeochemical elimination and marginal deposits. The latter loss occurs when the lake decreases in size, isolating near-shore ponds that dry up, and the less soluble dissolved salts precipitate. On the other hand, when the lake size increases, these salts are partially recovered. We admit that, for a long period, the precipitation–dissolution phenomena are balanced and, consequently, the only effective losses are due to infiltration and biogeochemical sedimentations.*

Thus for the period 1954–1972, the mean annual losses, $(\bar{\imath})_p$, were similar to the mean annual input.

$$(\overline{Ca^{++}})_P = 4.3 \times 10^{12} mM; \ (\overline{Mg^{++}})_P = 3.3 \times 10^{12} mM; \ (\overline{Na^+})_P$$
$$= 5.6 \times 10^{12} mM;$$
$$(\overline{K^+})_P = 2.05 \times 10^{12} mM; \ (\overline{HCO_3^-})_P = 22.4 \times 10^{12} mM: \ (\overline{SiO_4H_4})_P$$
$$= 16.1 \times 10^{12} mM.$$

Due to its physico-chemical properties and the hydrochemical environment, Na sedimentation may be unimportant. Therefore the Na supplies from the river $(\overline{Na})_F$ are mainly eliminated by infiltration: $(\overline{Na})_F = (\overline{Na})_I$ $= 5.60 \times 10^{12} mM$ per year. The infiltration volume has been calculated by this hypothesis in para. 2.1.2 to be $(v)_I = 3.85 \times 10^9 m^3$. Also calculated was the increasing salinity factor of 10.8 (para. 2.1.4) in the lake caused by the water currents.

This factor is lower for dissolved constituents other than Na^+ because they are subject to biogeochemical sedimentations. It is evaluated as follows: without sedimentation, the 'climato-geographical' lake concentration of i, $[i]_{L\ cg}$, would be: $[i]_{L\ cg} = 10.8[i]_F$. The annual loss by infiltration would be equal to $[i]_{L\ cg} \times (\bar{v})_I$. But as the lake concentration of i is equal to $[i]_{L\ cg}$, the effective infiltration of i is equal to $[i]_L \times (\bar{v})_I$. The difference represents the sedimentation, $(\bar{\imath})_S = \{[i]_{L\ cg} - [i]_L\} (\bar{v})_I$.

The mean annual losses by infiltration can then be calculated:

$$(\overline{Ca^{++}})_I = 2.4 \times 10^{12} mM; \ (\overline{Mg^{++}})_I = 1.95 \times 10^{12} mM; \ (\overline{Na^+})_I$$
$$= 5.6 \times 10^{12} mM;$$
$$(\overline{K^+})_I = 1.65 \times 10^{12} mM; \ (\overline{HCO_3^-})_I = 15.6 \times 10^{12} mM; \ (\overline{SiO_4H_4})_I$$
$$= 3.0 \times 10^{12} mM$$

and those by sedimentation:

$$(\overline{Ca^{++}})_S = 1.95 \times 10^{12} mM; \ (\overline{Mg^{++}})_S = 1.35 \times 10^{12} mM; \ (\overline{Na^+})_S$$
$$= 0.0;$$

---

* By neglecting the border precipitation–dissolution of salts, the losses by infiltration and biogeochemical pathways are overestimated because the redissolution of salts deposited on the border is certainly not complete.

$$(\overline{K^+})_S = 0.40 \times 10^{12}\text{mM}; \ (\overline{HCO_3^-})_S = 6.8 \times 10^{12}\text{mM}; \ (SiO_4H_4)_S$$
$$= 13.1 \times 10^{12}\text{mM}.$$

4.3.1.3 *Mean hydrochemical characteristics of the lake*. Let us recall that our purpose is to describe the mean hydrochemical state of the lake during 1954–1972, but the chemical data are only available for 1968–1972. Fortunately, the hydrochemistry depends on the hydrology and the dissolved salt content is largely proportional to the volume (Carmouze 1976). Although 1967 represented the average state of the lake, there are no data available for it. thus the calculations are made by extrapolation from the 1968–72 data.

Using this procedure, the mean salt stocks of the different constituents have been calculated:

$$(\overline{Na^+})_L = 105.2 \times 10^9\text{mM}; \ (\overline{Mg^{++}})_L = 36.65 \times 10^{12}\text{mM}; \ (\overline{Ca^{++}})_L$$
$$= 45.05 \times 10^{12}\text{mM};$$
$$(\overline{K^+})_L = 31.10 \times 10^9\text{mM}; \ (\overline{HCO_3^-})_L = 292.15 \times 10^{12}\text{mM}; \ (SiO_4H_4)_L$$
$$= 56.1 \times 10^{12}\text{mM}.$$

The absolute concentrations were calculated using $(\bar{v})_L = 72 \times 10^9 \text{m}^3$:

$$[\overline{Na^+}]_L = 1.46\text{mM } 1^{-1}; \ [\overline{Mg^{++}}]_L = 0.509\text{mM } 1^{-1}; \ [\overline{Ca^{++}}]_L$$
$$= 0.625\text{mM } 1^{-1};$$
$$[\overline{K^+}]_L = 0.432\text{mM } 1^{-1}; \ [\overline{HCO_3^-}]_L = 4.057\text{mM } 1^{-1}; \ [\overline{SiO_4H_4}]_L$$
$$= 0.779\text{mM } 1^{-1}.$$

4.3.1.4 *Conclusion: the hydrochemical regulation and biogeochemical sedimentation of dissolved salts*. The mean annual dynamic equilibria of the different mineral salts are shown in Figs. 5A and 5B, considering the lake as a homogeneous, single environment.

Comparing these equilibria, we note that each component has its own regulation. The geographical regulation alone would have the same concentration factor of 10.8 for each. So the concentration of Ca, Mg, K, HCO_3 and SiO_4H_4 would have been respectively equal to 1.14, 0.862, 0.527, 5.85 and 4.18mM 1^{-1}, but their actual values are 0.625, 0.509, 0.432, 4.06 and 0.779mM 1^{-1}. These observed decreases are caused by intervention of biogeochemical sedimentations.

Considering only the two extremes of Na and SiO_4H_4, we note that the turnover of the Na stock represents 5.3% while the turnover of the SiO_4H_4 stock is 78.7%. In other words the lake is a transit place for Na where the regimes of river supplies and losses by infiltration maintain a relatively high stock of this element. However for SiO_4H_4 the lake is primarily a place of chemical sedimentation where the regime of river supplies and losses by biogeochemical means and to a lesser degree by infiltration maintain it in a relatively low stock.

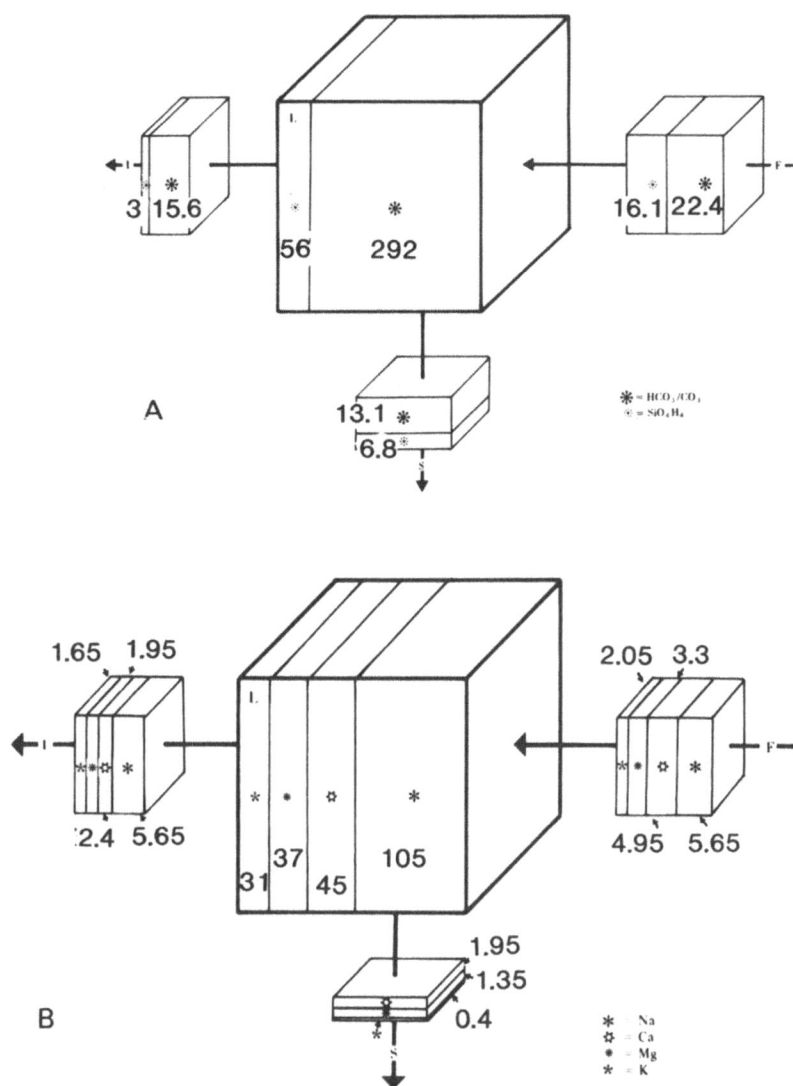

Fig. 5 Mean annual salt balances in the whole lake. A) HCO_3^-/CO_3^{--} and SiO_4H_4; B) Na^+, Ca^{++}, Mg^{++} and K^+.
F = fluvial input; S = sedimentation; I = infiltration. The flows and the lake stocks (L) are expressed in moles × 10⁹.

In summary, there are three main reasons for the low salinity of the water:
— the salinity concentration factor in this arid closed lake is only 10.8, a low value due to the considerable infiltrations;
— the low river salinity is on average twice as low as the mean salinity of continental waters;

— considerable biogeochemical sedimentation occurs in the lake, which reduces the salinity by 45%.

The varying importance of biogeochemical sedimentation according to the constituents causes the relative chemical composition of the river to tend towards the $Na-HCO_3/CO_2$ side.

The mean relative chemical composition of the rivers and the lake is:

$$\overline{[Ca^{++}]}_F^r = 29.5\% \quad \overline{[Ca^{++}]}_L^r = 20.8\%; \quad \overline{[Mg^{++}]}_F^r = 22.3\%, \quad \overline{[Mg^{++}]}_L^r = 16.8\%$$

$$\overline{[Na^{+}]}_F^r = 34.3\%, \quad [Na^{+}]_L^r = 47.8\%; \quad \overline{[\bar{K}^{+}]}_F^r = 13.9\%, \quad [\bar{K}^{+}]_L^r = 14.5\%$$

$$\overline{[HCO_3^-]}_F^r = 58.7\%, \quad \overline{[HCO_3^-]}_L^r = 84\%; \quad \overline{[SiO_4H_4]}_F^r = 41.3\%, \quad \overline{[SiO_4H_4]}_L^r = 16\%.*$$

4.3.2 Regional hydrochemistry: average salt balance in the north and south basins

The mean hydrochemical balance of the dissolved constituents has also been established for the two basins (Carmouze 1976). They provide good information on the salt transfers on a regional scale. The results are shown in Table 1 and Figs. 6A and 6B.

The south basin is mainly a transit zone for dissolved salts, with very short residence times (18 months for Na, 11 months for SiO_4H_4). About 73% of the river input flows into the north basin where the residence time ranges from 19 years for Na to 4.5 years for SiO_4H_4. Consequently the north basin is the principal reservoir of dissolved salts with 88.5% of the total stock compared with 11.5% in the other basin.

Sedimentation is equally important in both basins. In the south, for SiO_4H_4, the sedimentation is 48% compared with 52% in the north: for Ca 51% S/49% N; for Mg 44.5% S/55.5% N for K 37% S/63% N and for HCo_3^- 47.5% S/52.5% N.

On the contrary the salt loss by infiltration is much less important in the south basin being 11% of the total as compared with 89% in the north.

As a consequence of these different hydrological and hydrochemical balances between the basins, the salinity of the north basin is 4 times higher than that of the south basin (10.7 mM 1^{-1} against 2.55 mM 1^{-1}). This increase is accompanied by a higher relative concentration of Na and HCO_3/CO_3 to the detriment of other elements.

$$*[\text{Cation i}]^r = \frac{[\text{Cation i}]}{[\Sigma \text{ cations}]}; [HCO_3^-]^r = \frac{[HCO_3^-]}{[HCO_3^-] + [SiO_4H_4]}; [SiO_4H_4]^r$$

$$= \frac{[SiO_4H_4]}{[HCO_3^-] + [SiO_4H_4]}.$$

Table 1   Characteristics of the mean salt balances in the north and south basins (N.B. and S.B.); M = moles.

	Na$^+$		Ca^{++}		Mg^{++}		K$^+$		HCO$_3^-$/CO$_3^-$		SiO$_4$H$_4$	
	S.B.	N.B.	S.B.	N.B.	S.B.	N.B.	S.B.	N.B.	S.B.	N.B.	S.B.	N.B.
River inputs M × 10^9	5.60	5.14	4.35	3.09	3.30	2.50	2.05	1.76	22.35	17.6	16.1	9.03
Infiltrations M × 10^9	0.46	5.15	0.27	2.13	0.20	1.75	0.14	1.51	1.50	14.05	0.77	2.19
Sedimentations M × 10^9	—	—	0.99	0.96	0.60	0.75	0.15	0.25	3.23	3.60	3.30	6.83
Lake stocks M × 10^9	8.95	96.2	5.22	39.8	3.86	32.8	2.82	28.3	29.2	2.83	15.0	41.1
Absolute concentrations mM × 1	0.355	2.05	2.205	0.85	0.15	0.70	0.110	0.605	1.15	5.65	0.595	0.88
Relative concentrations %	42.9	48.8	25	20.2	18.5	16.6	13.5	14.3	66	86,5	34	13.5
Turnover %	62.5	5.35	83.9	8.0	85.5	7.6	88.4	7.20	76.7	6.2	107	22

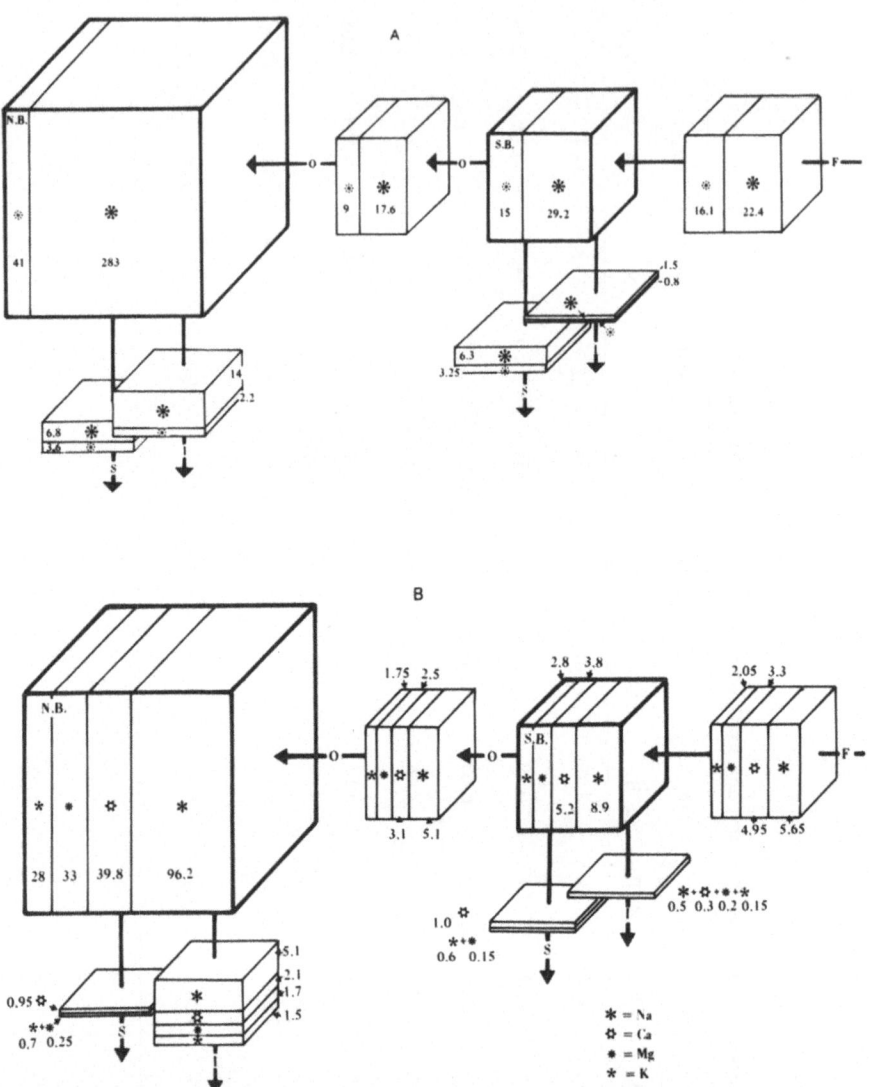

Fig. 6 Mean annual salt balances in the south and the north basins. A) HCO_3^-/CO_3^{--} and SiO_4H_4; B) Na^+, Ca^{++}, Mg^{++} and K^+.

F = fluvial input; S = sedimentation; I = infiltration; O = north basin input coming from the south basin. All the flows and the north (N.B.) and south (S.B.) basin stocks are expressed in moles $\times 10^9$.

4.4 Nature of the biogeochemical sedimentation

4.4.1 *The geochemical sedimentation: neoformation of smectites and calcite precipitation*

4.4.1.1 *Geochemical changes in the lake.* Each year the Shari transports an average of 2.8×10^6 t of suspended material to the lake, 69% of which is discharged between June and September. Its average mineralogical composition is given in Table 4.2. Kaolinite is the major component and the smectite is a ferriferous beidellite. $[Si_{3.47} Al_{0.53}] O_{10} [Al_{0.33} Fe^{II}_{1.17} Ti_{0.29} Mg_{0.21}] (OH)_2$ $Mg_{0.06} Ca_{0.16} Na_{0.025}$.

The lake sediments contain some minerals similar to those transported by the Shari but with different composition. The relative percentage of smectite is greater in the sediments.

To evaluate these modifications, the sum of the relative percentages of kaolinite, illite and feldspars was taken as a reference. These minerals have the advantage of being uniformly distributed throughout the lake, without mineralogical modifications and all showing identical composition in the solid load of the Shari and in the sediments (Carmouze 1976). Thus we calculated the gains and losses of their mineralogical constituents after sedimentation in relative and absolute values based on a mean annual transport for the period 1954–1972.

Thus 30% and 80% of Fe_2O_3 and Al_2O_3, supplied annually are recombined while there is an enrichment of sediments by amorphous silica and smectite in quantities representing 200% and 45% of the annual supplies of these constituents. One such enrichment of smectite comes largely if not completely from neoformation in the middle of the lake. As the mineralogical supplies of the Shari are less varied over the period of surface sediment formation, 35% of the smectites in the sediments can be estimated to be of allochthonous origin against 65% of autochthonous origin (Carmouze 1976).

4.4.1.2 *Characterization of neoformed smectites: formation, reaction and resulting hydrochemical modifications.* Three kinds of smectites were identified (Carmouze 1976; Carmouze et al. 1976–1977; Pedro et al. 1978). They were:
— a nontronite in the pseudo-sand sediments near the delta;
— a ferrous beidellite in the central region of the lake;
— a magnesium montmorillonite in the northern region of the lake.

In addition, sediments located in the north contained 5% to 10% calcite.

(a) *The nontronites.* The sediments near the Shari delta are composed of granules (so-called pseudo-sands) whose sizes range from 0.125 to 0.500 mm. They present different forms, colours and hardness, changing from black, manually unfriable and spherical granules to greenish-yellow, manually friable and oval pellets. They are constituted of goethite Fe O(OH) and nontronite whose average composition is $[Si_{3.83} Al_{0.06} Fe^{3+}_{0.11}] O_{10} [Fe^{3+}_{1.76} Mg_{0.20}] (OH)_2$ $Ca_{0.22} Na_{0.02} K_{0.02}$.

Black granules which derive from an agglomeration of monocristalline microparticles of ferric oxide of fluvial origin change into greenish-yellow ones in the delta. This change corresponds to a progressive silicification of goethite into nontronite: 1.87 Fe $O(OH) + 0.06$ $Al(OH)_3 + 3.83$ $SiO_4H_4 + 0.20$ $Mg^{++} + 0.22$ $Ca^{++} + 0.02$ $Na^+ + 0.02$ $K^+ \rightarrow$ nontronite $+ 0.88$ $H^+ + 7.2$ H_2O, followed by: 0.88 H^+ 0.88 $HCO_3^- \rightarrow 0.88$ $CO_2 + 0.88$ H_2O.

The dissolved silica and the magnesium in the nontronite structure come from the dissolved phase. The aluminium is the amorphous aluminium of fluvial origin or that contained in the goethite. This reaction produces an important hydrochemical elimination of silica (4 moles are necessary to form a mole of nontronite) and to a lesser extent, an elimination of Ca^{++} and Mg^{++} (0.20–0.22 moles). Besides, the reaction transforms a HCO_3^- fraction into Co_2 which is liberated, and thus the pH of the water decreases. Geochemically, there is an impoverishment of the environment, mainly of ferric oxide.

(b) *The ferriferous beidellite.* 65% of the smectite contained in the sediments located in the central zone is of autochthonous origin and 35% of fluvial origin, the formula of the former is: $[5Si_{4.0}] O_{10} [Al_{1.32} Fe^{3+}_{0.43} Ti_{0.02} Mg_{0.18}] (OH)_2$ $Ca_{0.14} Na_{0.01}$. The iron and the aluminium constituting this smectite come respectively from microcrystalline amorphous ferric hydrate and amorphous aluminium from the Shari. The titanium originates from the hydrate of titanium of fluvial origin in a free form. Finally the silica and the magnesium of the system and the exchangeable cations are extracted from the dissolved phase.

The formation reaction is: 4.0 $SiO_4H_4 + 1.32$ $Al(OH)_3 + 0.43$ $FeO(OH) + 0.02$ Ti $O(OH)_{2s} + 0.18$ $Mg^{++} + 0.13$ $Ca^{++} + 0.01$ $Na^+ \rightarrow$ ferriferous beidellite $+ 0.63H^+ + 8.9$ H_2O followed by 0.63 $H^+ + 0.63$ $HCO_3^- \rightarrow 0.63$ $CO_2 + 0.63$ H_2O. Thus, the formation of a mole of beidellite causes an impoverishment in the sediment of Al $(OH)_3$ (1.32 moles) and of Fe $O(OH)$ (0.43 moles) and in the aqeuous solution of SiO_4H_4 (4.0 moles), Mg^{++} (0.18 moles) and Ca^{++} (0.13 moles). CO_2, is also liberated, decreasing the pH of water.

(c) *The magneseous montmorillonite.* In the sediments of the north there is a mixing of smectites of fluvial origin, and smectites neoformed in the central zone with a locally formed magneseous smectite which formula is $[Si_4]O_{10} [Al_{0.24}$ $Fe^{3+}_{0.12} Ti_{0.11} Mg_{1.99}] (OH)_2 Ca_{0.195} Na_{0.07}$.

The aluminium component is derived from amorphous aluminium and the iron and the titanium from their corresponding oxides in free form. The silica and magnesium as well as their exchangeable cations come from the solution. Schematically the reaction is: 4.0 $SiO_4H_4 + 0.24$ Al $(OH)_3 + 0.12$ Fe $O(OH)_2 + 0.11$ Ti $O(OH)_2 + 1.99$ $Mg^{++} + 0.195$ $Ca^{++} + 0.07$ $Na^+ \rightarrow$ magneseous montmorillonite $+ 4.44$ $H^+ + 4.4$ H_2O, followed by 4.44 $H^+ + 4.44$ $HCO_3^- \rightarrow$ 4.44 $CO_2 + 4.44$ H_2O.

The neoformation of a mole of this smectite causes a major elimination of

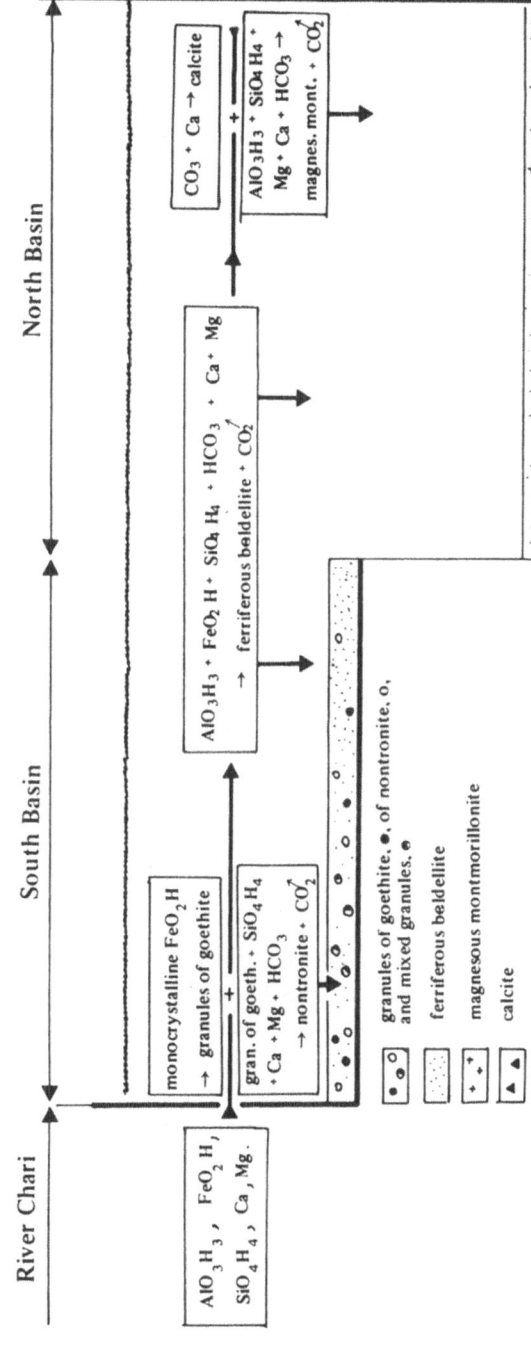

Fig. 7 Simplified geochemical reactions occurring in the lake and their localizations.

111

SiO_4H_4 (4 moles) and of Mg^{++} (2 moles) and to a lesser extent Ca^{++} (0.20 moles) from the aqueous solution. The large quantity of CO_2 liberated (4.44 moles) attenuates the pH increase due to evaporation.

(d) *In addition* to the neoformation of clays, precipitation of calcium carbonate occurs in the north, where the sediments contain about 5–10% calcite. A detailed study of this phenomenon has been made (Carmouze 1976) showing that the precipitation occurs when the product of the activities of free Ca^{++} and free CO_3^{--}, $\{Ca^{++}\}.\{CO_3^{--}\}$ exceeds 18 times the solubility product of calcite.

The calcite formation results in a reduction in the environment of Ca^{++} and CO_3^{--} and either in a decrease in alkalinity of the water or a lessening restraining its increase due to evaporation.

Figure 7 shows in a simplified form, these four geochemical reactions.

4.4.2 *The biochemical sedimentation*

The biological activity may profoundly affect the hydrochemistry. A given element is more affected when it participates largely in biological cycles of organisms whose biomass and production are high and when its release rate from dead organisms is low. It is indirectly affected when it is in direct relation with one or more elements which are modified by biological activity. With this general consideration, molluscs must have an influence on Ca regulation, the macrophytes on K and SiO_4H_4 and to a lesser extent on Mg and Ca and the diatoms on SiO_4H_4 regulation.

A few observations can be made on the biochemical sedimentation by the molluscs and the macrophytes. Although the sedimentation caused by the diatoms has been studied locally (Lemoalle 1978), it is difficult to draw any conclusion from it.

4.4.2.1 *Influence of molluscs.* Lévêque (1972) has shown the importance of molluscs on Ca regulation by estimating that a population equal to that of 1970 needs 700 000 t of Ca each year. This amount represents 4 times the annual river input of Ca or half the dissolved stock of Ca in the lake. Evidently, without recycling the Ca concentration would have been rapidly diminished in the lake and become zero in the zones furthest from the delta. If the biomass and production of molluscs remains constant and if the liberation of Ca from dying organisms is rapid, the regulation of Ca is not affected by the molluscs. In fact, the populations and the biomass change with lake fluctuations and the dissolution rate of shells varies regionally with the water pH.

The Ca transfers relative to variations in mollusc biomass are difficult to describe because we have little data on this subject. The total biomass changes from 560 000 t to 745 000 t in fresh weight from 1968 to 1970 while the lake level

112

decreased from 281.8 to 281.1 m. In 1972 it fell from a mean of 280.7 m to a value 5 to 10 times lower than that of 1970. At this time the calcium demand must have strongly decreased otherwise the dissolution rate of shells would have changed. This imbalance probably enriched the environment in one or two years from 100 000 to 125 000 tons of calcium which corresponded to 9.3 to 11.7% of the dissolved calcium stock in 1971. There was no increase in Ca stock at this time but as noted in para. 5.3 during the subsidence the increase of Mg and K sedimentation was greater than that of Ca. It may be explained by the influence of molluscs.

The rate of Ca liberation is indicated by the percentage of 'dead shells/live shells'. Lévêque (1972) showed that this percentage is low in the south basin as well as in the southern half of the north basin, proving that the recycling of Ca is largely assured during the period of relative stability of population or of the water level. In the northern half of the north basin the dead shells are relatively abundant and less eroded. They tend to be buried by the actual deposition and then the calcium loss is permanent. But this loss stays low because the mollusc biomass is low in this region. Moreover, at the beginning of their burial, the shells are in contact with the interstitial solutions aggressive against the aragonite, and are then dissolved. These shell levels are not found in recent sediments.

In summary, the permanent calcium losses are greater with increasing distance from the delta; they decrease with a decrease of biomass and increase with an increase in biomass. While the Ca stock tends to increase when the biomass decreases sharply (the fixation rate of Ca by the organisms decreases faster than the rate of Ca release by dissolution of dead shells) it tends to decrease when the biomass grows sharply for the opposite reasons.

4.4.2.2 *Influence of macrophytes.* The particulate materials of fluvial origin are enriched by organic matter during their sedimentation by a factor of 5. The organic matter largely derives from the degradation of macrophytes (Chevery 1974).

The principal macrophytes, *Phragmites australis*, *Cyperus papyrus*, *Vossia cuspidata* and *Typha angustiflora* represent 85%, 10%, 4% and 1% of the occupied surface and mostly hold dissolved silica and potassium.

The approximate quantities of K^+, Ca^{++}, Mg^{++}, Na^+ and SiO_4H_4 contained in the macrophytes represent respectively $138 \times 10^3 t$, $7.4 \times 10^3 t$, $6.6 \times 10^3 t$, $2.3 \times 10^3 t$ and $230 \times 10^3 t$ or 170%, 4%, 8.5%, 2.2% and 23.5% of the mean annual inputs of these elements. This represents 11.2%, 0.4%, 0.8%, 0.1% and 6.8% of their corresponding dissolved lake stocks. The amount of K stocked in the plants is relatively higher than that supplied each year by them. Considering that the ratio of production/biomass is about one, these quantities are similar to those which are annually taken up by the macrophytes themselves.

As organic accumulations are registered in some parts of the lake we can infer, from these data, that substantial amounts of K and, to a lesser extent, of

SiO_4H_4 are not recycled and thus represent permanent losses.

The annual amount of dissolved salts fixed by the macrophytes change in the same proportion as the biomass. It may vary greatly because the vegetation area which is about 2500 km^2 at the lake level of 281.5 m can double or be reduced by half respectively at lake levels of 280 and 283 m.

In conclusion, during the flood the biomass is reduced, the mineral salt requirements decrease, the rate of mineralization surpasses that of production and there is a partial restoration of mineral salts to the aqueous phase from the sedimented materials. The chemical sedimentation dependent on the activity of macrophytes must decrease.

During the lake recession the opposite phenomenon occurs. However, during a severe drought, some zones dry up causing the disappearance of macrophytic vegetation. When the waters reflood these regions they tend to be enriched in K and to a lesser degree in Mg (Chantraine 1978).

4.5 The hydrochemical regulation during the period of flood and subsidence

4.5.1 *Environmental instability from hydrological regulation*

(a) Instability of the entire lake against the annual supply–loss imbalance.

Due to the characteristics of its hydrological regulation, Lake Chad at first appears to be very unstable hydrochemically in the face of the irregularity of supplies. This aspect can be shown by a simple example:

Supposing that in an 'average' state, the lake is subjected to an annual supply–loss imbalance of water equal to:

$$\Delta(v)_A = \Delta(v)_F + \Delta(v)_M = \Delta(v)_F$$

For simplification

$$\Delta(v)_M = 0$$

Thus

$$\Delta(v)_A = \Delta(v)_F = \Delta(v)_L$$

Placing $\Delta(v)_F = x(\bar{v})_F$, we know that $(\bar{v})_F = 0.576\,(\bar{v})_L$.
Thus

$$\Delta(v)_F = \Delta(v)_L = 0.576\,x(\bar{v})_L$$

The new lake volume is thus

$$(v)'_L = (\bar{v})_L\,(1 + 0.576\,x)$$

The lacustrine stock of i also changes:

$$\Delta(i)'_L = \Delta(v)_F \times [i]_F = 0.576\,x(\bar{v})_L \times [i]_F$$

114

Now

$$[\tilde{i}]_F = 1/10.8 \ [\tilde{i}]_L$$

Thus

$$\Delta(i)'_L = 0.053 \ x(\bar{v})_L \times [i]_F \quad \text{and} \quad (i)'_L = (\bar{v})_L \times [\tilde{i}]_L \ (1 + 0.053 \ x)$$

The new value of i

$$[i]'_L \text{ is} = \frac{(i)'_L}{(v)'_L} = [\tilde{i}]_L \ \frac{(1 + 0.053 \ x)}{(1 + 0.576 \ x)}$$

Based on this formula when there is an annual excess of 33% in the fluvial supplies ($x = 1/3$), the salt concentration of the lake decreases by 15%. However, when there is a deficit of 33% ($x = 1/3$) the salt concentration increases by 21.5%. Thus considering only the geographical regulation, the variations in salt concentration, corresponding to frequent supply–loss imbalances, are relatively important.

(b) Instability of the south and north basins against the annual supply–loss disequilibria imbalance.

The previous example was used to calculate the instability of the two basins: 50.6% of the total supplies enter the north basin. Thus the volume variation of this basin is:

$$\Delta(v)_{CN} = 0.506 \Delta(v)_F = 0.506 \ x(\bar{v})_F$$

We know that

$$(\bar{v})_F = 0.888 \ (\bar{v})_{CN}$$

Thus

$$\Delta(v)_{CN} = 0.506 x \times 0.888 \ (\bar{v})_{CN} = 0.449 x \ (\bar{v})_{CN}$$

The new volume is:

$$(v)'_{CN} = (\bar{v})_{CN} \ (1 + 0.449 \ x)$$

The lacustrine variations of i in this basin are:

$$\Delta(i)_{CN} = 0.918 \ \Delta(v)_F \ x[i]_F$$

In fact 91.8% of the fluvial supplies arrive in the north basin (cf. para. 2.2). Thus

$$\Delta(i)_{CN} = 0.918 \ x(\bar{v})_F \cdot [i]_F$$

Given that $(\bar{v})_F = 0.888 \ (\bar{v})_{CN}$ and that $[i]_F = \dfrac{1}{15.24}[i]_{CN}$

$$(i)_{CN} = 0.918 x \times 0.888 \ (\bar{v})_{CN} \times \frac{1}{15.24} \ [\tilde{i}]_{CN} = 0.053 \ x \ (\bar{v})_{CN} \ [\tilde{i}]_{CN}$$

From this

$$(i)'_{CN} = (i)_{CN} + \Delta(i)_{CN} = (\bar{v})_{CN}[i]_{BN} \ (1 + 0.053 \ x)$$

Finally,

$$[i]'_{CN} = \frac{(i)'_{CN}}{(v)'_{CN}} = \frac{[i]_{CN} \ (1 + 0.053 \ x)}{(1 + 0.449 \ x)}$$

The remaining 49.4% of the annual water suplies are in the south basin. The volume variation is:

$$(v)_{CS} = 0.494 \ \Delta(v)_F = 0.494 \ x(\bar{v})_F$$

We know that

$$(\bar{v})_F = 1.64 \ (\bar{v})_{CS}$$

Then,

$$\Delta(v)_{CS} = 0.494x \times 1.64 \ (\bar{v})_{CS} = 0.81 \ x \ (\bar{v})_{CS}$$

and

$$(v)'_{CS} = (\bar{v})_{CS} + \Delta(v)_{CS} = (\bar{v})_{CS} \ (1 + 0.81 \ x)$$

The variation in the stock of i in the south basin $\Delta(i)_{CS}$ is:

$$\Delta(i)_{CS} = 0.082 \ x(\bar{v})_F \times [i]_F$$

Being given that $(v)_F = 1.64 \ (\bar{v})_{CS}$ and that $[i]_F = \dfrac{1}{2.61}[i]_{CS}$, we obtain:

$$\Delta(i)_{CS} = 0.082 \ x \times 1.64 \ (\bar{v})_{CS} \times \frac{1}{2.61}[i]_{CS} = 0.051 \times (\bar{v})_{CS} \times [i]_{CS}$$

and

$$(i)'_{CS} = (i)_{CS} + \Delta(i)_{CS} = (v)_{CS} = (v)_{CS} \ [i]_{CS} \ (1 + 0.051 \ x)$$

Finally,

$$[i]'_{CS} = \frac{(i)'_{CS}}{(v)'_{CS}} = \frac{[i]_{CS} \ (1 + 0.051 \ x)}{(1 + 0.81 \ x)}$$

Based on these two statements, when there is an excess of 33% ($x = 1/3$) of the fluvial supplies, the salt concentration decreases by 11.5% in the north basin and by 20% in the south basin. However, for a deficit of 33% ($x = 1/3$) the salt concentration increases by 15% in the north basin and by 34% in the south basin. The instability of the south basin is thus much greater.

4.5.2 *Hydrochemical regulation during a lake reduction*

From 1967 to 1972, the annual supplies were lower than the annual losses and the lake level has diminished by 1.35 m. In such a case the lake volume with a higher turnover rate than that of the saline stocks decreased faster; therefore the salt concentrations as well as the dissolved elements increased. This increase was more marked if the element decreased a little or if the removal coefficient of this stock was low.

The salinity changes can be calculated from the water and salt input deficits. During the period under consideration, the salt input deficit, $\Delta(i)_F$ which is equal to the difference between the supplies of an average regime (see para. 3.1.1) and those of 1967 to 1972, is as follows:

$$\Delta(Ca^{++})_F = 5 \times 10^{12} mM; \quad \Delta(Mg^{++})_F = 3.8 \times 10^{12} mM;$$
$$\Delta(Na^+)_F = 6.7 \times 10^{12} mM;$$
$$\Delta(K^+)_F = 2.4 \times 10^{12} mM; \quad \Delta(HCO_3^-)_F = 26 \times 10^{12} mM;$$
$$\Delta(SiO_4H_4)_F = 19.4 \times 10^{12} mM$$

or, in % of the corresponding initial stocks

$$\Delta(Ca^{++})_F = 11.1\%; \quad \Delta(Mg^{++})_F = 10.5\%; \quad \Delta(Na^+)_F = 6.5\%$$

$$\Delta(K^+)_F = 7.7\%; \quad \Delta(HCO_3^-)_F = 9\%; \quad \Delta(SiO_4H_4)_F = 34.6\%$$

These values indicate that the stock changes are proportional to the stock turnovers.

Assuming that annual outputs have not changed and that the salt stock variations arise only from input deficits, the salt stocks of 1972 could be calculated:

$$(i)'_L = (i)_L - \Delta(i)_F$$
$$(Ca^{++})'_L = 40 \times 10^{12} mM; \quad (Mg^{++})'_L = 32.8 \times 10^{12} mM;$$
$$(Na^+)'_L = 98.5 \times 10^{12} mM;$$
$$(K^+)'_L = 28.8 \times 10^{12} mM; \quad (HCO_3^-)'_L = 266 \times 10^{12} mM;$$
$$(SiO_4H_4)'_L = 36 \times 10^{12} mM.$$

The lake volume has changed from 72.10^9 in 1967 to $45,5.10^6 m^3$ in 1972, so the constituent concentration would be:

$$(Ca^+)'_L = 0.88 mM\ 1^{-1}; \quad (Mg^{++})'_L = 0.72 mM\ 1^{-1};$$
$$(Na^+)'_L = 2.16 mM\ 1^{-1};$$
$$(K^+)'_L = 0.63 mM\ 1^{-1}; \quad (HCO_3)'_L = 5.84 mM\ 1^{-1};$$
$$(SiO_4H_4)'_L = 0.80\ mM\ 1^{-1}.$$

In fact, the actual total salinity represents 67% of the forecasted value. The values in 1972 were lower than those we have just calculated:

$$(Ca^{++})''_L = 18.8 \times 10^{12} mM; \quad (Mg^{++})''_L = 20.2 \times 10^{12} mM;$$
$$(Na^+)''_L = 78.6 \times 10^{12} mM;$$

$(K^+)_L'' = 19.7 \times 10^{12} mM;$

$(CO_3H^-/CO_3^{--})_L'' = 172 \times 10^{12} mM;$

$(SiO_4H_4)_L'' = 25.85 \times 10^{12} mM.$

Initially this difference $(i)' - (i)''$ can be explained by an increase in the infiltration and sedimentation losses, taking as reference the outputs which occur when the lake is in the steady-state. But the infiltration can hardly increase because, although the waters which infiltrate are more concentrated during the subsidence, the infiltration volume diminishes because the infiltration zones are reduced and more impermeable when they are located toward the middle of the lake. In the case of sodium, the increased loss by sedimentation cannot be taken into account.

A very flat relief of the lake basin actually favours the isolation of bordering ponds. A small quantity of salt is then trapped and deposited when these ponds dry up.

Thus, for sodium, we assume that the increase in losses $\Delta(Na)_p$ is due only to these marginal losses $\Delta(Na)_{Pm}$ which represent 18.7% of the initial stock of 1967. The fraction corresponding to the marginal losses of the other elements is slightly different, depending on the stock distributions in space. For Ca^{++}, Mg^{++}, K^+, Na^+, HCO_3^- and SiO_4H_4 the marginal losses are respectively equal to 17.3, 18.1, 18.6, 18.7, 18.2 and 16.4% of the initial stocks.

Thus:

$$\Delta(Ca^{++})_{Pm} = 7.8 \times 10^{12} mM; \ \Delta(Mg^{++})_{Pm} = 6.65 \times 10^{12} mM; \ \Delta(Na^+)_{Pm}$$
$$= 19.7 \times 10^{12} mM;$$
$$\Delta(K^+)_{Pm} = 5.8 \times 10^{12} mM; \ \Delta(HCO_3^-/CO_3^{--})_{Pm}$$
$$= 53 \times 10^{12} mM; \ \Delta(SiO_4H_4)_{Pm} = 9.2 \times 10^{12} mM.$$

The higher the stocks are, the greater the losses.

The losses by chemical sedimentation $\Delta(i)_{Ps}$ are calculated from the difference: $(i)_{Ps} = \{(i)' - (i)''\} - (i)_{Pm}$.

$$\Delta(Ca^{++})_{Ps} = 13.45 \times 10^{12} mM;$$
$$\Delta(Mg^{++})_{Ps} = 6.0 \times 10^{12} mM; \ \Delta(Na^+)_{Ps} = 0;$$
$$\Delta(K^+)_{Ps} = 3.2 \times 10^{12} mM;$$
$$\Delta(HCO_3^-/CO_3^{--})_{Ps} = 41.0 \times 10^{12} mM; \ \Delta(SiO_4H_4)_{Ps} = 1.70 \times 10^{12} mM$$

or in % of the corresponding initial stocks:

$$\Delta(Ca^{++})_{Ps} = 28.9\%; \ \Delta(Mg^{++})_{Ps} = 16.3\%;$$
$$\Delta(K^+)_{Ps} = 10.3\%; \ \Delta(HCO_3^-/CO_3^{--})_{Ps} = 13.9\%; \ \Delta(SiO_4H_4)_{Ps} = 2.9\%$$

Charges in the hydrochemistry from 1967 to 1972 are shown in Fig. 8.

In summary:

— the decrease of the dissolved silica stock is due to a deficit of fluvial supplies (65% of the total decrease of SiO_4H_4 and to marginal deposits (30.5%),

118

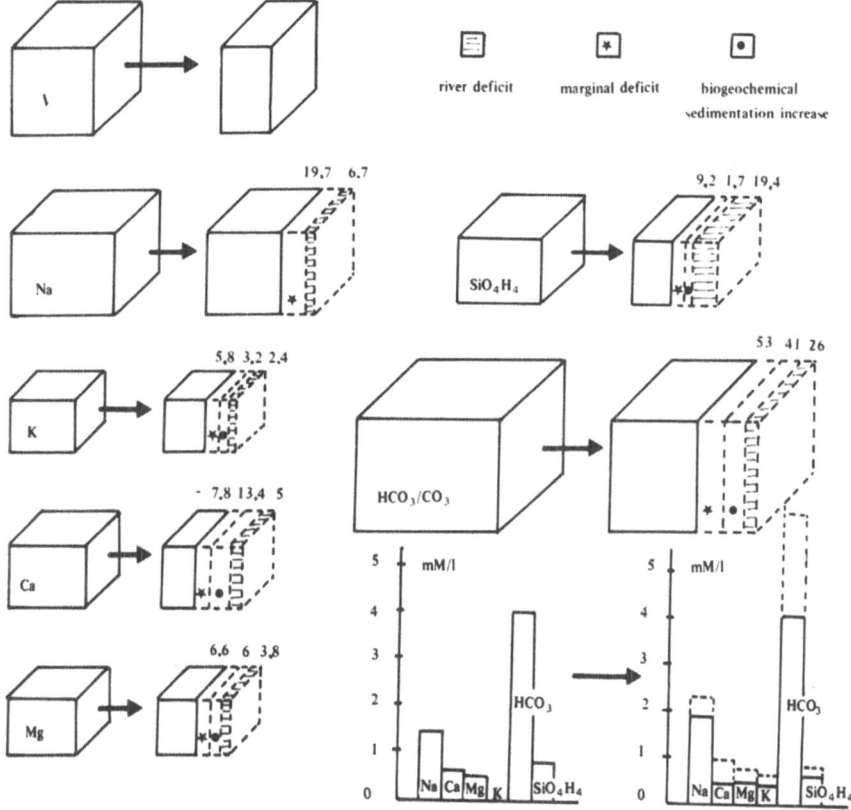

Fig. 8 Changes in the lake salt stocks and the chemical composition of the lake waters during the 1967–1972 lake contractions. The salt stocks variations are expressed in moles × 10⁹.

while the increase in losses by biogeochemical sedimentation is only 5.5%;
— the decrease in the sodium stock is caused by marginal deposits (14%) and decrease in fluvial supplies (25.5%);
— the decrease of the calcium stock is mainly caused by the increase in chemical sedimentation (51.2%). Marginal losses and those caused by a decrease in fluvial supplies are respectively 29.7% and 19%.

Finally, the decrease of magnesium, potassium and carbonate-bicarbonate stocks are due to marginal losses (of 40%, 50.9% and 44.1%) and to the increase in chemical sedimentation (respectively 36.5%, 28% and 34.0%). Deficits from fluvial supplies are 21 to 23%.

Thus the decrease in stocks in proportions similar to those of the volume is caused by marginal losses and an increase in chemical sedimentation. By comparison with the values of 1967 Na^+ increased by a third; $[K^+]$ $[HCO_3^-/CO_3^{--}]$ and $[Mg^{++}]$ were less varied while $[Ca^{++}]$ and $[SiO_4H_4]$

decreased significantly. The total salt content of the waters did not change with the sum of dissolved salts was 7.90mM 1^{-1} in 1967 and 7.86mM 1^{-1} in 1972.

The chemical composition changed following the flow of matter towards the sodium-bicarbonate pole, which was:

— in 1972

$$[Ca^{++}]_r = 13.7\%; [Mg^{++}]_r = 14.6\%; [Na^+]_r = 57.3\%;$$
$$[HCO_3^-/CO_3^{--}]_r = 87.1\%; [SiO_4H_4]_r = 12.9\%.$$

— in 1967

$$[Ca^{++}]_r = 20.8\%; [Mg^{++}]_r = 16.8\%; [Na^+]_r = 47.8\%;$$
$$[HCO_3^-/CO_3^{--}]_r = 89.4\%; [SiO_4H_4]_r = 10.6\%.$$

To understand the rhythm of the decrease in stocks from one basin to another, and for a given basin, of one element to another, the ratio $(i)_r/(i)_s$ is supposed to represent the stocks of i in the north and south basins. The relative change of the two basins from 1967 to 1972 is shown by Fig. 9.

The Na^+ and SiO_4H_4 stocks of the south basin decreased more quickly than those of the north basin. The HCO_3^-/CO_3^{--} and K^+ stocks decreased according to a rhythm similar to that of their homologues in the north basin. Finally the

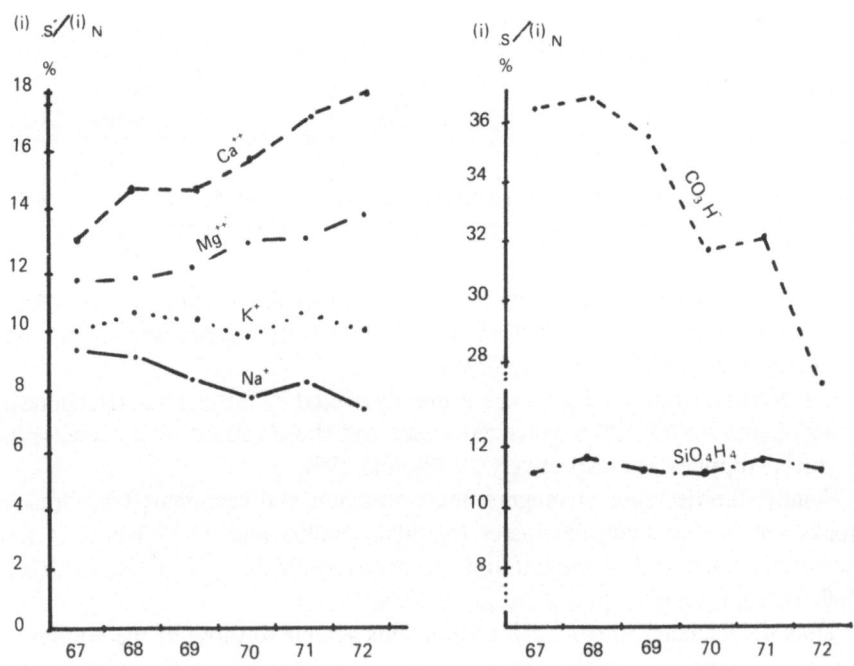

Fig. 9 Relative changes in the salt stocks of the north $(i)_N$ and the south $(i)_S$ basins from 1967 to 1972.

Ca^{++} and Mg^{++} stocks of the south basin decreased less quickly than the corresponding ones in the north basin.

These results show that the factors of the stock decline (deficit of fluvial supplies, increase of sedimentation, marginal losses) have a variable importance from one basin to another not only for a given element but also from one element to another. Unfortunately, it is not possible to calculate satisfactorily the contribution of each of these factors in each of the basins as has been done for the whole lake.

The following changes occurred in the chemical composition of the waters:
— in the south basin, ionic concentrations increased by a third, and the SiO_4H_4 concentrations did not change;
— in the north basin, only the Na^+ concentration increased by 30%; the other elements remained relatively constant (K^+, HCO_3^-/CO_3^{--} and SiO_4H_4) or decreased (Ca^{++} and Mg^{++}).

If relative concentrations are considered, the waters of the south basin developed towards the sodium-bicarbonate pole and those of the north basin towards the sodium pole, ($[HCO_3^-/CO_3^{--}]$ and $[SiO_4H_4]$ remained relatively constant between them).

4.5.3 *Hydrochemical regulation during a flood*

Only two reference points are available for the hydrochemical development of the lake when the water level was increasing: one from 1945 and the other from 1957.

The lake volume in 1945 was similar to that of 1972, attained after a similar lake decrease. Therefore we can assume that the dissolved stocks of 1945 were also similar to those of 1972, because they changed in approximatively the same proportions as volume. Consequently, the total ionized salt stock of the lake in 1945, evaluated with the data of 1972, was equal to 175×10^{12}mé. In 1957 the electrolyte stock, estimated from conductivity data (Guichard 1957) was about 265×10^{12}mé. So, the increase was 90×10^{12}mé, while the lake level rose 2.20m.

This increase arose from excess annual supplies and from partial redissolution of salt deposited on the marginal zone during the preceding contraction of the lake and possibility from a decrease in the annual chemical sedimentation rate. Losses by infiltration were not diminished; on the contrary they may have increased.

The increase of the electrolyte stock caused by excess river supplies between 1945 and 1957 was about 45×10^{12}mé (Carmouze 1976). This represents 50% of the stock increase. Consequently, the remainder, i.e. 45×10^{12}mé, was attributed to the redissolution of the marginal salt deposits* when the lake

* The redissolution of salt deposits is about 60% (Carmouze 1976).

121

increased, and, to a lesser extent, to a diminution of the chemical sedimentation rate.

During the period 1945–1957, the volume of the lake doubled: from 42×10^9 to $84 \times 10^9 m^3$, while the ionized salt stock increase was 50%. Consequently, the mean electrolyte concentration diminished from 4.15mé l^{-1} to 3.25mé l^{-1} (or from 365 µS cm^{-1} to 290 µS cm^{-1}). However, if the salt stock had only been dependent on the fluctuations of inputs during this period it would have been equal to about 2,60mé l^{-1} or 235 µS cm^{-1} (80% of the actual value).

This difference points out the importance of the dissolution process of the marginal salt deposits and may be, to a lesser extent, the diminution of the chemical sedimentation rate. They both help to reduce the decrease in salinity caused by geographical factors when the lake increases in size.

4.6 Conclusion

In spite of its endorheic nature and its subarid environment *Lake Chad is not a lake of high salinity*. Three main reasons are responsible for this:
— the river salinity is *low* (about 60 mg l^{-1} or half of the mean salinity of continental waters);
— the climato-geographical regulation of the salinity determines a concentration factor of the river water of about 10.8. This value is not very high for a closed lake situated in an arid zone because the infiltrations are considerable. This factor for the south basin, which is a transit zone, is only 2.60 against 8.45 for the north one;
— the biogeochemical regulation is characterized by major sedimentation of SiO_4H_4, Ca, Mg, HCO_3/CO_3 and to a lesser extent K. The geochemical sedimentation is due to neoformations of smectites and precipitation of calcite. The neoformations of smectites are facilitated by a relatively high concentration of SiO_4H_4 in the river Shari. The salinity of water where calcite precipitation takes place is not very high because the anions are only CO_3H and CO_3. The mollusc and macrophyte contributions to the biochemical sedimentation are surely important because the biomasses and the productions of these two groups are substantial. Then the 'climato-geographical' value of the salinity is reduced by 45%.

The salinity is relatively stable in spite of the high instability of the lake caused by the turnover of water and the dissolved salts. This is mainly due to the existence of marginal salt deposits formed when the lake decreases and partial redissolution of these deposits when the lake increases. These phenomena produce in the first case an additional output of salts, in the second case an additional input, in such a way that dissolved salt stocks vary in approximatively the same proportion as the volume. They are favoured by the morphology of the lake, which is flat and marked by a dune system. For this reason when the

122

lake contracts many ponds are isolated and rapidly dry up. Also, the biochemical sedimentation rate increases when the lake decreases because the biomasses of molluscs and macrophytes increase (this is true only when the decrease is not very strong). The inverse phenomenon takes place when the lake increases.

References

Billon, B. et al., 1969. Monographie hydrologique du Chari. ORSTROM Paris, 5 parts in 6 vols., 610 pp., mimeo.

Carmouze, J. P., 1969. Salures globales et spécifiques des eaux du lac Tchad en 1968. Cah. ORSTOM Sér. Hydrobiol. 3: 3–14.

Carmouze, J. P., 1972. Originalité de la régulation saline du Lac Tchad. C. r. Acad. Sci. Paris 275: 1871–1874.

Carmouze, J. P., 1976. La régulation hydrogéochimique du Lac Tchad. Trav. Doc. ORSTOM No. 58, 418 pp.

Carmouze, J. P., Golterman, H. P. and Pedro, G., 1976. The neoformation of sediments in Lake Chad. Their influence on the salinity control. Interaction between sediments and freshwater. Amsterdam, September 6–10: 33–39.

Carmouze, J. P., Pedro, G. and Berrier, J., 1977. Sur la nature des smectites de néoformation du lac Tchad et leur distribution spatiale en fonction des conditions hydrogéochimiques. C. r. Acad. Paris 284 Sér. D: 615–618.

Carmouze J. P. and Pedro, G., 1977. Contribution des facteurs géographiques et sédimentologiques à la régulation saline d'un milieu lacustre. Cah. ORSTOM Sér. Hydrobiol. 11: 231–237.

Chantraine, J. M., 1978. Evolution hydrochimique du lac Tchad de septembre 1973 à septembre 1975 lors d'une phase de décrue. Cah. ORSTOM Sér. Hydrobiol. 12: 3–18.

Dieleman, J. P. and De Ridder, N. A., 1963. Expertise sur les mouvements des eaux et du sel dans le polder de Bol-Guini. Int. Inst. Land. Rech. Wageningen, 50 pp.

Fontes, J., Maglione, G. and Roche, M. A., Données isotopiques préliminaires sur les rapports du lac Tchad avec les nappes de la bordure nord-est. Cah. ORSTOM Sér. Hydrol. 6: 17–34.

Guichard, E., 1957. Sédimentation du lac Tchad. ORSTOM Fort-Lamy, 46 pp., mimeo.

Lemoalle, J., 1978a. Application des images Landsat à la courbe bathymétrique du lac Tchad. Cah. ORSTOM Sér. Hydrobiol. 12: 83–87.

Lemoalle, J., 1978b. Relations silice-diatomées dans le lac Tchad. Cah. ORSTOM Sér. Hydrobiol. 12: 135–142.

Lévêque, C., 1972. Mollusques benthiques du lac Tchad; écologie, production et bilans énergétiques. Thèse d'Etat, Univ. Paris VI. 225 pp., mimeo.

Maglione, G., 1974. Géochimie des évaporites et silicates néoformés en milieu continental confiné. Thèse d'Etat Univ. Paris VI, 331 pp., mimeo.

Pedro, G., Carmouze, J. P. and Velde, B., 1978. Peloïdal nontronite formation in recent sediments of the lake Chad. Chem. Geol. 23: 139–149.

Pirard, F., 1963. Reconnaissance hydrogéologique du Niger Oriental. BRGM, Orléans.

Roche, M. A., 1969. Evolution dans l'espace et le temps de la conductivité électrique des eaux du Tchad. Cah. ORSTOM Sér. Hydrol. 6: 35–78.

Roche, M. A., 1973. Traçage naturel salin et isotopique des eaux du système hydrologique du lac Tchad. Thèse d'Etat, Univ. Paris VI, 385 pp., mimeo.

Schneider, J. L., 1967. Relation entre le lac Tchad et la nappe phréatique. A.I.H.S. Symposium de Garda, Publ. No. 70: 122–131.

Toucheboeuf de Lussigny, P., 1969. Monographie hydrologique du lac Tchad. ORSTOM Paris, 169 pp., mimeo.

II. The main types of communities and their evolution during a drought period

5. The aquatic vegetation of Lake Chad

André Iltis and Jacques Lemoalle

The phanerogamic vegetation of the lake was studied by Leonard (1969, 1974) during two surveys carried out in December 1964 and January–February 1968. The species composition and the location of the populations were determined and some observations on the aquatic vegetation of the then reduced lake area were made by Chouret and Lemoalle (1974), Fotius (1974), Fotius and Lemoalle (1976). These observations follow the changes in the lake basin which are related to the continuing slow recession of water since 1968. Due to travelling difficulties in the marshy or recently dried zones, the last surveys were made only in the south basin, while only some aerial surveys of the north basin could be made.

The existence, composition and population density are influenced by different ecological factors of which the principal ones are:
— the variations in water level which determine total or partial dryness during the period of lake contraction and determine the height and the duration or submersion at the time of high water;
— the nature and the slope of the substrate;
— the exposure to wind and its force; wave action determining the presence or absence of certain floating species highly sensitive to the disturbance of water;
— the physico-chemical composition of the water which is progressively loaded with salt with increasing distance from the Shari delta.

During the stage 'Normal Chad' until the end of 1972, the aquatic vegetation could be considered to be highly developed in comparison to that existing in the shallow European or African lakes. However the vegetation cover was limited to the deltaic zones; to the borders of the archipelago islands and to the floating islands which detach from it; to the shallower areas of the eastern part of the lake; and to the shallows. After this period, the stage 'Lesser Chad' can be distinguished, when this vegetation cover extended over approximately 50% of the normal lake basin. The drying up of vast areas situated below the former shore line and the appearance of extensive marshy zones with some temporary drying up gave rise to the development of considerable aquatic vegetation relatively poor in species.

J.-P. Carmouze et al. (eds.) Lake Chad
© *1983, Dr W. Junk Publishers, The Hague/Boston/Lancaster*
ISBN 978-94-009-7268-1

Photo 7 Aerial photograph of an island showing the fringe of aquatic macrophytes.

5.1 'Normal Chad' stage

5.1.1 *Main species associations*

— *Potamogeton* spp. and *Vallisneria* sp. association
Found in aquatic vegetation in weak currents, it is well represented east of the Shari delta and scattered through the lake. It also often includes two genera, *Najas* sp. and *Ceratophyllum demersum*. Several different groups can be distinguished which are characterized by the dominance, sometimes exclusive, of each constituent of this association.
— *Nymphaea* spp and *Utricularia* spp. (Nymphaeïdae association)
This is found in the sheltered bays, the leeward shores and the ponds.
— *Pistia stratiotes* and *Lemna perpusilla* association
This is an association of floating plants, characteristic of sheltered zones and more particularly of sheltered and open bays with semi-aquatic vegetation. It includes some *Spirodela polyrhiza*, *Azolla africana* (on the surface) and often *Ceratophyllum demersum*. It is found almost everywhere in the lake but especially in the south, and in the El Beïd and Yobé deltas. It is absent from the Shari delta.

126

Photo 8 Palm tree ('doum').

— *Ludwigia adscendens* subsp. *diffusa* association

This semi-aquatic vegetation, spread over the water's edge, is found in the sheltered bays and in the fringes of the aquatic meadows and reed beds. It includes some *Ipomoea aquatica* and *Neptunia oleracea* and dynamically follows the *Pistia stratiotes* and *Lemna perpusilla* association. It is widely spread throughout the lake especially in the south and particularly well represented in the deltas of the El Beïd and the Yobé.

— *Cyperus laevigatus* association

This association is characteristic of the fringes of the salt ponds of the Sahel zone and is frequently found around the Kanem lakes, a region bordering northern Lake Chad. It also appears very sporadically in the extreme north of the lake, where the waters have the highest dissolved salt concentration.

— Aquatic meadows of *Vossia cuspidata*

This is a semi-aquatic, well-spread grassy plant which is dominant in the Shari and El Beïd deltas. The series often includes the reed fringes in some other parts of the lake. Though abundant in the south and the east, the *Vossia* meadows are progressively scarce close to the 'Great Barrier', north of which the papyrus become yellowish and less developed. They progressively disappear and are almost absent from the northern border of the lake.

127

— *Phragmites australis* subsp. *altissimus* (Phragmitidae) association

The established populations form reed belts with a dense surface root mat and some tropical creepers. It is a very frequent association throughout the lake and is present in the south (Shari and El Beïd deltas), well developed in the east and the center ('Great Barrier') and clearly dominant in the northern part.

— *Typha australis* (Typhidae) association

The established population forms reed belts but without floating root mats and thus only very rarely with tropical creepers. This association, which forms some scattered groups, is absent in the south, fairly well represented in the east and the center and well developed in the north.

— *Aeschynomene elaphroxylon* (local name 'ambatch') group

This bush can reach five to six meters high and is scattered throughout the zone of the reed islands and on the border of the islands of the archipelago but appears to be absent in the south of the lake.

— Vegetation of the archipelago

Two different zones can be distinguished: (a) the zones of flat islands with long submerged aquatic meadows of *Leersia hexandra* and *Echinochloa* sp., or with long dried meadows of *Panicum repens* and *Sporobolus spicatus*, or with a belt of *Hyphaene thebaïca* ('doum' palm) in the high waters. These flat islands are especially abundant in the Northeastern Archipelago; (b) the

Photo 9 Fishermen camp on a reed island.

zones of islands which have prominent relief and so are never submerged, with a vegetation of Sahelian trees: *Acacia* sp., *Calotropis procera*, *Leptadenia* sp. and a belt of *Hyphaene thebaïca*. These islands constitute the major part of the Southeastern and Eastern Archipelago.

5.1.2 *Morphology of the vegetation*

The lake vegetation can be grouped according to species composition and the zone in which it is developed. Thus several major types of communities can be distinguished.

5.1.2.1 *Aquatic meadows*. Found especially in the deltaic zones, the aquatic meadows cover a triangular zone about 15 to 20 km at the outlet of the Shari. The river waters flow through it by some well-defined channels during recession and by filtering through the vegetation during high waters. *Vossia cuspidata*, which occurs in one to three meters of water is the most widespread species.

On the borders of the lake, the *Vossia cuspidata* meadows are replaced by meadows of *Cyperus papyrus* or *Phragmites australis*. The co-existing species are identical in the three areas.

Photo 10 Papyrus float used by local people to cross channels between islands.

The surveys made in the Shari delta showed the following species[a]:

	Vossia cuspidata meadow		Cyperus papyrus meadow	Phragmites australis meadow
	1st survey	2nd survey		
Vossia cuspidata	5.5	5.5	1.1	1.1
Ludwigia adscendens subsp. *diffusa*	+0.1	1.2	+0.1	
Cyperus papyrus subsp. *miliaceus*	+0.1	+0.2	5.5	
Phragmites australis	+0.1			5.5
Oxystelma bornouense		+0.2		2.1
Luffa cylindrica		1.3	2.2	
Cayratia sp.		+0.2	1.2	1.1
Thelypteris totta			+0.2	
Ipomoea rubens			1.2	1.1
Ipomoea aquatica			1.2	
Vigna sp.			+0.2	
Cyperus sp.			+0.1	

[a]Abundance and dominance notation (Braun-Blanquet):
+ = rare or very rare individuals, covering very low percentage of the area
1 = fairly abundant individuals but covering low percentage of the area
2 = very abundant individuals or found in 1/20 of the area
3 = individuals covering from $\frac{1}{4}$ to $\frac{1}{2}$ of the area
4 = individuals covering from $\frac{1}{2}$ to $\frac{3}{4}$ of the area
5 = individuals covering from more than $\frac{3}{4}$ of the area
Arrangement of the individuals of the same species in comparison to each other: 1. isolated individuals; 2. in group; 3. in set; 4. in small colonies; 5. in populations.

5.1.2.2 *Reed islands*. These are the vegetation islands established on the shallows and associated with *Cyperus papyrus* and *Phragmites* as well as other less abundant species. The latter develop on a compost of rhizoïdes and roots forming a thick mat of one to two meters under the water surface. The islands are anchored in the shallows whose form and size determines that of the reed island. These vegetation formations which can be several hundred meters long often include some clear gaps and some small closed ponds where some associations characteristic of sheltered waters develop. The reed islands occupy all the eastern part of the lake and are abundant in the region of the 'Great Barrier'. They also form a large border up a few kilometers between the open waters and the archipelago.

The group of the Tarara reed islands, situated a few kilometers northeast of the Shari delta are composed of the following species:

Central zone, about 50 to 100 m wide		
	Phragmites australis	5.5
	Cayratia sp.	1.1
Circular zone about 30 m wide		
	Cyperus papyrus ssp. *miliaceus*	5.5
	Ipomoea rubens	1.2
	Melothria sp.	+0.2
	Luffa cylindrica	1.2
	Vigna sp.	1.2
Circular zone about 5 to 10 m wide		
	Vossia cuspidata	5.5
	Cayratia sp.	1.1
	Ipomoea rubens	+0.1
	Polygonum sp.	+0.2
	Ludwigia adscendens subsp. *diffusa*	+0.2
	Cyperus imbricatus	+0.1
	Alternanthera sessilis	+0.2
	Cyperus papyrus subsp. *miliaceus*	1.1
Floating layer on the surface, about 2 m wide, sometimes absent		
	Pycreus mundtii	4.4
	Ludwigia stolonifera	1.1
	Cyperus nudicaulis	2.2
	Ipomoea aquatica	+0.1
	Neptunia oleracea	+0.1

In the north basin of the lake, the numbers of species on the reed islands decreases and *Vossia cuspidata* and *Cyperus papyrus* disappear towards the northern coast.

5.1.2.3 *Floating islands.* Some floating islands detach from the reed islands or from the vegetation borders of the islands and archipelago peninsulas and float away as they are pushed by the wind. Consisting most often of *Cyperus papyrus* and more rarely of *Phragmites* or *Vossia*, these islands are usually circular and are locally called 'kirtas'. Their size varies from a few meters to several hundred meters. At the time of the reversal of the dominant winds, during June and October and generally in the rainy season when the wind shifts are frequent during tornadoes, these islands move back and forth, modifying the aspect of the reed islands and closing the channels of the archipelagoes. The composition of the vegetation on these islands is similar to that seen on the reed islands.

The 'kirtas' were very numerous especially in the archipelago of the south basin until 1967 and the boats en route to Bol were often blocked for several days by the floating papyrus. These islands disappeared fairly quickly when the average level of the lake decreased below 282 m.

5.1.2.4 *Vegetation borders of the islands of the archipelago.* All the islands and peninsulas of the archipelago possess a vegetation fringe several meters wide and three to four meters high. *Cyperus papyrus, Vossia cuspidata* and especially *Phragmites australis* are the most abundant and largest species. At the base of the reeds, there are floating plants such as *Lemna perpusilla, Spirodela polyrhiza, Ceratophyllum demersum, Pistia stratiotes.* In the northern part of the lake, *Cyperus papyrus* is replaced by *Typha australis.* This vegetation barrier is interrupted in only a few places by narrow passages which allow local boats to land.

Three surveys made in three different borders showed the following species:

	1st survey	2nd survey	3rd survey
Phragmites australis	5.5	1.1	1.1
Cyperus papyrus subsp. *miliaceus*		5.5	1.2
Vossia cuspidata		1.2	5.5
Typha australis	+0.2		
Ipomoea aquatica	+0.1		+0.1
Cayratia sp.	1.1		
Oxystelma bornouense	1.1		
Ipomoea rubens	1.1	1.2	
Ludwigia adscendens subsp. *diffusa*	+0.1		
Alternanthera sessilis	+0.1	+0.1	
Polygonum sp.	+0.1	+0.1	
Pycreus mundtii	+0.2		
Lemna perpusilla	+0.1	+0.2	
Spirodela polyrhiza	+0.1		
Melothria sp.	+0.1		
Cyperus sp.	+0.1		
Oldenlandia sp.	+0.1	+0.1	
Luffa cylindrica		2.3	
Eclipta prostrata		+0.1	+0.1
Ceratophyllum demersum	2.2		

Inside this marginal fringe, a zone on the edge of the islands, 2–3 meters wide, is colonized by *Pycreus mundtii* (dominant) with *Leersia hexandra* and *Cyperus articulatus.* This is followed by a belt one to two meters wide of *Cynodon dactylon* circling a zone of doum palms (*Hyphaene thebaïca*) which marks the base of the

132

slope of the dunes. If the island has fairly marked relief, the summit and the slopes have bushes or trees of *Leptadenia hastata*, *Tinospora bakis*, *Salvadora persica*, *Phyllanthus reticulatus*, *Balanites aegyptiaca*, *Commiphora* sp., *Cassia occidentalis*. A grassy carpet develops during the rainy season (Fig. 1).

5.1.2.5 *Submerged vegetation banks*. These banks are very extensive in the shallow marshy zones situated to the east of the Shari delta, at the foot of the Hadjer el Hamis rock hills. In this region, the coast appears to be poorly delimited and is masked by vegetation. The wider channels are overgrown by beds of *Potamogeton schweinfurthii*, *Vallisneria* sp. and *Ceratophyllum demersum* alternating with emerged vegetation islands of *Vossia* and *Phragmites*.

Some less extensive submerged vegetation banks exist in the southern part of the lake near the El Beïd delta and especially in most of the extremities of the archipelago branches, that is in the Southeastern or Northeastern Archipelago or the channels of open water which penetrate into the vegetation border of the northeast coast.

5.1.3 *Conclusions*

The observations on the aquatic flora covering 2400 km² or 12% of the lake surface during the period of 'Normal Chad' (Carmouze et al. 1978) lead to the following conclusions on the distribution of species in the lake.

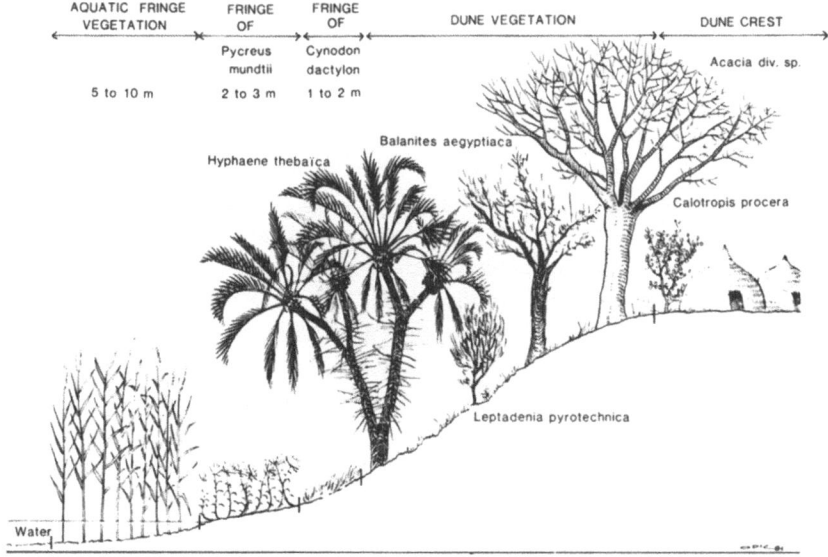

Fig. 1 Zonation of aquatic and terrestrial vegetation on a dune slope in the archipelago.

133

— the abundance of *Vossia cuspidata* in the entire Shari delta, before its progressive disappearance and removal from this zone;
— the abundance of *Cyperus papyrus* in the south basin of the lake and the 'Great Barrier' and its progressive disappearance in the north basin after an intermediate zone of plants in poor condition;
— the scarcity of *Typha australis* in the south basin and its relative abundance in the north basin illustrated by the local boats, called 'kadeï' being made of *Papyrus* in the south basin and central part of the lake and of *Typha* in the northern part;
— the abundance of *Potamogeton schweinfurthii* as submerged vegetation in the marshy zone to the east of the Shari delta;
— the sporadic but significant appearance in the extreme north of *Cyperus laevigatus*, a species exclusive to the edges of salt ponds.

The structure of reed island vegetation belts where the following concentric zones are found in centripetal order summarizes the development of the vegetation passing from the delta towards the north of the lake:
— to the south, *Vossia cuspidata, Cyperus papyrus, Phragmites australis*;
— towards the center, *Cyperus papyrus, Phragmites australis, Typha australis*;
— to the north, *Phragmites australis, Typha australis*.

Towards the north, there is a progressive disappearance of several aquatic species and groups. The impoverishment of the flora is directly related to the increasing salt concentration of the water.

5.2 'Lesser Chad' stage

5.2.1 *General remarks*

The observations and descriptions of the vegetation during the period of 'Lesser Chad' appear to be more complex than for the preceding stage because of the partition of the lake into two basins, each evolving independently of each other. On the other hand, during 'Lesser Chad' the lake appears to be very unstable, the effect of the floods being much less dampened in the south basin than in the entire lake during a normal period. On the whole, two periods can be distinguished.
— a period of drought, from 1972 to mid-1975 for the north basin and until the end of 1974 for the south basin. It led to total drought of the northern part of the lake and in the southern basin to a reduction of the open water to area 1500 km² and to the isolation of smaller ponds subsisting in the archipelago;
— a period of 'Lesser Chad' from the end of 1974. The south basin filled up to nearly the level of 1972. The north basin remained dry or contained little more than temporary ponds, which filled during the rains or with surplus water from the Shari flood which overflowed from the south basin and emptied through the 'Great Barrier' into the north basin.

The collection of data and observations were hindered by the impossibility or the difficulty of travelling through the invading marshy zones of vegetation which prohibit access to some very extensive regions. The aerial surveys could not solve this problem effectively because the flights needed to be supported by ground surveys. In the future, the use of satellite images may provide some more complete and much more directly utilizable data.

5.2.2 *Different population types*

5.2.2.1 *South basin.* Based on the surveys made in March 1974 (Fotius 1974), successions of plant species can be seen, following the open waters towards the dry ground.
— Shari delta: *Nymphaea* sp., *Ipomoea aquatica, Ludwigia adscendens* subsp. *diffusa, Sacciolepis africana, Vossia cuspidata, Aeschynomene elaphroxylon*;
— sandy islands north of the open waters in the southern basin:
  1st survey: *Aeschynomene elaphroxylon, Cyperus papyrus, Vossia cuspidata, Diplachne fusca, Cyperus articulatus, C. maculatus*;
  2nd survey: *Diplachne fusca, Cyperus articulatus, C. papyrus, Aeschynomene elaphroxylon*;
  3rd survey: *Cyperus articulatus, Typha australis, Phragmites australis, Aeschynomene elaphroxylon*;
— the archipelago island towards Bol: *Cyperus papyrus, Typha australis, Aeschynomene elaphroxylon, C. articulatus, Sesbania sesban, C. maculatus, Phyla nodiflora, Cassia occidentalis, Calotropis procera.*
From the partial observations made over the first phase of the period of 'Lesser Chad', the development of the lake can be outlined by noting:
— the massive development of *Aeschynomene elaphroxylon* (Ambatch) which existed only sporadically during the period of 'Normal Chad'. These ambatch forests developed on the dried sediments in 1973 and invaded the zones in the process of drying. Some similar species like *Aeschynomene pfundii* or *A. afraspira* are observed mixed with the *A. elaphroxylon* group;
— the extension of meadows of *Vossia cuspidata*, a species whose growth was earlier limited especially in the deltaic regions and to a lesser degree to some points of the Southeastern Archipelago. This species is now very marked in nearly all the surveys of the south basin;
— the recession of the *Phragmites australis* populations. Where the decrease is not obvious the population remaining stable, the % cover has decreased with the development of areas occupied by other aquatic macrophytes;
— the relatively limited development of *Typha australis* which was observed in several points of the archipelago, but never on very large areas;
— the development on dry land of species that were less abundant earlier, such as *Cassia occidentalis, Cyperus maculatus, Sesbania sesban, Phyla nodiflora*

and especially *Calotropis procera*. These plants which are particularly resistant to the drought and are not consumed by the animals, multiply rapidly on the sandy or clayey substrate when they are not flooded.

In the course of the following period, which began in 1974 with the filling up of the southern basin up to about its 1972 level, the aquatic vegetation underwent some modifications. This was partly due to the inundation of areas which had been dry for one to two years and partly due to seasonal variations in the water level caused by the Shari flood. The following observations were made in June 1976 (Fotius and Lemoalle 1976) in different parts of the lake:

— Shari delta

1. *Vossia cuspidata* (in the water).
2. *Vossia cuspidata, Cyperus articulatus, Cardiospermum halicacabum* (on the edge of the bank).
3. *Eragrostis barteri, Vossia cuspidata, Echinochloa pyramidalis, Phragmites australis, Fimbristylis cioniana, F. bi-umbellata, Cyperus* cf. *clavinus, C.* cf. *alopecoroïdes, Sphenoclea zeylanica, Mariscus* sp., *Rhamphicarpa fistulosa, Ludwigia leptocarpa, Ludwigia* sp. (cf. *perennis*), *Polycarpon prostratum* (level lower than the edge of the bank).

— Tarara reed island

1. Central part with *Phragmites australis* surrounded by some groups with *Vossia cuspidata, Aeschynomene elaphroxylon* and *Typha australis*.
2. In a sheltered position, *Cyperus papyrus, Ludwigia leptocarpa, L. adscendens* subsp. *diffusa, Cyperus* sp. (floating), *Nymphaea lotus* and *Ceratophyllum demersum*.

— Kalom reed island

Vossia cuspidata, Typha australis, Cyperus sp. (floating), *Leersia hexandra, Cyperus nudicaulis, Ludwigia leptocarpa* and *Phragmites australis*.

— Baga Sola channel in the archipelago

1. *Aeschynomene elaphroxylon* forest with an outer border of *Vossia cuspidata*
2. Other species found: *Ludwigia leptocarpa, Cyperus nudicaulis, Polygonum* cf. *limbatum, P. senegalense, P. albotomentosum, Nymphaea lotus, Ipomoea rubens, Pistia stratiotes, Leersia hexandra, Echinochloa stagnina, Cyperus* sp. (floating), *Commelina* sp.

The development of the vegetation in the south basin between 1974 and 1976 can be summarized in the following way:

— almost complete disappearance of *Ipomoea aquatica, Aeschynomene afraspera, A. pfundii, Lemna perpusilla, Diplachne fusca, Sesbania sesban* var. *nubica, Ludwigia adscendens* subsp. *diffusa, Sacciolepis africana*;
— considerable regression of stations and/or areas occupied by *Cyperus articulatus, C. papyrus, Typha australis, Polygonum senegalense*;
— stabilization of *Aeschynomene elaphroxylon* populations;
— massive development of *Pistia stratiotes, Cyperus nudicaulis, Nymphaea lotus*

136

Photo 11 Papyrus canoe ('Kadei') used by fishermen.

and increase in area covered by *Vossia cuspidata, Leersia hexandra* and *Ludwigia leptocarpa*;
— thus two species make up the lacustrine vegetation carpet: *Vossia cuspidata* and *Aeschynomene elaphroxylon*.

5.2.2.2 *North basin.* The water level became lower from year to year in the northern basin until 1975 which corresponds to a complete drying up. The rapidity of the water's retreat hindered vegetation development and plant formations which existed during the period of 'Normal Chad' disappeared over the course of the recession periods. Moreover, the young shoots were heavily grazed by the herds. Around most of the islands, no aquatic vegetation existed, especially on the windward shores. *Typha australis* was only present along some leeward flat shores or the eastern part of the basin. The following survey was made from the water towards the shore near Baga Kiskra (Fotius 1974):
— water;
— open uncovered beach;
— highly grazed cover of various Cyperaceae;
— several small plants of *Aeschynomene elaphroxylon* (5 cm high), similarly grazed;

137

— strip of decaying *Typha australis* with numerous little shoots mixed with *Phyla nodiflora, Pluchea ovalis, Sacciolepsis africana, Polygonum senegalense, Ipomoea rubens, Luffa* sp., crassulescent Rubiaceae (*Oldenlandia?*);

— *Calotropis procera* developed on a cover of highly grazed grasses.

Over the course of the period which followed 1975, the depressions in the north basin filled up during the rainy season and as a result of water supplies coming through the 'Great Barrier' acted like some temporary ponds. The information on the vegetation of this period is very limited because of the difficulty of travelling through this zone. Ambatches developed towards the north starting from the 'Great Barrier' and at several points in the eastern part of the basin where this species was found near groups of *Typha australis*.

5.2.3 Conclusions

Two periods can be observed, that of the evolution of the macrophytic vegetation during the period of 'Lesser Chad' from 1973 until 1976 and a later one. At first, there was a very marked impoverishment of the vegetation in the north basin while the south basin was overgrown by *Aeschynomene elaphroxylon*, associated with *Vossia cuspidata* and *Ipomoea aquatica* which developed during the low waters.

In the second phase, there was a development of ambatches and to a lesser degree *Typha australis* in the north basin. In the south basin, some considerable modifications were apparent due to the reflooding. The non-perennial plants could be maintained because of the drying of the previously occupied zones while some perennial species were destroyed by submersion, the annual water level oscillations having a higher amplitude than during the period of 'Normal Chad'.

The *Aeschynomene elaphroxylon* forests had their bases drowned in the deepest zones (more than one meter) and in many places they were blown down by the wind and only a few branches continued to develop as shoots. Their populations appeared stable. It is likely that *Vossia cuspidata* and *Cyperus papyrus* became dominant in the south basin, as in several places the establishment and development of these two species was observed in some channels which were cut in the ambatch forests. The zones of open water which appeared inside the extensive group were often colonized by *Pistia stratiotes*.

In May 1976, in the south basin (east of a line Baga Kawa–Baga Sola), the area occupied by the macrophytes was estimated as 3270 km² for a total inundated area of 5960 km² (Lemoalle 1978) (Fig. 2). At this time, the extension of the vegetation in the north basin was of the same order of magnitude as in the south basin, or a total of 6000 to 7000 km².

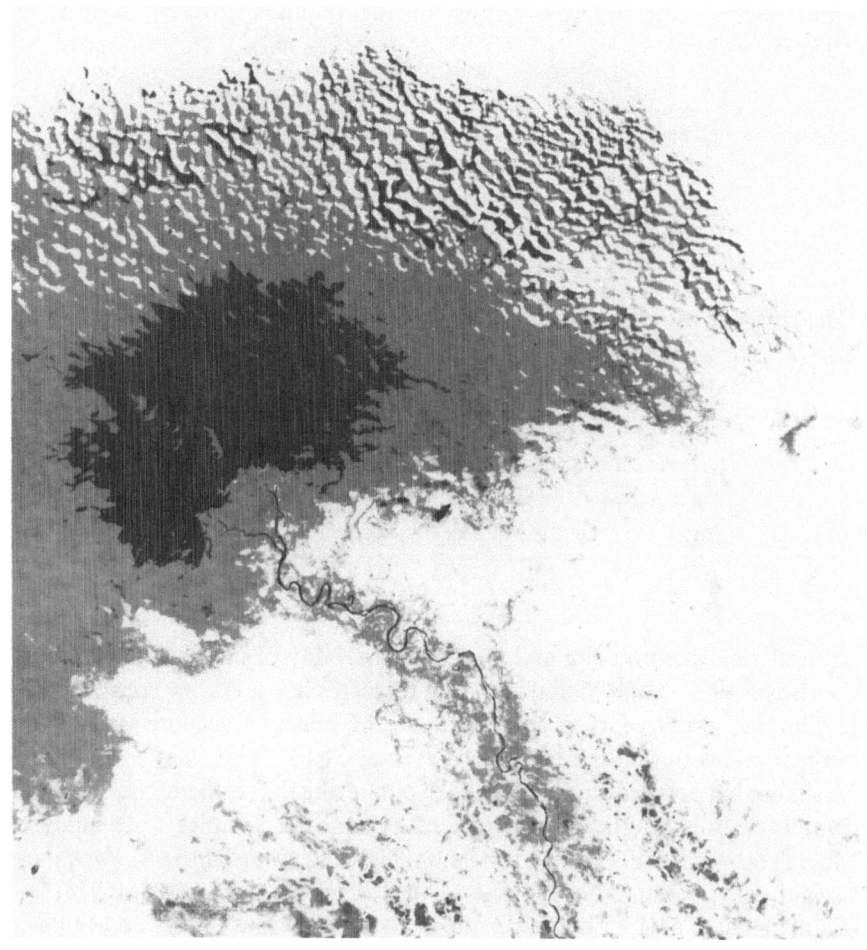

Fig. 2 Landsat image of the southern basin on 29 May, 1976. The open water areas are in black, the marshes in grey (corresponding to flooded areas). In total, the area under water is very similar to that in 1972. The marshy areas are relatively stable and are modified little between 1974 and 1979.

5.3 Mineral composition of macrophytes and its influence on water chemistry

5.3.1 *Analysis of macrophytes*

From the first study in 1970, both the biomass of emergent macrophytes in the lake and their mineral composition could be calculated corresponding to a period lower 'Normal Chad' (Carmouze et al. 1978).

Four species constituted the bulk of the biomass covering a total of 2400 km²:

Phragmites	6355×10^3 tons d.w. covering 2000 km²
Papyrus	674×10^3 tons d.w. covering 240 km²
Vossia	168×10^3 tons d.w. covering 100 km²
Typha	13×10^3 tons d.w. covering 25 km²

The amount of salts accumulated was similarly estimated (in thousands tons):

	Na	K	Ca	Mg	SiO₂
Aerial parts	2.3	138	7.4	6.6	230
Roots	8	62.5	56	31.5	2110
Total	10.3	200.5	63.4	38.1	2340

Considering the intra-site and inter-site variability of macrophyte composition (Boyd 1969, 1971; Gaudet 1975), these results must be considered as providing an order of magnitude estimate of minerals accumulated by the emergent plants during the period of 'Normal Chad'.

A similar but detailed study was carried out during the period of 'Lesser Chad' (1974) near Bol (Southeastern Archipelago) and 12 samples were analyzed (Table 1) (Lemoalle 1979). The results on *Typha, Cyperus papyrus, Vossia* and *Phragmites* are comparable to those published by Carmouze et al. (1978).

Salt content (Table 2) shows that the importance of ions (K > Ca > Mg > Na) is generally the reverse of their average composition in the water during 'Normal Chad'. Chloride was not measured in the macrophyte samples but published data give a range from 0.3 to 3% of the dry weight. If we choose an arbitrary value of 1% (30 mé 100 g^{-1}) that is a concentration close to that of Ca, the chloride content in the water is 10 to 30 times less.

Laboratory experiments on decomposition of these macrophytes are in agreement with the *in situ* observations made elsewhere (Gaudet 1977). More than half of the inorganic elements are returned into solution after some days of flooding. For the plants of Chad, the experiments using a limited volume of water provided the following results: (Chantraine, personal communication):

— potassium represents more than half of the dissolved cations;
— the ionic ratios observed are Ca/Mg < 1 and Na/K < 0.5;
— chloride equals 5 to 15% of the concentration of cations;
— silica rapidly reaches concentrations close to saturation (amorphous silica) especially for *Cyperus papyrus* and *Vossia cuspidata*.

Table 1 Dry weight composition (%) of main plants of the Bol region (Eastern Archipelago) in 1974. The symbols Vg, Fl and Fr followed by a number (1 to 3) indicate respectively the state of the development of vegetation, flowering, and fruiting.

Samples	Dry weight composition (%)								
	N	Na	P	K	Ca	Mg	Al_2O_3	SiO_2	C
1 *Typha australis* stem and leaves (Vg2)	1.60	0.571	0.348	3.70	0.37	0.24	0.029	0.07	43.8
2 *Cyperus articulatus* complete without roots (Fr3 Fl1)	0.91	0.418	0.296	2.08	0.42	0.18	0.087	3.30	42.1
3 *Phragmites australis* stem and complete leaves	1.63	0.018	0.167	1.60	0.24	0.14	0.037	7.19	42.8
4 *Vossia cuspidata* aerial part and stem (Vg2)	1.26	0.011	0.251	1.91	0.16	0.15	0.060	8.45	40.7
5 *Ipomoea aquatica* complete plant (Fl, Fr)	1.44	0.253	0.321	2.98	0.58	0.37	0.086	0.94	43.2
6 *Ludwigia* sp. complete plant (Fl, Fr)	2.48	0.218	0.590	3.63	1.15	0.49	0.053	0.58	42.0
7 *Cyperus* sp. (flottant) complete plant (Fl1)	1.21	0.100	9.279	2.52	0.63	0.30	0.076	4.99	40.8
8 *Cyperus maculatus* complete plant (Fl, Fr)	0.95	0.172	0.219	1.00	0.39	0.15	0.051	5.52	42.9
9 *Cyperus papyrus* complete plant with roots (Vg1)	0.62	0.200	0.234	2.83	0.29	0.16	0.022	1.85	43.7
10 *Aeschynomene elaphroxylon* (branches)	1.94	0.032	0.229	1.56	1.20	0.26	0.370	2.16	44.9
11 *Aeschynomene elaphroxylon* (emerged trunk)	0.80	0.072	0.183	1.43	0.66	0.16	0.110	0.58	44.6
12 *Aeschynomene elaphroxylon* (immersed trunk)	0.86	0.164	0.249	1.99	0.50	0.12	0.197	1.05	44.2

5.3.2 *Influence of macrophytes on the hydrochemistry of the lake*

The influence of macrophytes on the water chemistry is both direct, by the transfer of salts from the water to the plant tissues and the sediments and indirect, by a modification of the physical conditions of the environment. The comparison of two periods of isolation of the Bol region in 1973 and in 1974, respectively without and with the influence of macrophytes (Lemoalle 1979), is a good illustration of that.

When the macrophytes were abundant in the period of 'Lesser Chad', the

Table 2 Salt content (mé 100 g^{-1} of dried plants).

	Na$^+$	K$^+$	Ca^{++}	Mg^{++}
1 Typha australis	24.8	95	18.5	20
2 Cyperus articulatus	18.2	53	21	15
3 Phragmites australis	0.8	41	12	11.7
4 Vossia cuspidata	0.5	49	8	12.5
5 Ipomoea aquatica	11.0	76	29	30
6 Ludwigia adscendens subsp. diffusa	9.5	93	57.5	40
7 Cyperus sp. (floating)	4.3	65	31.5	25
8 Cyperus maculatus	7.5	26	19.5	12.5
9 Cyperus papyrus	8.7	73	14.5	13.3
10 Aeschynomene elaphroxylon	1.4	40	60	21.7
11 Aeschynomene elaphroxylon	3.1	37	33	13.3
12 Aeschynomene elaphroxylon	7.1	31	25	10

indirect effects noted were: (1) a dampening effect of water level oscillations and a diminution in the fetch of the wind; (2) a lowering of pH and of the dissolved oxygen with an increase of the CO_2 tension. These new environmental conditions limited the neoformation of clay which participates in the salinity regulation in the lake.

One of the direct effects was the variation of ion concentrations, especially of potassium and chloride which were assimilated by plants during the growth periods in flooded environment and redissolved during submersions at the time of lacustrine floods. The phosphate concentrations (dissolved reactive phosphorus) also become greater during these periods.

It is actually difficult to estimate the balance of transfer for a longer time. The salts accumulated in the plants initially originate from dried sediments on which they have developed considerably. Later on, the macrophytes remove salts from the sediments and the water in unknown proportions. Inversely, decomposition rapidly provides a considerable amount of dissolved elements while a smaller fraction remains trapped for a time in the organic matter of the sediments.

During the period of 'Normal Chad', the influence of macrophytes is relatively weak even if the quantity of salts involved is already important. During the period of 'Lesser Chad'; the proliferation of vegetation allows a better understanding of their interactions with the physico-chemical environment.

Acknowledgement

We are grateful to Mr. G. Fotius, ORSTOM botanist, who made a critical review of this chapter.

References

Boyd, C., 1969. Production, mineral nutrient absorption and biochemical assimilation by *Justicia americana* and *Alternanthera philoxeroïdes*. Arch. Hydrobiol. 66: 139–160.

Boyd, C., 1971. The dynamics of dry matter and chemical substances in a *Juncuns effusus* population. Am. Midl. Nat. 86: 28–45.

Carmouze, J. P., Dejoux, C., Durand, J. R., Gras, R., Iltis, A., Lauzanne, L., Lemoalle, J., Lèvêque, C., Loubens, G. and Saint-Jean, L., 1972. Grandes zones écologiques du Lac Tchad. Cah. ORSTOM Sér. Hydrobiol. 6: 103–169.

Carmouze, J. P., Fotius, G. and Lévêque, C., 1978. Influence qualitative des macrophytes sur la régulation hydrochimique du lac Tchad. Cah. ORSTOM Sér. Hydrobiol. 12: 65–69.

Chouret, A. and Lemoalle, J., 1974. Evolution hydrologique du Lac Tchad durant la sécheresse 1972–1974. ORSTOM N'Djamena, 12 pp., mimeo.

Fotius, G., 1974. Problèmes posés par l'évolution de la végétation liée à la baisse du lac Tchad. ORSTOM N'Djaména, 30 pp., mimeo.

Fotius, G. and Lemoalle, J., 1976. Reconnaissance de l'évolution de la végétation du lac Tchad entre janvier 1974 et juin 1976. ORSTOM N'Djaména, 13 pp., mimeo.

Gaudet, J., 1975. Mineral concentrations in *Papyrus* in various African swamps. J. Ecol. 63: 483–491.

Gaudet, J., 1977. Uptake and loss of mineral nutrients by *Papyrus*. Ecology 48: 415–422.

Lemoalle, J., 1978. Application des images Landsat à la courbe bathymétrique du lac Tchad. Cah. ORSTOM Sér. Hydrobiol. 12: 83–87.

Lemoalle, J., 1979. Biomasse et production phytoplanctonique du lac Tchad (1968–1976). Relations avec les conditions du milieu. Thèse d'Etat, Univ. Paris VI, 311 pp., mimeo.

Leonard, J., 1969. Aperçu sur la végétation aquatique. In: Monographie hydrologique du lac Tchad. ORSTOM Fort-Lamy, 11 pp., mimeo.

Leonard, J., 1974. Aperçu sur la végétation de la partie est du lac Tchad. ORSTOM N'Djaména, 14 pp., mimeo.

6. The phytoplankton

Pierre Compère and André Iltis

6.1 Qualitative composition of the algal flora (P. Compère)

About 1300 species and intraspecific taxa of algae and plankton were collected in Lake Chad, the lower course of its main tributaries such as the Shari, the El Beïd and the Yobé, and in various ponds of the flooded zones of these rivers (Iltis and Compère 1974; Compère 1974–1977). If those collected by J. Leonard in 1964 (Compère 1967) are added to these, the total number of algal species found in the Lake Chad region is 1500. Lake Chad itself had the greatest number of taxa (1042), but the collections there were much more numerous than in the tributaries and their flood plains which had 903 taxons. The southeastern part of Lake Chad appeared to have the most varied flora with 787 taxons compared with 628 for the northern part and only 461 in the southern part, where the collections were less numerous than in the other parts (Table 1).

The green algae were by far the most numerous within this flora since they represented 52% of the total taxa, with 12% Euchlorophyceae, 8% filamentous green algae such as Ulotrichophyceae and Zygnemataceae and 32% Desmidiaceae. The Diatoms followed with 27% and the Cyanophyceae with 13% and finally, all the other classes represented a little less than 8%. Of course, these data refer to the number of taxa and not to the number of individuals. Cyanophyceae often played a much more important part in the appearance and biomass of the plankton because of the mass development of some species such as *Microcystis aeruginosa* or *Anabaena flos-aquae*.

It appears that the qualitative composition of the flora was remarkably constant in the various subdivisions of the region under study (Table 1). Variations in the percentages observed were low or irregular. Diatoms were relatively more numerous in the plankton, while Desmidiaceae were much more abundant in the periphyton. Both groups contribute to about 30%, while Cyanophyceae ranged from 11 to 16%. In the northern part of the lake, these proportions were somewhat modified since Desmidiaceae represented only 21.7% of the whole, while the Diatoms reached 34%. If the regions furthest from the

© *1983, Dr W. Junk Publishers, The Hague/Boston/Lancaster*
ISBN 978-94-009-7268-1

Table 1 Distribution of the different algal types in the flora of the Lake Chad region.

	Lake Chad region		Tributaries		South-eastern part		South		North	
	N	%	N	%	N	%	N	%	N	%
Cyanophyceae	168	13.1	106	11.7	106	13.5	51	11.1	105	16.5
Diatoms	349	27.2	274	30.3	248	31.5	156	33.8	213	33.9
Green algae:										
. Euchlorophyceae	170	13.2	126	13.9	105	13.3	64	13.9	101	16.1
. Ulotrichophyceae + Zygnemataceae	101	7.9	65	7.2	44	5.6	16	3.5	35	5.6
. Desmidiaceae	398	31	270	29.9	251	31.9	140	30.4	136	21.7
Others	98	7.6	62	6.9	33	4.2	34	7.4	38	6.0
Total	1284		903		787		461		628	

delta and the tributaries are considered in this northern part where conductivity exceeds 500 μS cm^{-1}, this tendency clearly increased with Desmidiaceae representing only 17% of the flora while Diatoms represented 36% and Cyanophyceae nearly 20% (Fig. 1). In this last region, the qualitative composition of the algal flora tended to be more similar to that of the floras of North Africa.

Table 2 enables us to draw a parallel between the composition of the algal flora of the Lake Chad region and that of some tropical or subtropical regions. It was not very different from that observed previously in the lower Shari and the southeastern part of Lake Chad (Compère 1967). It is difficult to compare with the algae found in Mali (Bourrelly 1957; Couté and Rousselin 1975), in the Ivory Coast (Bourrelly 1961) or in Guinea (Bourrelly 1975) since there are no data on Diatoms on these countries (Fig. 2). In each region, the prevalence of the Desmidiaceae (from 40 to 75%) and the rather small amount of Cyanophyceae (from 3 to 10%) is emphasized. In Morocco (Gayral 1954) and Algeria (Gauthier-Lièvre 1931; Baudrimont 1974) the algal floras contained large numbers of Diatoms (37–40%) and Cyanophyceae (17–19%), while the green algae represented only 35% of the whole, among which Desmidiaceae represented only 11 to 13%. In Ennedi and Kanem, the floral composition was very similar to that of North Africa. The percentages of the various algal groups in the flora of the Lake Chad region seem to be situated between those of the North-African or Saharian floras and those of the floras from more tropical regions in Western Africa.

146

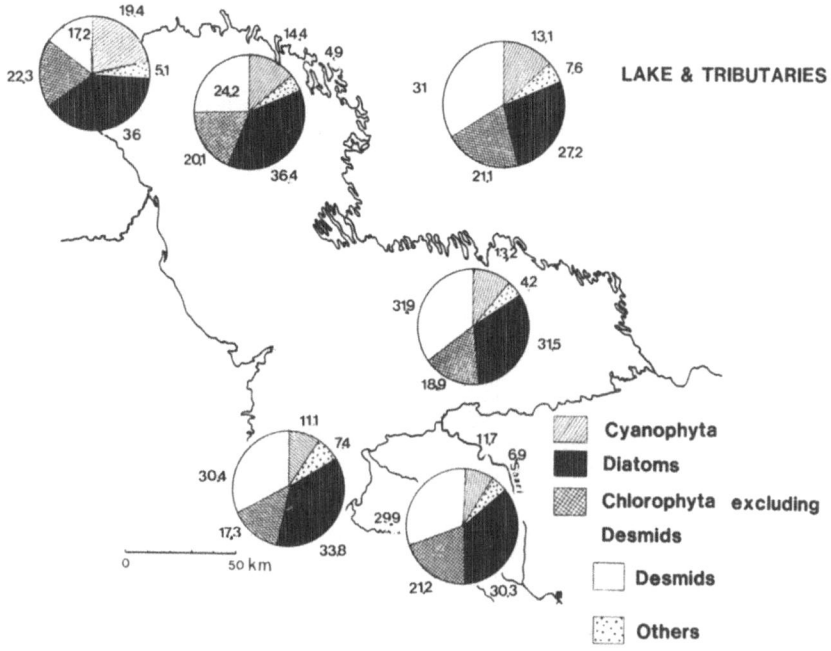

Fig. 1 Distribution of the different classes in the algal flora of Lake Chad. Upper right and outside the map for the total flora of the lake and its tributaries; inside the map, for the different local flora.

This floral composition is similar to that of the total flora of the great lakes of Eastern Africa (Van Meel 1954), although Diatoms were more abundant there than the green algae, among which the Desmidiaceae were less important. It must be pointed out that Van Meel's observations were made mainly on the plankton and in Lake Chad, we also observed that Diatoms were relatively more abundant in the plankton while Desmidiaceae were especially common in the periphyton. Finally the algal flora in Burma which was described by Skuja (1949) showed a qualitative composition similar to that of the Lake Chad region. Here Diatoms seemed to play a more discrete role, but it must be noted that the author appeared to be less interested in this class than in the others since he did not even devote even one plate out of the 37 included in his work to Diatoms.

The geographic distribution of the flora is given in Table 3 which shows that cosmopolitan or subcosmopolitan species predominated and represented more than 65% of the flora (Fig. 3). Tropical species represented more than 28% of the flora, of which the endemic or African species represented 14% and the tropical species with a wider distribution also represented 14%. The remaining 7% were algae known especially from temperate regions. The tropical compo-

147

Table 2 Distribution of the different algal types in a few tropical floras (in %).

Region	Lake (Lake Chad) (Table 1)	Lake (Lake Chad)	Macina (Mali)	Middle Niger (Mali)	Ivory Coast	Guinea	Kanem	Ennedi	Algeria	Morocco	Eastern Africa	Burma
		(1)	(2)	(3)	(4)	(5)	(6)	(7)	(8)	(9)	(10)	(11)
Cyanophyceae	13.1	10.1	3	3.5	9.2	6.3	18.9	19.5	17.4	19.5	11.8	20.0
Diatomophyceae	27.2	32.0	–	30.2	–	–	40.7	36.3	40.6	37.3	45.2	17.4
Green algae												
Desmidiaceae	31	41.4	75	49.8	38.6	65.4	8.7	11.1	11.5	12.8	22.1	37.0
Others	21.1	14.8	17	9.8	16.5	10.0	20.2	19.8	23.5	21.4	18.2	16.8
Others class	7.6	1.7	5	6.7	35.7	18.2	11.5	13.4	7	9	2.6	8.6
Total number of taxa	1284	444	213	550	313	220	514	324	1025	686	1216	841

From: (1) Compère 1967; (2) Bourrelly 1957; (3) Couté and Rousselin 1975; Maillard 1977; (4) Bourrelly 1961; (5) Bourrelly 1975; (6) Iltis 1972; (7) Compère 1970; (8) Gauthier-Lièvre 1931; Baudrimont 1974; (9) Gayral 1954; (10) Van Meel 1954; (11) Skuja 1949.

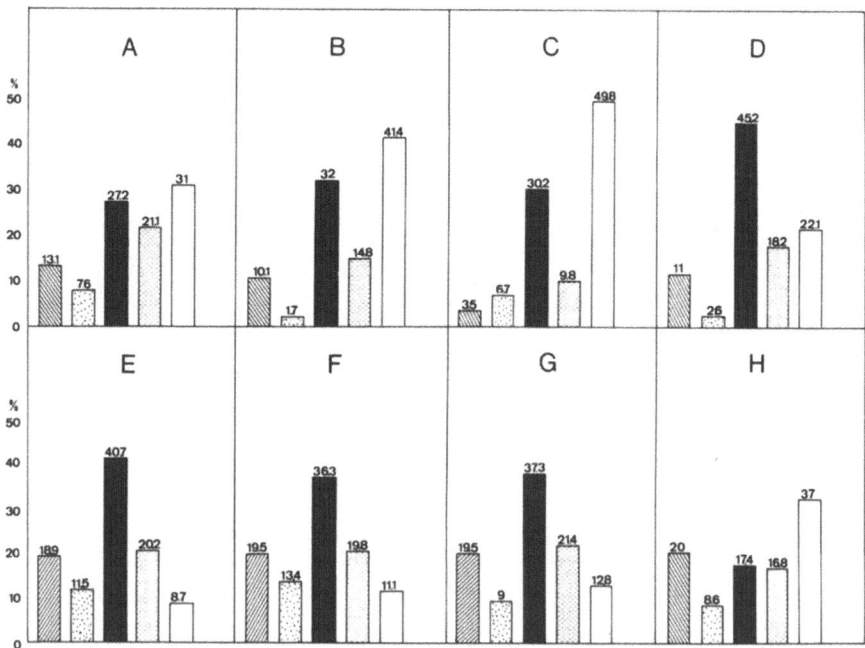

Fig. 2 Distribution of the different classes in some tropical algal flora. A, Lake Chad region; B, Lake Chad region in 1964, according to Compère (1967); C, Mali, Middle Niger, according to Coutè and Rousselin (1975) and Maillard (1977); D, Great lakes in Eastern Africa, according to Van Meel (1954); E, Kanem, according to Iltis (1972); F, Ennedi, according to Compère (1970); G, Morocco, according to Gayral (1954); H, Burma, according to Skuja (1949). Key to histogram: See Fig. 1.

nent was especially important in the Desmidiaceae where it represented 38% of the taxa and in the Diatoms where it was 33% of the taxons; it was average in the Cyanophyceae and almost zero in the Euchlorophyceae. Table 4 shows that the tropical component was in its broad sense less important in the flora of Lake Chad and its tributaries than in the floras of the various regions in Western Africa which were studied by Bourrelly and his assistants (Mali, Macina: 53%, Bourrelly 1957; Mali, Middle Niger: 44.5%, Coutè and Rousselin 1975; Guinea: 36.3%, Bourrelly 1975; Ivory Coast: 34.1%, Bourrelly 1961), Fig. 4 Nevertheless, the tropical African character of our flora was far from being insignificant and was much more pronounced than in the flora of Ennedi where the tropical component *sensu lato* represented only 18% of the flora (Compère 1970).

Thus the algal flora in the Lake Chad region showed a rather pronounced tropical African character, between that of the Saharian or Sahelian algal floras and that of the floras in the Sudano-Guinean regions of Western Africa.

149

Table 3 Simplified geographic distribution of the algal flora in Lake Chad.

	Lake Chad region		Tributaries		Lake Chad SE part		Lake Chad S part		Lake Chad N part	
	N	%	N	%	N	%	N	%	N	%
Cosmopolitan + subcosmopolitan	816	63.9	593	66.2	529	67.7	335	73.8	449	72.4
Tropical + subtropical	186	14.6	133	14.9	102	13.1	50	11.0	76	12.2
African (with endemic species and new taxa)	179	14.0	111	12.4	105	13.4	52	11.5	63	10.2
Temperate	95	7.5	58	6.5	45	5.8	17	3.7	32	5.2
Total	1276	100	895	100	781	100	454	100	620	100

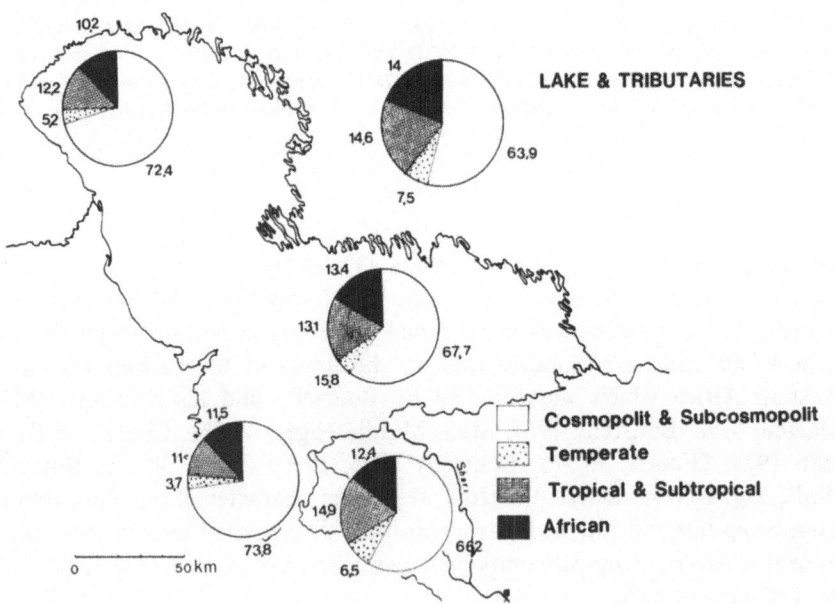

Fig. 3 Simplified geographical composition of the algal flora in the region of Lake Chad. Upper right and outside the map, for the whole lake and its tributaries; inside the map, for the different local flora.

150

Table 4 Simplified geographic distribution of a few African algal floras (in %).

| | Lake Chad region | Macina (Mali) | Middle Niger (Mali) | Ivory Coast | Guinea | Ennedi |
		(1)	(2)	(3)	(4)	(5)
Cosmopolitan + subcosmopolitan + temperate	71.4	47	61.3	61.8	63.7	82
Tropical + subtropical	14.6	25	20.0	15.9	22.7	10
African (with endemics)	14.0	28	18.7	22.3	13.6	8

From: (1) Bourrelly 1957; (2) Couté and Rousselin 1975; Maillard 1977; (3) Bourrelly 1961; (4) Bourrelly 1975; (5) Compère 1970.

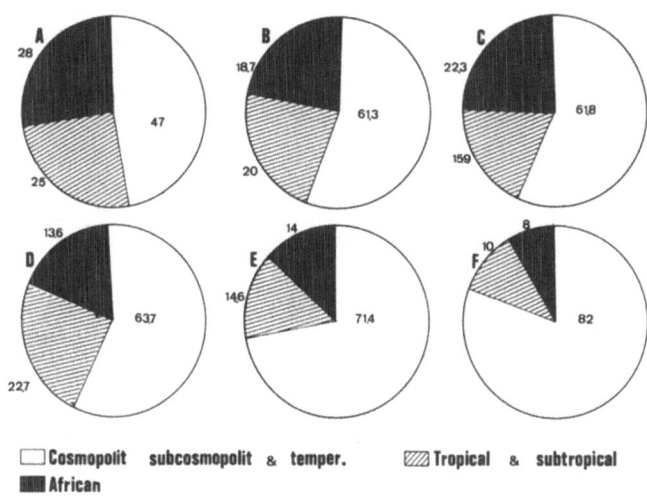

Fig. 4 Simplified geographical compositions of a few African algal florules. A, Macina, Mali, according to Bourrelly (1957); B, Middle Niger, Mali according to Couté and Rousselin (1975) and Maillard (1977); C, Ivory Coast, according to Bourrelly (1961); D, Guinea according to Bourrelly (1975); E, region of Lake Chad; F, Ennedi, according to Compère (1970).

151

This intermediate character was still apparent when Bourrelly's empirical indices (1957, 1961) were applied to Desmidiaceae of the flora under study. This index calculates the percentages of all the Desmidiaceae represented respectively by the total number of *Pleurotaenium* and filamentous Desmidiaceae and by the total number of *Pleurotaenium, Euastrum* and filamentous Desmidiaceae. Table 5 gives these indices for the Desmidiaceae belonging to our flora as well as a few tropical desmidial floras.

The resulting values are a little lower than those calculated previously for the lower Shari and the southeastern part of Lake Chad (Compère 1967). They are also lower than those calculated for various other tropical African floras and for Amazonia, but they are generally higher than those for temperate floras (Bourrelly 1957, 1961). They are very close to those calculated from Skuja's data (1949) for Burma and for Middle Niger (Couté and Rousselin 1975), and a little higher than those calculated from Van Meel's data (1954) for the plankton in the great lakes of Western Africa.

In summary, Desmids and Diatoms were qualitatively the dominant species in the algal flora of the Lake Chad region. Although they were represented by fewer taxa, Cyanophyceae played an important quantitative role. The composition of this flora clearly shows a tropical African character that was more pronounced than in the North-African floras or in those of the South Sahara massifs but certainly less pronounced than in those of the Sudano-guinean regions of Western Africa. In the region studied, this tropical African character was greater around the Shari and the southeastern part of the lake. By its floral composition and the distribution of its geographic elements the flora in the far northwest of the lake tended to approach the Sahelian and Saharian floras.

Table 5 Bourrelly's empirical indices (1957) for a few tropical desmidial floras.

	Region (Lake Chad)	Lake (Lake Chad) (1)	Macina (Mali) (2)	Middle Niger (Mali) (3)	Ivory Coast (4)	Guinea (5)	Sierra Leone (6)	Eastern Africa (7)	Amazonia (8)	Burma (9)
Pleurotaenium + fil. Desm. total Desmids	7	8.5	13.5	5.4	15.0	11.7	11.9	6.0	13.0	8.6
Pl. + *Euastrum* + fil. Desm. total Desmids	14	19.2	27.7	17.0	24.1	19.4	24.5	12.3	23.2	14.6

From: (1) Compère 1967; (2) Bourrelly 1957; (3) Couté and Rousselin 1975; (4) Bourrelly 1961; (5) Bourrelly 1975; (6) Grönblad et al. 1968; (7) Van Meel 1954; (8) Thomasson 1971; (9) Skuja 1949.

6.2 Phytoplanktonic communities and biomasses (A. Iltis)

The study of the Lake Chad phytoplankton was difficult for the following reasons. It is a very wide lacustrine basin with varied ecological zones and aquatic vegetation developed to varying extents depending upon the topography of the lake bottom and banks. Considerable variations in physico-chemical conditions occur in the regions, and the variety of lake zones make it necessary to sample all of them to analyze the plankton and estimate total biomass. Given the possible isolation of some regions by vegetation barriers or swampy shoals according to the variations in water level, sampling was fairly difficult in these shallow biotopes. The studies were designed to analyze the following points: seasonal variations, delimitation of the different ecological zones, species associations, plankton biomass and development during a period of drought. Studies on the primary production and its estimates were later added to these observations.

Given the shallowness of the lake, the euphotic layer included, with few exceptions, the total water column. Moreover, the samples taken below the surface could be considered as representative of the limnoplankton of the entire water surface. The analysis of samples taken in the archipelago from the surface to a depth of 5 to 6 meters showed a maximum variation in the composition and density of 14% according to the depth in the water column (Gras et al., 1967). This value is lower than the error generally occurring in plankton counts. Therefore, it can be accepted that for this study of the phytoplankton the algae were distributed homogeneously throughout the water column.

The phytoplankton biomass was estimated by counting the organisms under an inverted microscope from sub-samples taken in the lake about ten centimeters below the surface and immediately fixed in 10% formalin. The results were expressed either as the number of cells, cenobes or colonies per liter and transformed into biovolumes of living algae per liter after calculating the mean volume of each species. From 1974, titrations of chlorophyll *a* were conducted by spectrophotometry.

The studies on the phytoplankton were at first limited only to the Shari and the eastern part of the lake (Gras et al. 1967) but later extended to the whole lake. The main biotopes were then delimited for the first time (Carmouze et al. 1972), and from 1971 to 1975, general surveys provided biomass estimates and detailed analysis of the communities (Iltis 1977). Studies on primary production were also conducted in various parts of the lacustrine basin (Lemoalle 1979).

6.2.1 *Characteristics of the communities*

6.2.1.1 *Composition.* In the phytoplankton observed in the entire lake from 1964 to 1975, Cyanophyceae, Chlorophyceae and Diatomophyceae were the most important groups quantitatively, while the Pyrrhophyta, Euglenophyta, Chrysophyceae and Xanthophyceae seldom appeared in large proportions.

Cyanophyceae occurred throughout the year in the far east and the northeast where they represented 80% of the biomass and were generally the most important constituent of the phytoplankton in the archipelago zones. During the warm season, they extended into the open water and especially the Southern Open Water. *Microcystis* and *Anabaena* were the most frequent genera among which species *Microcystis aeruginosa*, *M. holsatica*, *M. elachista* and *Anabaena flos aquae* were very abundant. Moreover, *Lyngbya limnetica* and *L. contorta* occurred, forming up to 18 and 8% in the archipelago and the southern open water during certain periods. *Oscillatoria laxissima* and *Anabaenopsis tanganii-kae* also appeared occasionally in considerable number.

To show the importance of Cyanophyceae in the phytoplankton biomass, it should be pointed out that out of 35 samples taken over the whole lake in February 1971, 32% of the algal biovolume was composed of Cyanophyta and in January 1972, the percentage to 49% in 40 samples. In April 1974, the mean percentage of Myxophyceae was 25% (38 samples) in the lake and decreased to 11% in November 1974 (17 samples) and to 6% in February 1975 (17 samples) after the occurrence of the drought (Table 6).

In the Diatomophyceae, *Melosira granulata* var. *angustissima*, *Nitzschia spiculum* and an unspecified *Navicula* were the three species with the highest densities in the lake during the period of 'Normal Chad'. *Melosira granulata* predominated during certain periods in both the archipelago and the open water of the southern part of the lake where considerable changes occurred during the first quarter of the year before the establishment of dense populations of Cyanophyceae in the warm season. *M. granulata* was sometimes replaced by *Nitzschia spiculum* (Iltis and Lemoalle 1976), and *Navicula* sp. formed up 25 to 30% of the biomass in the Northern Open Water in January 1972. During the period of 'Lesser Chad', *Cyclotella meneghiniana*, *Coscino-discus rudolfii*, *Synedra berolinensis*, *Surirella linearis*, *S. muelleri* and *Nitzschia*

Table 6 Mean percentages of the different algal groups calculated from the phytoplanktonic biovolumes (analysis conducted on 147 samples collected during five surveys in Lake Chad from 1971 to 1975).

	Cya. %	Diat. %	Chlor. %	Eugl. %	Others %
February 1971 (*N*=35)	32	23	39	0	6
January 1972 (*N*=40)	49	27	22	0	2
April 1974 (*N*=33)	25	56	12	6	1
November 1974 (*N*=17)	11	47	28	10	4
February 1975 (*N*=17)	6	67	13	11	3

sp. occurred in high densities, especially in the northern part where they predominated.

In the whole lake, the average percentages of Diatomophyceae in the algal biovolume were as follows: 23% in 35 samples in February 1971, 27% in 40 samples in January 1972, 56% in April 1974, 47% in November 1974 and 67% in February 1975. Therefore, the percentage contribution of Diatoms to biomass considerably increased after the change towards the period of 'Lesser Chad'. In the northern part of the lake which was still occupied by water, 60% in April 1974, 49% in November 1974 and 89% of the phytoplankton biomass was composed of Diatoms.

Chlorophyceae represented 39% of the phytoplankton volume in the 35 samples taken in February 1971 and 22% in January 1972. In 1974 and 1975 during the period of 'Lesser Chad', this percentage decreased to 12%, 28% and 13%. The most abundant species in the phytoplankton were *Scenedesmus quadricauda, Oocystis* sp., *Closterium aciculare* and *C. strigosum, Coelastrum cambricum* and *C. microporum, Gonatozygon monotaenium, Pediastrum clathratum* and *Botryococcus braunii*. It appears that during the period of 'Normal Chad', the open water of the northern part of the lake was the richest in Chlorophyta such as *Scenedesmus quadricauda* or *Closterium aciculare*. Dense populations of *Closterium strigosum* appeared during certain periods in the archipelago of the southern part of the lake, but in most cases, the populations of Chlorophyceae were much more diverdified than those of Cyanophyceae or Diatomophyceae.

Euglenoïds only represented a considerable proportion of the biomass (up to 80%) during the period of 'Lesser Chad' in the region of the Eastern Archipelago which was divided into numerous ponds overgrown with vegetation.

Among the Pyrrhophyta, *Cryptomonas erosa* constituted more than 20% of the algal biovolume in some samples from the northern part of the lake. Moreover, it should be pointed out that an unidentified species of the genus *Peridinium* was abundant during certain periods and that *Mallomonas portae ferrae*, a Chrysophyceae, was sometimes abundant.

6.2.1.2 *Density and biomass.* The difference between the lowest and the highest phytoplankton densities was considerable. Algal biovolumes of 0.005 μ 1^{-1} were found in the Southern Open Water of the lake in February 1971 and biovolumes of 305 μl 1^{-1} were found in the part of the northern basin still occupied by water in November 1974. Therefore, these extreme values represent respectively the poorest waters during the period of 'Normal Chad' and the richest waters during the period of 'Lesser Chad'.

The magnitude of these differences was due to the yearly variations in the lake and the change from 1973 towards a Lake Chad with pond-like characteristics. The observations in time however showed that there were profound differences between the various lake zones and that the archipelagoes were much richer than the open water (Fig. 5). The minimum, mean and maximum values observed from the analysis of samples taken during five surveys from 1971 to 1975 are given in Table 7.

During the 'Normal Chad' period, the mean algal biomass ranged from 1 to 2 mg l^{-1} and it increased considerably (up to 74 mg l^{-1}, when the lake turned into 'Lesser Chad', as a result of the drought. More specifically, the mean

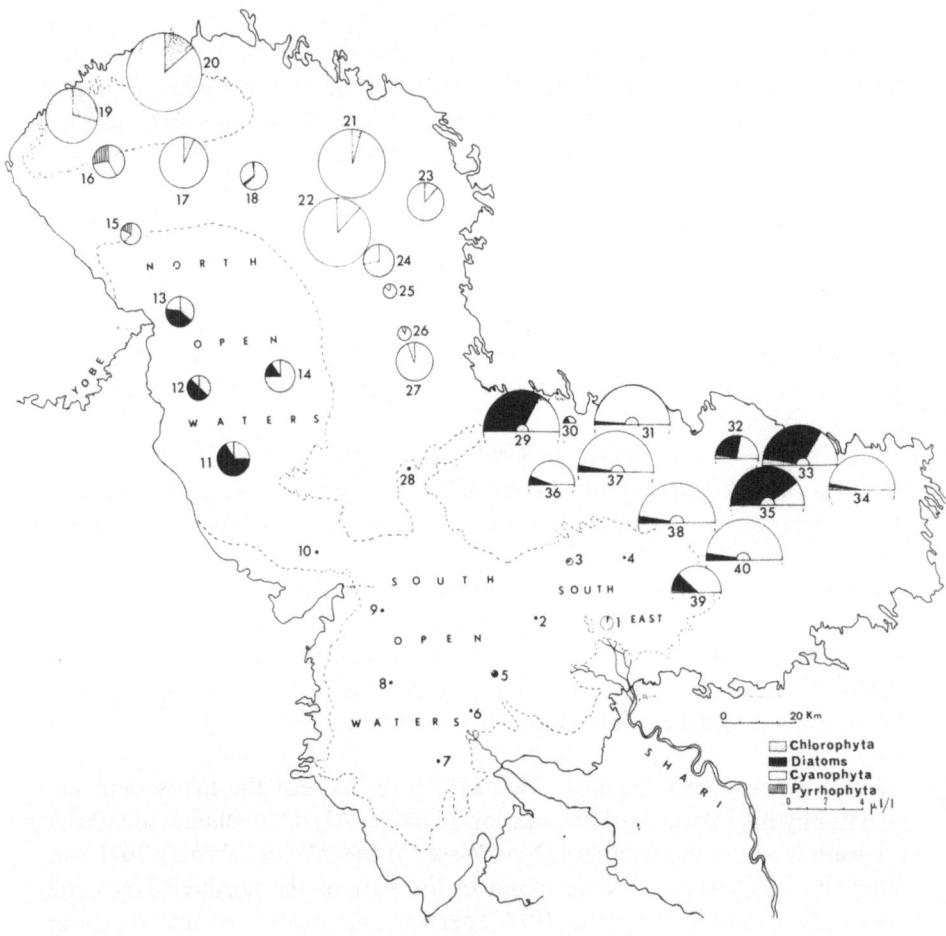

Fig. 5 Location and composition of the phytoplanktonic samples in January 1972. The diameter of the circles is proportional to the phytoplankton density except in points 29, 31, 33, 35, 37, 38 and 40 where the density is very high. The algal biovolumes which are lower than 0.2 µl l^{-1} are represented by a point. The approximate limits of the zones of open waters represented with broken lines.

Table 7 Algal biovolumes ($\mu l \ l^{-1}$) observed in Lake Chad during five surveys from 1971 to 1975.

	Minimum	Mean	S.D.	Maximum	
2/1971	0.005	1.1	2.11	12.1	} 'Normal Chad'
1/1972	0.06	2.4	2.45	10.5	
4/1974	0.74	21.8	30.79	183.1	
11/1974	0.11	74.1	102.31	305.6	} 'Lesser Chad'
2/1975	0.16	32.3	55.93	219.9	

biomass at this stage was obtained from the biomass of three individual waters bodies which then represented Lake Chad. The northern basin was the richest of the three and the average algal biovolume was 25.3 $\mu l \ l^{-1}$ in April 1974, 179 $\mu l \ l^{-1}$ in November 1974 and 74.5 $\mu l \ l^{-1}$ in February 1975. These densities were the highest ever observed and it can be considered that from November 1974, when the maximum depth was about one meter, these biotopes were closer to the natron ponds situated in the north of Lake Chad than to lacustrine waters.

On the contrary, the open water in front of the Shari delta remained poor, with mean biovolumes of 7.8 $\mu l \ l^{-1}$ in April 1974, 0.8 in November and 0.7 in February 1975. In the archipelago, the biovolumes observed during the same period were respectively 8.2, 0.4 and 7.6 $\mu l \ l^{-1}$.

Measurements of the chlorophyll *a* concentration were made by Lemoalle (1979) in different parts of the lake during various periods. During the period of 'Normal Chad', only the southern basin was sampled and two series of samples were taken in December 1970 and in June 1971 respectively. Minimum values of 5 mg m^{-3} were observed in the open water close to the delta in December 1970, while values ranging from 60 to 70 mg m^{-3} were found in the eastern part of the archipelago in June 1971. The values ranged from 15 to 70 mg m^{-3} in June 71 (period of low water) and from 5 to 36 in December (period of flood).

After 1973 during the period of 'Lesser Chad', concentrations remained very low (10 mg m^{-3}) during the flood in the southern basin in front of the delta. In that part of the open water furthest from the delta, they normally ranged from 80 to 120 mg m^{-3} in April, May and June. In the archipelago, which was divided by vegetation into a very heterogeneous group of patches water not always connected with each other, the values ranged from 2 to 325 mg m^{-3} according to the periods and stations. In the northern part of the lake which was surveyed only during the period of 'Lesser Chad', the mean chlorophyll *a* concentration, based on 7 to 11 stations around the islet of Kindjéria, was 141 mg m^{-3} in April 1974, 1658 in November 1974 and 204 in February 1975.

It seems that comparison with the other tropical environments must be made cautiously, as the available data are not very abundant or not always complete.

157

Moreover, they seldom represent annual averages so that values are over-estimates or underestimates according to the sampling seasons. In the case of Chad, it can be assumed that the samples and the calculations from February 1971 and January 1972 underestimated the mean annual biomass. As a matter of fact, the open water in the lake was composed of the Shari flood water which was still not colonized by the plankton. Given the extent and the shallowness of this zone in relation to the whole lake, it can be estimated that the differences between the values found and those observed when the river was at its lowest ranged from 12 to 13% (Iltis 1977). Comparison of results of the analysis of algal biomass in tropical or temperate lakes is also difficult as they may be expressed in different terms: biomasses expressed per unit of area, as number of cells per liter, as dry weight, as quantity of pigments or in microliters or grams of living algae per liter or per cubic meter. The last expression (living matter per unit volume of the euphotic layer) seems to be the most practical one. Moreover, it is pointed out that in very heterogeneous environments the existing biomass in each zone or group of regions is often more interesting than a single mean value, e.g. in Lake Chad after 1974 when the three water bodies which then constituted the lake were virtually isolated from each other and developed individually. Therefore, a mean value obtained from estimates made in different environmental zones has little significant use for lake comparisons. With the exception of very shallow lakes (Lake George in Uganda, natron lakes north of Lake Chad and certain lagoons) (Lewis 1978), only the Lake Chad values obtained during the period of 'Normal Chad' can be used for comparisons as the water depth was almost always less than two meters during the period of 'Lesser Chad'.

According to the results given by Lewis (1978), Lake Chad during the normal period (values ranging from 1.2 to 2.4 mg l^{-1}) is similar to tropical lakes rich in phytoplankton (Lake Lanao, the Philippines, 1.6 mg l^{-1}; Lake Lamongan, Java, 4.2 mg l^{-1}; Lake Lagartijo, Venezuela, 2.7 mg l^{-1}) especially when considering that the calculations, made when the southeastern part was submerged by the Shari flood, were underestimated (12 to 13%). These values are only averages compared with the biomass observed between Canada and the United States in the very eutrophic Lake Erie (Munawar and Munawar 1976) where the mean values range from 1.5 to 7.1 g l^{-1} according to season and lake region.

6.2.1.3 *Species diversity and environmental constant.* The species diversity (determined by the number of species present and the distribution of the total biomass between them) was calculated according to Shannon's formula. Only the taxa representing at least 0.07% of the total biomass present in the sample were used to calculate the diversity index.

The mean diversity values for the whole lake during five surveys conducted from 1971 to 1975 seem to be very similar and ranged from 2.14 to 2.75 bits

Table 8 Diversities (in bits) observed in Lake Chad during five surveys from 1971 to 1975.

	Minimum	Mean	S.D.	Maximum	Taxa number	
					Mean	S.D.
2/1971	0.611	2.752	0.81	4.186	15.2	7.15
1/1972	0.528	2.141	0.86	3.441	12.3	3.81
4/1974	0.422	2.328	0.93	3.835	21.0	6.24
11/1974	0.378	2.553	0.76	3.938	19.9	5.19
2/1975	1.182	2.324	0.86	3.778	18.7	7.71

(Table 8). These observations were too scattered in time to provide a thorough information on the changes on the biocenoses, but they were sufficient to allow the delimitation and characterization of the different ecological zones in terms of their phytoplankton communities. Thus, for the samples taken in 1972 (Fig. 6), the mean diversity was 1.775 in the Southern Open Water and 3.078 in the Northern Open Water, while the Northern Archipelago and the Eastern Archipelago showed diversities of 2.569 and 1.705 bits respectively. The communities in the northernmost part and in the eastern part of the Eastern Archipelago which were the richest in phytoplankton had a mean diversity of about 1.942 bits.

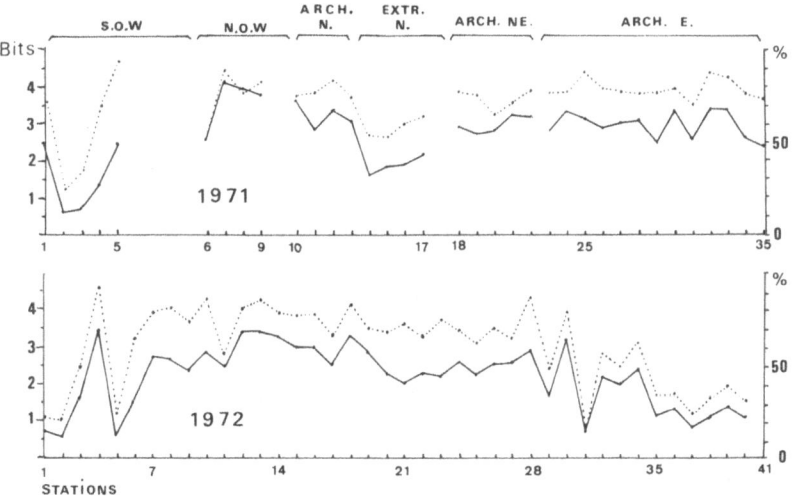

Fig. 6 Variations in the values of the specific diversity index (represented with a continuous line) and of the equitability (represented with dotted lines). (S.O.W. = southern open waters; N.O.W. = northern open waters; ARCH. N = northern archipelago; EXTR. N = far north, etc.).

159

The values of diversity and equitability (calculated as the percentage value of the ratio of the observed diversity to the maximum diversity) were represented on a graph in relation to the ecological zones where samples were taken. In February 1971, diversity was low in the Southern Open Water except at station 1 (perideltaic zone). Diversity increased in the Northern Open Water where it reached the maximum value (i = 4.186) and decreased abruptly in the north of the lake, especially in the northernmost stations (14 to 17) where the algal biomass was composed of only some species of Cyanophyceae. In the archipelago, the diversity index remained rather high in the north (station 18 to 21) as well as in the east (stations 23 to 35), while the easternmost stations generally had diversity indices which were slightly lower. The variations in equitability closely followed those of diversity with a high correlation between them (r = 0.80). In January 1972, the profile according to the ecological zones was almost identical to that of February 1971.

Diversity was low in the Southern Open Water except at the easternmost point 4 whose phytoplankton communities (with a rather high percentage of Cyanophyceae) showed similarities to that of the neighbouring stations 38 and 39. Maximum values were observed in the Northern Open Water (stations 12 and 13) and in the archipelago situated north of them (station 18), and lower values were observed in the northernmost part of the lake (stations 20 to 23). In the archipelago, diversity was rather high in the northeastern part of the lake (station 24 to 28) but very variable in the eastern zone, all the communities situated close to the open water (35 to 40) generally having diversities lower than those bordering the lake (29, 30, 32, 33, 34). As in 1971, equitability followed the variations in diversity (the correlation coefficient between these two series of values was 0.95) and provided little further information.

Several successive stages could be distinguished in the phytoplankton of the flood wave which moved from the delta towards the northwest where the open water was most extensive. Initially there was a stage with abundant Diatoms and low algal density and species diversity in the Southern Open Water. The second stage was characterized by the community of the Northern Open Water having an average phytoplankton density and a maximum diversity while Diatoms decreased in percentage contribution of total. The third stage which was related to the phytoplankton of the reed islands north of the previous zone was characterized by increasing density and decreasing diversity while the Diatom population practically disappeared to be replaced by Cryptophyceae. Finally a fourth stage could be defined in the northern part of the lake where there was a community characterized by maximum density and low diversity which was composed mainly of Cyanophyta. It is difficult to consider these stages as parts of succession cycles in phytoplankton development as shown by Margalef (1958–1967) for populations of marine algae. The changes in the phytoplankton were influenced by numerous local factors such as the presence of residual waters which were more or less mixed with flood waters when

160

moving through the northern basin, the heterogeneity of the zones or the existing biotopes, and the increase in dissolved salt content with increasing distance from the Shari delta. These factors upset the succession of developmental cycles which would normally exist in a water body subject to relatively homogeneous environmental factors. In 1974 and 1975, when the lake consisted of only three individual water bodies, the mean species diversity varied from one to another.

For a better knowledge of the algal population structure, the distribution of species abundance was analyzed in each sample. Motomura's log-linear model (Motomura 1932; Inagaki 1967) often appeared to be the most accurate representation of the existing distributions. From an analysis of 147 samples, Motomura's law of log-linear species distribution was perfectly corroborated in 37% of the cases and fairly well in 45% of them according to the thresholds suggested by Inagaki and Lenoir (1974). The adjustment was inadequate in 18% of the samples.

As in the case of the diversity index (Table 9), the different ecological zones could be characterized by the values found during five surveys. In 1971 and 1972, the highest values were observed in the Northern Open Water and the Eastern Archipelago (Fig. 7), with the minimum values being found in the Southern Open Water, in the far north and in certain points of the Eastern Archipelago. The environmental constant was related to the number of species comprising the biomass and the correlation coefficient between these two series on values was 0.78 in February 1971 and 0.82 in January 1972. After 1973, during the period of 'Lesser Chad', the values of the environmental constant were fairly high in the open water facing the Shari delta (0.71 to 0.85) except at station 3 in April and November 1974 when Diatoms developed on a large scale. In the archipelago close to Bol, the high environmental constant was close to or above 0.8 except in November 1974 when it ranged from 0.70 to 0.80 at the different sampling points. The lowest values were found in the far northwest, north and northeast of the northern basin where the salt content was highest and at station 12 where a species of Centric Diatom predominated in the algal population. In November 1974 when variations were low, the highest

Table 9 Values of the Motomura's constant for the algal communities in Lake Chad.

	Minimum	Mean	S.D.	Maximum
2/1971	0.308	0.681	0.14	0.861
1/1972	0.256	0.628	0.13	0.794
4/1974	0.376	0.754	0.09	0.854
11/1974	0.682	0.745	0.06	0.852
2/1975	0.399	0.689	0.14	0.831

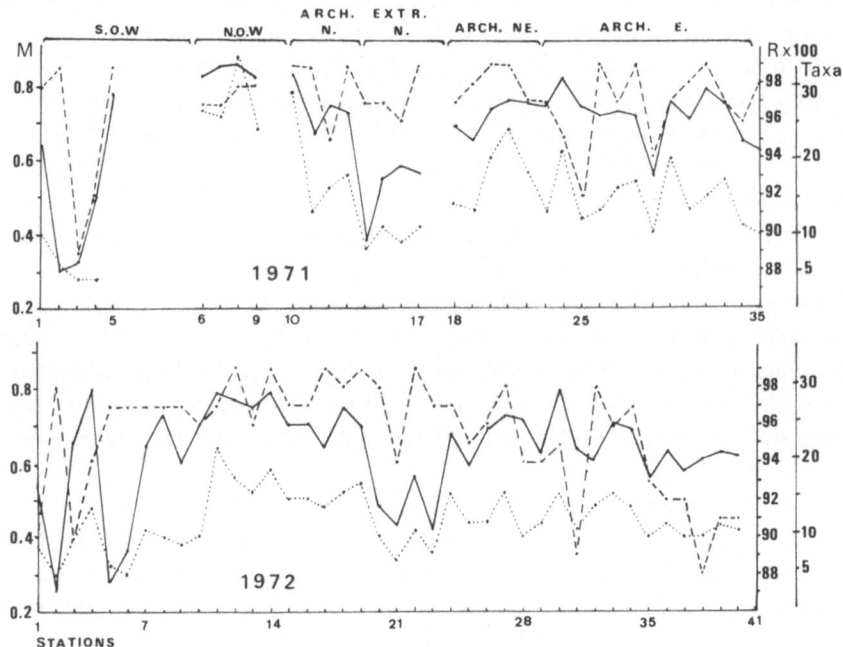

Fig. 7 Variations in the values of the 'R' coefficient between the logarithm of the biomasses of each species ranged in decreasing order of abundance and their rank (represented with broken lines); in the values of the 'constante de milieu' (represented with continuous lines); and of the number of species representing more than 0.07% of the total biovolume (represented with dotted lines). The abbreviations used are similar to those of the previous figure.

value was found at station 15 (0.776) in the far west. The lowest environmental constant values, ranging from 0.4 to 0.6, were found in February 1975.

6.2.2   General distribution of the phytoplankton

6.2.2.1   *Delimitation of the major ecological zones.* Despite the variety and complexity of the ecological zones present in the lacustrine system, the observations made from 1964 to 1972 during the period of 'Normal Chad' allowed the entire lake to be divided into a certain number of regions according to phytoplankton characteristics.

This zonation was determined from direct observation of the samples collected between 1964 and 1970 and after this time, and from calculation of the degrees of similarity (Spearman's correlation coefficient) existing between the quantitative surveys of the species. The correlation tables resulting from the samples collected during the two complete surveys conducted during 'Normal

162

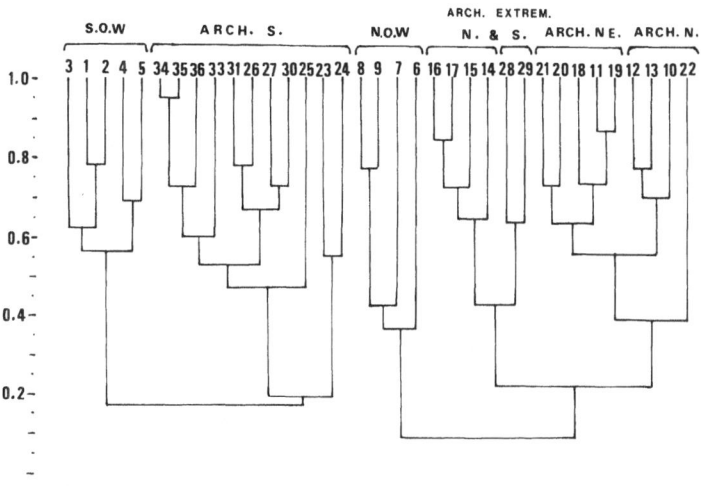

Fig. 8 Dendrogram for the interpretation of the Spearman correlation matrix between the records made in the lake in February 1971.

Chad' in 1971 and 1972 were analyzed in the form of a dendrogram. In 1971 (Fig. 8), the surveys made in the 35 stations were distributed in the following six groups: the Southern and Southeastern Open Waters, the archipelago south of the lake except the easternmost part (stations 28 and 29), the open water of the northern basin, the furthest parts of the archipelago (northeast and far east of the lake), the eastern part of the northern archipelago and the archipelago situated north of the Northern Open Water. In 1972 (Fig. 9), the resulting

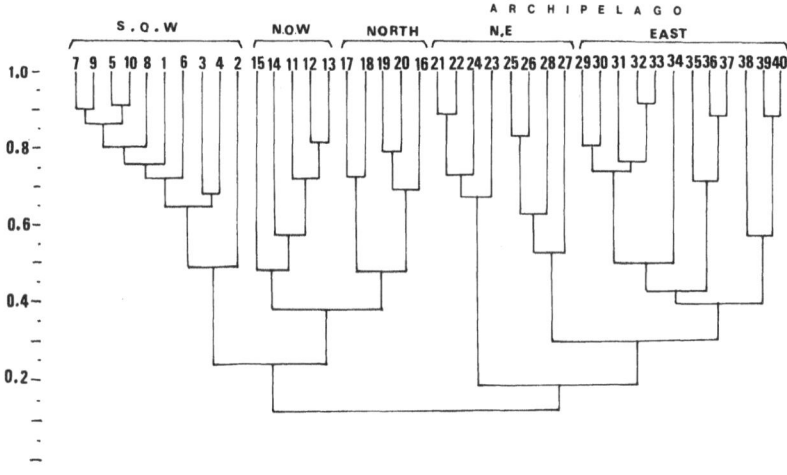

Fig. 9 Dendrogram for the interpretation of the Spearman correlation matrix between the records made in the lake in January 1972.

163

dendrogram was divided into five groups of surveys: those from the entire southern open water, those from the northern open waters, those from the vegetation fringe situated north of the previous group, those from the Northern Archipelago which includes a northeastern group (21 to 24) and an eastern group (25 to 28) and finally all the samples taken in the archipelago of the southern basin.

Given the phytoplankton distribution, the lake could be divided into two parts of nearly equal area by a line from Baga Kawa on the Niger coast to Baga Sola in the north of the Chad coast during the period of 'Normal Chad'. This line delimited a northern basin separated from the southern basin by a series of shallows. The northern part was characterized by high dissolved salt content (in the far north, the conductivity varied from 200 to 1300 μS cm^{-1} at 25°C). The depth was comparatively greater than in the southern and eastern parts so that resuspension of the bottom sediments resulted in lower turbidity, and finally the influence of the Shari flood was less. In the southern basin, the conductivity varied from 50 to 250 μS cm^{-1} at 25°C and exceeded this value only in the far northeast of the archipelago. The depth here was generally 2 to 4 meters resulting in a higher turbidity and the hydrological regime of the Shari and its tributaries caused considerable seasonal disturbances in the greater part of this zone. Each of the two basins was divided into a zone of open water and an archipelago, so that there were four major zones; two zones of open water, and two more heterogeneous archipelagoes subdivisible into two or three sub-zones

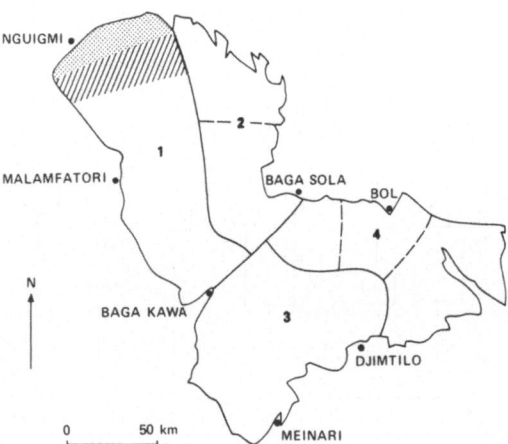

Fig. 10 Ecological zonation of Lake Chad. In 1, the northern open waters zone with, in the north, the region of the reeds islands and the vegetation fringe of the northern coast divided into two sub-zones; in 2, the northern archipelago zone divided into two zones; in 3, the southern open waters; and in 4, the southeastern archipelago divided into three sub-zones.

(Fig. 10). Particular biotopes such as the peri-deltaic zone, the immersed plant communities or the bottoms of the archipelago coves were distributed among the major zones but were so scattered and reduced in area that it was impossible to classify them into geographic regions.

The Southern Open Water included the open water situated between the Nigerian and the Cameroon banks (southern open water) and that spreading north of the Shari delta (southeastern open waters). It formed a shallow zone (3 to 4 meters) where there was frequent resuspension of the bottom sediments, especially during the rainy season (July–August) when a strong west wind blew. It was directly exposed to the disturbances caused by the outlet of the Shari waters; the flood waters which were very poor in plankton submerged the whole zone from the end of July to the beginning of January. The following species, belonging either to the Cyanophyceae, such as *Anabaena flos-aquae*, *Lyngbya limnetica* (especially in low waters) or to the Diatomophyceae, such as *Navicula* sp., *Melosira granulata*, *Gyrosigma kutzingii* and *Surirella muelleri* (especially in high waters) were the most important members of the phytoplankton. A Chlorophyceae, *Closterium aciculare* was very abundant in the northern basin and became fairly abundant in the western part of the southern basin. The diversity indices ranged from 0.53 to 3.44 bits, with an average of 1.523 in February 1971 and 1.775 in January 1972. The average environmental constant was 0.512 during the first year and 0.541 during the second year, while the mean algal biomasses were respectively 0.03 and 0.22 mg l^{-1}, the lowest values found in relation to the other regions of the lake.

The Southern Archipelago includes the whole northern zone of open water from the region of Baga Sola up to the easternmost part of the lake and the fringe of reed islands bordering the open water on the north and the east. The mean algal density was 1.4 mg l^{-1} in February 1971 and 4.5 mg l^{-1} in January 1972. The mean species diversity was 2.990 bits, in February 1971 and 1.558 in January 1972, while the average environmental constant was 0.714 and 0.639. Generally, Cyanophyceae predominated in the algae especially in the easternmost regions; Diatoms and Chlorophyceae were also abundant and formed blooms at certain periods. The whole zone was rather heterogeneous and divisible into three parts. The western region had an average plankton density (0.10 to 0.25 mg l^{-1}) and the phytoplankton was composed mainly of Chlorophyceae in January 1971 (*Oocystis* sp., *Pediastrum duplex*, *Coelastrum cambricum*). This region spread roughly from the 'Great Barrier' up to about ten kilometers east of Baga Sola. The central region around Bol was the poorest in phytoplankton and was the part of the archipelago most affected by the Shari flood during part of the year. The phytoplankton composed mainly of Diatoms (*Nitzschia spiculum*, *Melosira granulata*) and Chlorophyceae (*Oocystis* sp., *Coelastrum microporum*, *Pediastrum duplex*) in February 1971. Finally, the eastern region spread about ten kilometers west of Iseirom up to the eastern bank of the lake. It was characterized by dense phytoplankton (3 to 12 mg l^{-1})

where Cyanophyceae were predominant and the phytoplankton community was very similar to that present in the northern part of the archipelago of the northern basin.

In the northern part of the lake, the zone of open water included the Northern Open Water, the reed islands north of them and the small zone of open water spreading up to the vegetation fringe of the Niger coast. It was a rather deep zone where the highest values in the lake for species diversity were found (3.41 bits on average in February 1971 and 2.97 in January 1972). The average environmental constant reached 0.792 during the first year and 0.728 during the second year. The algal biomasses were 0.7 and 1.6 mg l^{-1} on average according to the year. While Cyanophyceae were abundant only in the northernmost part, the communities were generally very diverse and composed mainly of Chlorophyceae (*Oocystis* sp., *Closterium aciculare, Scenedesmus quadricauda, Coelastrum cambricum,* various *Pediastrum*) and Diatomophyceae (*Navicula* sp., *Synedra berolinensis, Fragilaria construens*). The waters in the reed islands north of the open water had a slightly different nature, characterized by abundant Pyrrhophyta belonging to the genera *Cryptomonas* and *Chroomonas.* Finally, in the northernmost part, near the plant fringe of the Niger coast, Cyanophyceae (*Microcystis delicatissima, Oscillatoria laxissima, Anabaenopsis tanganiikae*) become abundant and often predominant; the algal density was higher than further south (3 to 4 mg l^{-1}).

The archipelago of the northern basin was characterized by high phytoplankton biomass, equal to or higher than that of the Northern Open Water, especially in the northeastern part. In February 1971, the average was 1.4 mg l^{-1} and 2.0 mg l^{-1} in January 1972. The mean species diversity indices were 2.507 and 2.366 bits respectively, while the environmental constants were 0.628 and 0.588. The predominant groups of algae were the Cyanophyceae, with the following species: *Microcystis delicatissima, Oscillatoria laxissima, Lyngbya limnetica, L. contorta, Anabaenopsis tanganiikae*; and the Chlorophyceae with *Oocystis* sp., *Closterium aciculare, Scenedesmus quadricauda.* This zone was subdivided into two parts by a horizontal line a few kilometers north of Liwa. The region situated north of this was characterized by dense phytoplankton (1 to 4 mg l^{-1}) composed mainly of Cyanophyceae, while the southern part contained more varied and less dense populations (0.6 to 2 mg l^{-1}, where Chlorophyta were largely predominant. The characteristics of the previous zones are summarized in Table 10.

Various distinct smaller biotopes were scattered through some of these zones, e.g. the peri-deltaic regions covering an area of 5 to 8 km around the Shari delta and a lesser area near the deltas of El Beïd and the Yobé, which was a biotope having very pronounced seasonal variations in the phytoplankton and subject to the direct action of the affluents. In this latter region, a period from February to July of rich plankton alternated with a period of high water when the algae were reduced. This alternation was even more pronounced in the peri-deltaic

Table 10 Characteristics of the four ecological zones of the lake.

	Southern basin		Northern basin	
	Open waters	Archipelago	Open waters	Archipelago
Algal biovolume	0.03 & 0.22 µl l^{-1}	1.4 µl l^{-1}	0.7 & 1.6 µl l^{-1}	1.4 & 2.0 µl l^{-1}
Specific diversity	1.52 & 1.77	2.99	3.41 & 2.97	2.51 & 2.37
Motomura's constant	0.512 & 0.541	0.714	0.792 & 0.728	0.628 & 0.588
Prevailing algal groups	Diatoms Cyanophyceae	Chlorophyceae Diatoms Cyanophyceae	Chlorophyceae Diatoms	Cyanophyceae Chlorophyceae
Subdivisions	1 facies	3 sub-zones	2 facies	2 sub-zones
Seasonal variations	very pronounced	mean	low	low

zones where, in addition to river inflows during low water, dense lacustrine phytoplankton came and went with the flood. Consequently, an algal flora appeared which was more varied than that of neighbouring zones because of the different species carried over by the rivers and their flood zones.

A second distinct biotope was that of the submerged plant communities which consisted mainly of *Potamogeton*, and more rarely *Ceratophyllum*, *Vallisneria* or *Najas*. The phytoplankton was initially characterized by high diversity (the greatest number of taxa), low density, which was apparent from the high water transparency among the plants, and finally by the presence of abundant periphyton with *Gomphonema*, *Rivularia*, *Calothrix*, *Microchaete*, *Oedogonium*, etc. These biotopes existed all around the lake, in the archipelagoes and the reed island zones. They were most often found on the southeastern coast of the lake, east of the Shari delta, in the region of the 'Great Barrier' and in the northern archipelago, in the arms entering the plant fringe of the northern coast of the lake and generally in the ends of the archipelago channels.

A third distinct biotope was composed of the ends of the archipelago arms. The emergent vegetation along the banks was abundant and several meters wide. The water was transparent and often brown. The phytoplanktonic density was low and seasonal variations were insignificant. The dominant species belonged to the genera *Microcystis* and *Synechocystis*, the latter being found especially in the northern part of the lake. This biotope was often associated with the previous one, numerous cove bottoms being colonized partially or totally by submerged plant communities.

The limits of the different zones or biotopes were not always stable and water movements caused by the Shari flood or a change in the direction of the prevailing winds caused them to shift by several kilometers. For example, in 1972 (Fig. 5), the southern part of the Northern Open Water (stations 10 and 28) was similar to the Southern Open Water, so the separation between the northern plankton and the southern plankton 10 to 15 kilometers further northwest. The graphs made here, therefore, correspond to the calculation of their average position.

Moreover, this configuration of the lake was the one which existed during the period of 'Normal Chad' but it underwent profound modifications with the annual variations in water level which led to the periods of 'Greater Chad' or 'Lesser Chad'. Such was the case in 1973 when after several years of low rainfall, the lake had become separated into several different regions, thus assuming a different aspect which will be described later.

6.2.2.2 *Species associations.* In order to define the species associations related to the existing ecological zones, a factorial analysis (correspondence analysis) was applied to the surveys. This had the major advantage of illustrating the survey and the species on the same graph and it showed the reciprocal relations which existed between the surveys and their species components. Thus Fig. 11 shows the relative position of the stations and the species for the surveys made in January 1972. In this analysis, the surveys and the taxa were distributed on both sides of the axes 1, 2 and 3 which accounted for 23, 17 and 16% of the total variation respectively (a total of 56%), while 74% was represented by the first five axes. The different zones of the lake appear clearly on axes 1 and 2 (Fig. 11, top), and five groups can be distinguished:
— the group composed of the Southern and Eastern Open Water (surveys 1 to 10 and 28). The samples taken along the line separating the northern and the southern basin were related to the Southern Open Water. The following species represented the algal biocenosis: *Melosira granulata*, typical species and its variety *angustissima, Gyrosigma kutzingii, Surirella muelleri, Navicula* sp., *Scenedesmus acutus, Closterium acutum* var. *variabile, Lyngbya contorta* and *Anabaena flos-aquae. Lyngbya limnetica* was relatively abundant in the samples;
— the group composed of the Northern Open Water (stations 11 to 14) included some planktonic Diatoms, *Synedra berolinensis, Fragilaria construens* and

---

Fig. 11 Factorial analysis positions of the samples and the species occurring in the lake in relation to the axis 1 and 2 (top) and 1 and 3 (bottom) in January 1972. The long dashes delimit the groups of the southern and southeastern open waters, the dotted line delimits the groups belonging to the northern open waters; the continuous line delimits the groups of the archipelago of the southern basin; the broken and dotted line delimits the groups of the archipelago of the northern basin and the short dashes delimit the groups of the far northeastern and eastern archipelago (see Appendix I for the taxa code used).

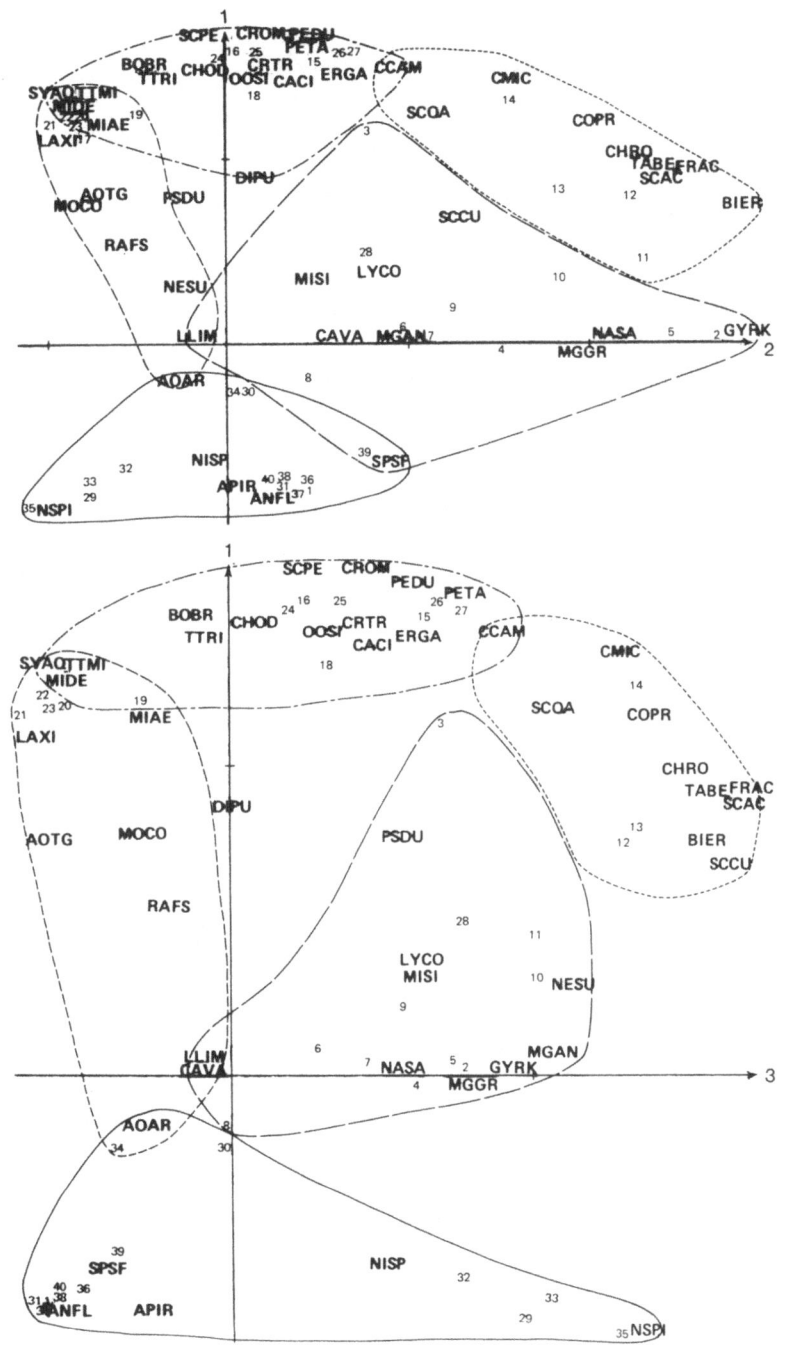

more especially Chlorophyceae, *Scenedesmus quadricauda, Scenedesmus acuminatus, Coelastrum microporum, C. cambrium, C. proboscideum, Binuclearia eriensis.* The species *Navicula* sp., *Oocystis* sp., *Lyngbya contorta* and *Chroococcus limneticus* were also present;
— the group formed by the Southeastern Archipelago whose samples represented a large zone with two dominant species constituting two extremes: *Nitzschia spiculum* for stations 29, 32, 33 and 35 and *Anabaena flos-aquae* for stations 31, 36, 37, 38, 39 and 40, the latter having similarities with the populations of the Southeastern Open Water. In all the populations, there were several small species: *Nitzschia, Surirella muelleri, Anabaena spiroïdes* and *Anabaenopsis arnoldii.* Stations 30 and 34 (end coves of the archipelago) showed similarities with the following group;
— the group formed by the northernmost and northeastern parts of the lake (surveys 17 and 19 to 23) corresponded to the northern part of the archipelago in the northern basin. The related species group consisted of mainly Cyanophyceae: *Synechocystis minuscula, Microcystis delicatissima, M. aeruginosa, Oscillatoria laxissima, Anabaenopsis tanganiikae, A. arnoldii, Lyngbya limnetica, Raphidiopsis* sp. and a few Chlorophyceae: *Nephrochlamys subsolitaria, Monoraphidium contortum, Tetraedron minimum* and *Oocystis* sp.;
— the last group included the reed islands and the archipelago bordering the northern open water to the north (stations 15, 16 and 18) and east (stations 24 to 27). The following species, mainly Chlorophyceae, were present *Oocystis* sp., *Crucigenia triangularis, Botryococcus braunii, Tetraedron trigonum, T. minimum, Scenedesmus perforatus, Chodatella* sp., *Pediastrum tetras, P. duplex, Eremosphaera gigas, Coelastrum cambricum, Closterium aciculare, Chroomonas* sp., *Microcystis delicatissima* and *Synechocystis minuscula.*

The plot of axes 1 and 3 (Fig. 11, bottom) provided very little additional information. The five above-mentioned groups were clearly separated. The group of the Southeastern Archipelago was extended more along the horizontal line than in the previous plot and the sub-groups dominated by *Anabaena flos-aquae* and by *Nitzschia spiculum* were well separated.*

6.2.2.3 *Type communities.* From the correlations observed between the species and the calculated species biovolumes, a type community schematically showing the relative species composition of the existing populations was determined for each zone or sub-zone. In each region, the mean percentage of each well represented species was calculated from the samples from the different stations in this region. After summing these mean percentages and adjusting them to 100%, the relative composition of the existing community was determined.

---

* Refer to the taxa code in Appendix I for abbreviations used in the text.

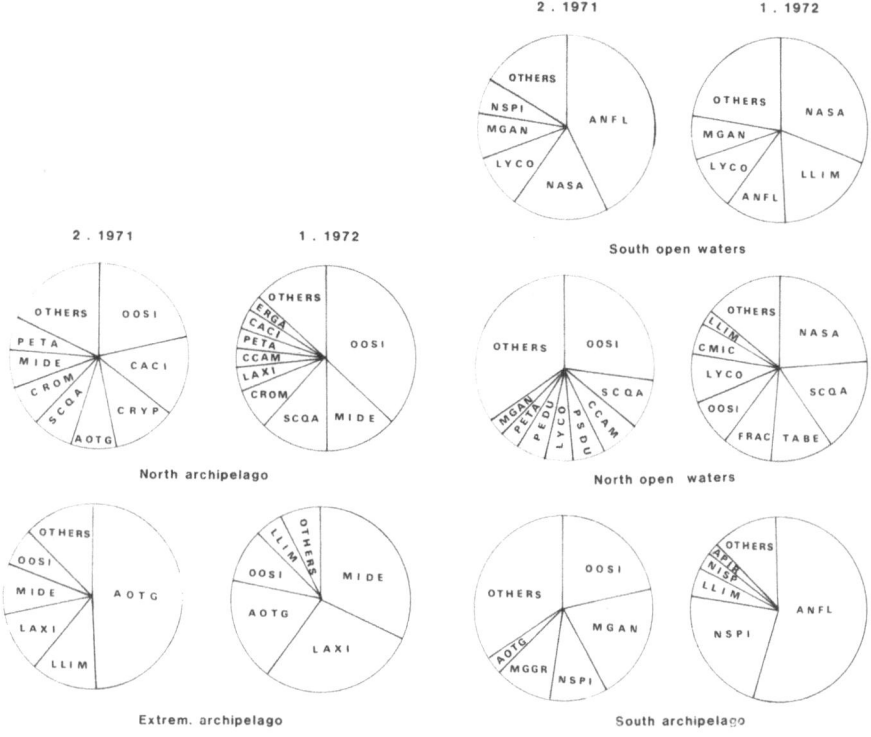

Fig. 12 Type-communities existing in the different zones of the lake in February 1971 and January 1972 (cf. Appendix I for the taxa code).

Only those taxa representing more than 2% of the biovolume were considered, the others being placed in the category 'various' (Fig. 12).

The phytoplankton community of the Southern and Southeastern Open Waters can be defined as follows:

	February 1971	January 1972
Melosira granulata var. *angustissima*	8%	7%
Navicula sp.	18%	31%
Nitzschia spiculum	6%	
Anabaena flos-aquae	43%	11%
Lyngbya contorta	9%	9%
Lyngbya limnetica		18%
Various	16%	24%

171

At the beginning of 1971, half the biomass was composed of Cyanophyceae, with Diatoms representing more than a third of the populations, while Chlorophyta were not very abundant. In January 1972, Diatoms and Cyanophyceae were almost equal in proportion.

In the open water of the northern basin, the biomass was extremely different in composition and included the following species during the two periods under consideration:

	February 1971	January 1972
Oocystis sp.	28%	8%
Scenedesmus quadricauda	8%	17%
Coelastrum cambricum	7%	
Coelastrum microporum		5%
Pediastrum clathratum	6%	
Pediastrum tetras	4%	
Pediastrum duplex	5%	
Melosira granulata var. *angustissima*	3%	
Synedra berolinensis		11%
Fragilaria construens		9%
Navicula sp.		24%
Lyngbya contorta	6%	8%
Lyngbya limnetica		2%
Various	33%	16%

While Chlorophyta represented three quarters of the community in February 1971, they formed only a third of the biomass in 1972. The Diatoms represented more than 50% of the algae.

In the whole Northern Archipelago, the algal species were distributed as follows:

	February 1971	January 1972
Oocystis sp.	23%	37%
Eremosphaera gigas		2%
Scenedesmus quadricauda	7%	12%
Coelastrum cambricum		4%
Pediastrum tetras	5%	4%
Closterium aciculare	14%	4%

	February 1971	January 1972
Microcystis delicatissima	7%	13%
Anabaenopsis tanganiikae	8%	
Oscillatoria laxissima		4%
Cryptomonas erosa	12%	
Chroomonas sp.	7%	7%
Various	17%	13%

The communities were varied and Chlorophyta formed from half to two thirds of the biomass.

The two type communities calculated for the archipelago in the southern basin are:

	February 1971	January 1972
Oocystis sp.	22%	
Melosira granulata	10%	
Melosira granulata var. *angustissima*	21%	
Nitzschia spiculum	11%	23%
Nitzschia sp.		3%
Anabaena flos-aquae		55%
Anabaena spiroïdes		2%
Anabaenopsis tanganiikae	2%	
Lyngbya limnetica		5%
Various	34%	12%

While there was a clear prevalence of Diatoms in February 1971, at the beginning of 1972 two thirds of the biomass was composed of Cyanophyta and only about one third was of Diatoms.

Finally, in the northernmost and easternmost parts of the archipelago, particular communities developed with a prevalence of Cyanophyceae and included:

	February 1971	January 1972
Oocystis sp.	6%	9%
Microcystis delicatissima	8%	32%
Anabaenopsis tanganiikae	49%	18%
Oscillatoria laxissima	11%	28%
Lyngbya limnetica	12%	5%
Various	14%	8%

173

Therefore, it was observed that Cyanophyceae and Diatomophyceae were abundant in the southern basin during the period of 'Normal Chad', while the Chlorophyceae appeared to dominate in the northern basin. The furthest parts of the archipelago, whether in the northern or southern basin, were almost totally colonized by Cyanophyceae.

6.2.3 *Temporal changes in the phytoplankton*

6.2.3.1 *Seasonal variations.* A cool season was distinguishable in December and January when water temperatures could drop below 20°C and a warm season occurred in April, May and June when water temperatures exceeded 30°C. Winds blow from the northeast for about eight months, but from May to September, the strong humid monsoons blow from the southwest and are associated with rainfall (250 to 500 mm over the lake according to the latitude). Insolation was very high all the year round declining to a minimum in August–September. The different zones of the lake were not affected in the same way by the factors, as the northern part had the greatest temperature differences throughout the year. If the channels were always protected from the prevailing winds in the archipelago, the resulting turbulence led to a re-suspension of the muddy and pseudo-sandy bottom sediments in the open waters, especially in the southern basin where depth did not exceed four meters. Therefore, water transparency was about 25 cm in this zone during the rainy season.

Given the high annual rate of water renewal in the lake (55% on an average), the lacustrine plankton was most affected by the hydrological regime of the Shari. The flood begins in the deltaic region at the end of June and the flood wave with very turbid waters submerges the southeastern part in July and spreads to the whole southern part of the lake. The flood reaches its maximum in October when turbidity is high in the central part of the eastern archipelago towards Bol. Thus the southern and southeastern open waters, the central part of the eastern archipelago and the southern fringe of the northern open waters are subject to the greatest seasonal variations.

A study conducted from August 1964 to July 1965 (Gras et al. 1967) in different zones and biotopes of the eastern part of the lake provided partial knowledge of these variations (Table 11). Quantitative estimates were made each month at eleven stations distributed over the Shari delta, the Peri-deltaic Zone, the Southeastern Open Waters, the plant communities east of the delta, the central part of the archipelago towards Bol and the eastern part (Fig. 13). They were expressed as number of cells per liter with all the cells being counted whether they were isolated or arranged in cenobes or colonies. The density values in Table 11 need to be multiplied by 10^5 to obtain the number of plant cells present in each liter (Fig. 14).

Table 11 Algal biomasses expressed in number of cells per liter in the different stations of the eastern part of the lake within a year (from August 1964 to July 1965).

Biotops	Shari	Peri-deltaic	Immersed plants	Open Waters SE		Archipelago (Center)	Bol	Archipelago (East)			N.E. Arms Ends
Stations	Delta	St 1	Adjiléié	St 2	St 3	St 5	St 6	St 7	St 8	St 9	St 10
August	256	650	985	1099	1099	1178	5165	10 735	11 071	10 663	4890
September	73	75	535	638	946	2154	5801	11 936	12 777	10 658	5836
October	71	74	281	659	914	1247	5345	12 569	12 770	10 374	6079
November	43	64	814	629	1156	1080	6127	10 429	10 461	11 954	5530
December	43	64	800	792	1440	981	4857	10 996	11 840	12 809	5799
January	115	722	589	1783	1291	760	5472	12 164	12 821	12 291	6406
February	114	1415	614	1316	1279	582	5532	9370	12 454	10 772	5045
March	1410	3813	811	1987	1012	955	5691	10 507	11 669	10 831	5252
April	2075	4366	1175	2807	3638	3067	10 167	12 837	10 866	11 361	4930
May	1748	5454	3606	2625	2285	3545	9516	11 513	11 759	10 667	5216
June	1955	5626	3981	3360	3117	3103	8395	10 803	10 340	11 295	6030
July	318	4677	4471	3166	1366	1218	6146	10 220	10 837	11 008	8713
Annual mean	685	2250	1555	1738	1628	1657	6518	11 173	11 638	11 224	5810

175

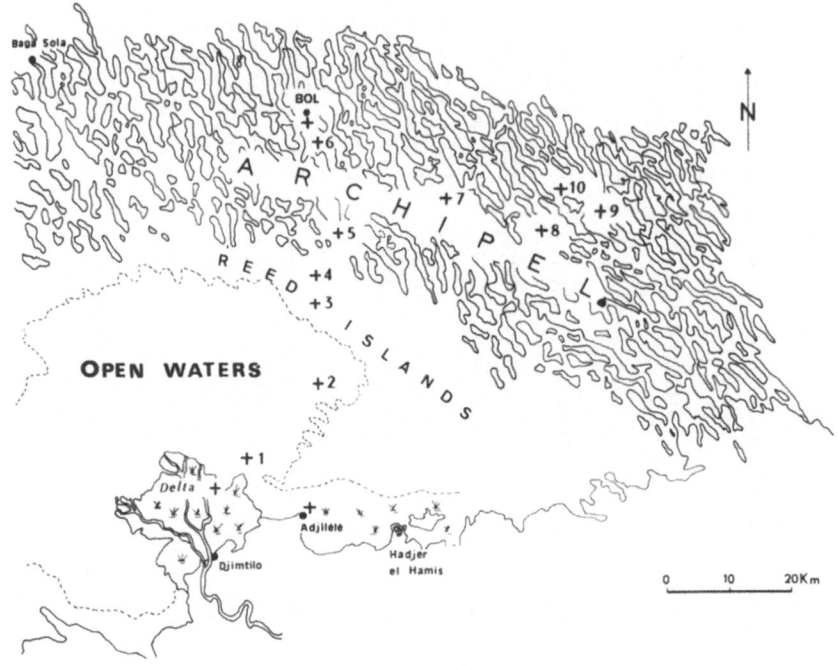

Fig. 13 Location of the sampling points which were regularly examined to analyze the seasonal variations of the plankton (1964–65).

If the seasonal development of the phytoplankton density is analyzed in each of these zones or biotopes, a fairly rich period in the Shari delta and peri-deltaic region can be observed in April, May and June when the water was low and warm. An impoverished period occurred from September to March with minimum values in November–December when the flood reached its peak. In the peri-deltaic zones, the algal density which was very low during the flood became higher than in the neighbouring open water during low water periods. However, the phytoplankton density was generally lower in the plant community zones situated east of the Shari delta towards Adjilele. In the Southeastern Open Water (stations 2 and 3), the minimum density resulting from the flood was very apparent during the last quarter of the year, while in the central part of the archipelago (stations 5 and 6), the low density was still observable from December to February especially in the southernmost part (station 5). In the eastern part of the archipelago (stations 7, 8 and 9), the algal density was very high (more than one million cells per milliliter) and remained stable throughout the year. At station 10 which was representative of the zones in the end coves of the archipelago, the biomass was only the half of that in the surrounding archipelago and there were virtually no variations throughout the year.

176

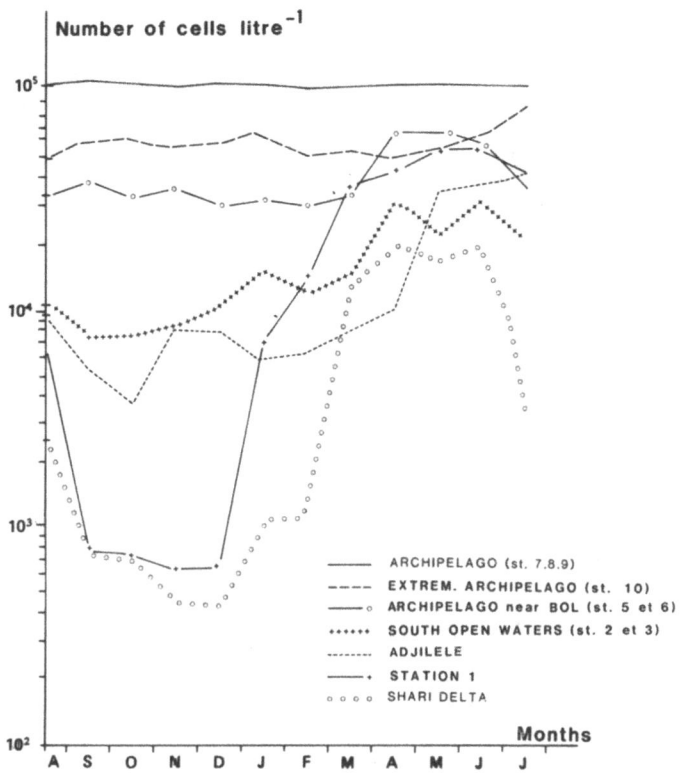

Fig. 14 Seasonal changes in the number of cells per liter in the different parts of the eastern zone of the lake. The mean related to 2 or 3 stations was represented in the Southern Open Water, in the archipelago towards Bol, and in the eastern archipelago.

The proportions of the three algal groups present in the samples from the river near the delta up to the end coves of the archipelago are presented on six graphs (Fig. 15). The Cyanophyceae expressed as the number of cells per liter, were generally composed of colonial species such as *Microcystis*, with numerous small cells and the percentages were always high. The proportion of Diatoms was equal to or slightly higher than that of the Cyanophyceae in the Shari delta and the peri-deltaic zone in September. Then Chlorophyceae became predominant in the Shari in December at the beginning of the flood when the plankton from the flood zones flowed back towards the river bed. A maximum number of Chlorophyceae also appeared in the peri-deltaic regions in November, but from January, Cyanophyceae predominated (more than 90% of the cells) until the time of the next flood. In the communities situated east of the delta, Diatoms and Chlorophyceae contributed 10 and 20% respectively in

177

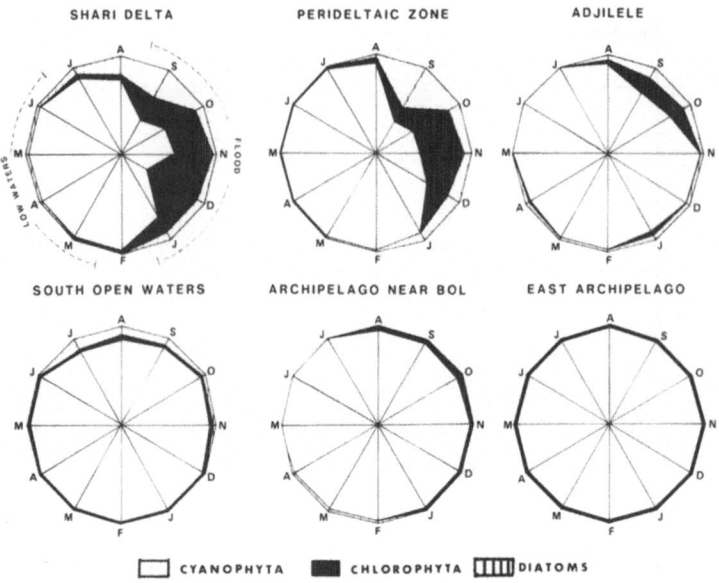

Fig. 15 Seasonal variations in the percentage of each of the three main groups of algae calculated from the number of cells per liter in the eastern part of the lake (1964–1965).

September and October but Cyanophyceae comprised more than 90% of the cells during the rest of the year. According to the average of stations 2 and 3, the percentage of Diatoms was about 10% (13% max.) in the southern open waters in July, August and September when the resuspension of the Diatom rich bottom film was related to the waves formed by the west winds. 5% of the cells were composed of Chlorophyta from August to December, but Cyanophyceae contributed 90% of the cells for nine months of the year. Finally 5% of Chlorophyceae were observed in the center of the archipelago from September to January followed by the development of *Melosira granulata* in February, while the Cyanophyceae, still forming 90% of the cells reached 98% between March and July. Diatoms were always insignificant in the eastern region and in the distinct biotopes of the end channels, while Chlorophyceae represented 2% of the cells and Cyanophyceae 98%, throughout the year.

During the period of 'Normal Chad', considerable seasonal variations in the phytoplankton occurred only in the peri-deltaic regions and the Southern and Southeastern Open Waters of the eastern part of the lake. In these regions, there was an alternation of a six-monthly period with considerable variations, with six stable months. Few observations were made in the northern part of the lake but it appeared that only the southern part of the open water and the region of the Yobé delta were affected by the flood.

178

6.2.3.2 *Annual variations.* Since the water level of the lake depended upon the balance between the inflows and evaporation, climatic variations which were similar for several years led to changes in the level, with rainfall excesses resulting in a period of 'Greater Chad' and rainfall deficiencies resulting in 'Lesser Chad', as happened from 1972 onwards. In this case the phytoplankton varied considerably from year to year, and in the following pages, the changes which occurred between 1971 and 1975 when the lake underwent considerable contraction are analyzed.

During 'Normal Chad', there were practically no long-term observations made on the plankton over several consecutive years, but from a few data obtained between 1964 and 1968 when the lake was slowly contracting, it would seem that the phytoplankton varied little from one year to another.

Apart from the regular changes resulting from the occurrence of the annual Shari flood in the southern basin, the southern and southwestern open waters were occupied by populations of *Anabaena flos aquae* each year during low water. Species such as *Melosira granulata, Micractinium pusillum* or *Mallomonas portae-ferrae* developed on a large scale in the archipelago of the southern basin from March to April. In the Northern basin, *Closterium aciculare* was predominant in the open water each year.

But current knowledge of the algal communities is still too limited in time to predict the composition and changes of the plankton over a long period during which the mean level of Chad remains stable.

6.2.4 *Dynamics of the phytoplankton during a dry period*

6.2.4.1 *History of the revolution of the lake towards the period of 'Lesser Chad'.* Towards the middle of 1972, the constant and regular lowering of the lake that had been occurring since 1964 led to a decrease in depth, to the exposure of a 3 to 20 kilometers wide fringe along the southern and western coasts of the lake and the occurrence of shelves in the archipelago channels whose aspect underwent changes. However, all the regions of the lake were connected with each other and the lake was still considered to be in the period of 'Normal Chad' up to the first quarter of 1973. But, since the 1972 flood was deficient (17.5 billion m^3 against 40 billion on average), a separation occurred between the southern and the northern basin during the second quarter of 1973 as a result of the exposure of the shallows situated in the place of the former 'Great Barrier' which existed at the time of Tilho's survey. In July 1973, the archipelago of the southern basin became separated from the zone of open water facing the delta; the shallow arms dried up and abundant vegetation developed. During the second half of 1973, this situation was temporarily modified by the inflows of the Shari flood, the Southeastern Archipelago was again occupied by water and very low inflows reached the northern basin. But, from March 1974, the lake was again divided into three parts: the Southeastern

Open Water connected to the river system and situated north of the delta (about 1300 km²), the Southeastern Archipelago reduced to a few channels which were usually covered with aquatic vegetation (about 150 to 200 km²) and the very shallow northern basin (about 6000 km²) separated from the southern basin by a wide vegetation barrier from Baga Sola to Baga Kawa. In July 1974, in the northern basin, the northern and eastern parts dried up and numerous islands appeared in the remaining water bodies.

In November 1974, the southern basin received considerable inflows from the Shari and from September, the Southeastern Archipelago was again occupied by water and connected to the open waters. However, extensive vegetation consisting of *Cyperus papyrus* and *Aeschynomene* covered the flooded regions. The northern part of the lake remained isolated by the vegetation barrier between the northern and southern basins; the water area was reduced and the depth barely exceeded one meter. In January 1975, the whole of the southern basin reached a level which was higher than that of the lake in 1972, but the southern coastal zone of the lake and the archipelago were always covered with vegetation although again occupied by water. In the northern basin, water inflows passed through the 'Great Barrier' and in February 1975, a slight increase (from 30 to 40 cm) was observed in the water level together with a decrease in the dissolved salt content. In June 1975, there were only a few ponds left in the northern part and almost total drying up occurred in August.

The chapter on the hydrological development of the lake could be referred to for locating this contraction period in the complete hydrology of the lake from 1960 to 1976, which is considered to provide an accurate description of the Chad situation when the phytoplankton samples were taken.

6.2.4.2 *Development of the phytoplankton.* During a dry period, starting from a period of 'Normal Chad', two stages can be observed in the development of the algal populations. During the first stage, the water cover still represented an hydrological unit as in 1971 and 1972, although its level decreased along with a considerable increase in the mean algal biovolume of the lake from 1 $\mu l\, l^{-1}$ in February 1971 to 2.4 $\mu l\, l^{-1}$ a year later. This increase was felt in all the zones of the lake and mainly in the archipelago of the southern basin. It was accompanied by a decrease in the species diversity index from an average of 2.752 to 2.141 and by a decrease in equitability from 72 to 59%. The mean number of taxa contributing to the biomass (above a threshold of 0.07% of the sample biovolume) decreased from 15.2 to 12.3. The environmental constant was 0.681 in February 1971 and 0.628 in January 1972.

The phytoplankton as a whole underwent few modifications in composition during this period. Almost the same taxa were found but the major algal groups were sometimes present in different proportions in the various zones. Cyanophyceae and Diatoms represented most of the algae in the Southern and Southeastern Open Waters. In the eastern archipelago, Diatoms were dominant in February 1971 with a rather high percentage of Chlorophyta, while Cyanophyta represent two thirds of the much higher biomass the following year.

In the northern basin, Cyanophyceae predominated in the northernmost part, while further south Chlorophyceae predominate in February 1971 and Diatoms in January 1972. In the Northern Archipelago, Chlorophyta were always dominant except in the far northeast. Therefore it is possible that, from 1972 onwards, given the decrease in the mean level of the lake by 30 to 40 cm as compared with the previous year, the flood waters were checked at the level of the reed islands bordering the Southeastern Open Water in the north. The flooding of the archipelago was smaller and slower, allowing the Cyanophyceae to quickly colonize. The flood waters which were rich in Diatoms moved far beyond the Great Reef increasing their percentage in the open water of the northern basin where they become dominant. In the parts of the archipelago furthest from the Shari delta, whether in the northern basin or in the southern basin, Cyanophyceae always represent about 90% of the biomass.

Thus this first stage was particularly characterized by an increase in the mean phytoplankton biomass, with a decrease in the species diversity values and in the environmental constant. The proportions of the different algal groups changed, mainly in the archipelago of the southern basin and in the Northern Open Water.

From 1974 onwards Lake Chad was no longer an hydrological unit. Three individual water bodies remained in the entire lacustrine basin. During this second stage, changes in the algal communities could only be observed by examining developments in each individual water body (Fig. 16).

In the Southeastern Open Water, the phytoplankton density was higher than in previous stages: 0.8 μl of algae per liter in November 1974; 0.7 in February 1975 and 7.8 μl l^{-1} in April 1974 when water level was low. The community diversity and the values of the diversity index, which were lower than 2 bits on average during the period of 'Normal Chad', were greater than 2.5 in 1974 and over 3 in February 1975. The mean number of taxa contributing more than 0.07% of the biomass was 6 in February 1971 and 25 in February 1975. The algal community which was composed mainly of Diatoms and Cyanophyceae during the period of 'Normal Chad' increased with large numbers of Euglenoids occurring during the warm season, and with Chlorophyceae and Chrysophyceae present from November 1974 onwards.

The plankton of this region thus evolved towards a more pronounced river type which was characterized by greater variety and by the existence of considerable differences between the poorer flood waters and the much richer low waters. If the water inflows from the Shari were sufficient to keep the southern and southeastern open water zones submerged, the communities in the southern and western parts were different from those in the eastern part but these subdivisions disappear with the flood. If the water inflows from the Shari

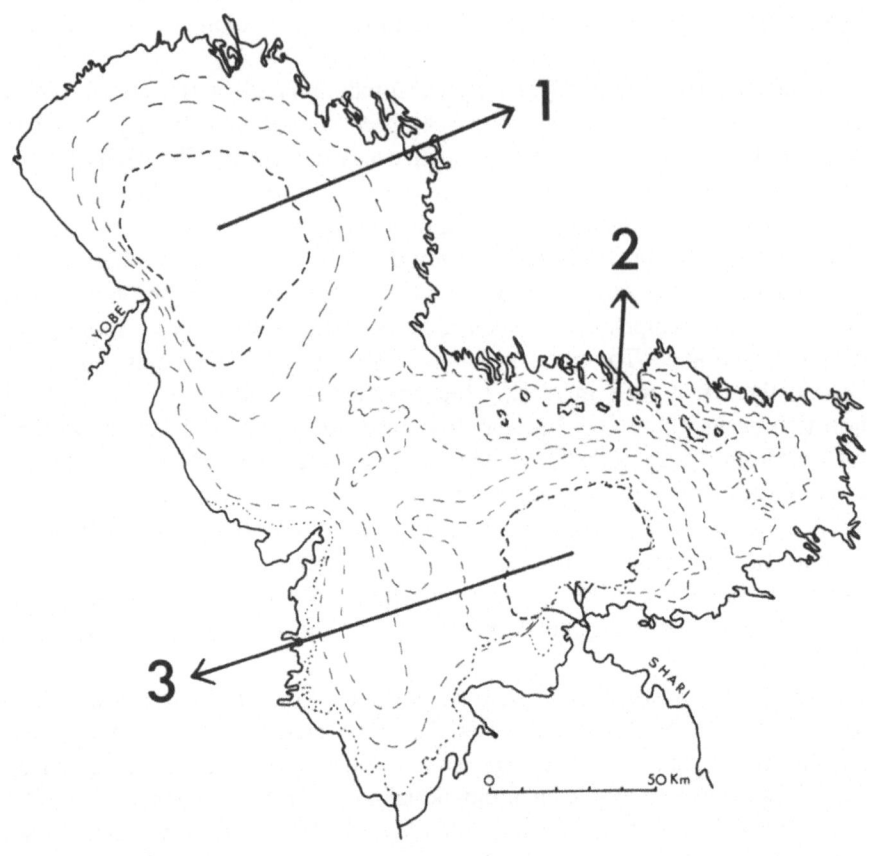

Fig. 16 Representation of the drying up of Lake Chad from 1972 to 1975 and development of the different parts of the lake; 1, towards a facies of natron pond; 2, towards a facies of swamp; 3, towards a facies of river.

remained very low, the pond-like character of this region was more marked. Cyanophyceae and Euglenoids increased during time of low water, while large numbers of Diatoms were always present during other seasons.

On the whole, apart from the development of the Euglenoids and the abundance of a Diatom *Synedra berolinensis* which was previously non-existent or rare in the samples, the phytoplankton has varied little in its components and this zone of the lake could be considered part of the lake which changed least during a dry period.

In the archipelago of the lake, after an increase in biomass in the period before the change towards 'Lesser Chad' (average 1.4 mg l^{-1} in April 1974 and 7.6 in February 1975) biomasses as a whole were still higher during this period (average 8.2 mg l^{-1} in April 1974 and 7.6 mg l^{-1} in February 1975). However, the flooding of this zone by water from the south which filtered through the reed islands surrounding the open water considerably impoverished the environment ($0.4 \, \text{mg} \, l^{-1}$ in November 1974) during certain periods (for instance October–November 1974). In this group of fairly large biotopes which were either isolated or connected with one other, and has a varied water supply during high water, the rather diversified algal populations (2.3 to 3.2 bits on average depending on period) were generally composed mainly of Euglenoïds with Chlorophyta and Diatoms reaching considerable proportions locally; Cyanophyceae were always low in proportion.

This group of ponds became established during the period of 'Lesser Chad' and evolved towards a swamp dominated by Euglenophyta, and occupied by aquatic macrophytes, whose importance varied with the water inflow from the south which affected the depth of these habitats. The chlorophyll measurements made by Lemoalle (1979) corroborated the differences which existed between the remaining ponds. After a concentration stage in 1972 and 1973 when values were high (more than $250 \, \text{mg} \, \text{m}^{-3}$), water circulation resulting from infiltrations through the vegetation caused a decrease in chlorophyll content which rarely exceeded $100 \, \text{mg} \, \text{m}^{-3}$. At the end of 1975, the chlorophyll content did not exceed $2 \, \text{mg} \, \text{m}^{-3}$ in most places.

As in the rest of the lake, the last stages of the 'Normal Chad' period in the northern basin were characterized by an increase in algal biomass from an average of 1.09 to $1.73 \, \text{mg} \, l^{-1}$ between 1971 and 1972. The evolution of the water body towards the period of 'Lesser Chad' then became rapid as a result of the low water inflows from the Yobé river during the flood period and from the southern basin through the restored 'Great Barrier'. As soon as this part became isolated, the average algal biovolume increased considerably to $179 \, \mu l \, l^{-1}$ in November 1974. This change was well reflected in the chlorophyll

Table 12 Chlorophyll a contents observed in the central part of the northern basin from March 1973 to April 1976.

Time	Station number	Chl ($\text{mg} \, \text{m}^{-3}$)	S.D.
21/24 March 1973	9	39	5.1
24/29 April 1973	11	40.7	13.5
18 September 1973	8	82.5	20
11/22 November 1973	14	102.9	16
19/27 April 1974	11	141.2	37
26/31 July 1974	21	343	214
25/30 November 1974	7	1658	360
16/18 February 1975	9	204	107
10/13 April 1975	16	464	257
16/21 April 1976	6	208	41

content of the central part of the basin (Table 12, from Lemoalle 1979). After reaching a maximum in November 1974 (1658 mg m^{-3}), the concentration became reduced to 204 mg m^{-3} by water inflows in February 1975. The dry period resumed and the chlorophyll content reached 3600 mg m^{-3} in August before total drying up occurred. In 1976 concentrations of phytoplankton were observed in ponds that were temporarily refilled by water.

The conductivity of the waters went up at the end of the period of 'Lesser Chad', showing a dissolved salt concentration which related these environments to continental salt water. It thus ranged from 1900 to 4200 µS cm^{-1} in November 1974 and from 486 to 1330 in February 1975 after the low inflows from the flood. The pH varied from 9.1 to 9.5 in November 1974. Therefore it can be estimated that the water remaining in the northern basin changed rapidly towards a temporary or permanent pond containing sodium such as existed in the region of Kanem bordering Lake Chad in the north. Species such as *Synechocystis minuscula, Anabaenopsis arnoldii, Oscillatoria platensis f. minor* which are typical of mesocarbonated environments in ponds with sodium (Iltis 1974) appeared in large numbers towards Kindjéria in November 1974. The algal populations in which Chlorophyta, Cyanophyta, Diatomophycea and Pyrrophyta were well represented in 1972 were reduced to Chlorophyceae and Diatoms in 1975, the latter representing 80% of the biomass.

Based on the study of populations in natron ponds, the Diatom population was replaced by Cyanophyceae as soon as the total dissolved salt content in water exceeded one to two grams per liter. These modifications in the community accompanied a decrease in the average species diversity from 2.931 bits in January 1971 to 1.758 in February 1975. When the total dissolved salt content became greater than 2 g l^{-1}, there was a highly significant inverse correlation between diversity and dissolved salt content. Diversity became practically zero when the salt content of the waters exceeded twenty grams per liter in the permanent natron ponds. The change in the environmental constant was similar to the change in diversity and became lower as the salt content increased. In fact, when the salt content exceeded two grams per liter and the salinity increased, a smaller number of halophilous species made up most of the biomass. Below 2 g l^{-1}, the phenomena resulting from variations in salinity were more diffused and could be masked by the action of other factors, either biological or physico-chemical.

6.2.4.3 *Phytoplankton characteristics in different parts of 'Lesser Chad'*. There was no correlation between the algal communities in the three water bodies or groups of individual water bodies which represented 'Lesser Chad'. Spearman's rank correlations calculated on samples taken during three complete surveys conducted in 1974 and 1975 always showed three separate groups of communities corresponding to the plankton of the Southeastern Open Waters, the Southeastern Archipelago and finally the northern basin. Even if there were temporary hydrological connections between the archipelago and the southeastern waters during certain periods, the phytoplankton in each zone maintained particular characteristics different to these of the other two zones.

Therefore the composition of the plankton in the three water bodies will be examined individually.

6.2.4.3.1 *Southeastern Open Water.* In the Southern Open Waters, the community was composed mainly of Diatoms during the Shari flood and of Diatoms and Cyanophyceae during the period of low water. Euglenoïds could amount to 20% of the biomass in certain zones during the warm season. The most abundant species were: *Melosira granulata, M. granulata* var. *angustissima, Synedra ulna, S. berolinensis, Surirella linearis, S. muelleri, Nitzschia spiculum, Gonatozygon monotaenium, Eudorina elegans, Anabaena flos-aquae, A. spiroides, Lyngbya limnetica, L. contorta, Microcystis elachista, Euglena* sp. and *Mallomonas portae-ferrae.* The phytoplankton densities remained low during the flood of the Shari (0.8 μl l^{-1} of algae on an average in November 1974 and 0.7 in February 1975) and were much higher in low water during the warm season (7.8 μl l^{-1} on average in April 1974) with higher biovolumes than in 'Normal Chad' in the same zone.

The average species diversity was 2.51 bits in April 1974 and it increased to 2.70 in November and 3.10 in February 1975, while the values of the environmental constant were respectively 0.717, 0.786 and 0.801.

The type-communities which were defined included the following species in April 1974 when the waters was low (Fig. 17).

30%	*Nitzschia spiculum*
17%	*Anabaena flos-aquae*
9%	*Euglena* sp. which are red on the surface
8%	*Melosira granulata*
7%	*Melosira granulata* var. *angustissima*
5%	*Lyngbya contorta*
5%	*Fragilaria contruens*
3%	*Anabaenopsis tanganiikae*
13%	various species

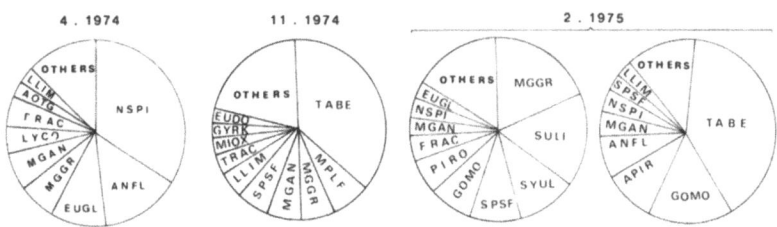

Fig. 17 Type-communities occurring in the southern open waters from April 1974 to February 1975 (cf. Appendix I for taxa code).

The community was composed of about a third Cyanophyceae, more than 50% Diatomophyceae and 10% Euglenophyta.

During the flood in November 1974, the algal biomass was composed roughly of:

35%	*Synedra berolinensis*
8%	*Melosira granulata*
8%	*Mallomonas portae-ferrae*
6%	*Melosira granulata* var. *angustissima*
6%	*Surirella muelleri*
5%	*Lyngbya limnetica*
4%	*Trachelomonas* plur. sp.
3%	*Microcystis elachista*
2%	*Gyrosigma kutzingii*
2%	*Eudorina elegans*
21%	various species

There were about two thirds Diatomophyceae, 10% of Cyanophyceae and 10% Chrysophyceae.

In February 1975 when the decrease of water ended, two type-communities could be distinguished in the zone of open water which then occupied the southern and southeastern regions of the lake.

	Western part	Eastern part
Gonatozygon monotaenium	8%	15%
Spirogyra sp.	6%	
Melosira granulata	19%	
Melosira granulata var. *angustissima*	4%	4%
Synedra ulna	10%	
Synedra berolinensis		41%
Fragilaria construens	5%	
Surirella muelleri	9%	3%
Surirella linearis	17%	
Nitzschia spiculum	4%	4%
Anabaena spiroïdes		9%
Anabaena flos-aquae		8%
Lyngbya limnetica		3%
Euglena sp.	2%	
Various sp.	16%	13%

There were about 75% Diatoms and 15% Chlorophyceae in the western part and more than 50% Diatoms, slightly over 20% Cyanophyceae and 15% Chlorophyceae in the eastern part.

During the period of 'Lesser Chad', Diatoms dominated the open water of the southern basin, while Cyanophyceae held respectively second and third position. So a considerable amount of Euglenophyta and to a lesser extent Chrysophyceae were included in the biomass. On the whole, the communities were very varied.

6.2.4.3.2 *Archipelago of the southern basin.* During the period of 'Lesser Chad', the archipelago in the eastern part of the lake no longer represented a perfectly homogeneous zone. The channels partially dried up and a dense macrophytic vegetation developed over the dried up areas and covered the flooded reaches. Only a group of ponds or small channels were left either cleared or more or less covered with vegetation and separated by exposed shelves. When the flood occurred, the complex system was again fed by water from the south which filtered through vegetation barriers leading to a temporary flooding of part of the vegetation and to a partial separation in the floating plant beds.

Only three stations could be explored around Bol because the region was not easily accessible. Generally the algal community was composed mainly of Euglenoids but Diatoms and to a lesser extent Chlorophyceae were well represented according to the stations. The most common species were *Euglena* sp., *Trachelomonas* sp., *Strombomonas fluviatilis*, *Lepocinclis* sp., *Melosira granulata* var. *angustissima*, *Synedra ulna*, *Dictyosphaerium pulchellum* and *Closterium strigosum*. A Pyrrhophyta such as *Cryptomonas erosa* was sometimes abundant in the samples.

Depending on station and period, the algal biovolumes varied greatly from 8.8 $\mu l \, l^{-1}$ on average in April 1974 to only 0.4 $\mu l \, l^{-1}$ in November 1974 when the archipelago was resubmerged. In February 1975, the phytoplankton density was again high and very varied (minimum of 2.3 $\mu l \, l^{-1}$ and maximum of 13.8 $\mu l \, l^{-1}$) with an average of 7.6 $\mu l \, l^{-1}$.

The average species diversity was 3.152 bits in April 1974, 2.506 in November

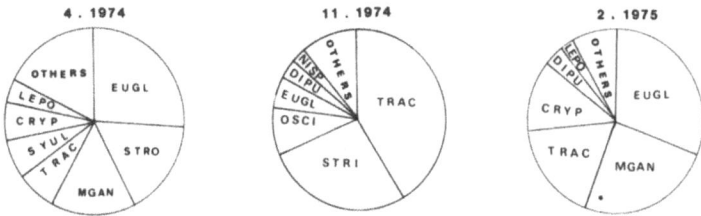

Fig. 18 Type-communities occurring in the archipelago of the southern basin from April 1974 to February 1975 (cf. Appendix I for taxa code).

1974 and 2.365 in February 1975. The values of the environmental constant was respectively 0.848, 0.726 and 0.796.

Despite the heterogeneity of the algae present in these environments with infrequent connections, type communities were defined for the three periods under study (Fig. 18).

	April 1974	November 1974	February 1975
Dictyosphaerium pulchellum		4%	4%
Closterium strigosum		27%	
Melosira granulata			
var. *angustissima*	16%		24%
Synedra ulna	7%		
Nitzschia sp.		2%	
Oscillatoria sp.		9%	
Cryptomonas erosa	7%		13%
Euglena plur. sp.	26%	6%	31%
Trachelomonas plur. sp.	7%	41%	18%
Strombomonas fluviatilis	16%		
Lepocinclis sp.	4%		2%
Various	17%	11%	8%

In this case, Euglenophyta represented 50 to 60%, Diatoms about 25% in April 1974 and February 1975 and Pyrrhophyta 7 to 13% during the same periods.

Generally, Euglenoids represented at least half the biomass in the ponds remaining in the archipelago during the period of 'Lesser Chad'. Although Chlorophyta and Diatoms reached considerable numbers locally, Cyanophyta were always low in proportion. The populations were very variable in density according to water movements through this region. The biomass enrichment in relation to the period of the larger lake resulted from the development of Euglenoïds and was interrupted by total reflooding of the whole zone in November 1974. This led to an increase of 2.50 m in the water level, and this flood wave was quickly colonized by very dense phytoplankton in February 1975.

6.2.4.3.3   *Northern basin.* During the period of 'Lesser Chad', Diatomophyceae generally dominated in the phytoplankton with the two abundant species being *Coscinodiscus rudolfii* and *Cyclotella meneghiniana*. Chlorophyceae were most usually abundant and the main species were *Scenedesmus quadricauda, S. opoliensis, S. intermedius, Pediastrum tetras,* and *Oocystis.* Finally, in April 1974, Cyanophyceae were abundant in the northern and northeastern parts of

188

the basin and in the part of the archipelago occupied by water; the most abundant species were *Anabaenopsis tanganiikae*, *Lyngbya contorta*, *L. limnetica*, *Oscillatoria laxissima* and *Chroococcus limneticus*.

The algal biomass was high with an average of $25.3 \, \text{mg} \, l^{-1}$ in the samples from April 1974 and $179 \, \text{mg} \, l^{-1}$ in the seven samples from November 1974 when the water level did not exceed one meter in the deepest zones. In February 1975 when the northern basin was again fed by low water infiltrations (30 to 40 cm) through the 'Great Barrier', the average biomass was $74 \, \text{mg} \, l^{-1}$. The chlorophyll *a* estimates around the island of Kindjéria in the center of the northern basin had means of $141 \, \text{mg} \, \text{m}^{-3}$ in April 1974, $1659 \, \text{mg} \, \text{m}^{-3}$ in November and $204 \, \text{mg} \, \text{m}^{-3}$ in February 1975.

The species diversity values were very variable in April 1974 and ranged from 0.4 to 3.8 bits (mean: 2.217). The lowest values were either from the populations in the northernmost and northeasternmost parts of the lake where *Anabaenopsis tanganiikae* was dominant, or from the populations in the center of the basin where *Coscinodiscus rudolfii* was largely dominant. In November 1974 and February 1975, the communities had homogeneous diversities of 2.422 and 1.536 bits on average. The mean values of the environmental constant were 0.741 in April 1974, 0.704 in November and 0.531 in February 1975.

The composition and proportions of the different species can be defined by two types of community in April 1974 (Fig. 19). The first corresponded to the plankton in the center, the south and west of the lake, while the second was representative of the surveys in the north, the northeast and east of the lake, i.e. the archipelago. In November 1974 and February 1975, the plankton of this region could be considered as homogeneous (Table 13).

If, in April 1974, two thirds of the population were still represented by Cyanophyceae, 25% by Diatoms and 10% by Chlorophyta in the north and the east of the northern basin, Diatoms (three quarters of the population and 10% of Chlorophyta) predominated in the southern and western parts. In November, the type community included about 50% Diatoms, 40% Chlorophycea and 10% Cyanophyceae and in February 1975, Diatomophyceae were almost 90%.

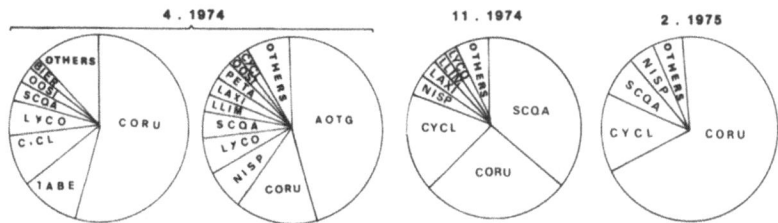

Fig. 19 Type-communities occurring in the northern basin from April 1974 to February 1975 (cf. Appendix I for taxa code).

Table 13 Changes in the relative composition of the algal communities in the northern basin of the lake from April 1974 to February 1975.

North basin	April 1974		November 1974	February 1975
	S and W parts	N and E parts		
	(%)	(%)	(%)	(%)
Oocystis sp.	3	2		
Scenedesmus quadricauda	3	5	36	7
Pediastrum tetras		3		
Binuclearia eriensis	2			
Cyclotella meneghiniana	9	2	18	14
Coscinodiscus rudolfii	54	15	27	69
Synedra berolinensis	11			
Nitzschia sp.		7	4	5
Anabaenopsis tanganiikae		46		
Oscillatoria laxissima		3	4	
Lyngbya limnetica		3	2	
Lyngbya contorta	6	6	2	
Others	12	8	7	5

Thus a very high algal density developed in the northern basin during the period of 'Lesser Chad' and Diatoms were particularly abundant. The partitioning of the northern basin into two different ecological zones such as the archipelago and the open water which existed during the period of 'Normal Chad', was still present in April 1974, so that with the permanent decrease in the water level, the algal populations become uniform in the reduced water cover existing at the end of 1974 and in early 1975.

In conclusion, after contraction of the lake, the composition of the algal communities seems to have changed little in the open water in the southern basin. In the archipelago the major change was caused by the Euglenoïds. Species composition in the Northern basin was very different because of the abundance of Centric diatoms and the occurrence prior to the drying up of the lake, of Cyanophyceae which were typical of the sodium ponds.

6.2.5 General conclusions

From the surveys made between 1971 and 1973, 'Normal Chad' can be classified as a tropical lake rich in phytoplankton compared with data from some lakes in the Philippines, Indonesia or Venezuela (Lewis 1978). Moreover, the surveys showed an increase in the algal biomass when the lake moved towards the stage of 'Lesser Chad' during the drought period. This increase

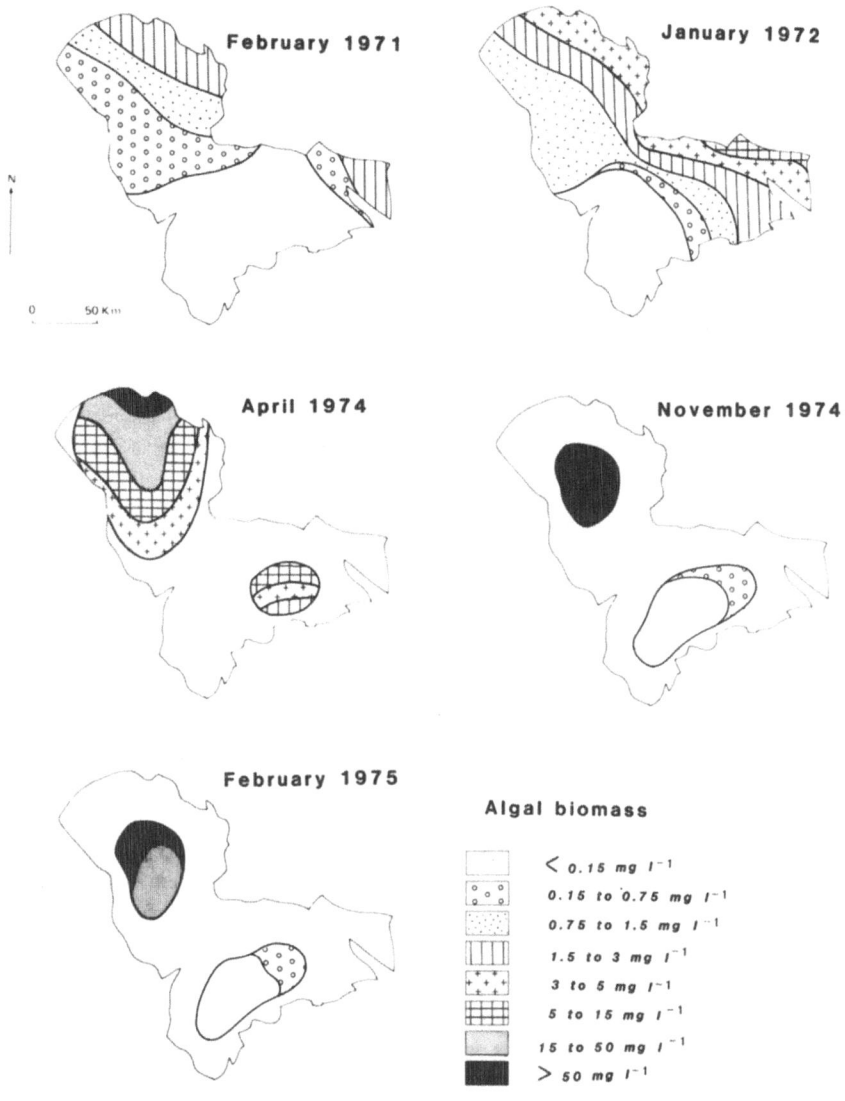

Fig. 20 Distribution of the phytoplankton biomass in Lake Chad: February 1971 (total biomass 40 800 t); January 1972 (total biomass 76 800 t); April 1974 (total biomass 183 500 t); November 1974 (total biomass 187 300 t); February 1975 (total biomass 244 100 t).

whose stages are represented in Fig. 20 was particularly marked in the northern basin which acquired characteristics of an eutrophic pond.

Vollenweider (1968) suggested the use of maximum biomasses to indicate the trophic level of European lakes. He thus classified the temperate lakes into

191

several groups, ranging from the ultra-oligotrophic lakes with maximum biomasses lower than one gram per cubic meter to the highly eutrophic lakes with maximum biomasses higher than ten grams per cubic meter. It seems difficult to apply this classification here since the observations were limited in time. Nevertheless, it can be assumed that starting with the analysis of the seasonal variations during the period of 'Normal Chad' (Gras et al. 1967), the open waters zones can be classified as mesotrophic environments, while the northern and southeastern archipelagoes were meso-eutrophic. On the contrary, the furthest parts from the deltas of the Shari, El Beïd and the Yobé (far east and far northeast where Cyanophyceae represented at least 90% of the algal biovolume throughout the year) ranged from eutrophic to highly eutrophic environments. During the period of 'Lesser Chad', the archipelago of the southern basin could be considered as eutrophic, while the waters of the northern basin and the zone of open water near the Shari delta were highly eutrophic.

Since the eutrophic zone extended to the bottom and the plankton was considered to be fairly well distributed in the water column, the total biomass was calculated from the data on surface algal density. The area and bathymetrical data used were those defined by Roche (1971) for the different regions. The results expressed in weight of living matter were in some way imprecise because of the numerous sources of errors likely to occur (methods of sampling and sub-sampling the phytoplankton, countings under the inverted microscope and conversion into biovolume calculation of areas, mean depths and dried up areas in relation to the level). However these estimates will be useful for calculating the general energy balance of the lake.

The algal biomass (Table 14) was estimated as 40 800 tons for a water area of 18 000 km^2 in February 1971. In February 1975, it reached 244 100 tons in a flooded area of about 11 000 km^2. It appears that phytoplankton increase varied greatly with seasons in the open waters of the southern basin, was greater in the archipelago and considerable in the remaining water of the northern basin.

However, these estimates only concerned the limnoplankton. In this type of shallow lake, the floating vegetation, whether submerged or not, covered considerable areas. In addition to floating 'Kirtas' and reed islands, there was submerged vegetation in all the sheltered zones of the archipelago and in the shallow border of the open water allowing abundant periphyton to develop. Wetzel (1964) showed in a Californian lake that the periphyton represented most of the primary production, limnoplankton contributing only 25%.

In Lake Chad, because of the hydrological instability leading to frequent modifications in water level and structure, the primary production was observed in three forms (limnoplankton, macrophytes, periphyton) which differed in proportions according to hydrological developments in the lake. Therefore, in the northern basin, the evolution towards 'Lesser Chad' under a

Table 14 Total algal biomasses (in tons, f.w.) for the different regions of the lake, from 1971 to 1975.

	2. 1971		1. 1972		4. 1974		11. 1974		2. 1975	
	Area (Km²)	Biomass (t)	Area (Km²)	Biomass (t)	Area (Km²)	Biomass (t)	Area (Km²)	Biomass (t)	Area (Km²)	Biomass (t)
Southern basin										
Open waters	5410	305	4910	1535	1300	14 195	4910	6140	4910	5030
Archipelago	3405	9375	3405	26 135	200	2295	3405	2135	3405	37 955
Northern basin										
Open waters	6160	15 780	5660	29 730	6000	166 980	2000	179 000	300	201 150
Archipelago	3160	15 340	3160	19 425						
Total	18 135	40 800	17 135	76 825	7500	183 470	10 315	187 275	11 315	244 135

dry climate leads to a reduction in the macrophyte belts covered by periphyton. The primary production was mainly due to a very dense limnoplankton or heleoplankton. In the southern basin, during the period of 'Lesser Chad' large developments of macrophytes and, to a lesser extent, of periphyton were observed in the archipelago while the limnoplankton remained at an average level. In the south and the west, the swampy zones towards the delta of El Beïd and the 'Great Barrier' were occupied by macrophytes.

So, limnoplankton studies do not give representative results for estimating quantitative richness of the whole plant level and establishing trophic classification of this lake. It is necessary to evaluate the importance of each class of primary producer (plankton, aquatic phanerogams, periphyton) in order to assess total primary production. The studies on the diets of fish populations in Chad (Lauzanne 1976) showed that the three forms of vegetation were used directly or indirectly by one group of fishes or another. The range of biomass values observed in the limnoplankton alone suggest that total plant production was particularly high.

References

Baudrimont, R., 1974. Recherches sur les Diatomées des eaux continentales de l'Algérie. Ecologie et Paleoécologie. Imp. Nord-Africaine. Alger 249 pp., 22 plates.
Bourrelly, P., 1957. Algues d'eau douce du Soudan français, région du Macina. Bull. IFAN Sér. A 19: 1047–1102.
Bourrelly, P., 1961. Algues d'eau douce de la République de Côte d'Ivoire. Bull. IFAN Sér. A 23: 283–374.
Bourrelly, P., 1975. Quelques algues d'eau douce de Guinée. Bull. Mus. Nat. Hist. Nat. 3 ème Sér. 276, Botanique 20: 1–71.
Carmouze, J. P., Dejoux, C., Durand, J. R., Gras, R., Iltis, A., Lauzanne, C., Lemoalle, J., Lévêque, C., Loubens, G. and Saint-Jean, L., 1972. Grandes zones écologiques du Lac Tchad. Cah. ORSTOM Sér. Hydrobiol. 6: 103–169.
Compère, P., 1967. Algues du Sahara et de la région du lac Tchad. Bull. jard. bot. nat. Belg. 37: 109–288.
Compère, P., 1970. Contribution à l'étude des eaux douces de l'Ennedi. VI Algues. Bull. IFAN Sér. A 32: 18–64.
Compère, P., 1974. Algues de la région du lac Tchad. II. Cyanophycées. Cah. ORSTOM Sér. Hydrobiol. 8: 165–198.
Compère, P., 1975a. Algues de la région du lac Tchad. III. Rhodophycées, Euglénophycées, Cryptophycées, Dinophycées, Chrysophycées, Xanthophycées, Cah. ORSTOM Sér. Hydrobiol. 9: 167–192.
Compère, P., 1975b. Algues de la région du lac Tchad. IV. Diatomophycées. Cah. ORSTOM Sér. Hydrobiol. 9: 203–290.
Compère, P., 1976a. Algues de la région du lac Tchad. V. Chlorophycophytes (1ère partie) Cah. ORSTOM Sér. Hydrobiol. 10: 77–118.
Compère, P., 1976b. Algues de la région du lac Tchad. VI. Chlorophycophytes (2ème partie). Cah. ORSTOM Sér. Hydrobiol. 10: 135–164.
Compère, P., 1977. Algues de la région du lac Tchad. VII. Chlorophycophytes (3ème partie: Desmidiées) Cah. ORSTOM Sér. Hydrobiol. 11: 77–177.

Couté, A. and Rousselin, G., 1975. Contribution à l'étude des algues d'eau douce du Moyen Niger (Mali). Bull. Mus. nat. Hist. Nat. 3ème Sér. 277, Botanique 21: 73–175.

Devaux, C., 1973. Contribution à l'étude des populations phytoplanctoniques du lac de Tazenat (Puy de Dôme) – Ann. Stat. biol. Besse-en-Chandesse, 7: 1–101.

Gauthier-Lièvre, L., 1931. Recherches sur la flore des eaux continentales de l'Algérie et de la Tunisie. Minerva, Alger, 299 pp.

Gayral, P., 1954. Recherches phytolimnologiques au Maroc. Travaux Inst. Scient. Cherifien. Sér. Botanique, No. 4, 306 pp., 14 plates.

Gras, R., Iltis, A. and Lévêque-Duwat, S., 1967. Le plancton du Bas Chari et de la partie est du lac Tchad. Cah. ORSTOM Sér. Hydrobiol. 1: 25–100.

Gronblad, R., Scott, A. M. and Croasdale, H., 1968. Desmids from Sierra Leone, Tropical West Africa. Acta Bot. Fennica. 78: 1–41, 159 figs.

Iltis, A., 1972. Algues des eaux natronées du Kanem (Tchad). 1st part. Cah. ORSTOM. Sér. Hydrobiol. 6: 173–246.

Iltis, A., 1974a. Phytoplancton des eaux natronées du Kanem (Tchad) VIII. Classification des milieux étudiés et espèces caractéristiques. Cah. ORSTOM Sér. Hydrobiol. 8: 81–91.

Iltis, A., 1974b. Le phytoplancton des eaux natronées du Kanem (Tchad). Influence de la teneur en sels dissous sur le peuplement algal. ORSTOM. Paris, Thèse AO 9523, 313 pp., mimeo.

Iltis, A., 1977a. Peuplements phytoplanctoniques du lac Tchad. I. Stade Tchad normal (février 1971 et janvier 1972). Cah. ORSTOM Sér. Hydrobiol. 11: 33–52.

Iltis, A., 1977b. Peuplements phytoplanctoniques du lac Tchad. II. Stade petit Tchad (avril 1974, novembre 1974 et février 1975). Cah. ORSTOM. Sér. Hydrobiol. 11: 53–72.

Iltis, A., 1977c. Peuplements phytoplanctoniques du lac Tchad. III. Remarques générales. Cah. ORSTOM Sér. Hydrobiol. 11: 189–199.

Iltis, A and Compère, P., 1974. Algues de la région du lac Tchad. I. Caractéristiques générales du milieu. Cah. ORSTOM Sér. Hydrobiol. 8: 141–164.

Iltis, A. and Lemoalle, J., 1976. Un plancton à Diatomées à Bol (lac Tchad) en 1973. ORSTOM N'Djaména, 11 pp., mimeo.

Inagaki, H., 1967. Mise au point de la loi de Motomura et essai d'une écologie évolutive. Vie et Milieu 18: 153–166.

Inagaki, H. and Lenoir, A., 1974. Une étude d'écologie évolutive: application de la loi de Motomura aux fourmis. Bull. Ecol. 5: 207–219.

Lauzanne, L., 1976. Régimes alimentaires et relations trophiques des poissons du lac Tchad. Cah. ORSTOM Sér. Hydrobiol. 10: 267–310.

Lemoalle, J., 1979. Biomasse et production phytoplanctoniques du lac Tchad (1968–1976). Relations avec les conditions de milieu. Thèse Univ. Paris 6, 287 pp.

Lewis, W. M., 1978. A compositional, phytogeographical and elementary structural analysis of the phytoplankton in a tropical lake: Lake Lanao. Philippines. J. Ecol. 66: 213–226.

Maillard, R., 1977. Diatomées d'eau douce du Mali, Afrique. Bull. Mus. nat. Hist. Nat. Paris 443: 17–45.

Margalef, R., 1958. Temporal succession and spatial heterogeneity in phytoplankton. In E. Buzzati-Traverso (ed.), Perspectives in Marine Biology, pp. 232–347. Univ. California Press, Berkeley, Los Angeles.

Margalef, R., 1967a. El ecosistema. Ritmos, fluctuaciones y sucesión. La vida suspendida en las aguas. In: Ecología Marina. Fund. La Salle de Ciencias nat., Est. Invest. mar. Margaríta, Caracas: 377–463, 454–492, 493–562.

Margalef, R., 1967b. Some concepts relative to the organization of plankton. Oceanogr. Mar. Biol. Ann. Rev. 5: 257–289.

Motomura, I., 1932. Etude statistique de la population écologique. Doobutagaku Zassi 44: 379–383.

Munawar, M. and Munawar, I. F., 1976. A lakewide study of phytoplankton biomass and its species composition in Lake Erie, April–December 1970. J. Fish. Res. Board Can. 33: 581–600.

Roche, M. A., 1971. Géographie et éléments numériques sur la superficie et la bathymétrie du lac Tchad. ORSTOM N'Djaména, 7 pp., mimeo.

Skuja, H., 1949. Zur Süsswasseralgen – Flora Burmas. Nova acta regiae soc. scient. Upsalla, Sér. 4, 14, 188 pp., 37 plates.

Thomasson, K., 1971. Amazonian algae. Mém. Inst. Roy. Sc. Nat. Belg., Sér. 2, 86, 57 pp., 24 plates.

Van Meel, L., 1954. Exploration hydrobiologique du lac Tanganika, 1946–1947. IV, 1, Le phytoplancton. Inst. Roy. Sc. Nat. Belg., 681 pp.

Vollenweider, R. A., 1968. Scientific fundamentals of the eutrophication of lakes and flowing waters, with particular references to nitrogen and phosphorus as factors in eutrophication. Organ. Econ. Coop. Dev. Rep., Paris.

Wetzel, R. G., 1964. A comparative study of the primary productivity of higher aquatic plants, periphyton and phytoplankton in a large shallow lake. Int. Rev. ges. Hydrobiol., 49: 1–61.

Appendix I

Taxa code

Melosira granulata	MGGR	*Monoraphidium contortum*	MOCO
Melosira granulata		*Binuclearia eriensis*	BIER
var. *angustissima*	MGAN	*Coelastrum cambricum*	CCAM
Coscinodiscus rudolfii	CORU	*Coelastrum microporum*	CMIC
Cyclotella meneghiniana	CYCL	*Coelastrum proboscideum*	COPR
Synedra ulna	SYUL	*Ankistrodesmus* sp.	ANKG
Fragilaria construens	FRAC	*Quadrigula quaternata*	QQUA
Synedra berolinensis	TABE	*Chodatella* sp.	CHOD
Gyrosigma kutzingii	GYRK	*Diclyosphaerium pulchellum*	DIPU
Navicula sp.	NASA	*Botryococcus braunii*	BOBR
Nitzschia sp. (petite forme)	NISP	*Micraclinium pusillum*	MAPU
Nitzschia spiculum	NSPI	*Eudorina elegans*	EUDO
Surirella linearis	SULI	*Gonatozygon monotaenium*	GOMO
Surirella muelleri	SPSF	*Closterium aciculare*	CACI
Tetraedron minimum	TTMI	*Closterium acutum* var. *variabile*	CAVA
Tetraedron trigonum	TTRI	*Closterium strigosum*	STRI
Nephrochlamys subsolitaria	NESU	*Spirogyra* sp.	PIRO
Crucigenia triangularis	CRTR	*Euglena* sp.	EUGL
Crucigeniella crucifera	CRUC	*Euglena oxyuris* f. *charkowiensis*	EOFC
Crucigenia tetrapedia	TETR	*Trachelomonas* sp.	TRAC
Oocystis sp.	OOSI	*Strombomonas* sp.	STRO
Eremosphaera gigas	ERGA	*Phacus* sp.	PHAC
Scenedesmus acuminatus	SCAC	*Lepocinclis* sp.	LEPO
Scenedesmus acutus	SCCU	*Synechocystis minuscula*	SYAQ
Scenedesmus intermedius	SINT	*Synechococcus leopoliensis*	ROME
Scenedesmus quadricauda	SCQA	*Chroococcus limneticus*	CHRO
Scenedesmus ecornis	SCEC	*Microcystis aeruginosa*	MIAE
Scenedesmus opoliensis	SCOP	*Microcystis delicatissima*	MIDE
Scenedesmus perforatus	SCPE	*Microcystis elachista*	MIOX
Pediastrum clathratum	PSDU	*Microcystis* sp.	MISI
Pediastrum duplex	PEDU	*Anabaena flos-aquae*	ANFL
Pediastrum tetras	PETA	*Anabaena spiroïdes*	APIR

Anabaena sp.	ANSA	*Oscillatoria laxissima*	LAXI
Anabaenopsis arnoldii	AOAR	*Oscillatoria platensis* f. *minor*	SPMI
Anabaenopsis tanganiikae	AOTG	*Cryptomonas erosa*	CRYP
Lyngbya contorta	LYCO	*Chroomonas* sp.	CROM
Lyngbya limnetica	LLIM	*Mallomonas portae-ferrae*	MLPF
Raphidiopsis sp.	RAFS	*Peridinium* sp.	PERD
Oscillatoria sp.	OSCI		

7. The zooplankton

Lucien Saint-Jean

The zooplankton community includes about 30 species or genera of Rotifers (Pourriot 1968; Robinson 1971) and several microcrustacea. During the high water period, the latter were represented mainly by 8 species of Cladocera (*Diaphanosoma excisum, Daphnia barbata, D. longispina, D. lumholtzi, Ceriodaphnia cornuta, C. affinis, Moina micrura dubia, Bosmina longirostris*), two Calanoids (*Tropodiaptomus incognitus, Thermodiaptomus galebi*) and three *Cyclopoids (Thermocyclops neglectus, Th. incisus circusi, Mesocyclops* cf. *leuckarti*). A Cyclopoid (*Thermocyclops tchadensis*) and two unidentified Calanoids of secondary importance can be added to the list. The littoral forms of the Cladocera and Copepods, mostly attached to the vegetation were much more numerous (Dussart and Gras 1966; Rey and Saint-Jean 1968, and 1969).

During the period of 'Normal Chad', the planktonic and littoral faunas were clearly separate. However, two exceptions were noted: *M. leuckarti*, normally found in grassbanks, and a species of *Alona* with a particular affinity for the center of grassbanks were also found in the open water in 1970–71. *Alona* became frequent in plankton samples during the period of 'Lesser Chad'. This species is very abundant in the open areas within grassbanks, in contrast to other littoral species which are actually attached to the vegetation (Dejoux and Saint-Jean 1972). This location causes this chydorid to be swept from the grassbanks by water currents and during storms and this could explain its incidental presence in the plankton.

Most of the species recorded up to the present from different parts of the lake are fairly widespread in Africa, or more generally in the intertropical regions. There were three cosmopolitan species among the planktonic forms: *B. longirostris, D. longispina,* and *M.* cf. *leuckarti*. The importance of the biomass of the two Cladocerans above, as well as the diversification of the genus Daphnia (3 species) must be emphasized.

The data collected on the zooplankton cover a period of lake recession, from 1964 (high waters) to 1975 ('Lesser Chad'). They are distributed very heterogeneously in space and time and are of three types: the annual cycles at one or several stations (1964–65, 1972 to 1975); observations covering a large part of

J.-P. Carmouze et al. (eds.) Lake Chad
© *1983, Dr W. Junk Publishers, The Hague/Boston/Lancaster*
ISBN 978-94-009-7268-1

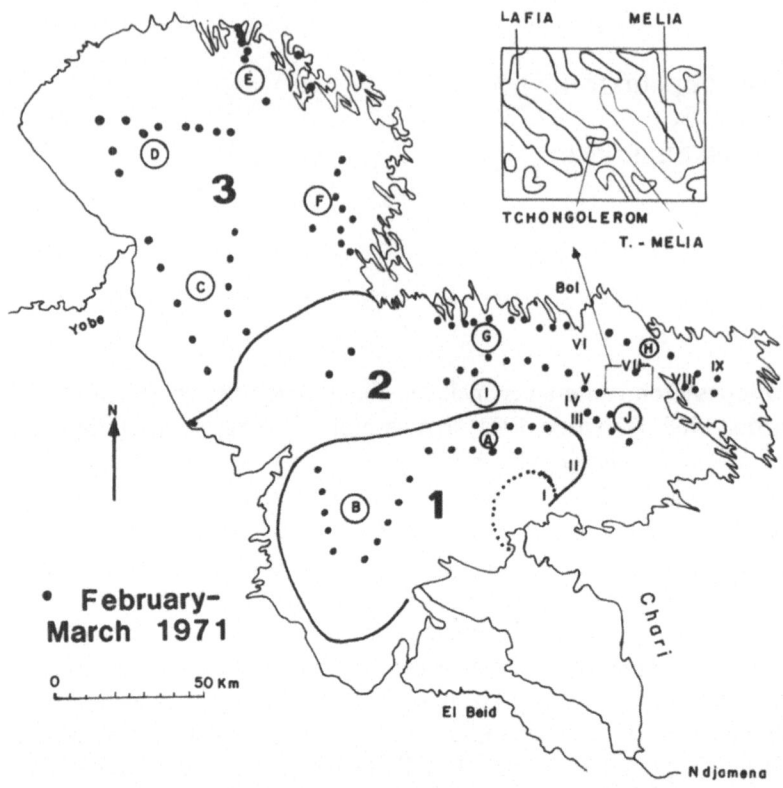

Fig. 1 The major ecological zones and stations sampled during the period of 'Normal Chad'. The Roman numerals designate the stations sampled over the course of the annual cycle of 1964–65; the dots and the letters correspond to the survey of February 1971; the stations sampled from 1968 to 1973 are framed.

the lake and collected at a given time of the year (February 1971); observations much more restricted in space and time (one station visited occasionally) (Fig. 1). We also have some information collected during 1967–68 by Robinson and Robinson (1971) on the annual cycle for the whole of the north basin.

All samples except those of Robinson and Robinson were taken by vertical tows from the bottom to the surface using one or two truncated nets of 1.5 m length, 30 cm opening diameter, and 60 μm mesh size. Robinson used oblique tows with a net of 60 or 200 μm mesh size and 30.5 cm diameter opening. The counts were made on subsamples taken with a calibrated syringe, representing 1 to 10% of the sample according to the abundance of the organisms. The individual weights by species or stage were used for biomass calculations and are shown in Table 1 as well as in Table 6 of Chapter 12.

200

7.1 The zooplankton of 'Normal Chad' (1964–71)

This period was characterized by a continuous mass of water, with a community that was relatively uniform in composition and structure. Excluding the Rotifer *Brachionus plicatilis*, which is well known in water of high salinity (Pourriot et al. 1967), and was only present in the regions of high conductivity at the extreme north of the lake, no variations appeared in the species composition in relation to water salinity. Nearly all the species recorded were present in the different regions sampled. Although very diverse the community was dominated by Calanoids, which represented about 46% of the biomass during 1964–65 in the eastern archipelago of the southern basin (37 and 17% for the Cladocera and Cyclopoids). There were, however, some regional and seasonal variations in biomass that appeared to be related to the hydrology of the Shari, or to other factors.

The Rotifers represented a negligible part of the biomass, particularly during the period of high waters, and will not be considered here.

7.1.1 *Regional and seasonal variations in abundance: the major ecological zones of the lake*

From the data of 1964–65 (Gras et al. 1967) and February 1971 (Carmouze et al. 1972) the lake can be divided into three zones defined in Fig. 1, differentiated on the basis of the frequency distributions of biomass observed in the February 1971 survey (Fig. 2).

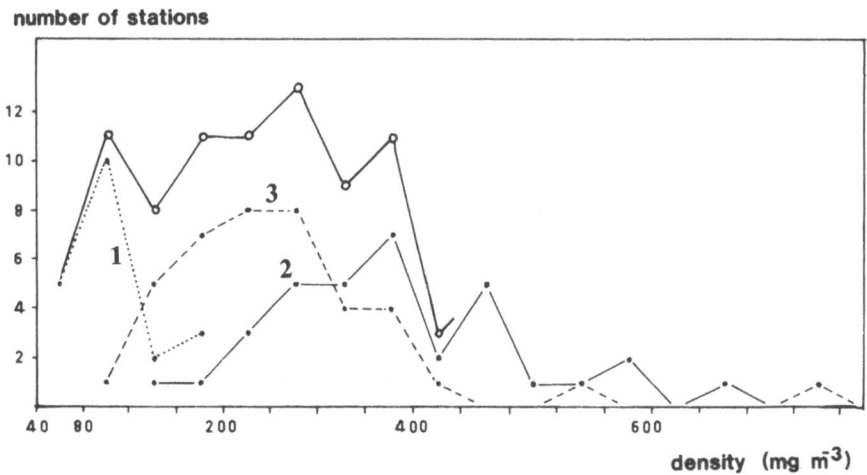

Fig. 2 Frequency distribution of biomass in samples taken in February 1971 in the north basin (zone 3, – – – –), the archipelago of the south basin (zone 2, ———), and the open waters of the south basin (zone 1, ...); total of 3 zones: O———O.

7.1.1.1 *The open water of the southern basin (zone 1)*. The open water was characterized by low densities for a long period, approximately June to February, and by a strong but brief increase during April (Fig. 3b, station II). During 1964–65 the average biomass between June and February was thus about 120 mg m^{-3} (96 individuals per liter), and the ratio between the annual maximum and minimum was 6.3 (biomass) and 10.3 (numbers). The values in 1971 (average 93 mg m^{-3} for 20 stations covering the whole zone) and from October–November 1969 (35 mg m^{-3}), confirm the low density that characterized this zone for most of the year. In the regions close to the delta represented by station I during 1964–65 (Fig. 3a), and sampled again during the flood of 1969, the minima were almost zero during October–November. These regions formed a 'perideltaic zone' which can be included in zone 1.

The comparison of the numbers of copepodites and adults of Cyclopoids and Calanoids, in stations O, I and II during April 1965 (maximum abundance) showed the differentiation of a river and deltaïc plankton (station 0) dominated by Cyclopoids, from plankton of stations I and II where the Calanoids were abundant. The ratio Cyclopoids/Calanoids was almost 10 at station 0, 0.7 at station I and 1.2 at station II. It was 1.2 in the reed islands and the archipelago (stations III to IX) during April 1965, with an annual average of 2. This difference suggests that the development of lake plankton in the peri-deltaïc region was independent of river inflows. These supplies were moreover very low at the time of the planktonic explosion observed in the region at the beginning of the year (Fig. 3a, station I).

7.1.1.2 *The archipelago and the reed islands of the southern basin (zone 2)*. The biomasses here were clearly much higher than those in zone 1, and the annual variations were low. In the archipelago the annual average in 1964–65 (station V to IX) was 333 mg m^{-3} (318 ind. l^{-1}) and the maximum/minimum ratio was 2.2 (biomass) and 2.0 (numbers). Two minima were seen, one corresponding to the low water of the lake and the rainy season (June–September) and the other to the cold season (December–March), as well as two maxima, during April–May and during October–November. The few complementary data reported in Fig. 3 confirmed the higher level of biomass in this region of the lake. The archipelago region also appeared to differ from zone 1 in having a greater abundance of the genus *Daphnia*. An average of one individual per liter was found in station I and II compared with an average of 10 individuals in stations V to IX, which corresponded to the density of the genera *Moina* and *Diaphanosoma*, whose distribution in the lake was more uniform. The average density of *Daphnia* in the reed island region (stations III and IV) was intermediate: 3 l^{-1}.

The transitory character of the reed islands (Chap. 2) was reflected at the level of community composition and abundance as shown above for the genus *Daphnia*. So, in 1971 (Table 1), some biomasses were much higher in the reed

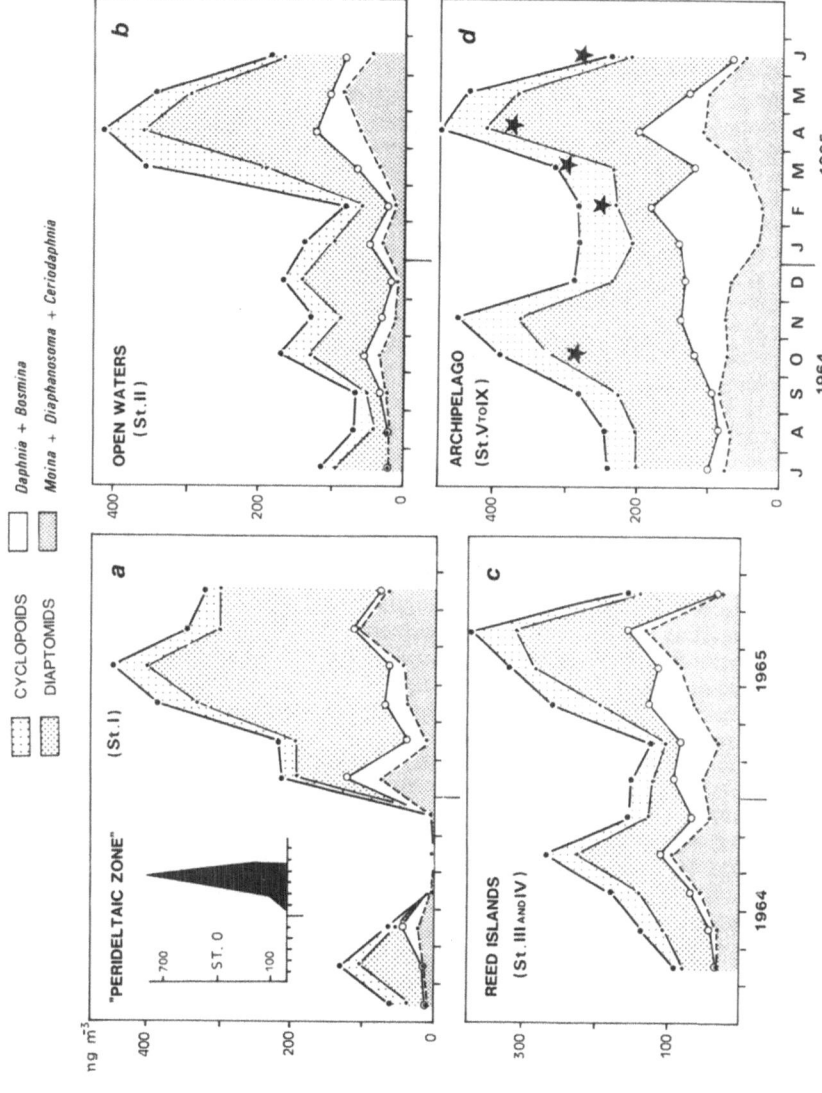

Fig. 3 Annual variations of principal group cumulative densities 1964–65 in the peri-deltaic and deltaic regions (a); in the open waters (b); in the reed islands (c); and in the archipelago (d) of the south basin; * other data.

203

Table 1 Density in mg dry weight m⁻³ and in number of individuals l⁻¹ in different regions of the lake during February 1971 (14/02/71 to 4/03/71). N, C, A, E: Nauplii, Copepodites, Adults, Eggs. The letters A to K refer to Fig. 1. The specific weight (µg) given in brackets have provided the biomass.

Species	Zone 1 Series		Zone 2 Series					Zone 3 Series			
	A	B	G	H	I	J	K	C	D	E	F
The Diaphanosoma (2.5 µg)	9.0	5.0	20.0	17.5	75.0	68.8	31.5	8.0	0.3	0.8	6.8
Moina (2.3)	12.0	3.7	20.0	3.2	49.0	47.2	24.4	6.4	1.8	0.5	3.9
Ceriodaphnia (1.0)	0.7	0.1	0	0	7.7	19.9	0	0	0	4.0	0
Daphnia (3.0)	6.0	1.5	20.7	0.6	21.6	16.8	137.4	13.8	74.1	59.1	71.1
Bosmina (0.75)	5.6	3.7	58.0	104.9	51.2	21.5	73.0	46.3	24.8	17.6	62.6
N. Cyclopoids (0.05)	3.5	3.8	6.9	8.0	11.0	7.9	5.3	4.9	4.5	5.3	4.4
C. Cyclopoids (0.55)	16.6	8.4	13.7	24.6	29.5	26.3	23.2	16.4	22.1	21.8	17.4
A. Th. neglectus (2.15)	4.9	11.6	19.6	34.2	37.6	19.1	21.3	10.3	9.5	11.4	9.2
A. Th. incisus (3.5)	1.1	0.7	4.9	4.6	3.2	1.8	6.3	3.2	4.2	0.7	4.6
A. Mesocyclops (5.0)	1.0	3.0	2.5	5.0	2.5	0.5	9.0	11.0	7.5	9.5	4.5
N. Tropodiaptomus (0.4)	1.4	0.7	3.2	5.1	2.9	1.5	4.5	1.0	0.9	3.5	1.4
C. Tropodiaptomus (3.8)	16.7	25.8	15.6	17.5	27.4	37.2	17.9	22.4	9.1	41.4	6.8
A. Tropodisptomus (11.0)	14.2	24.2	48.4	47.3	45.1	55.0	40.7	45.1	30.8	90.2	20.9
E. Cyclopoids (1.5% A. biom.)	0.1	0.2	0.4	0.7	0.6	0.3	0.5	0.4	0.3	0.3	0.3
E. Calanoids (1.5% A. biom.)	0.2	0.4	0.8	0.7	0.7	0.9	0.6	0.7	0.5	1.4	0.3
Cladocera											
weights	33.3	14.0	118.7	126.2	204.5	174.2	266.3	74.5	101.0	82.0	144.4
numbers	19.0	9.1	100.9	148.5	134.5	102.2	166.3	72.3	58.6	47.6	111.5

Cyclopoids											
weights	27.2	27.7	48.0	77.1	84.4	55.9	65.6	46.2	48.1	49.0	40.4
numbers	103.2	97.4	174.6	223.4	292.0	215.8	161.9	136.0	136.4	152.5	126.4
Calanoids											
weights	32.6	51.1	68.0	70.6	76.1	74.6	63.7	69.2	41.3	136.5	29.4
numbers	9.1	10.7	16.5	21.6	18.5	18.5	19.6	12.6	7.5	27.9	7.2
Total											
weights	93.1	92.8	234.7	273.9	365.0	304.7	395.6	189.9	190.4	267.5	214.2
numbers	131.3	117.3	292.0	393.5	445.0	336.5	347.8	220.9	202.5	228.0	245.1

islands (series I and J) than in the actual archipelago (series G and H, Fig. 1), and the opposite was seen during 1964–65, when the average biomass of stations III and IV (190 mg m^{-3}) was clearly lower than that of the archipelago, and in the same range as that of the open water (180 mg m^{-3}). Seasonal variations were identical in the two regions. This last fact shows a decrease in the influence of the hydrological regime of the Shari, linking the reed islands to the archipelago. The two form an heterogeneous complex from the point of view of densities, and this is clearly shown in the frequency histogram of biomasses (Fig. 2, zone 2).

Seasonal variations mainly affected the Cladocera and Calanoids. In the first, the genera *Bosmina* and *Daphnia* were dominant during the cold season, and the genera *Moîna*, *Diaphanosoma* and *Ceriodaphnia* were dominant for the rest of the year. As a whole, the Cladocera had their lowest density between June and September (less than 100 mg m^{-3} or 63 ind. l^{-1} on average), and the highest densities between October and May (114 to 199 mg m^{-3}, 102 ind. l^{-1}). The Calanoids were less abundant during the cold season (84 and 61 mg m^{-3} during January and February, 9 ind. l^{-1} on average without the nauplii), and showed two very high maxima, during October–November (220 and 224 mg m^{-3}, 34 ind. l^{-1}) and during April–May (238 and 273 mg m^{-3}, 35 ind. l^{-1}). This group was therefore responsible for most of the annual variation in zooplankton abundance.

The Cyclopoids as a whole only showed irregular small-scale variations. However, the *Mesocyclops* adults were less abundant during the cold season than during the warm season, and the *Thermocyclops* adults behaved in the opposite way (Fig. 4).

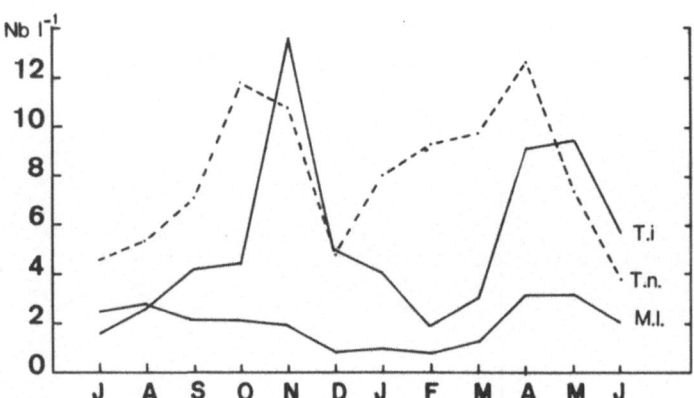

Fig. 4 Annual variations in the number of adults (Nb l^{-1}) of *Tropodiaptomus incognitus* (T.i) of *Thermocyclops neglectus* (T.n.) and *Mesocyclops leuckarti* (M.l.) in 1964–65 in the southern archipelago.

206

The winter plankton of the southern Archipelago (December to March), contained relatively few Calanoids (32% of the biomass) and more Cladocera (49%), two thirds of them belonging to the genera *Daphnia* and *Bosmina*. The corresponding annual averages were 46% for the Calanoids and 37% for the Cladocera, of which 46% *Daphnia* and *Bosmina*.

The preceding seasonal variations were confirmed by the winter nature of the community in February 1971 (Table 1), which contained 27% Calanoids, and 48% Cladocera (75% *Daphnia* and *Bosmina*), in the actual archipelago (series G and H). Similar variations were also observed in the reed islands. Finally, a similar development of the Calanoids was observed in the open water (station II), with a marked increase during April–May and a smaller increase during October–November, which may have been limited by the arrival of the Shari flood waters. It is necessary, however, to point out that the April 1968 population at Melia (Table 2) did not conform to the seasonal model, since among the Cladocera, the genera *Daphnia* and *Bosmina* were dominant, forming 75% of this group. This anomaly could be explained by the observed temperatures being lower than in 1965: 28°C during April 1965 and 24.5°C on average during April 1968.

7.1.1.3 *The northern basin (zone 3)*. Only the data of February 1971 allows a strict comparison of the zooplankton populations from the northern and southern regions of the lake. Based on these data, the northern basin was relatively homogeneous in density (Fig. 2 and Table 1). The average density (215 mg m^{-3}) was significantly lower than in the southern archipelago (254 mg m^{-3} for the series G and H) and of course in the entire zone 2 (315 mg m^{-3} for the series G to K). Except for the dominance of the genera *Daphnia* and *Bosmina* during the cold season, the observations by Robinson and Robinson (1971) for one annual cycle during 1967–68, showed that there was no seasonal variation in community abundance as in zone 2. Any such variation would not affect the Calanoids, but the observations of February 1971 were contradictory in this respect. The zooplankton then contained the winter population seen in zone 2, with 32% Calanoids and 47% Cladocera (92% *Daphnia* and *Bosmina*).

The densities observed by Robinson and Robinson, especially in April 1968, were 3 or 4 times lower (if considering numbers and biomass), than the values observed in the archipelago of the southern basin, the only region of this basin which could have been compared with the north*. Such a difference can only be explained by admitting that the sampling techniques used in the north (oblique tows of a small size net) caused net-avoidance and so underestimate zooplankton abundance.

The observations from 1967–68 thus confirmed the low level of density in the

---

* A misinterpretation of Robinson's data has led to a much higher annual biomass for the northern basin in an earlier work (Carmouze et al., 1972).

Table 2 Changes in zooplankton abundance (mg d.w. m^{-3}) between 1964 and 1975 in various stations of the archipelago of the south basin (cf. Fig. 1 and Table 1). 1971: mean of series G and H; 1968: estimated density of Cyclopoid (italics); 1972 and 1973: means of Tchongolerom and Tchongolerom-Melia stations (except May and August 1973); in 1972, the values in italics are the means of May and June series, and the others of 5 series sampled in May, July, August and October. N, C, A designate respectively the nauplii, the copepodites and the adults; NE not evaluated.

	1964–65 South Basin	1968 Mélia	1971 Ser. G and H	1972 Tchong. + T-Mélia	1973 Janu. Tchong. + T-Mélia	March Tchong. + T-Mélia	May Lafia	August Bol	1974–75 Bol
Daphnia	27.8	9.7	10.7	0.9	0	0	0	0	0
Ceriodaphnia	23.1	0	0	2.8	0	0.3	0	0	0.2
Moina	16.5	22.8	11.6	19.7	18.6	24.1	28.3	0.1	12.9
Diaphanosoma	27.5	18.6	18.8	28.0	3.9	6.7	29.0	0.1	10.9
Bosmina	29.6	90.0	81.5	1.9	3.5	0	0	0	0
N. Cyclopoids	7.5	8.2	7.5	6.2	5.4	29.6	58.2	N.E.	N.E.
C. Cyclopoids	22.4	16.3	19.2	14.2	21.4	43.4	31.4	0.5	11.1
A. *Th. neglectus*	14.5	24.5	26.9	10.8	42.5	124.3	81.7	3.1	21.0
A. *Th. incisus*	10.5	8.9	3.8	6.1	1.5	11.0	12.6	ε	0.2
A. *M. leuckarti*	1.5	7.3	4.8	1.6	1.6	ε	0	0	0
N. Calanoids	3.1	15.7	4.2	3.7	0.5	1.9	1.0	0	0
C. Calanoids	62.9	51.7	16.6	24.3	7.3	9.2	1.2	0	0
A. *Tropodiaptomus*	70.5	113.8	48.6	51.4	12.1	ε	0	0	0
A. *Thermodiaptomus*	15.4	4.3	ε	57.4	87.5	74.0	36.3	0	0
Cladocera	124.5	141.2	122.5	71.9	26.1	30.9	57.3	0.2	24.1
Cyclopoids	56.4	65.4	62.6	45.4	72.4	208.3	183.9	3.6	32.3
Calanoids	151.9	185.5	69.3	145.8	107.5	85.1	38.5	0	0
Total	332.8	392.0	254.4	263.1	205.9	324.3	279.7	3.9	56.4

northern basin. It is probable that the difference between zones 2 and 3 was well represented by the ratio (density in zone 2)/(density in zone 3) observed in February 1971: 1.5 for biomass and 1.6 for individuals.

7.1.2   Variation factors

The data presented here show that the densities were relatively homogeneous over the entire lake in April–May. These months corresponded to raised temperatures, lowering of the Shari, an average water level (by comparison to yearly variations) and the period of wind reversal from the harmattan of the NNE to the monsoons of the SSW. In addition to this short period, regional and seasonal variations already described were established, of which the most important occurred in the middle of the southern basin of the lake. Among the primary factors which appeared to regulate these variations were turbidity, the Shari flood, water depth, and the annual thermal gradient. Other factors such as the spatio-temporal variations affecting the zooplanctophagous fish species or the seston doubtless modified them, but the available data did not permit analysis of these probable interferences.

7.1.2.1   *The turbidity of the water.* The open water of the southern basin was differentiated from the rest of the lake mainly by the persistence of very low densities during most of the year. They were about 3 times lower than those of zone 2. The decrease in biomass in zone 1, represented by station II, probably commenced in June with the co-occurrence of a seasonal lowering of lake level and the presence of monsoon winds blowing onto the region at this time of year. They caused a major re-suspension of sediment, rich in mineral particles and increased turbidity, accentuating the decreasing transparency which started in January at this station (Fig. 5). Decrease of the phytoplankton followed after a slight delay so that the zooplankton decrease during June was difficult to explain with the available data, although it was highly likely that the sediment re-suspension was the main factor.

This mechanism persisted until about August–September and the low abundance of zooplankton at this station was thus maintained by the activity of the Shari flood and the lowered winter temperature.

7.1.2.2   *The flood.* In the delta and in the open water, the flood occurred on average from August to December, with a maximum in October–November. The water, muddy at first, was extremely poor in plankton. The massive arrival of this water at first pushed back and then diluted the lake water, resulting in a displacement in time and space of the minimum zooplankton abundance. The results of sampling along several transects starting from the delta during the 1969 maximum flood, give an illustration of this phenomenon (Fig. 6). Such mechanical effect of the floods, frequent in estuarine zones, is particularly

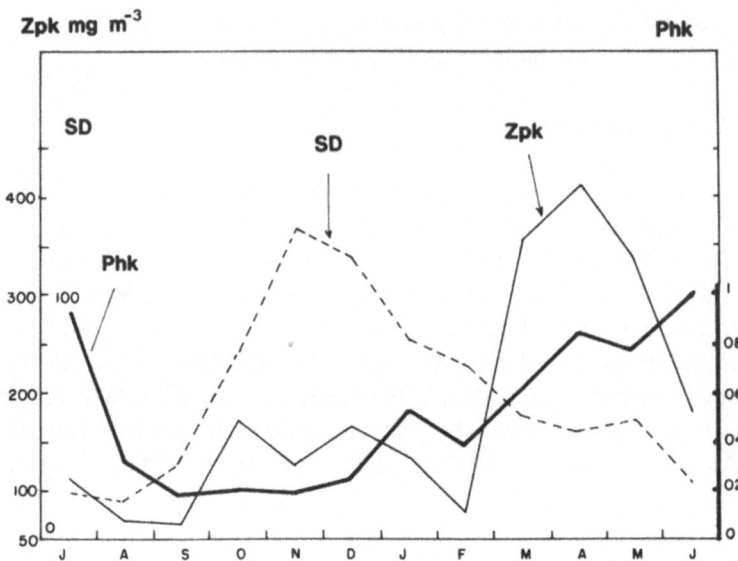

Fig. 5 Variation of Secchi disc transparency (SD, in cm), zooplankton densities (Zpk, in mg m^{-3}) and phytoplankton (Phk) expressed by ratio to a value 1 in June (after Gras et al. 1967) in the southern open waters (st. II) from July 1964 to June 1965.

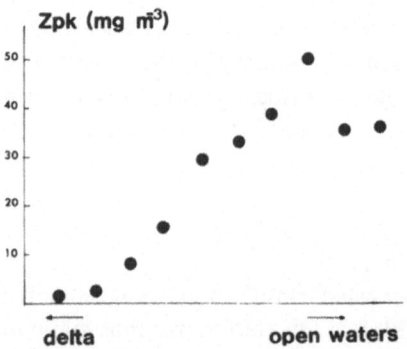

Fig. 6 Variation of density from the delta to the open water during the flood of 1969 (from 11 to 17 November).

comparable with that pointed out by Brandorff and Andrade (1978) for a lake in the Amazonian system. In Lake Chad, it probably had a fairly reduced importance spatially and temporally. As already seen, among the open water and the river system, the flood caused a differentiation of a perideltaïc zone in which the population densities reached zero values (station I, Fig. 3a).

The influence of the flood on the population of station II during October–

November is very problematic. In this station, the minimum densities did not occur during the passage of the flood, but during low water and at the beginning of the flood (June–September). The months of October and November were characterized by increased transparencies (reliable supplies from the Shari floods, increasing lake level, NNE winds) and low algal densities (Fig. 5). The dilution of the environment by the flood waters and low algal densities no doubt limited Calanoids and other zooplankton development. This limitation could have been due to very intense grazing as the ratio between the zooplankton stock and phytoplankton was then higher than at any other time of the year at this station, or throughout the year in the adjacent archipelago.

There was practically no effect of the flood in zone 2 with the maximum abundance occurring in October–November. The algal densities during the winter period, at the maximum lake level, were also relatively high, at least in the actual archipelago. The lesser influence of the Shari in zone 2, allowed the biological factors to be more important conferring to this zone the character of a closed lake.

7.1.2.3 *The depth.* Water depth was one of the most important differences between zones 2 and 3, and one can attempt to relate the lower densities of zooplankton observed in the north, to the increased depth which characterized this zone (about 2 m greater than the south basin). A possible explanation is that, at any one time, a quantitative ratio exists between the primary production of a homogeneous biotope (zone) and its zooplankton standing crop, the former operating in the surface layers whose thickness does not vary in proportion with the depth, so that an increase in depth corresponds to a decrease in the density of zooplankton under a given surface area. The reality is of course not so straight-forward, and the situations encountered are without doubt very diverse. Morever, it is difficult to know if such an explanation is acceptable for a productive shallow lake, which receives significant allochthonous supplies. Be that as it may, the observation of Brandl (1973) and other authors reported by him, tend to support this supposition. In Lake Lipno (6 m average depth), Brandl gives a relationship of the form $d = az^{-b}$ between the average biomass per m^3, observed under $1\ m^2$ of surface for the whole water column, and the depth z in meters at this site, a and b being two constants ($b \simeq 0.26$). This hypothesis is also supported by the ratio between the phytoplankton (in $mg\ m^{-3}$ fresh weight) (Iltis 1977) and zooplankton (in mg m^{-3} dry weight) in zones 2 and 3, being respectively 1400/315 (4.4) and 1059/216 (4.9) and thus close. On the other hand, the density found in series I and J in February, 1971 (335 $mg\ m^{-3}$ for an average depth of 1.60 m) was clearly higher than that of the much deeper series G and H (254 $mg\ m^{-3}$ for 2.60 m).

The preceding explanation would thus suggest two types of relationship, between the depth of a region and the quantity of plankton present. Below a certain depth threshold and in regions very exposed to wind action, this factor would have a negative effect, analogous to the biomass decrease recorded at station II at low waters. In the most general view, or above the threshold, the densities would tend to increase when the depth decreases. This criterion can be particularly important in shallow lakes.

7.1.2.4 *The temperature*. It is probable that the variations of the Cladocera (dominance) of the genera *Daphnia* and *Bosmina* in winter and the genera *Moina*, *Diaphanosoma* and *Ceriodaphnia* the rest of the year), were related to the strong annual variation in lake water temperature (12°C). Experimental data on development rates (Gras and Saint-Jean 1976b, 1978), suggest that the first two genera are less adapted to increased temperature than the other three. *Daphnia* and *Bosmina* thus have slow development rates, relatively low optimal embryonic development temperatures (22°C for *Daphnia barbata* and 21°C for *Bosmina*) and an optimal zone (the above optimal temperature ±6°C) which does not include temperatures of 28–30°C which existed in the lake during a major part of the year (from April to October in 1964–65). The three other genera, on the contrary, have rapid development rates, high optimal temperatures (24.6 25.3, 23.7°C, respectively) and an optimal zone clearly falling within the range of temperature found in the lake (18–30°C). Moreover, the duration of embryonic development of *Daphnia* and *Bosmina* is lower than that of the three other genera at lower temperatures (< 22°C), the situation being reversed above this threshold. The variation in abundance of these two groups of Cladocera could thus be the result of a passive competition, in which the development characteristics in relation to temperature play the major role.

No interpretation of the seasonal variation in abundance of the Calanoids in zone 2 is satisfactory. Temperature cannot be the cause as this does not explain the disappearance of this group during the rainy season. *Tropodiaptomus* has a high optimum temperature (24.1°C) and it appears, from other characteristics of the relationship of the duration of embryonic development to temperature, to be one of the species least sensitive to thermal factors. A general explanation can be put forward, according to which *Tropodiaptomus* is a pelagic form, thus well adapted to the north basin, deeper and not affected by the flood. In the south basin, the low temperatures occurring during the cold season and/or the low level of the lake at the period of low water, accompanied by a slight increase in turbidity could have provoked, by unknown factors, the reduction of this group during the cold season and between June and September. This would appear to be more plausible for this species, which progressively disappeared in 1973 with the lowering of the water level. Other intervening factors combining with each other or with the preceding factor may have been: variation in algal biomass, variation in the production rate and in predation pressure (Fig. 7). The decrease in phytoplankton biomass, production rate (or birth rate), and the high zooplanktophagous fish stock, would thus contribute to

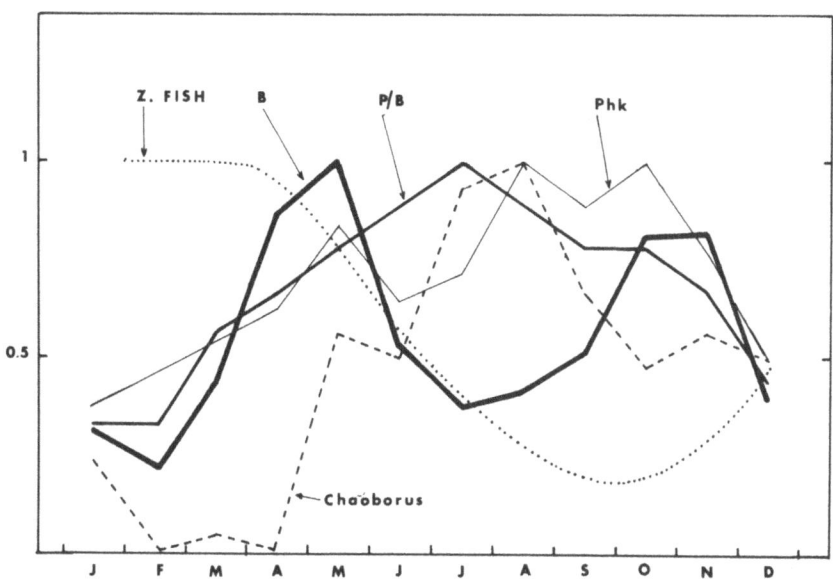

Fig. 7 Seasonal variations in the abundance of zooplanktophagous fish (z. fish), of *Chaoborus* larvae, of phytoplankton (Phk), of production rate (P/B) and biomass (B) of Diaptomids over an hypothetical annual cycle in the southern archipelago during the period of 'Normal Chad'. Every monthly value has been divided by the corresponding annual maximum.

the decrease in Calanoids in the cold season, but the reasons for the decrease between June and September are not clear.

More generally we must emphasize that a decrease in abundance occurred approximately between June and September throughout the south basin, and affected all the species, with the possible exception of *C. cornuta* (Fig. 3). At station II (zone 1) we saw that it was probably started by an increase in turbidity, due to intensive resuspension of bottom sediments by the waves, then maintained by some other factors in particular the reduction in algal density. This explanation would not appear to be valid in zone 2, because there was no decrease in transparency or algal density (Fig. 8), and no increase in turbidity was noticeable at this time. The only possible factor appeared at first to be an increase in the numbers of *Chaoborus* lsrvae (Fig. 7). However it would seem unlikely that this increase was sufficient to cause a considerable and rapid decrease in zooplankton, so that this major aspect of the seasonal variations in the zooplankton populations observed over all the south basin, has no satisfactory explanation.

In conclusion, no single factor appeared to govern the level of zoopiankton abundance in Lake Chad or to regulate its regional or seasonal variations, which remained low. Among the factors cited, the temperature, which was

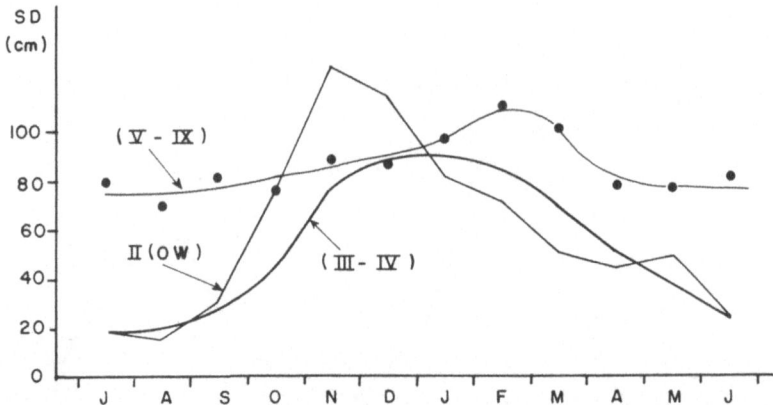

Fig. 8 Secchi disc transparency at station II (open waters), III–IV (reed islands) and V–IX (archipelago) in 1964–65 in the south basin (after Gras et al. 1967).

identical in all the biotopes considered, was only responsible for small quantitative variations affecting the Cladocera. On the contrary the most important factors seemed to be the Shari flood in the south basin and the greater depth in the north basin, as they were directly responsible for the lowest densities observed in the two regions.

7.2 Community changes during the establishment of 'Lesser Chad' (1972–1975)

7.2.1 *Structure and composition of the zooplankton community*

In spite of a progressive lowering of water level, it is reasonable to state that the zooplankton population changed little until 1971, or even 1972. After this period profound modifications occurred, that were well studied in the archipelago of the south basin.

7.2.1.1 *Changes in the populations in the archipelago of the southern basin.* Until 1972, the lake could be considered to be in a state of 'Normal Chad' from the point of view of water area, and there was a communication between the water masses inside the archipelago and the open water. After December, the water area decreased rapidly, with the emergence during April–May 1973 of all the regions of the reed islands. The open waters of the archipelago were isolated and then progressively reduced to several ponds fairly large and deep, most of which dried up before a partial recovery of the water occurred in October. The changes in the community were successively followed at 3 stations in the archipelago, Tchongolérom, Lafia and Bol (Fig. 1), but not until a final drying up stage.

214

Because of the agreement in the structures observed at these stations (Table 2) it is possible to describe the changes in comparison with 'Normal Chad.' The populations of 1972 (May to October) differed essentially in: (a) the low densities of *Bosmina*; (b) the return of large numbers of *Thermodiaptomus*; (c) the presence of a considerable number of littoral Cladocera (genera *Alona* and *(Macrothrix)*. In 1973 changes were rapid and one observed: (a) the quasi disappearance of Cladocera other than *Moina* and *Diaphanosoma*; (b) the disappearance of *Tropodiaptomus* and its replacement by *Thermodiaptomus* which also decreased during April–May; (c) the development of cyclopoids, particularly *Th. neglectus* and the disappearance of *Th. incisus circusi*, not very abundant during the period of 'Normal Chad'.

In addition to the modifications in species composition, a change in the demographic structure of all the populations was observed characterized by an increase in the proportion of adults, by weight. Littoral forms remained numerous and a great increase in Rotifer numbers was observed, evaluated only in three samples. Thus, 3836 rotifers per liter were counted, and 467 microcrustacea per liter, of which 82% were nauplii, on 9/4/73, at Tchongoleron. Similar values were obtained on 1st/2nd. May at Lafia, with rotifer counts of $2974\,l^{-1}$ and $2436\,l^{-1}$, and microcrustacean counts of $1151\,l^{-1}$ and $1216\,l^{-1}$ (92% and 83% nauplii). Rotifer numbers in May 1965 were only $47\,l^{-1}$ in the archipelago, rising to $111\,l^{-1}$ in February 1971. This change, which eliminated or masked the seasonal variations of 'Normal Chad', occurred, without major changes in total population density until May 1973, even July, when the qualitative observations made at Bol-Berim in a study on embryonic and juvenile development showed that, at that time, the population again had the same characteristics as in May. The considerable fall in biomass during August may have been caused by similar reasons to those that caused sporadic mortality of fishes at much the same time in the same region (Chapter 10).

After the return of water in October, the community showed some characteristics in agreement with the development shown above, with essentially, 3 species, *Moina*, *Diaphanosoma* and *Th. neglectus* (Table 3). Only *Thermodiaptomus* did not reappear and the zooplankton did not include any Calanoids. Three other important features were noted: (a) the densities were very low, with 56.4 mg m^{-3} on average for the non-zero values of 1974–75 and an absolute maximum of 142 mg m^{-3}; (b) *Mesocyclops*, a carnivorous species in 'Normal Chad', disappeared after the low water period of 1974; (c) the abundance of littoral forms was not as great as might have been expected with the considerable development of vegetation.

It is interesting to note that the two Cladoceran species subsisting at the end of this period were, at their maximum density, with some variation, similar to their densities in the high water period: 26.7 mg in February and April 1973; 38.3 mg in the same months in 1965 and 30.4 mg in February 1971. The void left by the Cladocera was thus not filled by species of the same group, but rather

Table 3 Densities of planktonic crustacea (mg dry weight m^{-3}) and some littoral forms of Cladocera (N m^{-3}) at Bol-Berim in 1973, 1974 and 1975. In total, for all the samples taken, representing 12.497 m^3 filtered water, one C_5 and one female of *Thermodiaptomus* were observed.

Dates	Crustacea							Cladocera (total %)	Littoral forms (N m^{-3})
	Moina	*Diaphanosoma*	*Ceriodaphnia*	*Bosmina*	*Thermocyclops neglectus*	*Mesocyclops leuckarti*	Total		
12/08/73	0.1	0.1	0	0	3.6	ε	3.9		107
7/11	0.4	ε	0	ε	0.1	ε	0.7		1253
12/12	0.1	0.4	ε	0	0.8	0.2	1.5		699
17/01/74	0.1	ε	0.3	ε	10.9	0.3	11.7	4.1	1605
2/03	0.4	0.1	0.2	0	7.0	0.2	7.9	9.7	81
18/04	41.1	0.1	0	0	26.9	3.0	71.2	57.9	137
16/05	17.9	18.6	0	0	38.9	ε	75.4	48.4	136
22/07	5.9	15.5	0	0	45.6	0	67.0	32.0	2877
15/09	8.2	6.5	0	0	18.5	0	33.2	44.1	491
21/10	0	0	0	0	0	0	0		0
11/12	ε	0	0	0	0	0	ε		0
23/05/75	16.4	36.0	0.2	0	89.8	0	142.4	36.9	0
2/09	13.5	10.7	1.1	0	17.2	ε	42.4	59.5	310
1/12	ε	ε	0	0	ε	0	ε		0

by *Th. neglectus*. The only change which occurred after October 1974 was the disappearance of *Mesocyclops* which was presumed to have kept to the carnivorous diet shown previously. Its disappearance was probably caused by the five-fold decrease in the microcrustaceans, especially Cladocerans, upon which it normally fed, particularly if one takes into consideration the almost certain frequent occurrence of cannibalism by the adults and later carnivorous stages upon the earlier stages.

The profile of variations in abundance after October 1973 (Fig. 9) can be compared with that of the peri-deltaic zone (station I, Fig. 3a), which had some zero values. It differed by having a very late and much slower increase in biomass, which, as already indicated, remained very low. The water entering the archipelago after October were filtered by the abundant vegetation which had become established between the open water and the archipelago; these waters lacked zooplankton, were very poor in oxygen, and contained dissolved CO_2 (Benech et al. 1975). These unfavourable conditions continued until almost February, and undoubtedly were a partial explanation of the slow development of the zooplankton in the cold season. Events occurred as if the development of phytoplankton and zooplankton were independent of one another (Fig. 9).

It will be noted that *Moina* and *Diaphanosoma* developed less rapidly than *Th. neglectus*, whose development cycle must nevertheless be much longer. It has to be compared to the fact that in the period of 'Normal Chad' the two Cladodera were summer forms whereas *Th. neglectus* was, among the Cyclopoids, relatively well developed in the cold season.

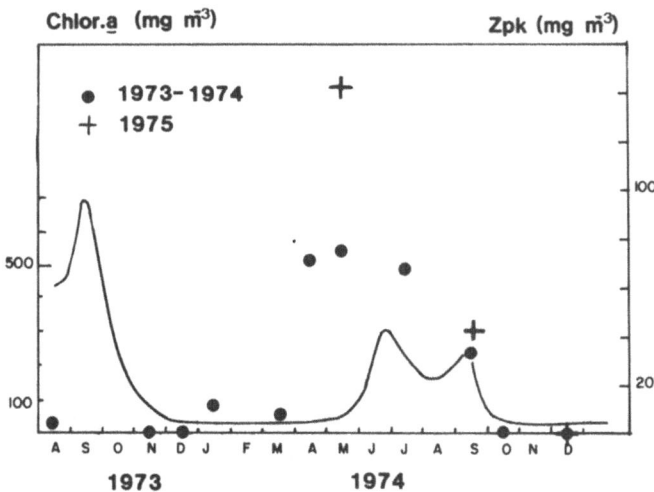

Fig. 9 Variation in chlorophyll concentration (plain line) and in zooplankton density (circles and crosses) at Bol-Berim in 1973–74.

None of the following factors — a change in salinity or in water composition (which were not so different in 1974–75 from those of 'Normal Chad'), insufficient food, or an over-abundance of zooplanktivorous species — appear to be the cause of low densities in 1974–75. The chlorophyll concentrations were indeed identical or even higher than those of 'Normal Chad', and the zooplanktivorous *Brachysynodontis batensoda*, still abundant, tended to change from a strict zooplanktonic diet in the high water period to a diet of detritus. Moreover, there was a growth check in 1973 and in 1974, the year during which the fish disappeared (Benech 1975; Im 1977). However a change in phytoplankton composition was noted. Euglenoids becoming dominant. These changes may have been important, but the most probable cause of low density levels was a permanent deficit of oxygen. Although oxygen tensions of 75–100% saturation were still noted in the whole water column during 'Normal Chad', they were much lower in 1974–75, being 5 to 60% saturation at the surface, with intermittent creation of almost anoxic conditions in the deeper layers of the entire water column (Chapter 10). This situation could have caused raised natural mortality rates, slowing down of reproduction, accumulation of organisms in the upper layers, or the three phenomena together.

7.2.1.2 *Changes in the other regions. The open water of the south basin.* Available data from the station nearer to the delta than station II in 1964–65 (Table 4) showed that as no profound changes in environmental conditions occurred in this region, the populations retained the same characteristics as those of 'Normal Chad'. This made it difficult to interpret the differences observed in 1973–74. One can however consider the densities in 1973–74 to be comparable with those obtained for the corresponding months in 1964–65, 1969 and 1971, although systematically lower. The zooplankton was moreover fairly well diversified with seven relatively abundant species, including *Tropodiaptomus incognitus*.

The north basin. A single series of data is available for this region from October 1973 a time, when the changes had ended in the south. These data refer to the east central part of the north basin in which 14 stations were sampled. They were situated in the three main types of natural zones already distinguished: open water, reed islands, and archipelago. The community was almost as diverse as during 'Normal Chad', with nine species, including *Tropodiaptomus*. Although the scarcity of *Bosmina longirostris* must be noted, it can't be attributed with some precise significance to the quasi disappearance of this species. The average population density (483 mg m^{-3}) was higher compared with the data of February 1971 (about 200 mg m^{-3}, cf. Table 1), but this corresponded to the seasonal maxima noted in zone 3 during 'Normal Chad' (514 mg m^{-3} in April 1965 and 440 mg m^{-3} in November 1964). The coefficient of variation of the mean population density were also greater than in 1971: 57% compared with 43% in 1971.

Table 4 Densities (in mg dry weight per m³) observed in the center part, east of the north basin, and in the open water of the south after the drying of the southern archipelago, Copepod nauplii excluded.

Zones	Dates	Z (m)	Densities (mg m⁻³) M.m.	D.e.	C.c.	D.b.	B.l.	T.n.	M.l.	T.ic.	T.i.	Total
Open water	18–25/	1.40	28.7	56.2	0.3	120.8	4.0	4.7	47.5	4.7	325.0	591.9
		1.44	62.9	102.5	0.9	92.5	2.7	6.2	42.0	11.4	187.8	508.9
		2.00	5.5	10.5	0.1	35.8	1.1	1.0	4.2	2.3	53.3	113.8
		2.18	22.3	33.2	1.5	30.4	0.1	2.7	11.5	3.0	125.8	230.5
Archipelago and Reed Islands	10/73	0.95	252.6	73.7	16.0	70.1	0.7	10.0	33.0	9.0	206.0	671.1
		1.37	63.5	111.0	9.4	144.1	2.1	7.3	22.2	15.4	314.5	689.5
		2.21	100.3	38.6	17.3	94.0	0.4	3.9	13.6	7.5	226.0	501.6
		1.93	46.1	53.1	11.0	72.7	1.0	3.0	22.3	7.9	265.4	482.5
		2.49	31.5	20.1	11.6	23.6	0.1	3.6	10.4	3.5	106.2	210.6
		2.06	18.8	12.2	6.1	12.1	0.1	3.3	21.5	3.2	114.8	192.1
		1.08	141.4	39.1	55.1	258.8	0.1	11.8	18.6	6.5	189.2	720.6
		0.97	309.2	43.3	87.8	309.2	0.3	16.4	31.5	7.8	268.3	1073.8
		0.82	221.5	36.0	33.8	200.9	1.7	17.0	23.1	2.2	167.5	703.7
		2.74	6.6	2.0	0.3	10.6	2.4	15.2	6.2	0.1	24.6	68.0
South Basin Open Water	18/04/73	1.72	42.6	18.0	0.3	0	0.5	136.4	7.0	0.3	14.6	219.7
	12/01/74	1.72	11.0	2.8	3.1	0.7	10.5	42.2	1.4	E	20.8	92.5
	4/02/74	1.74	14.8	7.3	1.6	0	9.6	46.4	0.9	0	15.4	96.0
	13/03/74	1.55	31.3	21.7	0.2	0	2.3	37.7	0.4	0	2.3	95.9
	6/10/74	2.61	11.9	42.7	3.5	0	2.9	12.8	3.5	0	5.0	82.3
	2/12/74	2.12	3.4	0.9	1.3	0	5.8	9.1	0.3	0	14.0	34.8

The population of this region which was similar to the normal stage changed with the lowering of the water level. It is possible that, in view of what happened to the phytoplankton, these changes did not exhibit the same features as in the southern archipelago, if one considers maximum levels of conductivity: 1120 and 3530 μS cm^{-1} respectively for the 1974 min. and max. values, while the 1973 maximum in the southern archipelago was 524 μS cm^{-1}.

7.2.2 Annual variations in density and stock (1964–73)

The zooplankton density remained almost constant at about 300 mg m^{-3} from 1964 to May 1973 in the southern archipelago (Fig. 10a, Table 2). This relative

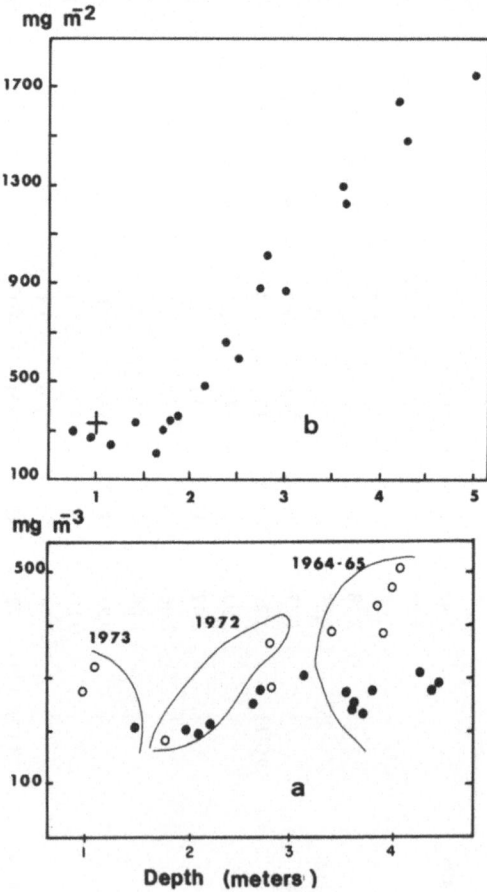

Fig. 10 Yearly variation of density and biomass per m^2 in the southern archipelago from 1964 to 1973. In (a) the plain circles represent the seasonal minimums.

constancy persisted in spite of the changes already described in structure and composition of the population in which the Copepods remained dominant (63% of the biomass in 1964–65, 87% and 90% in January and April 1973). However, two facts indicate a slight lowering of densities, which may have begun in 1972 with the quasi disappearance of the genera *Bosmina* and *Daphnia*. The higher densities observed in 1964–65 were never reached in following years, and the seasonal minima of the period 1972–73 were lower than the threshold of 240 mg m^{-3} above which all the values found during the period of 'Normal Chad' were situated.

The density remained about the same, the biomass per unit area decreased almost proportionally to the lowering of the water level (Fig. 10b), and the stock decreased in proportion to the decrease in volumes. This stock was distributed very unevenly between the three zones distinguished at the 'Normal Chad' stage, following the unequal distribution of water volume. In February 1971, for a water level height of about 281.30 m, the north basin thus contained 60% of the lake zooplankton stock estimated at 12 200 tons (Fig. 11). The biomass per hectare was evidently much higher in the north. If one extrapolates to the water level of 283 m, which was present in 1964–65, keeping the same densities and the same areas under water, one observes a small change in the stock distribution (55% in the north) and significant changes in the biomass per hectare (12.5 kg in zone 3, 12 kg in zone 2 and 4.2 kg in zone 1). These changes were the result of a relatively greater increase in depth (1.70 m) in the south than in the north.

At Bol station, which is supposed to be representative of the archipelago, the gross phytoplankton production (mgO$_2$m^{-2} h^{-1}) remained almost constant from 1968 to 1975 with some increases during the last year (Chapter 11). The chlorophyll concentrations or phytoplankton densities also markedly increased (Fig. 12). The decrease in grazing (lowering of the stock of zooplankton) would have contributed to this increase, without being the major cause (Chapter 12). Be that as it may, these data tend to show that zooplankton abundance was not proportional to that of phytoplankton. They support the fact that no clear dependence was found between the two biocoenoses, as seen in 1973–74 (although conditions differed, Fig. 9) or when comparing the annual cycles of 1964–65 for the zooplankton (Fig. 3) and 1969 for the phytoplankton (Fig. 7). In the first case, the curve is clearly bimodal, and it is almost unimodal in the second.

Thus, it appears that the phytoplankton biomass and its production were not the factors limiting the zooplankton, in any of the periods of change in the lake. This may not have been so during certain periods in zone 1 where algal densities were very low and the ratios of phytoplankton and zooplankton abundance were much lower than in the archipelago (Fig. 13). However, if one decides that phytoplankton was not limiting, variations in zooplankton abundance and diversity must have been regulated primarily by the presence of several

216 mg m^{-3},8.9 kg ha^{-1}

7400 t,8370 km^2

315 mg m^{-3},6.9 kg ha^{-1}

3950 t ,5934 km^2

93 mg m^{-3},2.5 kg ha^{-1}

850 t , 3770 km^2

0   50 km

Fig. 11 Stock and biomass per hectare in zone 1, 2 and 3 in February 1971, calculated for a of water level altitude 281.30 m.

zooplanktophagous species, or more generally by the essential importance of the pelagic food chain in 'Normal Chad'. Simplified examples of such regulation are offered by Lago di Anone (Bernardini and Giussani 1978). After the disappearance of the zooplanktivorous *Alburnus alburnus alborella* from a part of the lake, it results in strong oscillations in the *Daphnia hyalina* population, which was previously very stable. In such conditions, supposing that food is always in excess in the southern archipelago of the lake, the annual stability of density is explainable, in spite of the changes in structure or composition of the phytoplankton, the zooplankton and the zooplanktivorous species. Some experimental observations (Hall et al. 1970) suggest that an increase (or a decrease) in the predation pressure has no influence on the mean zooplankton abundance, but only on its composition or diversity, provided that the nutritional conditions are favorable (which was the case in enriched ponds studied by the authors). Other data appear to confirm the absence of a direct or close relationship between the

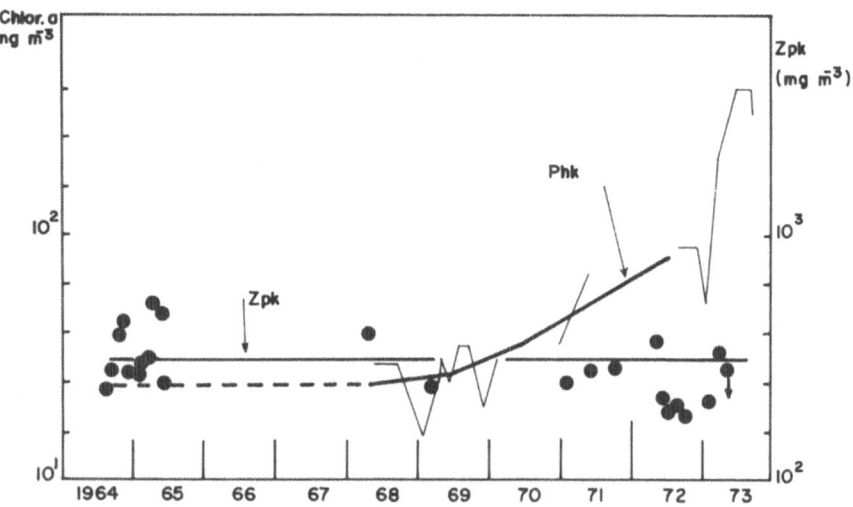

Fig. 12 Changes in chlorophyll concentration and zooplankton density, for the high water period until the drying up, in the southern archipelago. For the phytoplankton, the thick line shows an adjustment by eye of the data of Lemoalle (1973) representing the yearly evolution. The fine lines correspond to the short period variations corresponding to the measurements. For the zooplankton, the horizontal lines shows the general mean of annual values obtained.

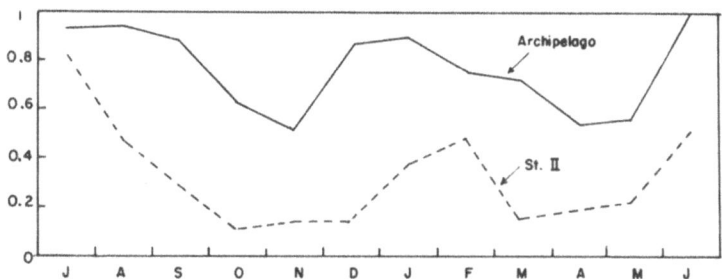

Fig. 13 Ratio between the numbers of cellular algae per liter and the zooplankton biomass ($mg\ m^{-3}$) in the southern archipelago and station II in 1964–65 (relative values, expressed by comparison to maximum archipelago value of June 1965: N/B value month X divided by N/B value in June).

phyto- and zooplankton biomasses, when the former exceeds a certain threshold. Apart from Lake Chad, where the zooplankton density was almost the same both in the high water period and in 1973 with very different chlorophyll concentrations (20 to 40 $mg\ m^{-3}$ in 1968–69; 60 to 280 $mg\ m^{-3}$ from January to April 1973), one can quote Lake George (370 $mg\ m^{-3}$ zooplankton for 177 $mg\ m^{-3}$ chlorophyll); Loch Leven (350 $mg\ m^{-3}$ average for 10 to 250 $mg\ m^{-3}$

223

chlorophyll) Lake Naivasha (202 mg m^{-3} for 20 to 50 mg m^{-3} chlorophyll), and two more eutrophic reservoirs (thus having *a priori* a high chlorophyll concentration) from South Africa, Lakes Hartbeespoort and Rietvlei, (about 400 mg m^{-3} zooplankton, Seaman 1979). Among the data considered here, low chlorophyll concentrations occur with low zooplankton: Lake Ototoa has a mean of 1 mg m^{-3} chlorophyll (from 0.04 to 4.6 mg m^{-3}) for 12 mg zooplankton, and two oligotrophic lakes of South Africa (Lake Lindleyspoort and Buffelspoort) studied with the two previous eutrophic lakes, support zooplankton densities of 30 and 50 mg^{-3}. These data are insufficient and too heterogeneous to predict whether or not a relationship exists between the mean zoo- and phytoplankton concentrations, and in particular if, as suggested by some previous data, there is a fixed limit for the high concentrations, by supposing that chlorophyll is a good indicator of phytoplankton abundance. In the range of the high concentrations, the variations in phytoplankton abundance would have been reciprocated by some changes in the structure or the production rate of the zooplankton population such as seen in Lake Chad (Chapter 12). In all hypotheses, it will be noted that the annual changes in zoo- and phytoplankton abundance express a total decrease in ecosystem efficiency which leads to the end of the preponderance of the pelagic transfer chain. From this point of view, the situation in the southern archipelago in 1973, is more similar to Lake George than to 'Normal Chad'.

7.3 General characteristics of the community

7.3.1 *Species richness*

Compared with some well-studied African lakes (Table 5), the Lake Chad community in the high water period would seem to be one of the most diverse in Africa, with 9 abundant species, of which seven (3 Cladocera, 1 Calanoid and 3 Cyclopoids) were simultaneously present and abundant throughout the year. Comparison with some other biotopes sampled occasionally, sometimes for faunal inventories, leads to the same conclusion. Among the large East-African lakes, the zooplankton of Lake Tanganyika in the pelagic zone consists of 3 Copepods (*Diaptomus simplex, M. leuckarti, Th. schurmanni*) (Lindberg 1951) and that of Lake Kivu, 3 Cladocera and 3 Cyclopoids, based on the data of Lindberg (1951) and Harding (1957). Collections in smaller water bodies have also been made. Jebel Aulia (Monakov 1969) with 11 species of which 3 are Copepods, is relatively diverse, and 4 small lakes of the Cameroon contained monospecific populations at the time of sampling, with either *M. leuckarti* (Lake Soden) or *Thermocyclops hyalinus* (Lake Mborombi Mbo, Kotto, Mboandong) present (Green 1972). Without attempting an exhaustive inventory, which is after all very difficult, it appears that the situations are all too diverse under other latitudes and emphasize the diversity of Lake Chad community (Weglenska 1971; Comita 1972; George and Edwards 1974; Kwik and Carter 1975; Salanki and Ponyi 1975; Rey and Capblancq 1975; Green 1976; Lair 1977, 1978; Gophen 1978). In Lake Lipno (temperate cold) (Brandl 1973), the population consists of eight species of which 5 are Cladocera. On the other hand the crustacean zooplankton communities of Loch Leven (cold temperate) or of Lake Kinneret (14–28°C) are almost monospecific, with *Cyclops strenuus abyssorum* and *M. leuckarti* respectively (Burgis and Walker 1972: Gophen 1978).

The diversity of the Lake Chad populations may be due to the large size of the lake, to the diversity as well as the abundance of food (about 25 mg m^{-3} of chlorophyll *a* in 1968–69 at Bol, or 1400 mg wet weight of phytoplankton per m^{-3} in the whole southern archipelago in 1971), and to the permanence of favourable temperature conditions. The strong thermal gradient (18–30°C) and the existence of relatively low temperatures over a fairly long period of the year were the causes of diversity, because they allowed the cold water forms *Bosmina* and *Daphnia* to establish themselves, as well as others which were well adapted to the thermal conditions of the environment. In Table 5, one notes that *B. longirostris* is present and the genus *Daphnia* is diverse in Lake Chad, Lake Midmar (South Africa) and Lake McIlwaine (Zimbabwe) which have the lowest minimum temperatures. The fact that the predators contribute to increased diversity and stability of the plankton community has been shown in some cases (Hall et al., 1970; Dodson 1970; Lane 1978; Zaret 1978). The existence in the lake of a large number of vertebrate and invertebrate predators is also favourable to species richness. In the archipelago, the zooplanktivorous fishes were numerous and represented a high fraction of the ichthyomass, about 44% according to calculations made from 1966 to 1970 (Lauzanne 1972). Two of these fish, *Alestes baremoze* and *Brachysynodontis batensoda* (27% of total ichthyomass) are respectively macro- and microphagous complementary predators. To these predators must be added other fish, *Chaoborus* larvae, 2 Cyclopoids, and the juveniles of several fish species. This situation can be contrasted with that of Lake George, where the zooplankton is less diverse and the zooplanktivorous species less numerous consisting of *Chaoborus larvae*, *M. leuckarti* and one fish of secondary importance, *Haplochromis poppenheimi* (Burgis et al. 1973).

The decrease in the number of species which began in about 1972 in the archipelago, was accompanied by a reduction in the number of zooplanktivorous species since *B. batensoda* alone existed to the end of the community changes, this species having mainly a detritivorous diet. It is thus possible that there was a cause and effect relationship between these two phenomena, but the hypothesis that the disappearance or scarcity of certain zooplankton species is due to a change in environmental conditions is quite valid. These two factors are not exclusive of each other as the Cladocera that

Table 5 Characteristics and composition of communities of planktonic microcrustacea from some well-sampled African lakes; * = abundant or dominant species, by numbers or by weight; A = artificial; N = natural; E = eutrophic; O = Oligotrophic or mesotrophic.

Lakes	Type	Latitude	Altitude (m)	Area (km²)	Mean depth (m)	Surface Temp. (°C)	Species	References
Chad	N E	12°30 N	282	18 000	3.9	18–30°C	*Diaphanosoma excisum*, Daphnia barbata*, D. longispina, D. lumholtzi, Ceriodaphnia affinis, C. cornuta*, Moina micrura*, Bosmina longirostris*, Tropodiaptomus incognitus*, Thermodiaptomus galebi, Thermocyclops neglectus*, Th. incisus circusi*, Th. tchadensis, Mesocyclops cf. leuckarti**	present work (high water)
George (Uganda)	N E	0°	913	250	2.25	25–26	*Thermocyclops hyalinus**, Mesocyclops cf. leuckarti*, M. micrura, C. cornuta, D. barbata*	Burgis (1974)
Chilwa (Malawi)	N	15°15 S	654	700	2	21–28	*D. excisum*, D. barbata, M. micrura, C. cornuta, Tropodiaptomus kraepelini*, M. cf. leuckarti**	Kalk (high water)
Naivasha (Kenya)	N E	0°46 S	1890	145	4.7	19.5–23	*D. excisum*, Simocephalus vetulus*, Thermocyclops* schurmanni, M. cf. leuckarti**	Litterick et al. (1979)
Mc. Ilwaine (Zimbabwe)	A		1370	26.3	9.5	16–28	*D. excisum, Daphnia laevis, D. lumholtzi, Ceriodaphnia dubia**, M. dubia, B. longirostris, Tropodiaptomus orientalis,*	Munro (1966)

								Akhurst et al. (1979)
Midmar (South Africa)	A O	29°S	1000	15.6	11.4	11–25	*Tropocyclops prasinus, Thermocyclops emini*, *D. excisum*, Daphnia pulex*, D. longispina*, D. tenuispina, D. barbata, M. micrura, B. longirostris, Tropodiaptomus spectabilis*, Metadiaptomus transvaalensis, Thermocyclops oblongatus, Th. tchadensis, Thermocyclops retrocurvus*	
Turkana (Kenya)	N	2–4° N	406	7200	~80 max.		*D. excisum*, D. barbata, C. cornuta, Moina brachiata, Tropodiaptomus banforanus*, M. leuckarti*, Th. hyalinus**	Ferguson (1974)

227

Table 6 Zooplankton abundance in some African and temperate lakes (* corresponds to less accurate data, generally through graphical estimates).

Lakes	Depth (m)	Surface temp. (°C)	Mean density $\text{dry weight mg m}^{-3}$	Nb l^{-1}	Maximum density	Maximum/ minimum	g m^{-2}	Remarks	References
Chad (Archipelago)	3.9	18–30°C	333	318	515 mg 453 indiv.	2.2	1.3	annual mean (a.m.)	present work
Chad (Zone 1)			181	152	413 mg	6.3		a.m.	present work
George	2.25	25–26	368		600 mg	1.8	0.8	a.m.	Burgis (1974)
Chilwa		21–28			100–800 ind.	~50 à 400		high water, limited data	Kalk (1975)
Naivasha	4.7	19.5–23	202*		364* mg	4.9	1.0	a.m.; DW/m² = DW/m³ × 4.7 m.	Litterick et al. (1979)
Turkana	80 max.			200*		3.5*			Ferguson (1975)
Loch Leven	3.9	0–19	350*		1030* mg	30*	1.37*	a.m. *Cyclops strenuus abyssorom*	Burgis and Walker (1972)
Eglwys Nynydd	3.5	0–18	445	50*	2030 mg	50*	1.6	a.m., 1970 and 1971; values related to *Daphnia hyalina* (dominant species in the community)	George and Edwards (1974)
Lipno	6.6	~18	160		—	—	1.25	May–October mean (30 values for several stations from 1960 to 1967); mg DW = mg N × 8.33	Brandl (1973)
Mikolajskie	11.7	~16	933		—	—	—	Summer mean (22/07–30/08/64) mg DW = 0.10 mg FW	Weglenska (1971)
Balaton			100 70		—	—	—	May–November 1965, 66 and 67; values for 2 regions of the lake:	Salanki and Ponyi (1975)
Ototoa	12–16ᵃ	10–25	11.7		14.2 mg	1.7	0.21	*Calamoecia lucasi* only (75% total number);	Green (1976)
Kinnereth		14–28	—		—	2.7*	1.5*	a.m.; *M. leuckarti*	Gophen (1978)

ᵃMetalimnion depth.

disappeared were the winter forms, of which two, *D. longispina* and *B. longirostris*, are cosmopolitan or widespread in temperate regions. These species must be at the limit of favourable thermal conditions in Lake Chad and they continued to develop as other environmental aspects (abundant food, considerable predation pressure, existence of appropriate conditions in a true pelagic zone), allowed them to compete with the other very ubiquitous species, which were well adapted to the thermal regime of the lake. Thus it is not surprising that these forms have not survived the rapid transformation of environmental conditions during the establishment of the 'Lesser Chad'.

If the size criteria of Dodson et al. (1976) are adopted the lake community could be placed in the category of small zooplankters (0.2–0.3 mm length) which are generally subjected to strong predation pressure.

7.3.2 *Abundance*

The mean biomass and the annual variations observed in the actual lacustrine part of Lake Chad (zones 2 and 3) were of the same order of magnitude as those of Lakes George, Naivasha, and Turkana which are among the best known African lakes (Table 6). Although other more partial data are similar, there are also data which differ. Monakov (1969) mentions 300 mg m^{-3} (April) and 150 mg m^{-3} (autumn) for Jebel Aulia; Seaman (1979) from 30 to 400 mg m^{-3} annual average for small reservoirs in South Africa, according to their trophic state: 30 and 50 mg m^{-3} for two oligo- and mesotrophic lakes and 400 mg m^{-3} for two eutrophic lakes. One can also include the values reported by Green (1972) for 4 smaller Cameroon lakes (from 2 to 53×10^3 individuals m^{-3} in April), and by Rufli (1976) for Lake Tanganyika (1890 ind. m^{-3} in May 1976), these numbers being low when compared with those of Lake Turkana or Lake Chad.

The insufficient number and the heterogeneous or incomplete character of data collected up to now on the African biotopes as well as from other continents or climatic zones; show that a valid comparison of average densities may not be made. The few data shown in Table 6 do not show any major differences between temperate and tropical environments from this point of view.

References

Akhurst, E. G. J., Breen, C. M., Johnson, I. M., Rayner, N. A. and Twinch, A. J., 1979. Lake Midmar: an oligotrophic impoundment in Natal. SIL Workshop on African Limnology, Nairobi, No. 14, 15 pp., mimeo.
Beadle, L. C., 1974. The inland waters of tropical Africa. An introduction to tropical Limnology. Longman, London, 365 pp.

Bénech, V., 1975. Croissance mortalité et production de Brachysynodontis batensoda (Pisces, Mochocidae) dans l'archipel sud-est du lac Tchad. Cah. ORSTOM Sér. Hydrobiol. 9: 91–103.

Bénech, V., Lemoalle, J. and Quensiere, J., 1976. Mortalités de poissons et conditions de milieu dans le lac Tchad au cours d'une période de sécheresse. Cah. ORSTOM Sér. Hydrobiol. 10: 119–130.

Bernardini, R. de and Giussani, G., 1978. Effect of mass fish mortality on zooplankton structure and dynamics in a small Italian Lake (Lago di Anone). Verh. int. Ver. Limnol. 20: 1045–1048.

Brandl, Z., 1973. Relation between the amount of net zooplankton and the depth of station in the shallow Lipno reservoir. Hydrobiol. Stud. 3: 7–51.

Brandorff, G. O. and Andrade, E. R. de., 1978. The relationship between the water level of the Amazon river and the fate of the zooplankton population in Lago Jacaratinga, a Varzea Lake in the Central Amazon. Stud. Neotrop. fauna. envir. 13: 68–70.

Burgis, M. J., 1974. Revised estimates for the biomass and production of zooplankton in Lake George, Uganda. Freshwat. Biol. 4: 535–541.

Burgis, M. J. and Walker, A. F., 1972. A preliminary comparison of the zooplankton in a tropical and a temperate lake (lake George, Uganda and Loch Leven, Scotland) Verh. int. Ver. Limnol. 18: 647–655.

Burgis, M. J., Darlington, J. P., Dunn, I. G., Ganf, G. G., Gwahaba, I. J. and McGowan, L. M., 1973. The biomass and distribution of organisms in Lake George, Uganda. Proc. r. Soc. Lond. B 184: 271–298.

Carmouze, J. P., Dejoux, C., Durand, J. R., Gras, R., Iltis, A., Lauzanne, L., Lemoalle, J., Lévêque, C., Loubens, G. and Saint-Jean, L., 1972. Grandes zones écologiques du lac Tchad. Cah. ORSTOM Sér. Hydrobiol. 6: 103–169.

Comita, G. W., 1972. The seasonal zooplankton cycles, production and transformation of energy in Severson Lake, Minnesota. Arch. Hydrobiol. 70: 14–66.

Dejoux, C., Lauzanne, L. and Lévêque, C., 1969. Evolution quantitative et qualitative de la faune benthique dans la partie est du lac Tchad. Cah. ORSTOM Sér. Hydrobiol. 3: 3–58.

Dejoux, C. and Saint-Jean, L., 1972. Etude des communautés d'invertébrés d'herbiers du lac Tchad: recherches préliminaires. Cah. ORSTOM Sér Hydrobiol. 6: 67–83.

Dodson, S. I., 1970. Complementary feeding niches sustained by size selective predation. Limnol. Oceanogr. 15: 131–137.

Dodson, S. I., Edwards, C., Wiman, F. and Normadin, J. C., 1976. Zooplankton: specific distribution and food abondance. Limnol. Oceanogr. 21: 309–313.

Dussart, B. and Gras, R., 1966. Faune planctonique du lac Tchad. I. Crustacés Copépodes. Cah. ORSTOM Sér. Océanogr. 4: 77–91.

Ferguson, A. J. D., 1975. Invertebrate production in lake Turkana (Rudolf). Symposium on the Hydrobiology and fisheries of Lake Turkana, 25–29th May 1975, 28 pp. mimeo.

George, D. G. and Edwards, R. W., 1974. Population dynamics and production of *Daphnia* in a eutrophic reservoir. Freshwat. Biol. 4: 445–465.

Gophen, M., 1978. The productivity of *Mesocyclops leukarti* (Claus) in Lake Kinneret, Israël. Hydrobiologia 4: 63–70.

Gras, R., Iltis, A. and Lévêque-Duwat, S., 1967. Le plancton du bas Chari et de la partie est du lac Tchad. Cah. ORSTOM Sér. Hydrobiol. 1: 25–96.

Gras, R. and Saint Jean, L., 1976a. Etude de la répartition spatiale du zooplancton dans le lac Tchad: variation de la dispersion en fonction des modalités d'échantillonnage et des conditions hydrodynamiques du milieu. Cah. ORSTOM Sér. Hydrobiol. 10: 201–229.

Gras, R. and Saint-Jean, L., 1976b. Durée du développement embryonnaire chez quelques espèces de Cladocères et de Copépodes du lac Tchad. Cah. ORSTOM Sér. Hydrobiol., 10: 233–254.

Gras, R. and Saint-Jean, L., 1978. Durée et caractéristiques du développement juvénile de quelques Cladocères du lac Tchad. Cah. ORSTOM Sér. Hydrobiol. 12: 119–136.

Gras, R. and Saint-Jean, L., 1981a. Durée du développement juvénile de quelques Copépodes du lac Tchad. Rev. Hydrobiol. trop. 14: 39–51.

Gras, R. and Saint-Jean, L., 1981b. Croissance en poids de quelques Copépodes planctoniques du lac Tchad. Rev. Hydrobiol. Trop. 14: 135–147.

Gras, R., Lauzanne, L. and Saint-Jean, L., 1981. Régime alimentaire et sélection des proies chez *Brachysynodontis batensoda* (Pisces, Mochodidae). Rev. Hydrobiol. Trop. 14: 223–231.

Gras, R. and Saint-Jean, L., 1983. Production du zooplancton du lac Tchad en période de hautes et de basses eaux. Rev. Hydrobiol. Trop. 16 (in press).

Green, J., 1972. Ecological studies on Crater Lakes in West Cameroun. Zooplankton of Borombi Mbo, Mboandong, lake Kotto and lake Soden. J. Zool. 166: 283–301.

Green, J. O., 1976. Population dynamics and production of the Calanoid Copepod *Calamoecia lucasi* in a northern New Zealand Lake. Arch. Hydrobiol. Suppl. 50: 313–400.

Hall, D. J., Cooper, W. E. and Werner, E. E., 1970. An experimental approach to the production, dynamic and structure of freshwater animal communities. Limnol. Oceanogr. 15: 839–928.

Harding, J. P., 1957, Crustacea, Cladocera. In: Explor. Hydrobiol. Lac Tanganyika (1946–47), 3,6: 55–89.

Iltis, A., 1977. Peuplement phytoplanctonique du lac Tchad III Remarques générales. Cah. ORSTOM Sér. Hydrobiol. 11: 189–199.

Im, B. H., 1977. Etude de l'alimentation de quelques espèces de *Synodontis* (poissons, Mochocidae) du Tchad. Thèse Paris, 150 pp., mimeo.

Kalk, M., 1979. Zooplankton in Lake Chilwa: adaptations to changes. In: M. Kalk, A. J. McLachlan and C. Howard-Williams (eds.), Lake Chilwa, studies of change in a tropical ecosystem, pp. 125–141, Monographiae Biologicae, Vol. 35, W. Junk Publishers, The Hague.

Kwik, J. K. and Carter, J. C. H., 1975. Population dynamics of Limnetic Cladocera in a Beaver pond. J. Fish. Res. Bb Can. 32: 341–346.

Lair, N., 1977. Biomasse et production dans deux lacs du Massif Central français. Arch. Hydrobiol. 79: 247–273.

Lair, N., 1978. Répartition spatio-temporelle, biomasse et production des populations zooplanctoniques du lac d'Aydat en période estivale. Hydrobiologia 61: 237–256.

Lane, P. A., 1978. Role of invertebrate predation in structuring zooplankton communities. Verh. int. Ver. Limnol. 20: 480–485.

Lauzanne, L., 1970. Sélection des proies chez *Alestes baremoze* (Pisces, Characidae). Cah. ORSTOM Sér. Hydrobiol. 3: 15–27.

Lauzanne, L., 1972. Régimes alimentaires des principales espèces de poissons de l'archipel oriental du lac Tchad. Verh. int. Ver. Limnol. 18: 636–646.

Lemoalle, J., 1979. Biomasse et production phytoplanctoniques du lac Tchad (1968–76). Relations avec les conditions de milieu. Thèse Univ. Paris VI, ORSTOM, 311 pp., mimeo.

Lindberg, K., 1951. Cyclopides (Crustacés, Copépodes) — Result. Scient. Explor. Hydrobiol. Lac Tanganyika (1946–47), 3, 2: 45–91.

Litterick, M. R., Gaudet, J. J., Kalff, J. and Melack, J. M., 1979. The Limnology of lake Naivasha. SIL Workshop on African Limnology, Nairobi, 16–23rd, December 1979, 72 pp. mimeo.

Monakov, A. V., 1969. The zooplankton and the zoo-benthos of the White Nile and adjoining waters in the Republic of Sudan. Hydrobiologia 33: 161–185.

Munro, J. L., 1966. A limnological survey of Lake MacIlwaine, Rhodesia. Hydrobiologia 28: 281–308.

Pourriot, R., 1968. Rotifères du lac Tchad. Bull. IFAN, A. 30: 471–496.

Pourriot, R., 1975. Relations proie-prédateur: réactions adaptatives et fluctuations des populations du zooplancton sous l'influence d'une prédation sélective. Ann. Biol. 14: 69–85.

Pourriot, R., Iltis, A. and Lévêque-Duwat, S., 1967. Le plancton des mares natronées du Tchad. Int. Rev. ges. Hydrobiol. 52: 535–543.

Rey, J. and Capblancq, J., 1975. Dynamique des populations et production du zooplancton du lac de Port-Biehl (Pyrénées Centrales). Ann. Limnol. 11: 1–45.

Rey, J. and Saint-Jean, L., 1968. Les Cladocères (Crustacés, Branchiopodes) du Tchad: 1ère note. Cah. ORSTOM Sér. Hydrobiol. 2: 79–118.

Rey, J. and Saint-Jean, L., 1969. Les Cladocères (Crustacés, Branchiopodes) du Tchad: 2ème note. Cah. ORSTOM Sér. Hydrobiol. 3: 21–42.

Robinson, A. H. and Robinson, P. K., 1971. Seasonal distribution of zooplankton in the northern basin of Lake Chad. J. Zool. 163: 25–61.

Rufli, H., 1976. Preliminary analysis of zooplankton sampling in Lake Tanganyika in May 1976. Lake Tanganyika Fishery Research and Development Project. W.P. 44, 9 pp.

Salanki, J. and Ponyi, J. E., 1975. The biomass of zooplankton in Lake Balaton. Symp. Biol. Hung. 15: 215–224.

Seaman, M. T., 1979. Zooplankton characteristics of some Transvaal dams. SIL Workshop on African Limnology, Nairobi, 16–23rd December 1979, 2 pp.

Weglenska, T., 1971. The influence of various concentrations of natural food on the development, fecundity and production of planktonic crustacean filtrators. Ekol. Pol. 19: 427–473.

Zaret, T. M., 1978. A predation model of zooplankton community structure. Verh. int. Ver. Limnol. 20: 2496–2500.

8. The benthic fauna: ecology, biomass and communities

Christian Lévêque, Claude Dejoux
and Laurent Lauzanne

Benthos consists of all those aquatic organisms which are associated in some way with the bottom sediments. We shall, therefore, consider here only the organisms and communities inhabiting the lake sediments, reserving a study of the periphyton found on higher aquatic plants for a further chapter. This periphyton was far from negligible in Lake Chad where fringing vegetation and submerged water grasses were numerous. It was usually composed of different species to the benthos although some were occasionally found among the benthos when it was sampled close to plant clumps.

Most of the observations on the benthic communities and their biomasses were made between 1968 and 1971, i.e. during the 'Normal Chad' period, and a general zonation was established for the Lake in 1970 (Carmouze et al. 1972). The distribution patterns obtained at this time were completely changed later, because of the rapid drop in the Lake level after 1972, which caused partial drought in some zones and major changes in the ecological conditions. The evolution of the benthic communities could only be partly studied during this drying phase of Lake Chad.

All samples were taken with an Ekman grab. For worms and insects, five samples were taken at each station with a 15 × 15 cm grab. The samples, washed on a 0.3 mm mesh sieve, were then fixed in 10% formol, and later sorted in the laboratory. At least six samples were collected for molluscs at each station, with a 30 × 30 cm grab, these washed on a 0.8 mm mesh sieve.

8.1 Composition of the benthic fauna

The bulk of the benthic fauna of Lake Chad was represented by three groups of macroinvertebrates — worms, molluscs insects — which have been the subject of intensive study during recent years.

We give here a list of the most common benthic species. More detailed systematic studies have been published by Dejoux (1968, 1969, 1970, 1971, 1973) for insects, Lauzanne (1968) for oligochaetes, and Lévêque (1968, 1974) for molluscs.

J.-P. Carmouze et al. (eds.) Lake Chad   233
© *1983, Dr W. Junk Publishers, The Hague/Boston/Lancaster*
ISBN 978-94-009-7268-1

OLIGOCHAETES

Alluroïdae
Alluroïdes tanganykae
Tubificidae
Aulodrilus remex, Euilodrilus sp.
Naïdidae
Branchiodrilus cleistochaeta, Allonais paraguayensis ghanensis, Pristina synclites, Naïs sp.

Only Alluroïdidae and Tubificidae were abundant in the bottom substrata, while Naïdidae were found only occasionally, for they preferred the water grasses.

MOLLUSCS

Prosobranchia
Melania tuberculata, Bellamya unicolor, Cleopatra bulimoïdes
Lamellibranchia
Corbicula africana, Caelatura aegyptiaca, Caelatura terestiuscula, Pisidium pirothi, Eupera parasitica, Mutela dubia, Mutela rostrata

The three species of Prosobranchs, as well as *C. africana* and *C. aegyptiaca*, were very abundant with a wide dspribution. The other species were less numerous and more localized.

INSECTS

Chironomids
Chironominae
Chironomus formosipennis, Cryptochironomus stilifer, Cryptochironomus nudiforceps, Cryptochironomus dawulfianus, Cryptochironomus diceras, Tanytarsus nigrocinctus, Polypedilum fuscipenne, Polydepilum griseoguttatum, Polypedilum abyssiniae, Polypedilum longicrus, Cladotanytarsus lewisi, Cladotanytarsus pseudomancus
Tanypodinae
Ablabesmyia pictipes, Ablabesmyia dusoleili, Clinotanypus claripennis, Procladius brevipetiolatus
Orthocladiinae
Cricotopus scottae
Ephemeroptera
Cloeon fraudulentum, Eatonica schoutedeni, Coenomedes brevipes, Povilla adusta
Trichoptera
Dipseudopsis capensis, Ecnomus dispar, Ecnomus sp.

Most of the species in these different invertebrate groups are widely distributed in Africa and none was endemic to Lake Chad.

8.2 Factors influencing species distribution and abundance

Temporal and spatial distribution of benthic organisms depends on various physical and chemical factors which favour or discourage the presence and

234

abundance of species. Other phenomena, such as fish predation, can also play an important role, but none of these processes have been evaluated in Lake Chad so far.

It is true that a number of ecological factors are more or less interrelated. The nature of the sediment, which serves as a universal basis for distinguishing benthic population, results in fact, from a complex group of physico-chemical factors which have not all been identified.

Under these circumstances it is often difficult to determine precisely which factors exercise a real influence on benthic species. It is generally considered sufficient to establish the correlation existing between the density of the species and certain environmental parameters, without confirming the direct relationship that exists between them. Only experimental studies can prove this.

In the case of Lake Chad, we have been able to define the role played by the sediment type, conductivity and temperature on the composition of the communities and the abundance of species. The generally very shallow depth and sufficient oxygenation at the bottom due to thorough mixing by fairly strong winds throughout the year, did not appear to limit the distribution of the benthic fauna as in deeper lakes. However, local vegetation barriers surrounding the islands acted as screens, attenuating the action of the dominant winds, and promoting the formation of swarms of chironomids with a gregarious tendency. The larvae of these chironomids were therefore more numerous near the shores than at the centre, but this phenomenon was relatively infrequent (Dejoux 1976).

8.2.1 *Bottom types*

Granulometry and the chemical characteristics of the sediment are known to play a very important role in the distribution of species and the composition of benthic communities. The influence of these factors has been well studied in Lake Chad (Dejoux et al. 1971; Dupont and Lévêque 1968; Carmouze et al., 1972).

In the region of Bol (southeastern archipelago), which may be considered relatively homogeneous as regards other factors affecting the distribution of species (salinity, type of landscape, etc.), a precise map of the bottom was established by core sampling and by dredging the zone under study. This represented a water surface of about $20 \, km^2$ and mineralogical analyses revealed five characteristic sediment types: sand mud, peat, soft clay and granular clay. Samples from January 1967, at various points in this zone, showed the influence of sediment type on the qualitative and quantitative composition of the malacological communities (Dupont and Lévêque 1968). This study was repeated in January 1970 for the entire benthic fauna (Dejoux et al. 1971). An analysis of variance showed that the 'bottom type' factor was highly significant for most of the worm and mollusc species, but that more than half of the insect larvae were unaffected by it. The factor under consideration was significant for three species only: *Cladotanytarsus lewisi, Polypedilum* sp. III and *Nilodorum rugosum*. The insects therefore were less dependent on a particular sediment than worms and molluscs. It will be noted that the greatest densities of worms and molluscs were to be found on clay bottoms, and those of insects on sandy bottoms (Table 1). Worms and insects were particularly scarce in peat.

Throughout Lake Chad, the bottom features also proved to be important in the distribution of species, although results were not always so clear as at Bol, due to interference from other ecological factors. If the mean densities and biomasses of the principal groups of worms are examined in the four types of bottoms sampled during March and November 1970 (Table 2), it can be seen that the Alluroïdidae were absent from mud, whereas the Tubificidae were abundant there. *Alluroides* was dominant in the 'pseudo-sand', whereas in peat there were no worms. These results confirmed the observations made in the Bol region, with sand being considered of the same texture and granulometry as 'pseudo-sand'. For molluscs the phenomenon was less clear, for it seemed that other factors in relation to geography also had a strong influence. However, in a particular zone, the type of community depended also on the nature of the sediment (Lévêque 1972). Finally for the insects, the results obtained for the whole lake were not comparable with those of the Bol region and other distribution factors probably intervened to a greater extent (Dejoux 1976).

8.2.2 *Conductivity of the water*

The ratio varied from 1 to 15 between the Shari delta and the north of the lake (Carmouze et al., 1972).

A fairly radical change was noticed in the structure of the oligochaete communities on clay beds of about 420 µS cm^{-1}. The Alluroididae, which had the greatest biomass below this threshold, disappeared completely beyond it and were replaced by Tubificidae which became very abundant. An ecological barrier related to the saltiness of the water seemed to exist for *Alluroides tanganyikae*, since this species disappeared from the clay sediments of Lake Chad when conductivity was over 420 µS cm^{-1} (Fig. 1).

An identical phenomenon was observed in the molluscs as the mean density of different species of prosobranchs diminished rapidly from 400 µS. *Cleopatra* and *Bellamya* disappeared above 550 µS and *Melania* beyond 600 µS (Fig. 2). Benthic molluscs were totally absent in samples from the north of the lake, where conductivities reached 750 µS, whereas environmental conditions of sediment, depth and landscape were apparently the same as for water with a lower conductivity. Moreover, in the zone devoid of molluscs, there were many

Table 1   Mean number of individuals (m^{-2}) of the different species of benthic invertebrates for each type of sediment in the Bol region in January 1970 (after Dejoux et al. 1971).

	Soft clay	Granular clay	Sand	Mud	Peat
INSECTS					
Ecnomus dispar	3.0	32.6	14.8		7.4
Cloeon fraudulentum	35.6	213.3	17.8		14.8
Povilla adusta		5.9			
Eatonica schoutedeni		3.0		1.8	
Orthotrichia	5.9	5.9	8.9		
Chaoborus ceratopogones	5.9	5.9			
Ceratopogonides	14.8			5.3	
Ablabesmyia dusoleili	3.0	41.5		16.0	55.6
Chironomus formisipennis	14.8	32.6	3.0	8.9	
Chironomus sp. I	17.8	3.0			
Cladotanytarsus lewisi			962.9		
Cladotanytarsus sp. I	3.0				
Clinotanypus claripennis		3.0		16.0	
Cryptochironomus stilifer	477.0	130.3	23.7	446.2	185.1
Cryptochironomus diceras	11.8	35.5	53.3	1.8	25.9
Cryptochironomus sp. I	5.9	14.8	26.7		
Cryptochironomus sp. II		8.9	204.4		
Cryptochironomus sp. III			11.9		
Cryptochironomus sp. IV					
Nilodorum rugosum	100.7	5.9			
Polypedilum fuscipenne	50.3		29.6	46.2	37.0
Polypedilum sp. I	23.7	5.9		3.6	11.1
Polypedilum sp. II	5.3				
Polypedilum sp. III			405.9		
Procladius brevipetiolatus		5.9			
Stictochironomus sp. I		3.0			
Stictochironomus sp. II		14.8			22.2
Tanytarsus nigrocinctus	3.0	11.8		3.6	
Tanytarsus sp. I	34.9	47.9			32.6
WORMS					
Aulodrilus remex		115.6	231.1	851.0	3.0
Euilyodrilus sp.	32.4	68.0	38.2	11.1	
Alluroïdes tanganikae	1434.0	1431.0	44.4		5.8
Branchiodrilus cleistochaeta	23.6	17.8	8.9	4.4	
Pristina synclites		35.6			
Naïs sp.			414.6		

237

Table 1 (continued).

	Soft clay	Granular clay	Sand	Mud	Peat
Aulophorus sp.			14.7		5.8
Nématodes	787.9	408.8	85.8	15.6	
MOLLUSCS					
Melania tuberculata	0.3	1.1	8.9	18.5	6.6
Cleopatra bulimoïdes	48.1	222.4	10.2	3.2	24.9
Bellamya unicolor	0.7	2.5	0.5	0.6	0.2
Corbicula africana	5.4	15.4	0.8	0.8	0.4
Caelatura aegyptiaca	1.6	3.1	0.5	0.1	0.1
Pisidium pirothi	0.1	0	0	0	0.1
Eupera parasitica	0.1	11.1	0	0	0

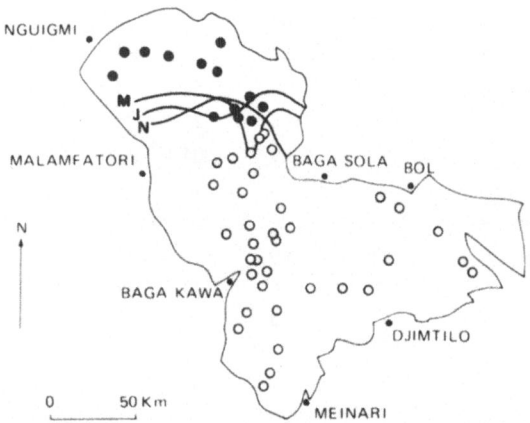

Fig. 1 Worms: distribution of Alluroïdidae on clay substrata (○ = presence; ● = absence). Lines M, J, N indicate the 420 µS cm^{-1} conductivity limits in March, July and November, 1970 (after Carmouze et al. 1972).

dead shells of *Melania* and *Bellamya*, indicating that these species must have prospered there. Finally, we never observed benthic molluscs in the ponds of Kanem with conductivities of 800 to 1000 µS, whereas Pulmonates were present in the vegetation.

The conductivity of the water did not appear to be a limiting factor in the distribution of numerous species of Chironomids (Dejoux 1976). Some of them however seemed to show a preference for high salinities (*Chironomus calipterus*,

Table 2 Mean densities and biomass for each group of benthic worms, according to the main substratum types investigated in March and November 1970 over the whole of the Lake Chad (after Carmouze et al. 1972).

March 1970		Alluroididae	Tubificidae	Naididae	Nématodes	Total
Mud	N m^{-2}	0	9134	54	229	9417
(20 stations)	mg m^{-2}	0	3197	19	14	3230
Clay	N m^{-2}	157	2462	0	1916	4535
(22 stations)	mg m^{-2}	535	866	0	119	1520
Pseudo-sand	N m^{-2}	626	219	30	177	1052
(8 stations)	mg m^{-2}	2132	77	10	11	2230
Peat						
(5 stations)		0	0	0	0	

November 1970		Alluroididae	Tubificidae	Naididae	Nématodes	Total
Mud	N m^{-2}	0	1089	2	70	1161
(22 stations)	mg m^{-2}	0	381	—	4	385
Clay	N m^{-2}	216	540	4	211	971
(27 stations)	mg m^{-2}	736	189	1	13	939
Pseudo-sand	N m^{-2}	369	61	0	50	480
(6 stations)	mg m^{-2}	1257	21	0	3	1281
Peat						
(2 stations)		0	0	0	0	0

Dicrotendipes polosimanus, Dicrotendipes fusconotatus, Tanytarsus nigrocintus, Cladotanytarsus lewisi, Cryptochironomus diceras, Cryptochironomus stilifer, Polypedilum laterale). Others were abundant especially in water with a low salinity (*Chironomus pulcher, Chrironomus acuminatus, Dicrotendipes peryngeyanus, Tanytarsus zariae, Tanytarsus flexibile, Nilodorum brevipalpis, Nilodorum fractilobus, Cryptochironomus miligenus, Cryptochironomus sinatus, Cryptochironomus melutensis, Polypedilum longicrus, Clinotanypus rugosus, Ablabesmyia nilotica*). Finally, in the Kanem ponds or in polders that were drying up, *Cryptochironomus deribae* was abundant, usually associated with *Chironomus calipterus. C. deribae,* known in the lagoons of the Camargue and the Baltic, may be considered a very halophilic species.

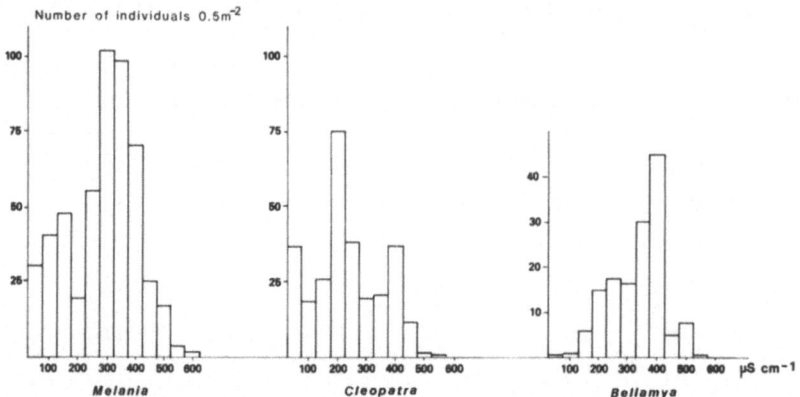

Fig. 2 Relationship between conductivity (μS cm⁻¹) and the mean density per sample for the three species of benthic Prosobranchs in Lake Chad (after Lévêque 1972).

8.2.3 *Seasonal abundance rhythms*

Monthly samples were taken between 1966 and 1967 at twelve stations within the eastern archipelago (Dejoux et al. 1969), and they revealed the existence of a seasonal abundance cycle in worms and insects, related to certain environmental factors. If the yearly change in water temperature and the lake level is compared with that of the mean density and biomass of organisms in the stations studied (Fig. 3), abundance was at a maximum for oligochaetes and insect larvae in the cool season (January to March) when the lake water was high. The minimum was found in the hot season (August to September) at the time of low water. The abundance of these two groups increased with lake level, but fell when the temperature of the water rose during the annual cycle.

The existence of a seasonal abundance rhythm has also been confirmed during missions undertaken over the whole lake in March, July and November, 1970. The greatest densities of oligochaetes (Carmouze et al. 1972) were observed in March, with the lowest in July (Table 4). This was also the case for the insects (Dejoux 1976) (Table 3).

Variation in seasonal abundance could not be shown for molluscs, with the exception of *Corbicula africana*, which reproduced in the cold season, and so had its maximum density (Dejoux et al. 1969; Lévêque 1972).

The low density of benthic organisms at the lowest water level could have been due to the greater disturbance of the sediments with decreasing depth when the wind stirred up the water, making conditions unsuitable for the establishment of dense communities. The tornadoes which occurred from April to June must have also altered the state of the bottom. Finally, the water reached its lowest level in the hot season when the biological cycles of the insects were much shorter than in the cool season.

240

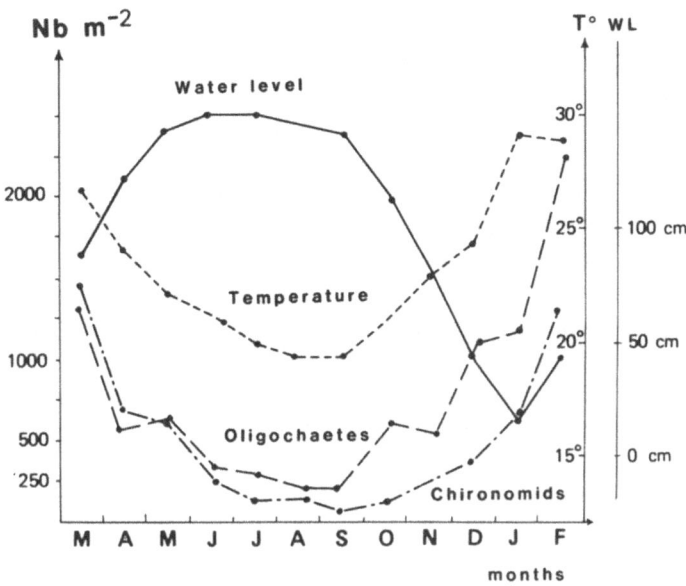

Fig. 3 Variation in mean density of benthic chironomids and oligochaetes for eastern Lake Chad in 1966–1967, with the temperature (T°) and water level (H) on the Bol gauge (after Dejoux et al. 1969, and Dejoux 1976).

Table 3 Variations in mean seasonal density of benthic insect larvae in numbers m^{-2} for three regions of Lake Chad: zone 1, north of the lake, above the Malamfatori parallel; zone 2, center of the lake between the preceding zone and the Great Barrier; zone 3, south and east of the lake (Dejoux 1976).

Date	Zone 1	Zone 2	Zone 3
March 1970	1574	482	369
July 1970	97	27	17
November 1970	540	298	119
March 1971	763	386	47

8.3 Communities and the major ecological zones

Samples from 1968 and 1970 from the whole lake revealed the main community types existing in each group studied. From these data it has been possible to distinguish according to each group, ecological zones within which the communities showed certain similarities regarding specific structure and species density (Carmouze et al. 1972). These main ecological zones clearly emphasized the original character of the different parts of the lake.

8.3.1 *Worms*

Four main ecological zones (Fig. 4) were determined, according to nature of the bottom and conductivity, the principal factors of species distribution. The mean densities and biomass of the worms in the diverse bottom types were calculated in each of these zones (Table 4).

Zone 1 — Open water of the north, consisting mainly of muddy substrate in a 180–420 μS cm^{-1} conductivity range. This zone was rich in Tubificidae which constituted almost the total community.

Zone 2 — Open water of the south and southeast in which the beds were made up of clay and 'pseudo-sand'. Here, the conductivity was below 180 μS, and Alluroididae largely dominated the biomass.

Zone 3 — Archipelago and reed islands of the north, where the conductivity was higher than 420 μS, Alluroïdidae were absent and Tubificidae represented 99% of the biomass in this zone.

Zone 4 — Archipelago and reed islands of the 'Great Barrier' and of the east. The sediments were more varied in this region and the communities were fairly heterogeneous. Tubificidae were found in the mud, whereas the Alluroïdidae were abundant in the clay bottoms.

8.3.2 *Molluscs*

The results of sampling the whole of the lake during 1968 and 1970 defined twenty-five biotopes (Fig. 5) according to three criteria: nature of the bottom, type region (open water, archipelago, reed island) and geographical position

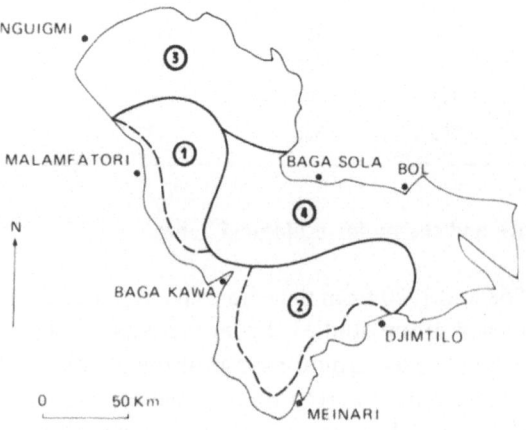

Fig. 4 Main community distribution zones of worms (after Carmouze et al. 1972).

Table 4 Mean densities and biomass of benthic worms in March, July and November, 1970, on the main substratum types and in the four large distribution zones defined for this group (Fig. 4) (after Carmouze et al. 1972).

March 1970		Zone 1	Zone 2	Zone 3	Zone 4
Mud	$N\ m^{-2}$	14353		2874	5009
	$kg\ ha^{-1}$	50.25		9.97	17.52
Clay	$N\ m^{-2}$	2583	500	5867	1083
	$kg\ ha^{-1}$	5.23	5.67	20.08	12.45
Pseudo-sand	$N\ m^{-2}$		1052		
	$kg\ ha^{-1}$		22.30		
Peat	$N\ m^{-2}$			0	0
	$kg\ ha^{-1}$			0	0

July 1970		Zone 1	Zone 2	Zone 3	Zone 4
Mud	$N\ m^{-2}$	279		0	0
	$kg\ ha^{-1}$	0.67		0	0
Clay	$N\ m^{-2}$		235	0	683
	$kg\ ha^{-1}$		6.45	0	8.61
Pseudo-sand	$N\ m^{-2}$		808		
	$kg\ ha^{-1}$		10.76		
Peat	$N\ m^{-2}$			0	0
	$kg\ ha^{-1}$			0	0

November 1970		Zone 1	Zone 2	Zone 3	Zone 4
Mud	$N\ m^{-2}$	2828		218	412
	$kg\ ha^{-1}$	9.44		0.75	0.70
Clay	$N\ m^{-2}$	1131	581	4750	372
	$kg\ ha^{-1}$	11.51	15.11	15.81	5.39
Pseudo-sand	$N\ m^{-2}$		526		
	$kg\ ha^{-1}$		12.83		
Peat	$N\ m^{-2}$			0	0
	$kg\ ha^{-1}$			0	0

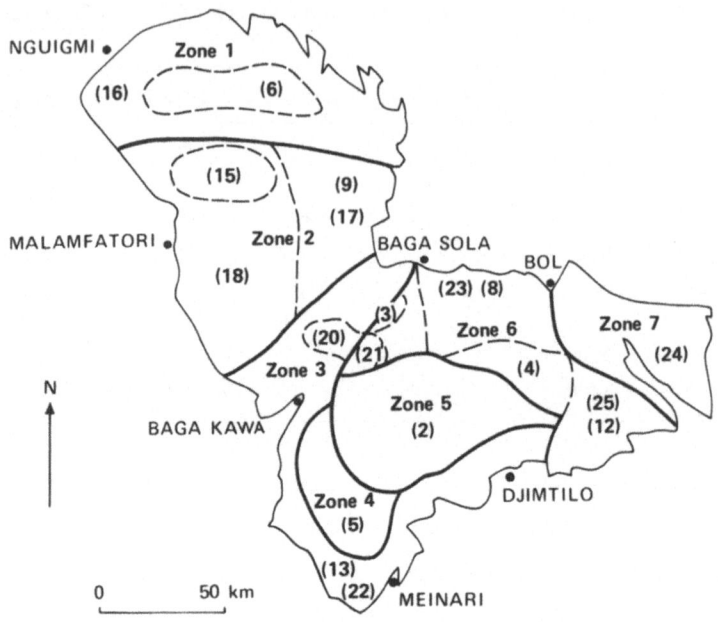

Fig. 5 Localization of the major geographical zones and main biotopes chosen for a study of benthic mollusc communities (after Lévêque 1972); 1 = sand (whole lake); 2 = pseudo-sand (west of the delta); 3 = pseudo-sand of the Great Barrier; 4 = blue clay (east); 5 = granular clay of the south; 6 = granular clay of the north; 7 = granular clay of the Great Barrier; 8 = granular clay of the eastern archipelago; 9 = granular clay of the central archipelago; 10 = peat of the eastern archipelago; 11 = peat of the Great Barrier; 12 = peat of the eastern reed islands; 13 = peat of the southern border; 14 = peat of the reed islands in the eastern open waters; 15 = blue mud of the central zone; 16 = mud of the northern zone; 17 = mud of the eastern archipelago; 18 = mud of the central open water; 19 = mud of the Great Barrier; 20 = mud of zone A; 21 = mud of zone B; 22 = mud of the southern reed islands; 23 = mud of the eastern archipelago (west of Bol); 24 = mud of the eastern archipelago (east of Bol); 25 = mud of the eastern reed islands.

(Lévêque 1972). Because of their reduced surface some of these biotopes were sampled once, either in 1968 or 1970.

A comparison of mean figures per biotope, using correspondence analysis (Lévêque and Gaborit 1972) enabled us to regroup those communities having fairly similar species composition and to define seven main ecological zones (Tables 5 and 6).

Zone 1 — Archipelago and reed islands of the north (biotopes 6 and 16). conductivity was higher than 500 µS cm^{-1} and molluscs were scarce, consisting mainly of *Melania*.

Zone 2 — Open water of the north and the southern part of the northeastern archipelago (biotopes 9, 15, 17, 18). *Melania* was dominant, but the abundance

Table 5 Mean densities of molluscs by number of individuals m^{-2} in the different biotopes (Fig. 5) sampled in 1968 (after Lévêque 1972).

Zones	1		2				3						4	5	6						7
Biotope No.	6	16	9	15	17	18	3	7	11	13	19	22	5	2	4	8	10	14	23	25	24
Nature of sediment	AG	V	AG	VB	V	V	PS	AG	T	T	V	V	AG	PS	AB	AG	T	T	V	V	V
Number of samples	16	9	1	9	12	22	2	2	2	8	17	6	24	11	7	5	5	3	14	8	6
Melania	13.3	7.6	314.8	51.5	119.1	160.6	202.8	22.2		14.4	102.3	9.3	197.6	180.4	18.5	0.4	2.6	11.1	1.3	6.3	
Cleopatra	0.9	0.6	75.9	1.2	39.8	4.7	171.8	199.1	37.0	21.8	96.2	12.0	55.0	102.7	64.8	8.1	19.3	130.9	6.1	33.3	1.9
Bellamya	4.6	0.8	174.1	20.4	73.7	21.4	235.7	45.4	5.6	8.3	25.3	1.5	3.9	0.3	0.3	4.5	3.0	1.9	1.9	2.6	
Corbicula					4.0	1.7	166.7	24.1	2.8	10.9	30.6	6.8	9.9	79.4	3.4	10.4	1.1	29.6	1.1	11.3	0.8
Caelatura			7.4		4.0	0.3	11.1	2.8	0.9		1.7	2.5	3.5	1.0	1.6	0.7	1.1		0.5	0.2	
Pisidium	0.6		1.9		1.9		3.2				1.7			6.4			6.4			2.6	
Eupera			3.7		2.3		11.6	37.0	0.9		3.8		0.5	0.5	0.3	0.4		0.9		0.5	
Total	19.4	9.0	577.8	73.1	244.8	168.7	802.9	330.6	47.2	55.4	261.6	32.1	270.4	370.7	88.9	24.5	33.5	174.4	20.9	56.8	2.7

245

Table 6 Mean density of benthic molluscs by number of individuals m^{-2} in the different biotopes (Fig. 5) sampled in 1970 (after Lévêque 1972).

Zones	1		2				3			4	5	6			7		
Biotope No.	6	16	9	15	17	18	11	19	20	5	2	4	10	21	23	25	24
Nature of sediment	AG	V	AG	VB	V	V	T	V	V	AG	PS	AB	T	V	V	V	V
Number of samples	11	7	2	5	30	11	4	9	3	16	27	10	15	4	14	13	20
Melania	1.9	3.5	259.0	185.6	159.3	250.9	34.3	263.9	332.0	94.1	125.9	34.8	4.4	75.6	8.2	10.9	7.2
Cleopatra			28.7	3.0	23.3	9.4	115.7	38.5	103.1	18.1	76.3	63.7	3.9	157.4	11.7	23.1	
Bellamya	0.2		173.1	40.3	41.7	101.9	63.9	6.5	171.7	1.1	0.5	0.9	0.5	4.6	1.7	1.5	0.1
Corbicula					7.0		27.8	22.6	77.2	18.0	73.1	11.1	0.6	29.1	1.5	8.7	0.1
Caelatura			2.8		0.7		6.5	12.6	25.9	2.8	2.2	3.0	0.4	7.4	0.6	0.4	0.1
Pisidium			8.3		0.6		0.9		1.9		1.3				0.1		
Eupera			2.8		0.2	0.4	5.6		3.2		0.4			0.5	0.2		
Total	2.1	3.5	474.7	228.9	232.8	362.6	254.7	344.1	715.0	134.1	279.7	113.5	9.8	274.6	23.7	44.8	7.6

of *Bellamya* in the communities was characteristic of this zone. Molluscs densities were very high.

Zone 3 — 'Great Barrier' and sand banks of the south (biotopes 3, 7, 11, 13, 19, 20, 22). This was a transitional zone between the north and the south basins and the communities were more heterogeneous. *Melania* was still abundant, but *Cleopatra* and *Corbicula* were far better represented than in zone 2, whereas *Bellamya* was scarce. In 1970, *Caelatura* was also seen to be abundant here. Certain biotopes of zone 3 showed some affinities with other zones. In particular biotopes 7 and 11 (granular clay and peat of the Great Barrier) resembled the community of zone 6.

Zone 4 — Open water of the southern basin (biotope 5) *Melania* was dominant, but *Cleopatra* and *Corbicula* were fairly abundant, and *Bellamya* very scarce.

Zone 5 — 'Pseudo-sand' of the southern open water (biotope 2). The 'pseudo-sand' occupied a large area of the southern open water and its community was well defined during 1968 and 1970. *Melania* dominated, but the high abundance of *Corbicula* was especially characteristic of this zone.

Zone 6 — Archipelagos, reed islands and open water of the southeast (biotopes 4, 18, 10, 14, 21, 23, 25). The communities were characterized by the dominance of *Cleopatra*. *Melania* remained well represented, but the rest of the fauna was poor in general, with the occasional exception of *Corbicula*. Density was not very great in most of the biotopes.

Zone 7 — Eastern archipelago (biotope 24). The community was very sparse and identical to that of zone 1 in 1970 (dominance of *Melania*), whereas in 1968 it was very similar to that of zone 6 in structure (dominance of *Cleopatra*).

According to previous observations, it appeared that in 1970 *Melania* was dominant almost everywhere, with the exception of zone 6 where *Cleopatra* dominated. *Bellamya* was well represented particularly in zones 2 and 3, *Corbicula* in zones 3, 4 and 5 and *Caelatura* in zone 3. Zones 1 and 7 were particularly low in molluscs, for which the highest densities were to be found in zones 2 and 3. A fairly clear zonation pattern of mollusc communities was therefore apparent for the whole lake. The geographical situation which may correspond to all the non-identified ecological factors seemed to have considerable importance.

8.3.3 *Insects*

8.3.3.1 *Chironomidae.* Samples taken in 1970, permitted identification of six main ecological zones in Lake Chad, according to which dominant species characterized the community. However, the limits of these zones were not constant throughout the year (Fig. 6).

Zone 1 — with *Cladotanytarsus lewisi* and *Tanytarsus nigrocinctus*. In

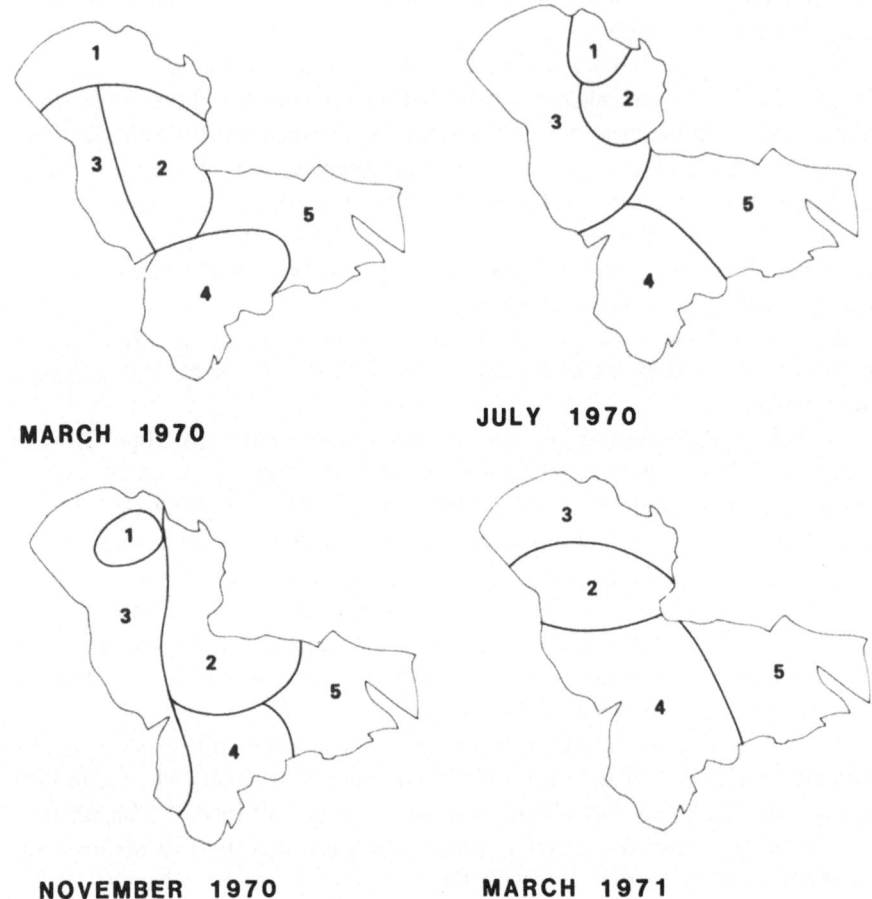

MARCH 1970

JULY 1970

NOVEMBER 1970　　　　MARCH 1971

Fig. 6 Zonation of Lake Chad based on the distribution of chironomids (after Dejoux 1976); 1 = zone with *Cladotanytarsus lewisi* and *Tanytarsus nigrocinctus*; 2 = zone with *Polypedilum fuscipenne*; 3 = zone with *Clinotanypus claripennis*; 4 = zone with *Cryptochironomus diceras*; 5 = zone with *Chironomus formosipennis*.

March, this zone covered the north of the lake and these two species, representing 82% of the community were absent from the rest of the lake. In July, *C. lewisi* disappeared, but *Polypedilum fuscipenne* and *Clinotanypus claripennis* were abundant. The presence of *T. nigrocinctus* distinguished this zone, although it was scarce. The zone became more clearly defined in November when *T. nigrocinctus* and *C. lewisi* represented 50% of the community.

Zone 2 — with *P. fuscipenne*. In March 1970 this zone covered the north-eastern archipelago and part of the open water of the north and the Great Barrier. *P. Fuscipenne* was very abundant (60% of total captures) with

Photo 12 Some benthic molluscs.

Photo 13 Boxes used to rear benthic molluscs.

Cryptochironomus stilifer also well represented (14%). In July the zone was reduced to the northeastern archipelago. *Clinotanypus claripennis* (48% of the community) was more abundant than *P. fuscipenne* (23%) and *C. stilifer* had disappeared. In November the zone spread once again to the Great Barrier. *P. fuscipenne* became dominant again (67%) and *C. stilifer* reappeared (27%).

Zone 3 — with *Clinotanypus claripennis*. In March the zone occupied the eastern part of the open water of the north, spreading north and south during the year. *C. claripennis* was abundant and *C. stilifer* was present in notable quantities.

Zone 4 — with *Cryptochironomus diceras* and *Ablabesmyia* sp. This zone, which was the poorest of the whole lake, fluctuated little during the year. *C. diceras* and *Ablabesmyia* sp. were found throughout the year at low densities rarely exceeding 100 individuals m^{-2}.

Zone 5 — with *Chironomus formosipennis*. No species was predominant in this zone which covered the archipelago and the reed islands of the east and southeast. The regular presence of *C. formosipennis* was however, characteristic, together with the three species of *Nilodorum*.

Zone 6 — with *Cricotopus scottae*. This shapeless and patchy zone all over the lake corresponded to the submerged water grasses. It had therefore no precise geographical location. *C. scottae* was largely predominant and *Dicrotendipes fusconotatus* was well represented.

If the changes observed in 1970 are considered characteristic of the seasonal changes affecting the chironomid population, a situation similar to that of March 1970 should have been found in March 1971. But, in fact, certain changes seem to have taken place:
— zone 1 disappeared and *T. nigrocinctus* became scarce;
— zone 2 spread distinctly westwards;
— zone 3 moved northwards, occupying the position of zone 1 in March 1970;
— zone 4 also spread northwards and a species of *Procladius* supplanted *Ablabesmyia* as the characteristic species;
— zone 5 changed little, but *P. fuscipenne*, absent before in this part of the lake, appeared for the first time.

It thus appeared that within one year a community transfer took place, occurring curiously in the same direction as that of the great water masses.

Even over a fairly short time, the chironomid communities appeared to be much less stable than those of worms and molluscs. The rapid development cycles of these insects, together with their great mobility and their great facility for adaptation, were certainly the main causes of the modifications observed in their distribution.

8.3.3.2 *Other insects*. Most species had a very extensive distribution, but the nature of the bottom was an important factor. Thus *Dipseudopsis capensis* and *Eatonica schoutedeni* were found principally on muddy bottoms rich in plant debris, whereas *Ecnomus* sp. and *Cloeon fraudulentum* always lived on clay and sandy bottoms. Generally the density of larvae was greater in the northern basin than in the southern basin of the lake (Carmouze et al. 1972). In the latter, the mean densities were respectively 9, 3 and 21 larvae m^{-2} in March, July and November 1970, whereas they were 36, 26 and 44 larvae m^{-2} in the northern basin. As for the sub-benthic species not living in the sediment, (*Chaoborus anomalus* and *Micronecta scutellaris*), density was also three times lower in the southern basin than in the northern basin.

8.4 Characteristics of the communities

8.4.1 *Distribution of organisms*

A study of the distribution of the molluscs has shown that they have a slight tendency to aggregate (Lévêque 1972). Here we have used Taylor Power law: $S^2 = a X^{-b}$. The relationship between the population mean ($\bar{x}$) and the variance (S^2) was obtained for each species from the series of samples from different biotopes and from different stations. The constant (b) was considered an aggregation index characteristic of the species under study. It is equal to 1 if the distribution is uniform and its value exceeds 1 as the individuals of the species studied tend to form aggregates.

In the case of the benthic molluscs of Lake Chad, the mobile prosobranchs ($1.12 < b < 1.84$) had a greater tendency to congregate than the lamellibranchs ($0.92 < b < 1.56$) whose distribution tend to be uniform. In the chironomids (Dejoux 1976), larval distribution was usually aggregated.

8.4.2 *Abundance distribution*

It has been shown that for most communities of benthic molluscs, the number of individuals in each species can be classified according to a geometrical progression law, the law of Motomura (Daget and Lévêque 1969; Lévêque 1972). This law holds (Inagaki 1967) when the correlation coefficient, calculated between the numbers and rank of each species (classified in decreasing order of numbers) is higher than 0.95. When it is not confirmed ($r < 0.95$), it may be concluded that the community was going through a period of change or else that the sampling was inadequate.

One interesting aspect of this law is that it permits the definition of 'nomocenoses' (Daget et al. 1972). They can be characterized by three parameters: the number of species; the environmental or Motumura (m), constant which is the linear regression slope between the numbers and ranks and which corresponds to a diversity index; the density of the community.

These parameters have been calculated for the communities of the Bol region and for the whole lake sampled in 1970 (Table 7). The law of Motomura proved true in most cases signifying that fairly stable communities existed and were well sampled.

8.4.3 Characteristic species of biotopes

These species may be investigated by the minimal deviation method (Bonnet 1964). We applied it to communities sampled at Bol in 1970 (Dejoux et al. 1971; Lévêque 1972; Dejoux 1976). Among the oligochaetes, *Naïs* and *Aulophorus* which were well represented in the periphyton, were always found on sandy bottoms near submerged vegetation as in the case of the zone under study. *Alluroides* and *Pristina* were abundant on clay beds and *Aulodrilus* in the mud and sand. Among the molluscs, *Melania* was typical of mud bottoms, *Bellamya* of granular clay; *Corbicula* and *Caelatura* of clays. There were no characteristic species of peat or sand. In comparison with the two preceding groups, the insects had relatively few characteristic species for any type of bottom. However, *E. dispar* and *C. fraudulentum* were found in granular clay and *N. rugosum* and *Chrionomus* sp. 7 in soft clay, *C. claripennis* in mud, *A. dusoleili*, *Stictochironomus* sp. 2 and *Tanytarsus* sp. 1 in peat, *C. lewisi*, *Cryptochironomus* sp. 1 and sp. 2 and *Polypedilum* sp. 3 in the sand.

The characteristic species of a bottom type did not necessarily constitute a biological association and were only valid for the zone studied at a given time, i.e. in well-defined ecological conditions. Thus, even though characteristic mollusc species were found in the biotopes sampled in 1970 over the whole lake (Lévêque 1972), the same results were not necessarily found at Bol at the same time. *Melania* was not characteristic of any biotope whereas *Corbicula* was characteristic of the blue clay sand and pseudo-sand beds and *Caelatura* of sand and mud bottoms of the Great Barrier. Many biotopes, had no particular typical species. This was also true of the insects, although *C. lewisi* was characteristic of sand over the whole lake and at Bol. Other species were found in the mud and granular clay, while peat and soft clay had no characteristic species (Dejoux 1976).

8.4.4 Interspecific relations

The study of affinities between species was based on the assumption that species from different samples whose numbers varied in a similar manner all had common ecological requirements. These affinities have been examined in mollusc samples taken at Bol in 1967 and 1970 and over the lake in 1970 (Fig. 7). With the exception of a strong affinity between *Corbicula* and

Table 7 Abundance distributions of benthic molluscs in different biotopes of the Bol Region and Lake Chad in 1970 (cf. Fig. 5 for the biotope numbers). Values of the parameters defining the nomocenoses: density, Motomura constant (m) and number of species; r = the correlation coefficient between the ranks and the abundance of species (after Lévêque 1972).

Biotopes	A at Bol	G at Bol	T at Bol	S at Bol	V at Bol	1	2	4	5	9	10	11	17	18	19	20	21	23	25
Sampled surface in m²	8	8	13	7	19	5.5	13.5	5	8	1	7.5	2	15	5.5	4.5	1.5	2	7	6.5
Density m⁻²	61.6	277	34.8	23	25	258.4	301.8	122.2	144.9	513	30.7	275	251.5	391.6	371.3	772	296.5	25.1	48.5
r	0.952	0.968	0.921	0.898	0.988	0.970	0.948	0.993	0.975	0.973	0.932	0.970	0.990	0.971	0.957	0.963	0.980	0.969	0.981
m	0.417	0.414	0.374	0.488	0.290	0.381	0.325	0.335	0.336	0.354	0.353	0.472	0.313	0.108	0.424	0.410	0.329	0.454	0.347
Number of species	7	6	6	5	5	7	7	5	5	6	5	7	7	4	5	7	6	5	6

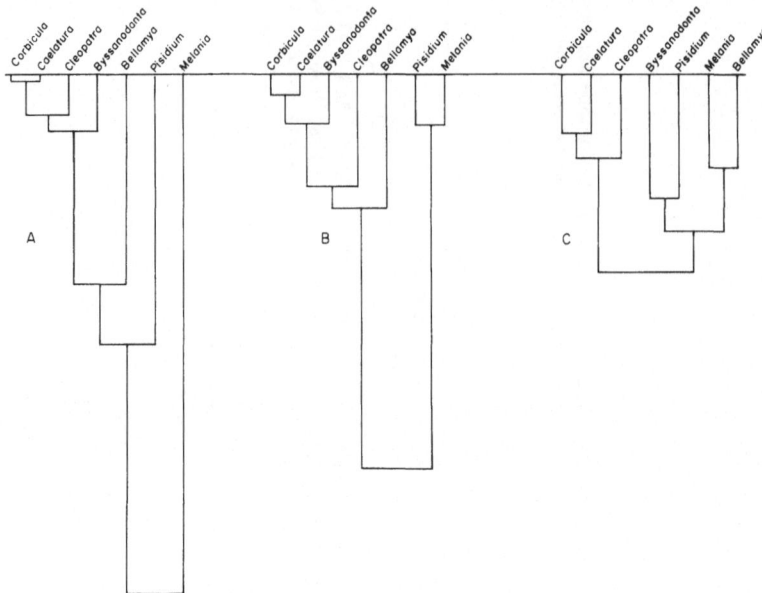

Fig. 7 Interpretation of the interspecific correlation matrices by the dendrogram method for molluscs; A = Bol, 1970; B = Bol 1967; C = whole lake, 1970.

Caelatura found in three series of observations, the other results did not agree. It may be considered then that relations revealed in a given situation are mere coincidences that must be verified elsewhere at other times to ensure their validity.

8.5 The benthic biomass

8.5.1 *Mean weight of the principal species and conversion factors*

The mean alcohol weight of the worms was evaluated as 3.41 mg for Alluroïdidae, 0.35 mg for Tubificidae and 0.06 mg for Nematodes. The dry weight was estimated at around 12% of the alcohol weight and the energy content at 5300 cal g^{-1} of the dry weight (Cumins and Wuycheck 1971).

For the principal species of benthic molluscs, the mean alcohol weight, the dry body weight, and the weight of the shell have been calculated (Table 8) from samples taken in 1970 over the whole lake. The energy content was 4000 cal g^{-1} for the different species (Lévêque 1973) (body without shell).

Finally, for the insects (Table 9), the fresh and dry weight of some species were determined by different methods (Dejoux 1976). The energy content was estimated as an average of 5300 cal g^{-1} dry weight, similar to the oligochaetes.

254

Table 8 Mean alcohol weight (in mg), dry body weight and shell weight for the main benthic species of Lake Chad collected in 1970.

	Alcohol weight (A.W.)	Dry body weight (D.B.W.)	A.W./D.B.W. (%)	Shell weight (S.W.)	S.W./A.W. (%)
Melania	88	6.6	7.5	48	55
Cleopatra	178	30.3	17	107	60
Bellamya	652	52.2	8	300	46
Corbicula	232	6.3	2.7	153	66
Caelatura	1571	78.6	5	1037	66

8.5.2 *Distribution of benthic biomass in Lake Chad*

8.5.2.1 *Worms.* The biomass distribution in March, July and November 1970 is shown in Fig. 8. It will be noted that the highest values were found in March and the lowest in July, corresponding to the seasonal abundance cycle observed in this group of invertebrates. Nevertheless, the biomass on the pseudo-sand and clay sediments of the southern open water, composed almost exclusively of Alluroïdidae, remained relatively important throughout the year. Variations were more marked in other zones where the Tubificids were dominant.

8.5.2.2 *Molluscs.* The number of individuals and the biomass have been calculated for the different species for each biotope sampled in 1970, using, the mean numbers per biotope (Tables 5 and 6), the surface occupied, and the mean individual weight. In order to obtain an estimate for the whole lake, the results obtained in 1968 from the biotopes 3, 7, 13 and 22, which were not sampled in 1970, were also considered.

The total number of benthic molluscs was 3.5×10^{12} individuals in 1970, made up of 61% *Melania*, 15% *Cleopatra*, 15% *Bellamya*, 8% Corbicula, and 1% *Caelatura*. A comparable value of 3×10^{12} individuals was obtained in 1968 (Lévêque 1972).

Expressed as dry body weight, the molluscs biomass totalled 64 000 tons in Lake Chad in 1970, for a water surface area estimated at 19 200 km^2. In terms of proportion of the biomass, 93% of the total consisted of three species: *Bellamya* (43%), *Cleopatra* (23%) and *Melania* (22%). the Lamellibranchs *Caelatura* (5%) and *Corbicula* (2%) were less common. The biomass of shells was 410 000 tons.

The biomass was not uniformly distributed in the lake (Fig. 9), since mean values between 0.2 and 200 kg ha^{-1} were observed, varying with biotopes. The highest values recorded, between 35 and 200 kg ha^{-1}, occurred in the open

Table 9 Mean wet and dry weight of larval insect species of Lake Chad.

Species	Fresh weight (mg)	Dry weight (mg)
Chironomidae		
Chironomus formosipennis	3.08	
Cryptochironomus stilifer	0.56	0.03
Cryptochironomus diceras	2.41	
Polypedilum fuscipenne	2.35	0.33
Tanytarsus nigrocinctus	0.45	
Cladotanytarsus lewisi	0.27	
Cladotanytarsus pseudomancus	0.25	
Rheotanytarsus ceratophylii	0.40	
Dicrotendipes fusconotatus	0.78	
Cricotopus scottae	0.84	
Ablabesmyia dusoleili	0.35	
Procladius brevipetiolatus	1.42	
Clinotanypus claripennis	2.50	0.39
Nilodorum rugosum	7.45	0.75
Chaoborus anomalus	1.89	0.04
Ceratopognus sp.	0.43	0.026
Ephemeroptera		
Eatonica schoutedeni		3.0
Cloeon fraudulentum		0.04
Povilla adusta		1.0
Trichoptera		
Ecnomus dispar		1.5
Dipseudopsis capensis		2.5
Hemiptera		
Micronecta scutellaris		0.07

water and archipelago of the north basin, and around the Great Barrier. In the extreme north and east of the lake much lower values were obtained.

8.5.2.3 *Chironomidae.* As with oligochaetes, a seasonal abundance cycle was observed for this group, with minimal biomass in July (Fig. 10). The northern basin was richer than the southern basin. This phenomenon could be related to the greater ecological stability of the slightly deeper northern basin, which was not directly affected by the flood waters of the Shari, as was so in the southern basin.

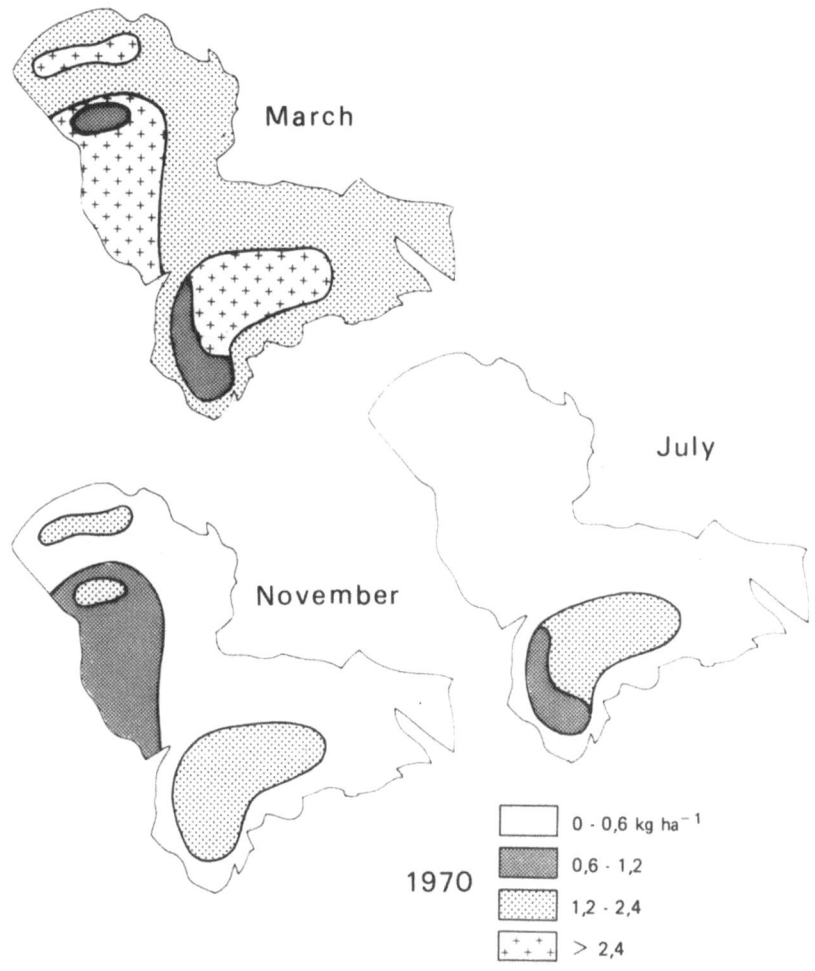

Fig. 8 Distribution of benthic oligochaete biomass (dry weight) in March, July and November, 1970, over the whole lake.

It should be noted that this group of invertebrates had a fairly rapid biological cycle and great facility of movement, in contrast to those mentioned earlier, and this probably accounted for the greater variations observed between two sampling periods.

8.5.3 *Estimation of benthic invertebrate biomass in 1970*

The zonation patterns proposed for the three groups of benthic invertebrates did not overlap exactly, because the distribution factors did not all have the

257

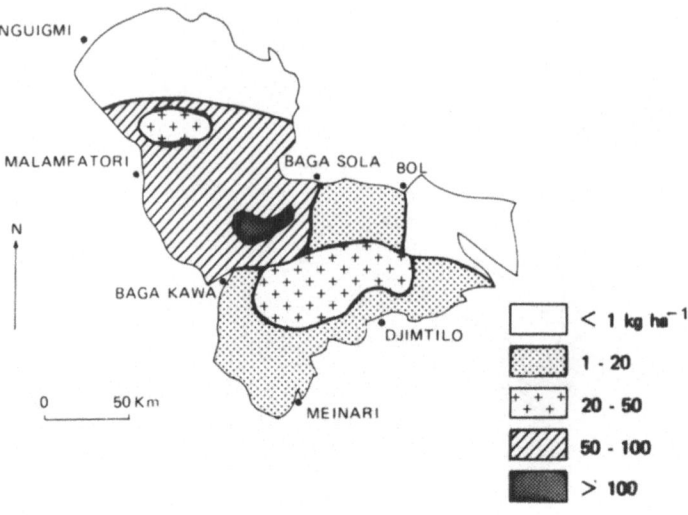

Fig. 9 Distribution of benthic mollusc biomass (dry weight, shell free) in March, 1970, over the whole lake.

same significance for each. It should be emphasized that a distinct difference seemed to exist between the north and south basins for all groups, as well as between the open water zone and the archipelago. On the basis of the distribution of the molluscs, which consititued the bulk of the biomass, seven zones could be clearly defined (Fig. 11). For each, the mean biomass (dry weight) of the three groups studied in March 1970 (Table 10) has been calculated. For the whole lake, the mean biomass of molluscs (33 kg ha^{-1} d.w.) was about 11 times greater than that of the worms (2.9 kg ha^{-1}) and 27 times greater than that of the insects (1.2 kg ha^{-1}). The mean biomass of shells was approximately 210 kg ha^{-1}. The total bentic biomass was estimated to be 71 100 tons composed of 5500 tons of worms, 2300 tons of insects and 63 000 tons of molluscs. Most of the biomass (74%) was concentrated in zones 2, 3 and 4, corresponding to the open water of the north, the northeastern archipelago and the Great Barrier, which represented no more than 40% of the lake surface. The energy value of the total benthic stock was 294 540 × 10^6 Kcal, of which 29 200 × 10^6 Kcal was contributed by the worms, 12 190 × 10^6 Kcal by the insects and 253 000 × 10^6 Kcal by the molluscs. The average energy value was 152.4 × 10^3 Kcal ha^{-1}. The above estimations only concerned the benthos and completely excluded the invertebrate communities associated with the aquatic macrophytes. This was a major omission in so far as aquatic vegetation stands in Lake Chad were estimated to cover about 2000 km^2 in 1970 (clumps of *Potamogeton* or *Ceratophyllum*, *Papyrus*, *Phragmites* or *Typha*). Such habitats were some-

258

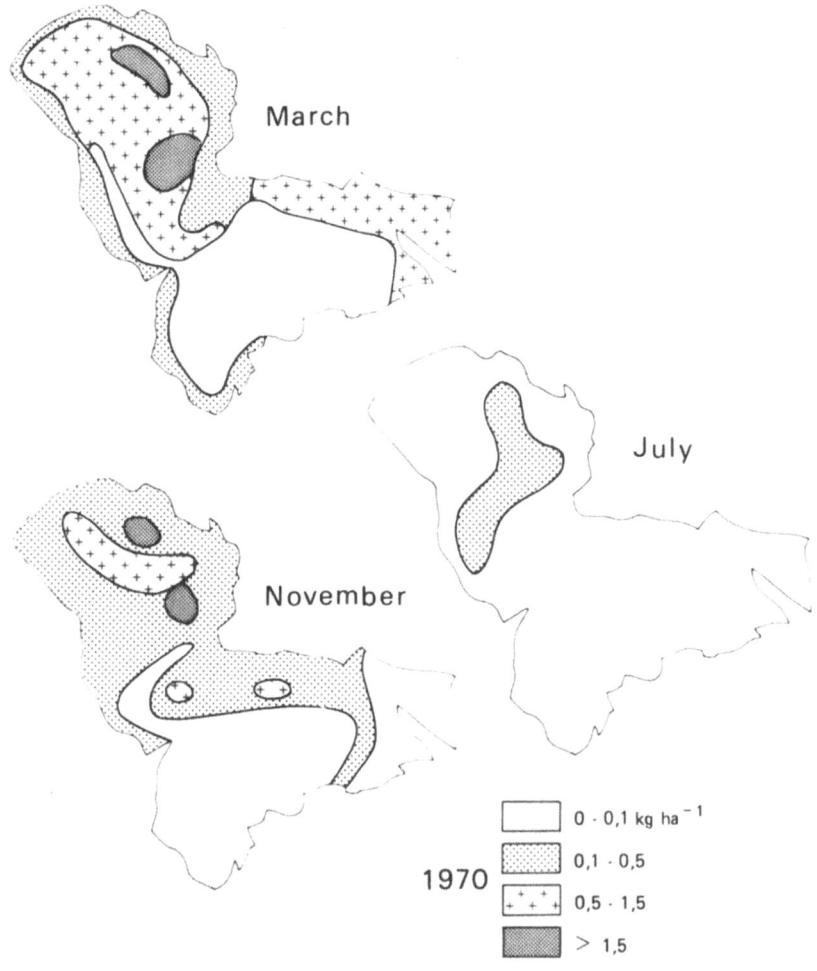

Fig. 10 Chironomids: distribution of biomass (dry weight) in March, July and November, 1970, over the whole lake.

times fairly rich in insects, pulmonate molluscs and worms (Naididae in particular).

8.6 Temporal community changes

Throughout the period during which the benthic communities were studied, the level of Lake Chad never ceased to drop. This phenomenon, fairly discreet until 1970, accelerated later until in 1973 the north and south basins separated. Later the north basin dried up (1975).

259

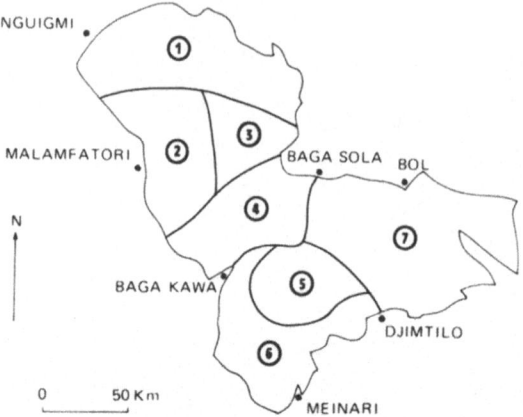

Fig. 11 Zonation pattern of the benthic fauna in Lake Chad in 1970 (cf. Table 10) (after Carmouze et al. 1972).

Unfortunately it was impossible to simultaneously study the changes within the benthic communities for all groups in the whole lake. We do however have some information showing that significant changes occurred in certain zones.

8.6.1 *Molluscs*

The temporal changes of the malacological communities had been observed in the four bottom types of the southeastern archipelago (Bol region) where sampling was carried out in January 1967, 1970 and 1972 (Carmouze et al. 1972). The density of all species in all biotopes diminished generally from 1967 to 1972 (Table 11), with the exception of *Cleopatra* and *Melania* which, in 1970, became abundant in granular clay substrata and in mud substrata respectively. Regular sampling from 1967 to 1970 made it possible to follow more precisely the changes occurring in the mollusc communities on clay and mud bottoms. The most important changes occurred in 1968 and included a rapid decrease in the density of *Cleopatra* in the mud, followed by a slow increase in *Melania*, and finally the dominance of this species in 1970. At the same time *Bellamya* and *Eupera* decreased in granular clay and *Eupera* declined in soft clay (Fig. 12).

The temporal changes revealed in the southeastern archipelago were also observed in other regions of the lake. The mean numbers of species in the biotopes investigated during 1968 and 1970 (Fig. 13) were compared by factorial analysis (Lévêque and Gaborit 1972). This shows that the position of some biotopes, tended to move towards the left and the base of the graph in relation to axes 1 and 2, between 1968 and 1970. This observation may be explained by the fact that the relative density of *Melania*, in comparison with

Table 10 Mean benthic invertebrate standing stocks during 1970 in the seven main ecological zones of Lake Chad (Fig. 11).

Zones	Surface (km^2)	Mean biomass in dry weight (kg ha^{-1})				Mean biomass in kcal ha^{-1} $\times 10^3$
		Worms	Insects	Mollusca	Total	
1	3082	2.1	1.4	0.2	3.7	19.4
2	3871	8.0	2.1	64.2	74.3	310.3
3	1501	1.1	2.9	47.8	51.8	2.2.4
4	2133	1.9	1.6	72.0	75.5	306.5
5	2290	1.5	0.1	38.6	40.2	162.9
6	2083	2.6	0.1	11.8	14.5	61.5
7	4259	0.8	0.6	10.6	12.0	49.8
Mean for the whole lake		2.9	1.2	33.0	37.1	153.0

that of *Cleopatra*, increased in the biotopes examined in 1970, just as that of *Corbicula* increased more than that of *Bellamya*. The most spectacular example was that of biotope 24 (mud of the southeastern archipelago west of Bol and peat of the Great Barrier).

Only a few fragmentary results were obtained in 1972 in two biotopes of the open water of the east (Table 12). The pseudo-sand community had changed little during the preceding four years, either in structure or the species density. In the blue clay (the biotope neighbouring the southeastern archipelago), however, densities which had been fairly comparable in 1968 and 1970, had decreased considerably by 1972, thus coinciding with observations from the Bol region.

It is difficult to accurately determine the factors which provoked the observed changes in the malacological communities between 1967 and 1972, but, either directly or indirectly, the lower level of the lake was certainly largely responsible for these modifications. The shallower the lake, the greater the effect of waves on bottom sediments. With loose beds, the surface sediment may become partly suspended greatly perturbing ecological conditions, since the environment becomes highly unstable. A layer of very fluid mud may even develop at the water–sediment interface, reducing the mollusc survival which have a tendency to sink into the mud because of their weight. Furthermore, the waters become loaded with silt, as noticed in the Bol region (Lemoalle 1974) and this can be harmful to the small lamellibranch filter-feeders. On bottoms such as pseudo-sand (Table 12), where larger and heavier particles predominated, this phenomenon must have been less evident. Thus the communities of this substratum altered little between 1968 and 1972.

Table 11 Changes in the mean density of benthic molluscs, as number of individuals m^{-2} in the Bol region (southeastern archipelago) in January 1967, 1970 and 1972 (after Carmouze et al. 1972).

Species	Sediment type											
	Soft clay			Granular clay			Mud			Peat		
	1967	1970	1972	1967	1970	1972	1967	1970	1972	1967	1970	1972
M. tuberculata	2.0	0.3	0	1.0	1.1	0.4	0.6	18.5	1.8	4.8	6.6	3.3
C. bulimoides	121.7	48.1	0	37.2	222.4	1.5	7.2	3.2	0	37.6	24.9	5.9
B. unicolor	3.1	0.7	0	14.8	2.5	0.6	0.6	0.6	0.2	5.1	0.2	0.4
C. africana	66.3	5.4	0.4	62.4	15.4	1.7	0.5	0.8	0.2	0.3	0.4	3.3
C. aegyptiaca	11.0	1.6	0.2	4.0	3.1	3.2	0	0.1	0.4	0.2	0.1	0.2
P. pirothi	5.7	0.1	0	3.0	0	0	0	0	0	3.5	0.1	0.6
E. parasitica	16.8	0.1	0	97.2	11.1	0	0.2	0	0	0	0	0
Total	226.6	56.3	0.6	219.6	255.6	7.4	9.1	23.2	2.6	51.5	32.3	13.7

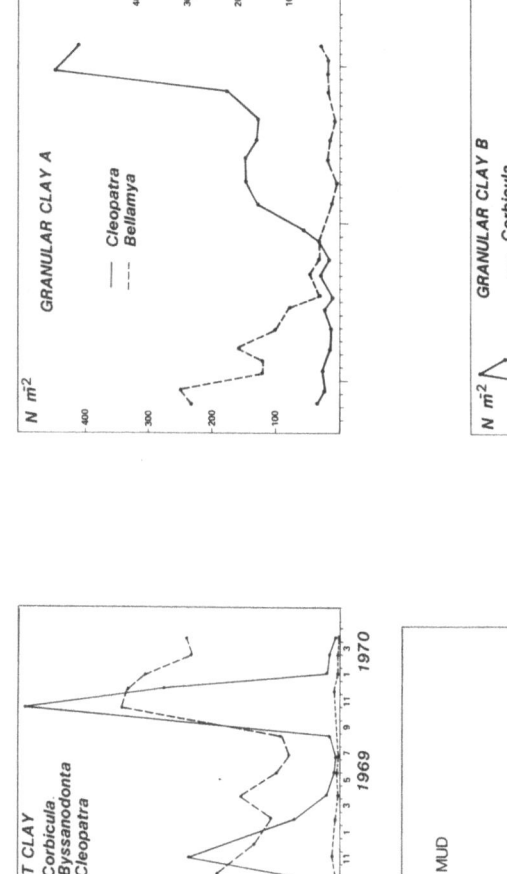

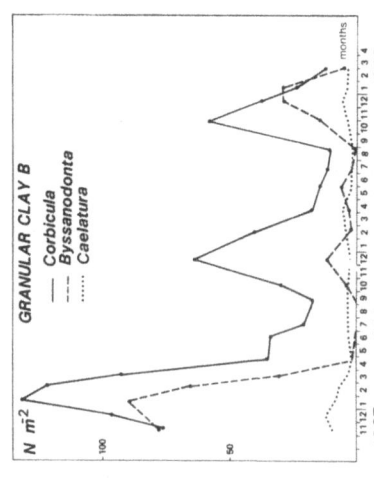

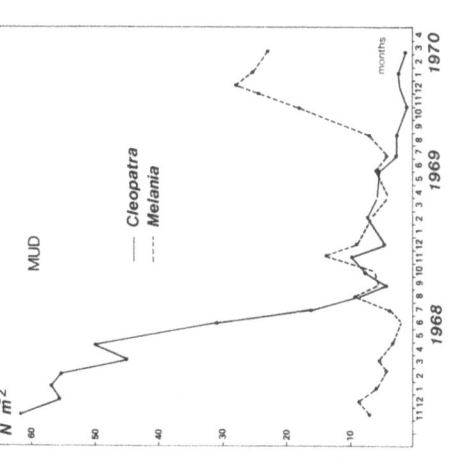

Fig. 12 Variations in the density of the main mollusc species on three bottom types in the Bol region (eastern archipelago), in November 1967, and March 1970 (after Lévêque 1973).

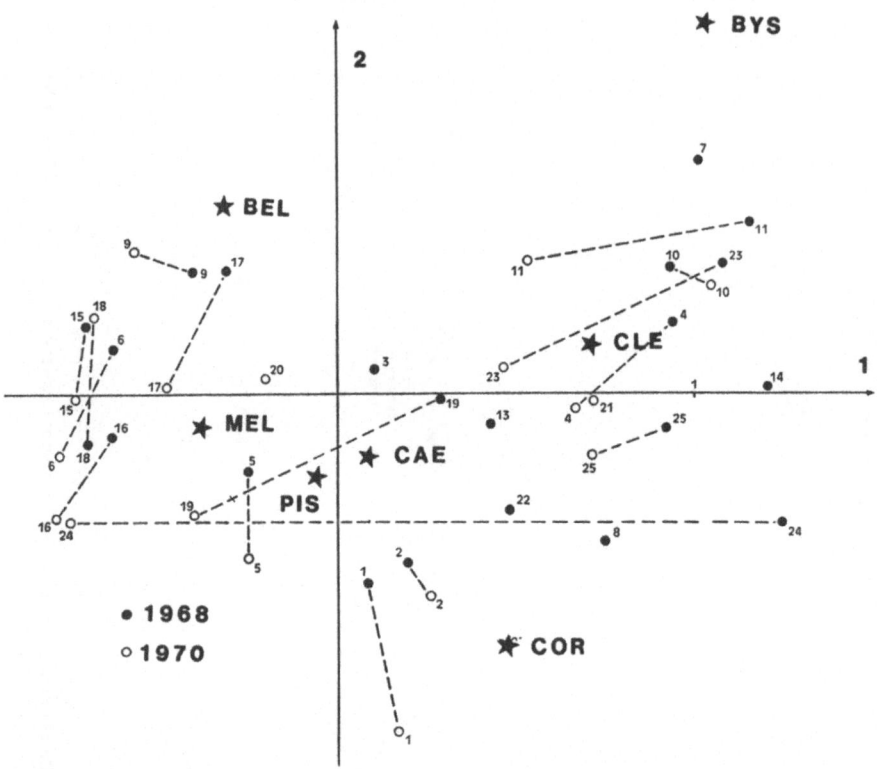

Fig. 13 Position on axes 1 and 2 of the mollusc species and mean plot readings in the biotopes sampled during 1968 and 1970 in Lake Chad (after Lévêque and Gaborit 1972).

8.6.2 *Chironomids*

We have already seen that the aquatic insect communities were fairly unstable over the course of a year. The same was true between years as shown by samples taken with a light trap between 1965 and 1974 at various points of the lake (Dejoux 1976).

In the southeastern archipelago (Bol region) a change in the dominant species was observed and particularly important was a decrease in the number of species collected between 1965 to 1974 (Table 13). On the other hand, species that were abundant and typical in the north of the lake in 1970 (*C. stilifer* and *T. nigrocinctus*) became dominant in 1972 and 1974 at Bol.

In the Shari delta (Table 14) the same phenomena were observed, but the decrease in the number of species was less spectacular than at Bol. In fact, at this station, lake species mingled with river species and the latter seemed to have been less affected than the others by the fall in lake level. This was for example

264

Table 12 Changes in mean density of the major species of benthic molluscs as number of individuals m^{-2} on the pseudo-sand and blue clay substrata of the southeastern open water in 1968, 1970 and 1972 (after Carmouze et al. 1972).

Species	Biotope 2			Biotope 4		
	pseudo-sand			blue clay		
	1968	1970	1972	1968	1970	1972
M. tuberculata	181	126	139	19	35	1
C. bulimoides	103	76	78	65	64	11
B. unicolor	1	1	0	0	1	0
C. africana	80	73	54	4	11	1
C. aegyptiaca	1	2	1	2	3	1
Total	366	278	272	90	114	14

the case, with *P. longicrus* and *P. griseoguttatum* which were collected fairly regularly. Among the lacustrine species, such as *P. maculosus* and *C. melutensis*, the same phenomenon occurred as at Bol, with sudden appearance or disappearance. At the same time *T. nigrocinctus* was also abundant in 1974.

8.7   Conclusions. Comparison with other lakes of the intertropical zone

Due to its shallow depth, Lake Chad, as a whole, can be considered as a vast littoral zone where the benthic communities are subjected to the combined action of several ecological features. One of the most important is the annual fluctuation of the water level under the influence of the Shari flood waters together with between year fluctuations dependent upon hydrological budget and the climatic 'past' of the system. During the period of our observations, a progressive fall in the level of the lake was evident, accompanied by modifications in the structure and density of the benthic populations. Thus, between 1967 and 1972, the malacological communities changed progressively and their densities decreased considerably on loose substrata. As the level of Lake Chad has fluctuated considerably since the beginning of the century, the majority of the benthic communities have probably been constantly disturbed, occasionally reaching extreme situations when they disappear completely (e.g. the north basin dried up in 1975). An identical phenomenon has recently occurred in another shallow African lake: Lake Chilwa.

During the period under study, which corresponded to the 'Normal Chad'

Table 13   Changes in the composition of the chironomid communities of the Bol region from 1965 to 1974, based on catches by light trap (Dejoux 1976).

Species collected (%)	1965	1966	1971	1972	1974
Polypedilum deletum	34	0.1	0	0	0
Polypedilum abyssiniae	42.1	0.5	0	0	0
Procladius brevipetiolatus	0	62.4	1.6	0	0.05
Polypedilum longicrus	3.4	6.8	15.5	0	0
Ablasbesmyia dusoleili	0.1	0.2	36.1	0	0.05
Cryptochironomus stilifer	0	0	34.9	88.1	0
Cladotanytarsus pseudomancus	0.3	0	0	0	14.3
Tanytarsus nigrocinctus	0.2	2.7	0	0.03	80.7
Total number of individuals collected in the year	69 143	4530	180	3399	5420
number of species collected	58	30	8	6	12
Equitability (%)	40.4	43.2	69.6	23.1	26.4

period, the mean biomass for the whole lake was approximately 3.7 g m^{-2} (dry weight). It will be recalled however, that vast zones of aquatic vegetation mats existed in Lake Chad, where insects were abundant. Nevertheless, the dominance of molluscs in the benthic biomass should be stressed, for it is a fairly infrequent phenomenon. Most of the stock was concentrated in the north basin and the Great Barrier. No definitive explanation was found for the relative paucity of this group in the south basin but it was probably related to the greater hydrological instability of this area. As the north basin was a little deeper than the south basin and did not receive the direct inflow of the Shari water, it could be considered as being truly lacustrine, whereas the south basin was merely an extension of the river system, influenced by disturbances from floods (Carmouze et al. 1972). It was also possible that the distribution of biomass was influenced by predation, which was not necessarily of the same intensity in different regions of the lake and on different species. The lack of quantitative data in this field, however, makes it impossible, at present, to estimate its importance.

The benthic communities and biomasses of tropical African lakes have been the subject of few quantitative research studies. Therefore it is worthwhile to examine the information at our disposal and to compare it with the results obtained in Lake Chad.

Table 14 Changes in the dominant species of the chironomid communities in the Shari delta from 1965 to 1974, based on catches by light trap (Dejoux 1976).

Species collected (%)	1965	1966	1968	1971	1972	1973	1974
Polypedilum abyssiniae	12.6	1.2	0	0	0	17.3	1.3
Ablabesmyia pictipes	18.0	6.6	0	0	0	0.5	0
Polypedilum griseoguttatum	23.3	2.7	23.3	0	0	2.4	8.6
Cryptochironomus dewulfianus	0.3	28.5	0	0	1.2	0.4	2.1
Polypedilum longicrus	3.4	52.6	0	40.2	0.7	9.6	22.6
Procladius maculosus	0	0	19.5	0	0	0	0
Clinotanypus claripennis	1.5	0.6	19.5	0	0	0	0
Ablabesmyia dusoleili	1.2	1.3	15.6	2.4	0.3	3.2	0.4
Cryptochironomus melutensis	0	0	0	0	73.1	0	0
Cryptochironomus nudiforceps	0.1	0	0	8.5	0.4	22.1	0.05
Cladotanytarsus pseudomancus	0.4	0	0	0	9.2	29.6	25.0
Tanytarsus nigrocinctus	0.3	0	0	0	0.2	3.1	34.5
Total number of individuals collected in the year	4914	834	71	82	1326	4718	6720
Number of species collected	49	20	11	18	14	23	18
Index of diversity	3.602	1.991	2.701	3.090	1.470	2.044	2.834
Equitability (%)	64.2	46.1	78.1	74.1	38.6	52.1	66.7

In Lake Léré (Chad), the mean biomass was estimated at $2.6 \, \text{gm}^{-2}$ (dry weight) for February 1970 (Dejoux et al. 1971). This value is comparable to that of Lake Chad. Further, the molluscs also dominated this system ($2 \, \text{g m}^{-2}$, shell-free dry weight).

A series of lakes in the region of the White Nile has been studied by Monakov (1969). The benthic standing crop varied between 0.1 and $0.5 \, \text{g m}^{-2}$ (estimation in dry weight). Oligochaetes were dominant but molluscs were sometimes well represented.

Lake George is a shallow lake which has a mean depth of 2.4 m. In this system, the ecological factors vary little with the seasons, due to its situation below the equator (Burgis et al. 1973). The superficial sediment consists essentially of very soft mud, rich in organic matter, and the first 6 to 10 cm of this are often resuspended by the action of the wind (Viner and Smith 1973). The benthic fauna (Darlington 1977) is composed mostly of Dipteran larvae (*Chaoborus*, chinoromids). Close to the shore, it is more diversified and the

267

gastropod *Melanoides tuberculata* is found. The absence of this species from most of the central regions of lake may be explained by the presence of unstable and the excessively fluid superficial sediments. The mean standing crop was approximately 1.2 g m^{-2} (dry weight) (Burgis et al. 1973).

Like Lake Chad, Lake Chilwa (Malawi) is a shallow endorheic lake with a level susceptible to considerable annual variations (McLachlan and McLachlan 1969; MacLachlan 1979). Observations on the benthic fauna were made in 1967, during a period of severe drought. At this time the fauna was very sparse and comprised only a few species of chironomids, Coleoptera and Hemiptera. However a study of the remains and debris of the fauna, thought to represent the benthic communities existing just before the fall in Lake level, showed that the number of species was originally far greater and that there were many molluscs. Among these, benthic species identical to those of Lake Chad (*Bellamya unicolor, Melanoides tuberculata*), as well as *Lanistes, Aspatharia* and diverse pulmonates were found. The benthic biomass, estimated in 1967 at 0.07 g m^{-2} (dry weight) in one region of the lake, must therefore have been considerably higher before the level fell and it may be supposed that molluscs constituted a large part of it, as in the case of Lake Chad. The disappearance of numerous benthic species could perhaps be due to an increased salinity of the water caused by the evaporation accompanying the drought. When the lake was flooded again in 1969, the mud substrata were rapidly colonized by *Chironomus transvaalensis*, but this species had practically disappeared ten weeks later (Maclachlan 1974). The benthic fauna, in 1969 and 1970, consisted mainly of insect larvae; the molluscs, apparently, not yet having had time to recolonize the sediments.

In Lake Turkana (ex-Rudolph), which reaches 80 m in depth, some observations have been made by Ferguson (1975). The benthic fauna was poor but was generally more abundant in the littoral zone than in deeper areas. Below 10 m, the community consisted mainly of two mollusc species (*Cleopatra pirothi* and *Malanoides tuberculata*), four species of chironomids and five species of ostracods. The abundance of these species diminished down to 80 m, where only *C. pirothi* and one chironomid species were found. The Ostracoda disappeared at 20 m. Oligochaetes were scarce beyond the littoral zone. Ferguson (1975) gives no estimation of the biomass. No explanation has been found so far for the scarcity of benthic fauna in Lake Turkana. It seems to have no connection, as in other lakes, with the oxygen content of the bottom water, which rarely decreases below 70% saturation at 80 m. It should be noted that the high conductivity of the Lake Turkana water (around $3000 \text{ }\mu\text{S cm}^{-1}$) did not appear to hinder the presence of molluscs, as in Lake Chad. One species (*M. tuberculata*) was found in both systems, and we must presume that the water composition differs or that the limit of conductivity observed in the north of Lake Chad is a mere coincidence and results from the intervention of other non-identified factors.

The benthic biomass of Lake McIlwaine (Rhodesia), a reservoir with a mean depth of 10 m (31 m maximum), can be estimated at about 3 g m^{-2} (dry weight), from Munro's data (1966). Insects and oligochaetes were dominant in 1962–1963. Later observations in 1968 (Falconer and Marshall 1970) showed that the 2–4 m deep zone was the richest and that below 5 m, no animals were found, probably due to the absence of oxygen at the bottom together with the high ammonia content. Moreover, Lamellibranchs were seen to disappear (especially *Sphaerium*) in 1968–1969. This was probably due to a very sharp drop in lake level during 1967–1968, which caused a large part of the littoral zone to dry up. Chironomid larvae were still dominant in the benthic fauna during 1968–1969, but less abundant than in 1962–1963 (similar to the oligochaetes).

The research by Petr (1972) in Lake Volta, during the years following the closing of Akosombo dam, deals essentially with the establishment of benthic communities. With the exception of *Pisidium*, molluscs were chiefly represented by the pulmonates, the abundance of which was linked to the presence of aquatic plants, as in Lake Chad (Lévêque 1975). The same was true for the oligochaetes, which were represented by the Naididae. Chironomids were largely dominant throughout the lake, at least during the first years of its filling and the Ephemeroptera, *Povilla adusta* was also an extremely important component of the benthic community during this filling phase in the development of the lake.

In Lake Kariba, a reservoir created in 1958, McLachlan and MacLachlan (1971) showed that chironomids largely dominated the benthos. Molluscs (represented by a pulmonate species) and oligochaetes were scarce. The total biomass was low, between 0.02 and 0.1 g m^{-2} (dry weight). However standing stocks recorded ponds along the shoreline were higher (1.4 g m^{-2}).

In comparison, the biomasses observed in certain lakes of Central Amazonia (Filtkau et al. 1975) are generally much lower (0.02 to 0.9 g m^{-2} and insects usually dominate.

It is impossible here to give an exhaustive account of the results obtained in other lakes around the world. It should be mentioned, however that Cole and Underhill (1965) give values for benthic biomass ranging between 4.3 and 8.6 g m^{-2} for seven lakes in North America, which they considered to be the most productive. The values are higher than the mean values observed in Lake Chad in 1970, but comparable to those of the north basin and the Great Barrier (Table 10). In Lake Esrom (Denmark), the mean biomass of 16.5 g m^{-2} is one of the highest ever observed (Jonasson 1972), but on the whole the benthic biomass of European lakes ranges from 0.2 to 2 g m^{-2}. We may therefore consider Lake Chad, and in particular the northern basin, to be one of the richest lakes as regards benthic biomass. Higher biomass values are generally found for invertebrates inhabiting aquatic vegetation mats. No data are available for Lake Chad, but in the south temperate coastal lake Smartvlei,

Davies (personal communication) observed mean standing stocks of 132 g dry weight m^{-2} (summer) and 34 g (winter), the mussel *Musculus virgiliae* being dominant too.

References

Bonnet, L., 1964. Le peuplement thecamoebien des sols. Rev. Ecol. Biol. sol 1: 123–408.

Burgis, M. J., Darlington, J. P. E. C., Dunn, J. G., Ganf, G. G., Gwahaba, J. J. and McGowan, L. M., 1973. Biomass and distribution of organisms in Lake George, Uganda. Proc. r. Soc. London B 184: 271–298.

Carmouze, J. P., Dejoux, C., Durand, J. R., Gras, R., Lauzanne, L., Lemoalle, J., Lévêque, C., Loubens, G. and Saint Jean, L., 1972. Grandes zones écologiques du lac Tchad. Cah. ORSTOM Sér. Hydrobiol. 6: 103–169.

Cole, G. A. and Underhill, J. C., 1965. The summer standing crop of sublittoral and profundal benthos in lake Itasca, Minnesota. Limnol. Oceanogr. 10: 591–597.

Cummins, K. W. and Wuycheck, J. C., 1971. Caloric equivalent for investigations in ecological energetics. Mitt. int. Ver. Limnol. 18, 158 pp.

Daget, J. and Lévêque, C., 1969. Application de la loi de Motomura aux mollusques du lac Tchad. Cah. ORSTOM Sér. Hydrobiol. 3: 81–85.

Daget, J., Lecordier, C. and Lévêque, C., 1972. Notion de nomocenose: ses applications en écologie. Bull. Soc. Ecol. 3: 448–462.

Darlington, J. P. E., 1977. Temporal and spatial variation in the benthic invertebrate fauna of Lake George, Uganda, Uganda. J. Zool. 181: 95–111.

Dejoux, C., 1968. Le lac Tchad et les Chironomides de sa partie est. Ann. Zool. fenn. 5: 27–32.

Dejoux, C., 1968. Contribution à l'étude des insectes aquatiques du Tchad. Catalogue des Chironomidae, Chaoboridae, Odonates, Trichoptères, Hémiptères, Ephéméroptères. Cah. ORSTOM Sér. Hydrobiol. 2: 51–78.

Dejoux, C., 1968a. Description d'une méthode d'élevage des Chironomides, adaptée aux pays tropicaux. Hydrobiologia 31: 435–441.

Dejoux, C., 1968b. Contribution à l'étude des premiers états des Chironomides du Tchad (1ère note). Description de *Tanytarsus nigrocinctus* et *Chironomus pulcher*. Hydrobiologia 31: 449–463.

Dejoux, C., 1969a. Les insectes aquatiques du lac Tchad. Aperçu systématique et bio-écologique. Verh. int. Ver. Limnol. 17: 900–906.

Dejoux, C., 1969b. Contribution à l'étude des premiers états des Chironomides du Tchad (2e note). Description de *Tanypus fuscus* et *Tanypus lacustris*. Bull. Mus. nat. Hist. nat. 2e sér. 41: 1152–1163.

Dejoux, C., 1970a. Contribution à l'étude des premiers états des Chironomides du Tchad (3e note). Description comparée des nymphes de *Nilodorus brevipalpis, N. brevibucca* et *N. fractilobus*. Bull. Mus. nation. Hist. nat. Paris 42: 175–184.

Dejoux, C., 1970b. Contribution à l'étude des premiers états des Chironomides du Tchad (4e note). Description de *Stictochironomus puripennis, Chironomus formosipennis, C. calipterus*. Cah. ORSTOM Sér. Hydrobiol. 4: 39–51.

Dejoux, C., 1971a. Recherches sur le cycle de développement de *Chironomus pulcher*. Can. Ent. 103: 465–470.

Dejoux, C., 1971b. Contribution à l'étude des premiers états des Chironomides du Tchad (5e note). Description de *Chironomus (Cryptochironomus) deribae, (Polypedilum) fuscipenne*. Cah. ORSTOM Sér. Hydrobiol. 5: 87–100.

Dejoux, C., 1973. Contribution à l'étude des premiers états des Chironomides du Tchad (6e note).

Description de *Tanytarsus (Rheotanytarsus) ceratophylli* n. sp. Cah. ORSTOM Sér. Hydrobiol. 7: 65–75.

Dejoux, C., 1976. Synécologie des Chironomides du lac Tchad (Diptères, Nématocères). Trav. Doc. ORSTOM n° 56, 161 p.

Dejoux, C., Lauzanne, L. and Lévêque, C., 1969. Evolution qualitative et quantitative de la faune benthique dans la partie est du lac Tchad. Cah. ORSTOM Sér. Hydrobiol. 3: 3–58.

Dejoux, C., Lauzanne, L. and Lévêque, C., 1971a. Prospection hydrobiologique du lac de Léré (Tchad). Cah. ORSTOM Sér. Hydrobiol. 5: 179–185.

Dejoux, C., Lauzanne, L. and Lévêque, C., 1971b. Nature des fonds et répartition des organismes benthiques dans la région de Bol (archipel est du lac Tchad). Cah. ORSTOM Sér. Hydrobiol. 5: 213–223.

Dupont, B. and Lévêque, C., 1968. Biomasse en mollusques et nature des fonds dans la zone est du lac Tchad. Cah. ORSTOM Sér. Hydrobiol. 2: 113–126.

Falconer, A. C. and Marshall, B. E., 1970. Limnological investigations on Lake McIlwaine. Newsl. limn. Soc. sthn. Afr. Suppl. 13: 66–69.

Ferguson, A. J. D., 1975. Invertebrate production in Lake Turkana (Rudolf) — Symposium on the hydrobiology and fisheries of Lake Turkana (Rudolf), 25–29th May 1975, 28 pp., mimeo.

Fittkau, E. J., Irmler, U., Junk, W. J., Reiss, F. and Schmidt, G. W., 1975. Productivity, biomass and population dynamics in Amazonian water bodies. In F. B. Golley and E. Modina (eds.), Tropical Ecological Systems — Trends in terrestrial and aquatic research, Chapter 20, pp. 289–311.

Inagaki, H., 1967. Mise au point de la loi de Motomura et essai d'une écologie évolutive. Vie Milieu B 18: 153–166.

Jonasson, P. M., 1972. Ecology and production of the profundal benthos in relation to phytoplankton in Lake Esrom. Oikos Suppl. 14: 1–148.

Lauzanne, L., 1968. Inventaire préliminaire des oligochètes du lac Tchad. Cah. ORSTOM Sér. Hydrobiol. 2: 83–110.

Lévêque, C., 1968, Mollusques aquatiques de la zone est du lac Tchad. Bull. IFAN, A, 29: 1494–1533

Lévêque, C., 1971. Equation de von Bertalanffy et croissance des mollusques benthiques du lac Tchad. Cah. ORSTOM Sér. Hydrobiol. 5: 263–283.

Lévêque, C., 1972. Mollusques benthiques du lac Tchad: écologie, étude des peuplements et estimation des biomasses. Cah. ORSTOM Sér. Hydrobiol. 6: 3–45.

Lévêque, C., 1973. Dynamique des peuplements, biologie et estimation de la production des mollusques benthiques du lac Tchad. Cah. ORSTOM Sér. Hydrobiol. 7: 117–147.

Lévêque, C., 1974. Etude systématique et biométrique des Lamellibranches Unionidés et Mutelidés du bassin tchadien. Cah. ORSTOM Sér. Hydrobiol. 8: 105–117.

Lévêque, C., 1975. Mollusques des herbiers à *Ceratophyllum* du lac Tchad. Biomasses et variations saisonnières de la densité. Cah. ORSTOM Sér. Hydrobiol. 9: 25–31.

Lévêque, C. and Gaborit, M., 1972. Utilisation de l'analyse factorielle des correspondances pour l'étude des peuplements en mollusques benthiques du lac Tchad. Cah. ORSTOM Sér. Hydrobiol. 6: 47–66.

McLachlan, A. J., 1970. Some effects of water level fluctuation on the benthic fauna of two central african lakes. Newsl. limn. Soc. sthn. Afr. 13: 15–19.

McLachlan, A. J., 1974a. Development of some lake ecosystems in tropical Africa with special reference to the invertebrates. Biol. Rev. 49: 365–397.

McLachlan, A. J., 1974b. Recovery of the mud substrate and its associated fauna following a dry phase in a tropical lake. Limnol. Oceanogr. 19: 74–83.

McLachlan, A. J., 1979. Decline and recovery of the benthic invertebrate communities. In M. Kalk, A. J. McLachlan, R. C. Howard-Williams (eds.), Lake Chilwa, Monographiae Biologicae 35: 145–160.

271

McLachlan, A. J. and McLachlan, S. M., 1969. The bottom fauna and sediments in a drying phase of a saline African lake (L. Chilwa, Malawi). Hydrobiologia 34: 401–413.

McLachlan, A. J. and McLachlan, S. M., 1971. Benthic fauna and sediments in the newly created Lake Kariba (Central Africa). Ecology 52: 800–809.

Monakov, A. V., 1969. The zooplankton and the zoobenthos of the White Nile and adjoining waters in the Republic of the Sudan. *Hydrobiologia* 33: 161–185.

Munro, J. L., 1966. A limnological Survey of Lake McIlwaine, Rhodesia. Hydrobiologia 28: 281–308.

Petr, T., 1969. Development of bottom fauna in the man-made Volta Lake in Ghana. Verh. Int. Ver. Limnol. 17: 273–282.

Petr, T., 1972. Benthic fauna of a tropical man-made lake (Volta Lake, Ghana 1965–1968). Arch. Hydrobiol. 70: 484–533.

Roche, M. A., 1973. Traçage naturel salin et isotopique des eaux du système hydrologique du lac Tchad. Thèse d'Etat, Univ. Paris VI, 385 pp., mimeo.

Viner, A. B. and Smith, I. R., 1973. Geographical, historical and physical aspects of Lake George. Proc. r. Soc. Lond. B 184: 235–270.

9. The fauna associated with the aquatic vegetation

Claude Dejoux

The flora of Lake Chad is relatively poor and consists of common species with no endemisms. These facts plus the area and position of the lake in the Sahelian zone led Leonard (1965) to propose the establishment of a phytogeographic district for Lake Chad.

The aquatic vegetation (submerged or semi-aquatic macrophytes) show considerable variation depending on the hydrology of the lake. In the period of 'Normal Chad' when an area of 12 to 18 000 km^2 is under water, the area covered by this vegetation can be very considerable. On the contrary, both the periods of very high water in the lake and relative dryness are generally accompanied by a decrease in vegetation covers.

Few studies were undertaken on this biological component of the lake ecosystem; the locations of these studies are shown in Fig. 1. This is generally true of studies in African lakes whose aquatic vegetation has usually been studied only when it has been a nuisance.

Generally, the aquatic macrophytes of Lake Chad are grouped in associations, which show some varied morphological aspect but are characterized by relatively low species diversity. In the following pages we try to show the faunal composition of the macrophyte populations as well as their dynamics.

9.1 Faunal composition

9.1.1 *The invertebrates*

The qualitative samples of the submerged macrophytes were taken by manually collecting the vegetation and washing thoroughly through different sized mesh sieves.

The quantitative samples were taken with a 'phyto-isolator', an apparatus specially designed for this purpose (Dejoux and Saint-Jean 1972). Some crustaceans, for example, *Caridina*, can escape from this device and therefore a different system was used for them (Troubat 1975). The submerged or semi-submerged vegetation of Lake Chad consisted of morphologically varied

© *1983, Dr W. Junk Publishers, The Hague/Boston/Lancaster*
ISBN 978-94-009-7268-1

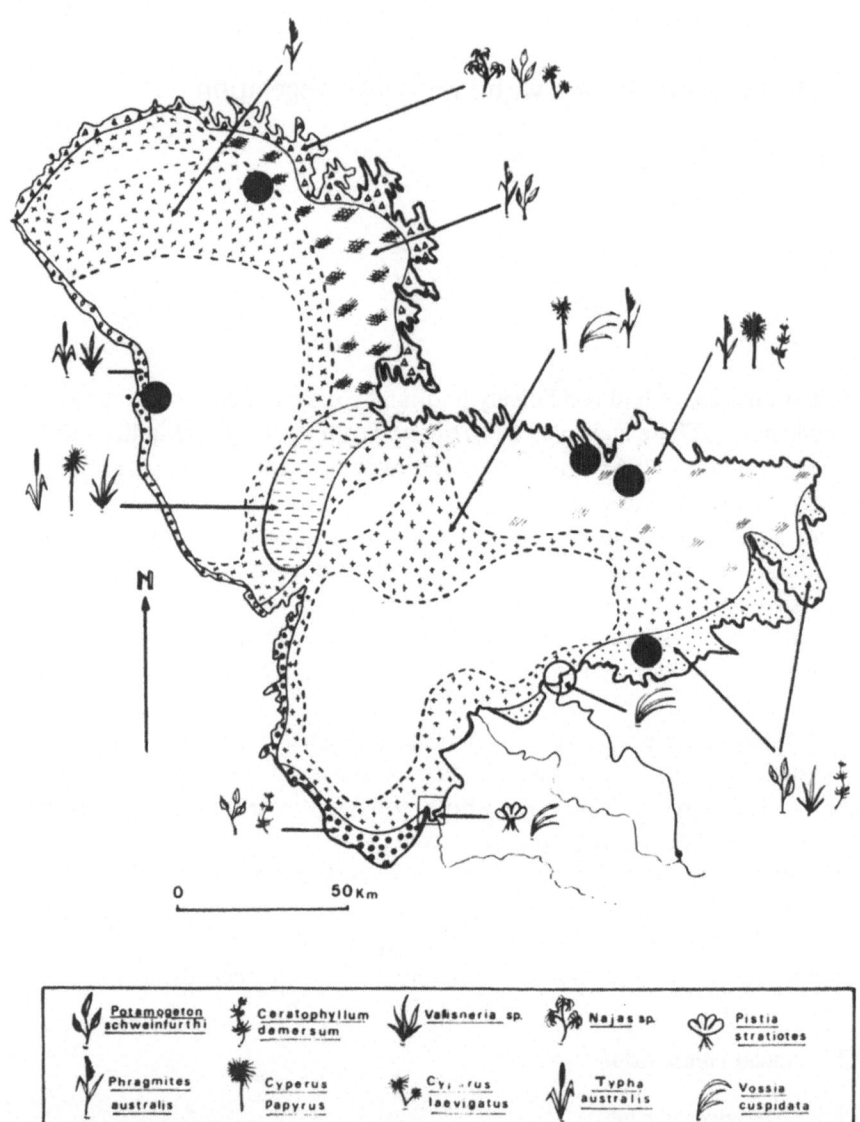

Fig. 1 Distribution of aquatic macrophytes in Lake Chad in 1968. Areas where special studies have been carried out are shown by black spots. For the different lake regions, decreasing dominances are drawn by ordination from left to right.

macrophytes which did not offer the same potential shelter for invertebrates. For this reason the list in Table 1 related invertebrates to the more important 'facies', a term used to designate a vegetation group where one species formed 90% or more.

Table 1 Banks of aquatic or semi-aquatic macrophytes.

Taxonomical groups	P	C	V	TP	RP
CILIATA					
Holotrichia sp.	+	+	–	–	–
Coleps hirtus	–	+	–	–	–
Didinium nasutum	–	–	–	+	–
Litonotus quadrinucleatus	–	+	–	–	–
Tetrahymena pyriformis	+	+	–	–	–
Colpidium campylum	+	+	–	–	–
Loxocephalus luridus	+	+	–	–	–
Neobursaridium gigas	–	–	–	+	–
Halteria grandinella	+	+	–	–	–
Urosoma acuta	–	–	–	+	–
COELENTERATA					
Hydra sp.	–	+	+	–	+
HIRUDINEA	–	+	–	+	+
OLIGOCHAETA					
Branchiodrillus cleistochaeta	–	–	+	–	+
Dero digitata	+	+	+	–	+
Aulophorus furcatus	+	+	–	–	–
Allonaïs paraguayensis ghanensis	+	+	–	–	–
NEMATODA	+	+	–	+	+
CRUSTACEA					
Entomostracea					
Chydorus globosus	+	+	–	–	–
Chydorus eurynotus	+	+	–	–	–
Alona diaphana	+	+	–	–	–
Alona monacantha	+	+	–	–	–
Alona pulchella	+	+	–	–	–
Alona verrucosa	+	+	–	–	–
Simocephalus sp.	–	+	–	–	–
Macrothrix triserialis	+	+	–	–	–
Macrothrix goeldii	–	+	–	–	–
Alona novae zelandiae	+	–	–	–	–
Alona kanna	+	+	–	–	–
A. guttata	+	–	–	–	–
Ilyocryptus spinifer	–	+	–	–	–
Decapoda					
Caridina africana	+	+	+	–	+
Macrobrachium niloticum	+	+	–	–	–

Table 1 (continued).

Taxonomical groups	P	C	V	TP	RP
INSECTA					
CHIRONOMIDAE					
C1 *Nilodorum brevibucca*	−	+	+	+	+
C2 *Polypedilum* sp.	+	+	+	+	+
C3 *Stictochironomus caffrarius*	−	−	−	+	+
C4 *Cryptochironomus* sp.	+	+	+	+	−
C5 *Dicrotendipes* sp.	−	+	−	−	−
C6 Orthocladiinae (*Nanocladius*?)	+	+	+	+	−
C7 *Cricotopus* sp.	+	+	+	+	+
C8 *Tanytarsus ceratophyllus*	+	+*	+	+	'
C9 *Cryptochironomus nudiforceps*	+	+	−	−	'
C10 *Cryptochironomus dewulfianus*	−	−	−	+	'
C11 *Cricotopus scottae*	+*	+	+	+	−
C12 *Polypedilum griseoguttatum*	+	+	+	+	−
C13 *Polypedilum* sp.	+	−	−	−	−
C14 *Dicrotendipes fusconotatus*	+*	+*	+	−	−
C15 *Stictochironomus puripennis*	+	+	+	−	−
C16 *Cricotopus* sp.	+	+	+	−	−
C17 *Tanytarsus nigrocinctus*	+	+	+	−	−
C18 *Ablabesmyia dusoleili*	+	+	+	+	−
C19 *Polypedilum* sp.	+	−	−	−	−
C20 *Polypedilum* sp.	+	+	+	−	+
C21 Orthocladiinae	+	+	−	−	−
C22 *Stictochironomus* sp.	−	−	−	−	+
C23 *Cladotanytarsus* sp.	+	+	−	−	+
C24 *Cricotopus* sp.	−	−	−	−	+
C25 *Tanytarsus* sp.	+	+	+	−	−
C26 *Tanytarsus* sp.	+	+	−	−	−
C27 *Polypedilum* sp.	+	+	+	−	−
C28 *Cryptochironomus* sp.	+	+	+	−	−
C29 *Dicrotendipes chloronotus*	+	−	−	−	−
C30 *Cryptochironomus* sp.	−	+	−	−	−
C31 *Tanytarsus bifurcus*	−	+	−	−	−
C32 *Dicrotendipes sudanicus*	−	+	+	−	−
C33 *Polypedilum melanophilus*	−	+	−	−	−
C34 *Cladotanytarsus* sp.	−	+	−	−	−
C35 *Ablabesmyia nilotica*	−	+	+	−	−
C36 Chironominae	+	−	+	−	−
C37 Chironominae	−	−	−	−	+

Table 1 (continued).

Taxonomical groups	P	C	V	TP	RP
C38 Chironominae	−	+	−	−	−
C39 *Polypedilum* sp.	+	+	+	−	−
C40 *Cryptochironomus*	+	+	+	+	−
C41 *Tanytarsus* sp.	+	−	+	−	−
C42 *Tanytarsus* sp.	+	+	+	+	−
C43 *Cricotopus albitibia*?	+	+	+	−	−
C44 *Cryptochironomus*	+	+	+	−	−
C45 Orthocladiinae (*Smittia*?)	+	+	+	−	−
C46 *Stictochironomus* sp.	−	−	−	−	+
C47 *Cryptochironomus* sp.	−	−	+	−	−
C48 Orthocladiinae sp.	+	−	−	−	−
C49 *Polypedilum* sp.	−	−	+	−	−
C50 Chironominae	−	−	+	−	−
C51 *Cryptochironomus* sp.	−	−	+	−	−
C52 Orthocladiinae	−	−	+	−	−
C53 Chironominae	−	−	+	−	−
C54 Chironominae	−	−	−	+	−
C55 *Polypedilum*	−	−	+	−	−
C56 Orthocladiinae	−	+	−	−	−
C57 Chironominae	+	−	−	−	−
C58 Orthocladiinae	+	+	+	−	−
C59 Chironominae	+	+	+	−	−
C60 *Cryptochironomus* sp.	+	−	−	−	−
ODONATA					
Agrionidae					
Libellulidae	+	+ *	+	−	−
EPHEMEROPTERA					
Baetidae	+	+	+	+	+
Povilla adusta	+	+	+	+	+
TRICHOPTERA					
Ecnomus					
Orthotrichia straeleni	+	+	+	+	+
Dipseudopsis sp.	+ *	+ *	+	+	+
HEMIPTERA					
Diplonychus grassei	+ *	+	−	−	−
Plea minuta	+	+	−	−	−
Hydrocyrius sp.	−	+ *	−	−	−
Micronecta scutellaris	+	+	−	−	−
Naboandelus sp.	−	+	−	−	−

Table 1 (continued).

Taxonomical groups	P	C	V	TP	RP
LEPIDOPTERA					
Pyralidae	+ *	+ *	+	–	–
COLEOPTERA					
Elmidae	+	–	–	–	–
DIPTERA					
Tipulidae	+	+	–	–	–
Muscidae	+	–	+	–	–
Culicidae	+	+	–	–	–
Ceratopogonidae	+	+	–	–	–
MOLLUSCS					
Bulinus truncatus and *Bulinus forskalii*	+ *	+ *	–	–	–
Ferrissia sp.	+ *	+	–	–	–
Biomphararia pfeifferi	+ *	+ *	+	–	–
Afrogyrus coretus	+ *	+	–	–	–
Gabiella tchadiensis	+	+	–	–	–
Bellamya unicolor	–	+	–	–	–
Gyraulus costulatus	+	+	–	–	–
Segmentina angustus	+	+	–	–	–
Limnea natalensis	–	–	–	–	–

P = *Potamogeton*; C = *Ceratophyllum*; V = *Vossia*; TP and RP = stems and roots of *Cyperus papyrus*.
+ * = highly characteristic; + = abundant; – = never collected.

In general, some of the major invertebrate groups were always abundant in aquatic vegetation; these were the Chironomidae, the Hemiptera, the Odonata (Libellulidae and Agrionidae), the Ephemeroptera (Baetidae and some Ephemeridae), the Lepidoptera (Pyralidae), the Ostracods, some Entomostraca and the pulmonate molluscs (Table 1).

These organisms were protected by the vegetation from predators. The vegetation also provided an abundant food source. The abundance of the invertebrates varied greatly with the different morphological structures of each facies.

Major ecological factors regulating the geographical distribution of invertebrates over the whole lake also determined to some extent the distribution patterns of the weedbeds.

9.1.2 *The fish community*

The term weedbed dwelling fish groups together several species whose degree of interrelationship with the submerged vegetation can be very different. In fact, some species were permanent inhabitants of weedbeds, while others lived in the spaces of open water between the clumps of vegetation. Others penetrate into the vegetation only to feed or to obtain shelter during a part of the day, while others were only found in these environments as juveniles.

The determination of the composition of these different groups in Lake Chad has been made easier by using ichthyotoxins (Loubens 1969, 1970). Four successive groups were distinguished in order of decreasing interrelationship with the presence of submerged vegetation stands.

(a) A group of species, mostly of small size strictly related to the presence of submerged vegetation and which were only rarely found elsewhere. In particular, there were several species of the genus *Barbus* as well as *Haplochromis bloyeti, Neolebias unifasciatius, Petersius intermedius, Paradistichodus dimidiatus, Aplocheilichtys* and *Epiplatys* ... (Table 2).

It is almost certain that the different regions of the weedbeds were populated by different species of fishes as a result of weedbed structure and location. As an example, we will point out the extreme abundance of *Petersius intermedius* in the mixed species vegetation stands of a secondary arm of the Logone, consisting of *Potamogeton* and *Nymphae* (Loubens 1969).

(b) A group of eurytopic species of small size which found temporary shelter in the vegetation but which could also be found in other environments. They include: *Alestes dageti* and *Micralestes acutidens* which were also found in the open water.

(c) A group of medium or large-sized fishes which over the course of their life and especially during the juvenile phase accomplished part of their growth in these environments. These species sought food, and protection against the predators of the open water. The most typical case was that of *Lates niloticus* whose juveniles concentrated in the shallow waters of the lake after their first pelagic phase (until about 2 cm) and became established in the beds of *Vallisneria spiralis, Potamogeton* and *Ceratophyllum* (Hopson 1972). Their development period among the aquatic vegetation ceased at a size of 15 to 20 cm. Their predation was exerted principally on the larvae of aquatic insects and crustacea (*Macrobrachium niloticum*) between 2 and 10 cm. After this they became mainly ichthyophagous, above 10 cm. This group included *Alestes baremoze, Alestes dentex* and *Distichodus rostratus*.

(d) A group of adult species of average or large size which were frequently found in the vegetation banks but also in other environments. Some, like the Polypterids or *Mastacembelus loennbergi* came there for shelter, others like several Mormyridae (insectivorous) or *Tetraodon fahaka* (malacophagous) fed there. Others, such as *Heterotis, Hyperopisus, Gymnarchus*, etc. built their nests

Table 2 Qualitative and quantitative composition of a fish sample collected by poisoning in a region covered with aquatic macrophytes (Matafo 11/8/68).

Weedbeds	kg	Open water	kg
Barbus pleuropholis	9767	*Distichodus rostratus*	134
Haplochromis wingatii	2640	*Lates niloticus*	87
Alestes dageti	1803	*Hyperopisus bebe*	86
Barbus calliplerus	419	*Synodontis frontosus*	83
Barbus leonensis	284	*Hydrocyon forskali*	54
Micralestes acutidens	279	*Synodontis schall* and *S. gambiensis*	35
Distichodus rostratus	173	*Gnathonemus cyprinoides*	31
Petersius intermedius	150	*Citharinus citharus*	30
Paradistichodus dimidiatus	133	*Marcusenius isidori*	25
Lates niloticus	100	*Tilapia galilaea*	19
Neolebias unifasciatus	91	*Alestes dentex*	18
Mochocus brevis	60	*Alestes baremoze*	12
Kribia nana	48	*Auchenoglanis occidentalis*	10
Alestes dentex	29	*Citharinus latus*	9
Hemichromis bimaculatus	24	*Labeo senegalensis*	6
Tetraodon fahaka	20	*Malaplerurus electricus*	5
Tilapia zilli	5	*Mormyrus rume*	4
Alestes baremoze	4	*Alestes macrolepidotus*	4
Aplocheilichthys sp.	4	*Tilapia zilli*	3
Hemichromis fasciatus	3	*Auchenoglanis biscutatus*	3
Auchenoglanis biscutatus	2	*Synodontis clarias*	3
Nannocharax sp.	2	*Synodontis batensoda*	2
Polypterus senegalus	1	*Tilapia nilotica*	2
		Distichodus brevipinnis	2
	16 041	*Gnathonemus senegalensis*	2
		Mormyrops deliciosus	1
		Heterobranchus bidorsalis	1
		Gymnarchus niloticus	1
			672
23 species		28 species	

in the vegetation and were thus found there at the time of reproduction.

Therefore generally the submerged vegetation of Lake Chad contained a specific fish fauna comprised of small species which lived there permanently plus larger species whose presence was temporary or sporadic.

9.1.3 *The other vertebrates*

There is little quantitative information on vertebrate groups other than the fishes associated with the vegetation. What is certain is that the scale of the lake makes them important.

The Batrachia for example (*Bufo* sp.) were particularly abundant in the reed islands and among the entire vegetation belt bordering the islands. The reptiles were less frequent although some snakes were found periodically (*Grayia*, *Python* ...). They could find various prey species (Batrachia and fishes) in the vegetation.

The crocodiles and the chelonians of the genus *Trionyx* were more numerous but their importance was only apparent after extreme lowering of the water level. They also exploited different trophic levels in the vegetation. The bird could be locally very dense finding temporary shelter (cormorants, grebes ...) or food (Anatidae, water hens, jacanas) in the vegetation. In the papyrus or reeds some species found favorable conditions for their nest building.

The mammals were less numerous over the large scale of the lake, whether the rare *Situtunga* or the hypothetical manatee. On the other hand the hippopotamus and the otters did show some small localized concentrations exploiting the submerged vegetation.

9.2 Factors causing abundance and distribution of communities

9.2.1 *Structure of some invertebrate populations*

The submerged vegetation was generally much richer in invertebrates than the portion of partially submerged macrophytes growing in the water, especially for the chironomids. Similarly, *Ceratophyllum* was seen to be more favorable to the establishment of the Diptera than *Potamogeton* or *C. papyrus*. The differences in density ranged from 10 to 400. By contrast, if the species richness is studied by the Margalef index $(D(n-1)/(\sqrt{}\log N))$ where n was the total number of species collected in an area, and N the total number of individuals of these n species, it would appear that *Valisneria* has the highest index (9.4) compared with *Potamogeton* (8.0) and *Ceratophyllum* (6.5), and finally the stems and roots of *C. papyrus* with respectively 3.9 and 3.3. These values were obtained from the eastern archipelago but these indices were clearly much lower in the delta region.

It appears that two factors were responsible for these differences: the oxygenation of the environment, related to the penetration of light, and the vegetation texture.

An analysis of some dominant Chironomid species was made in a similar region in two different weedbeds. The mean numbers for 10 grams dry vegetation weight were calculated and then the relative variance s^2/n (Table 3) was calculated. The value of s^2/n was mostly higher than the total for all the species considered and showed their distribution to be contagious. The extent of the error was moreover proportional to the density of the weedbeds and in the present case, was greater among *Ceratophyllum* than among *Potamogeton*. There is no doubt that mechanically the weedbed restrained larval dispersion and the denser the structure, the higher the aggregation of species.

A more complete analysis of chironomid populations from *Ceraphyllum* was made by considering all the species present, and the numbers were classed by frequency distribution after logarithmic transformation. In Fig. 2 we reported the top n species having the rank i for abcissa and the value of the logarithm of the abundance (log qi) for ordinate.

The abundance distribution approximates the log-normal type to one such species association appearing to correspond to 'nomocenoses', according to Daget et al. (1972).

By contrast, in a more open environment, there was a high chance of meeting some truncated and asymmetrical distributions. In comparing *Ceratophyllum* and *Potamogeton* the sampling in the first facies covered 31 species whereas the theoretical number present was 34 and in the second only 27 species were collected for a theoretical number of 48. These figures thus express the great homogeneity of the *Ceratophyllum* facies but by contrast its lesser richness was also confirmed by the calculation of shannon diversity index, S, in the two cases. We actually found an index of 1.523 bits/species for the *Ceratophyllum* against 2.78 in the *Potamogeton* which corresponded respectively to an equitability of 30.7% and 47%.

It would thus appear that *Ceratophyllum* represented a rather more closed environment than *Potamogeton*, explained at the time by the denser structure of *Ceratophyllum* but also by the relative isolation from the bottom. In comparison, *Potamogeton* cannot grow detached from the substrate and thus exchanges with the benthic fauna *sensu stricto* were certainly more numerous.

9.2.2 *Spatial variations*

The submerged weedbeds represented a tridimensional environment whose morphological structure could vary largely from one facies to another. In order to show the variations which existed between them and the homogeneity of a facies, we studied the distribution of two groups of well represented organisms, the Entomostraca and the Chironomids.

The mean densities (number of individuals/unit of plant weight) of the Entomostraca and Chironomid species were compared by Kendall's rank correlations method. The samples were taken from two *Potamogeton* vegetation stands, at a distance of 100 meters as well as from two *Ceratophyllum*

Table 3   Mean numbers and dispersion coefficient calculated for some species of Chironomids from submerged weed beds.

Species	Potamogeton I		Potamogeton II		Ceratophyllum I		Ceratophyllum II	
	$\bar{N}$	$s^2/\bar{N}$	$\bar{N}$	$s^2/\bar{N}$	$\bar{N}$	$s^2/\bar{N}$	$\bar{N}$	$s^2/\bar{N}$
Dicrotendipes fusconotatus	–	–	–	–	81	97.2	303	22.7
Stictochironomus puripennis	18	33.0	–	–	107	235.1	–	–
Rheotanytarsus ceratophyllus	197	72.4	133	39.2	7358	427.1	2606	306.0
Cricotopus sp.	22	15.0	42	24.0	63	132.1	73	67.5
Polypedilum sp. C 13	–	–	94	151.7	85	121.4	424	136.8
Polypedilum sp. C 20	–	–	37	9.4	43	49.0	134	51.2

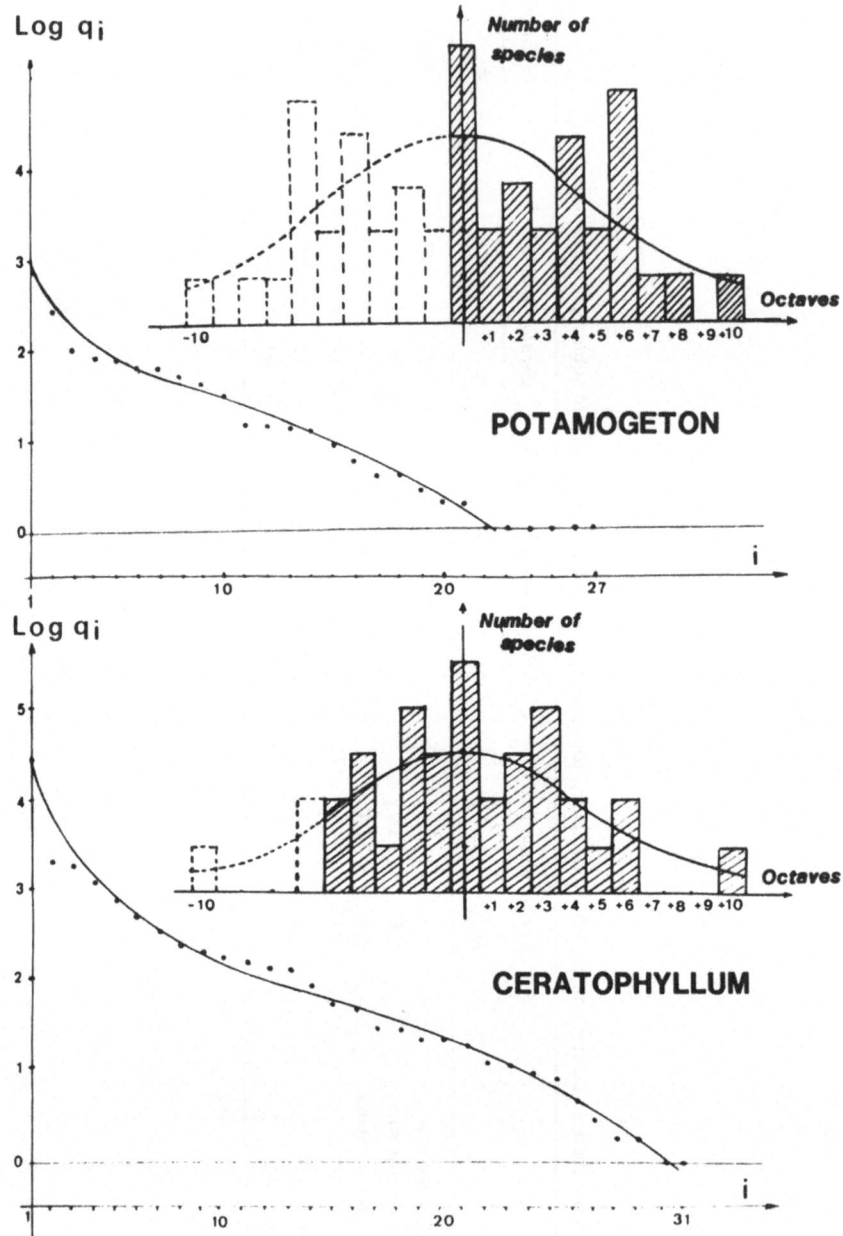

Fig. 2 Log-normal distribution of Chironomids from aquatic vegetation banks.

vegetation stands equally distanced from one another. A supplementary distinction was made at the time of sampling with some samples taken at the surface of the macrophytes and others in deeper water at about 60 cm.

The correlation coefficients were all positive which indicates some similarity between populations from different facies. Two vegetation banks of the same facies however had greater similarity although one hundred meters apart than two neighbouring vegetation banks of different facies.

Observations on changes in density of the organisms between the surface and a depth of 60 cm were variable from one species to another. They may have been due to physico-chemical factors of the environment but also to the distribution of the organisms in aggregations especially as the action of biotic factors related to the actual vegetation facies (vegetative cycle, morphology ...).

All the coefficients were much less than one, even if the coefficient involved was slightly underestimated by the assignment of a different rank to some species having low and similar densities. The value one must be approached if the community is perfectly homogenous vertically. Generally, when the interior of the same weedbeds or two of the same facies apart from each other are considered, community similarity is always greater for the Entomostracea than for the Chironomids or for other aquatic insects. This is almost certainly because Entomostraca are less attached to the vegetation support than the insects. The bracing of the vegetation banks by the wind and the waves contribute to an even distribution in the open water between the vegetation patches.

The morphological structure of each facies is shown in the distribution of densities of different invertebrates. A study of the Chironomid population of northern Lake Chad (Dejoux 1976) shows a very clear increase in densities related to a change from an open type of vegetation bank (*Potamogeton*) to a compact type (*Najas*) (Table 4).

One can now consider events on a larger scale, e.g. when comparing a facies situated in the zone of the Shari delta with the same facies in the southeastern archipelago, the very numerous invertebrates appeared to be ubiquitous and the dominant forms were found in the two places. The relative abundance could however vary as the greater abundance of *Alonella excisa* was shown in the delta zone in comparison with the zone of the southeastern archipelago. Similarly and although it was impossible to locate some groups which were found sporadically in one of the other stations, three facts can be shown: the absence of hydroid polyps in the delta when they were frequent in the archipelago; the concentration of Hemiptera in *Ceratophyllum* and *Potamogeton* in the two regions and finally, the greater abundance of Lepidoptera in the delta region. The adults of the family Pyrallidae were certainly affected by the more unfavourable climatic conditions which characterized the archipelago in comparison with the southeastern shore of the lake.

The comparison of Chironomid populations finally showed that the fauna was qualitatively richer in the southeastern archipelago region (49 species

Table 4 Comparative densities of Chironomid larvae inhabiting three different vegetation biotopes of northern Lake Chad (numbers rounded up and reduced to 10 g dry vegetation weight).

Species collected	Potamogeton						Ceratophyllum						Najas					
	1	2	3	4	5	$\bar{N}$	1	2	3	4	5	$\bar{N}$	1	2	3	4	5	$\bar{N}$
1 *Cricotopus scottae*	202*	137	93	177	144	151	326	151	387	251	196	262	281	335	554	376	212	352
2 *Cricotopus* sp. (cf. *albitibia*)	25	61	5	21	3	12	60	47	22	85	15	46	115	138	145	40	78	103
3 *Trichocladius* sp. (cf. *metallescens*)	4	–	3	–	3	12	14	10	3	24	5	11	16	2	13	3	6	8
4 *Dicrotendipes fusconotatus*	53	36	34	31	38	39	129	83	34	68	108	84	99	83	170	198	223	155
5 *Cryptochironomus dewulfianus*	72	49	10	59	51	48	385	265	363	201	314	305	532	313	471	297	342	392
6 *Tanytarsus nigrocinctus*	2	3	1	–	9	3	46	36	9	37	44	34	22	5	48	7	3	17
7 *Tanytarsus bifurcus*	4	6	–	–	13	5	5	–	–	24	15	9	13	7	39	23	38	24
8 *Stictochironomus* sp.	4	3	20	–	3	6	5	21	3	55	15	25	19	2	39	43	6	20
9 *Ablabesmyia* sp.	–	–	–	–	–	–	–	–	–	–	–	–	3	–	6	–	–	2
						T = 276						T = 775						T = 1073

collected) than in the delta (29 species collected) for all facies studied. Some species with a very wide distribution in one zone were totally absent in another.

9.2.3 *Seasonal variations*

The malacological fauna of the grass banks represented an important faunal component because of its diversity, rapid density, rapid growth and the role of the snails in the transmission of some tropical diseases. They were particularly abundant among *Ceratophyllum* and variations in their density were studied over an annual cycle in several regions of the lake (Lévêque 1975).

Some species (Fig. 3) such as *Gyraulus costulatus* and *Bulinus truncatus* did not show distinct variations in their densities throughout the year, or at least these variations did not appear to be related to those major ecological factors affecting the change in the lake. By contrast, some others, amongst which must be included *Gabbiella tchadiensis*, *Ferrisia* sp. *Biomphalaria* sp. and especially *Afrogyrus coretus* and *Segmentina augustus* showed a very clear maximum density during the high waters of the lake, from September to November. This period also corresponded to that of the maximum vegetation from different vegetation bank facies and it is certain that they were a factor favourable to the development of gastropods in providing abundant shelter and food.

We also studied the annual variations in density of several other invertebrate species in a nearby zone of the Shari delta and in a *Potamogeton* stand. Generally, this environment was fairly poor and some taxa were only found one or two times in the year. The oligochaetes and chironomids were by far the most abundant groups, followed by the Ephemeroptera, Baetidae (mainly *Procloeon fraudulentum*), the Hemiptera with mostly *Micronecta scutellaris*, and the Trichoptera with *Orthotrichia straeleni*.

The annual change in densities of the five dominant groups is outlined in Fig. 4: the Oligochaetes and *Orthotrichia* showed two maxima: one in January–February, the other, less marked in June–July at the beginning of the rise of the lake waters. *Pseudagrion*, *Micronecta* and the Baetidae showed only one maximum also during the high water period which partly corresponded to the rainy season. The particular abundance of the insect during this period was moreover a general phenomenon in the Sahelo-Soudanian zone. Several aquatic insects, found better climatic conditions for their dispersion during the rainy season than during the dry season when the humidity of the air frequently fell below 30%.

The annual change in densities of Chironomids showed a completely different profile and the particular abundance at the beginning of the rainy season was much less marked. Having taken into consideration the variation in density of the 11 species inhabiting the weedbeds studied and being given the short duration of the cycles of these insects, the emergence of adults throughout

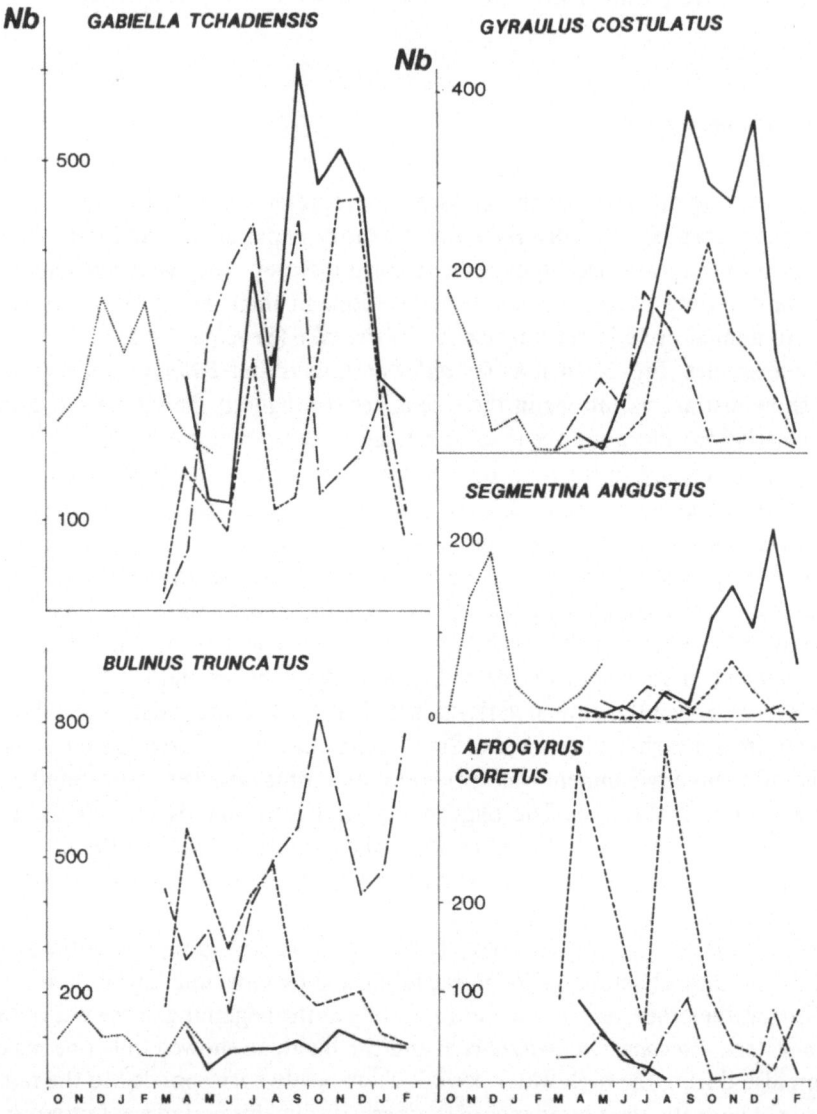

Fig. 3 Seasonal variations in density of main species of molluscs inhabiting *Ceratophyllum* banks. *N* = number of individuals in 100 g of dry weight vegetation.

the year closely followed the multiple population peaks (Fig. 5). If by contrast two of the most abundant species (*Cricotopus scottae* and *C. albitibia*) are considered separately, it appears that each of them showed a different type of population change. The first had two annual maxima whereas the second was characterized by several peaks distributed throughout the year (Fig. 5).

288

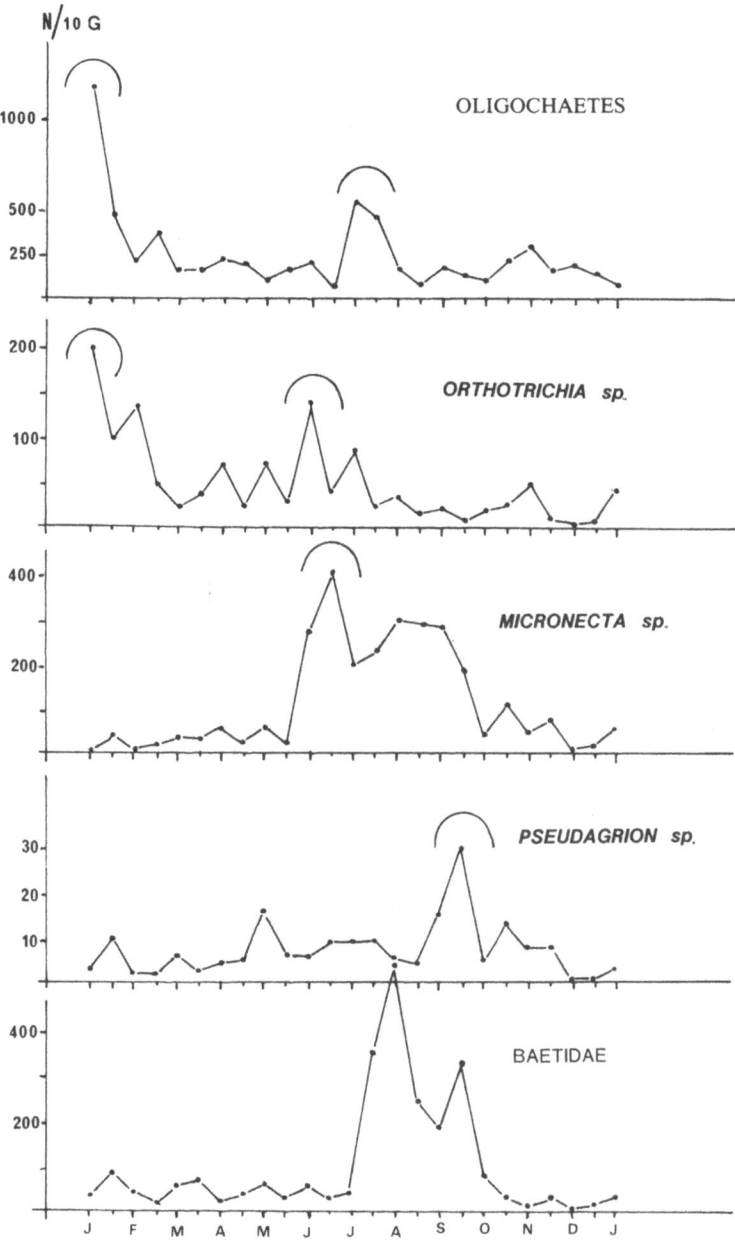

Fig. 4 Annual variations density of some invertebrates living in *Potamogeton* banks of the delta region. $N/10$ G = number of individuals for 10 g of dry vegetation.

289

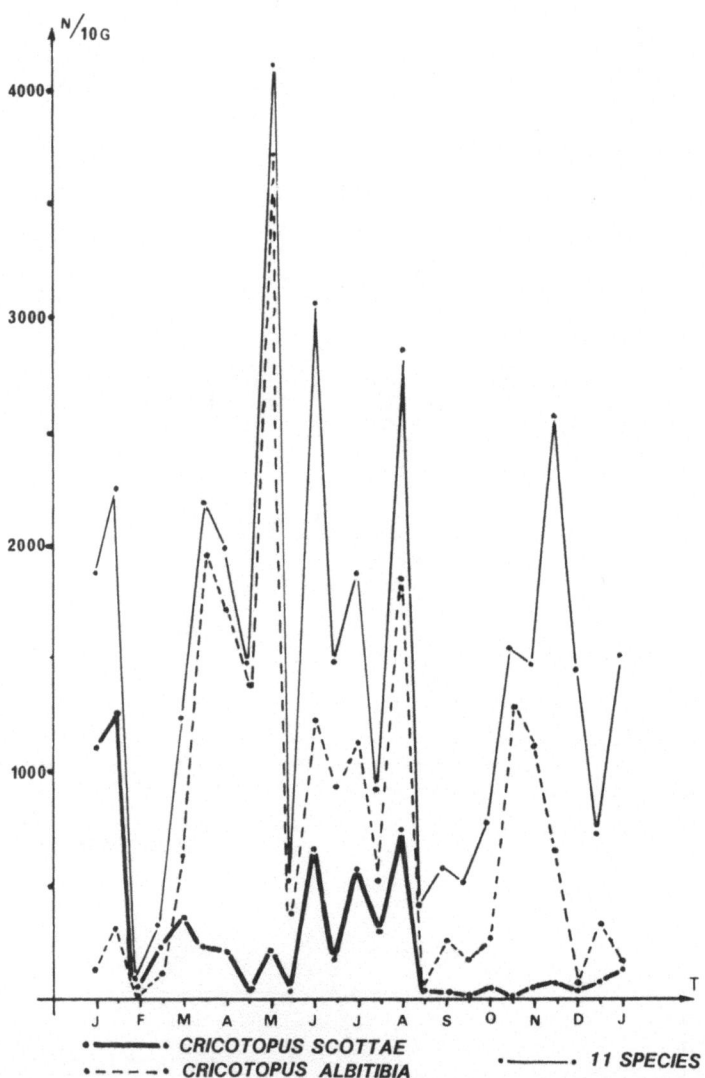

Fig. 5 Seasonal fluctuations of Chironomid density in *Potamogeton* banks of the delta region. *N*= number of individuals in 10 g of dry vegetation.

9.3 Discussions and conclusions

Although represented by heterogenous facies, the aquatic and semi-aquatic vegetation constituted among the biotopes of Lake Chad, an original and well-individualized environment whose importance in area varied as a function of the lake level. It was considerable in some parts of the lake basin during the period of mean or low level, and reduced during the period of the reflooding during the 'Greater Chad'. In the last case only, the vegetation banks, such as the floating islands (kirtas) or vegetation belts, could be an obstacle to navigation, limiting access to the emerged zones.

The aquatic vegetation contributed largely to the mineral cycling in the lake. At the mean water level of 28.5, the semi-submerged macrophytes covered about 2400 km^2, or about 12% of the lake surface (Carmouze 1976). They represented a considerable biomass of organic matter as well as an extremely important stock of mineral salts and ions from the lacustrine environment. As indicated, the quantity of potassium ions thus stocked in the aerial parts of the plants was about 138 000 tons, thus much higher than the annual river supplies (cf. Chapter 5).

The rate of mineralization of plant organic matter was lower than its rate of production. This resulted in a considerable accumulation of slightly degraded vegetation debris, especially in the zones of reed islands, and each year considerable quantities of dissolved elements (calcium, magnesium, potassium) were thus accumulated and so definitely lost.

In addition to elements common with the other biotopes, the vegetation was populated by one particular and characteristic fauna. The depth of the lake was essentially made up of detrital material and it is certain that the biotopes colonizable by the fauna would have been less diversified without the presence of aquatic and semi-aquatic vegetation.

References

Andrewartha, H. G. and Birch, L. C., 1964. The distribution and abundance of animal. University of Chicago Press, 785 pp.

Carmouze, J. P., 1976. La régulation hydrogéochimique du lac Tchad. Trav. Doc. ORSTOM, No. 58: 1–418.

Daget, J., Lecordier, C. and Lévêque, C., 1972. Notion de Nomocénose: ses applications en écologie. Bull. Soc. Ecol. 3: 448–462.

Dejoux, C., 1976. Synécologie des chironomides du lac Tchad. Trav. Doc. ORSTOM No. 56, 161 pp.

Dejoux, C. and Saint-Jean, L., 1972. Etude des communautés d'invertébrés d'herbiers du Lac Tchad: recherches préliminaires. Cah. ORSTOM Sér. Hydrobiol. 6: 67–83.

Hopson, A. J., 1972. A study of the nile perch (*Lates niloticus*, Pisces, Centropomidae) in Lake Chad. Overseas Research publication No. 19, London, 93 pp.

Leonard, J., 1969. Aperçu sur la végétation. In: Monographie hydrologique du lac Tchad, ORSTOM Paris, 11 pp., mimeo.

Lévêque, C., 1972. Mollusques benthiques du lac Tchad: Ecologie, production et bilans énergétiques. Thèse d'Etat, Univ. Paris VI, 225 pp.

Lévêque, C., 1975. Mollusques des herbiers à *Ceratophyllum* du lac Tchad: Biomasses et variations saisonnières de la densité. Cah. ORSTOM Sér. Hydrobiol. 9: 25–31.

Loubens, G., 1969. Etude de certains peuplements ichtyologiques par des pêches au poison (1ère note). Cah. ORSTOM Sér. Hydrobiol. 3: 45–73.

Loubens, G., 1970. Etude de certains peuplements ichtyologiques par des pêches au poison (2ème note). Cah. ORSTOM Sér. Hydrobiol. 4: 45–61.

Petr, T., 1968. Populations changes in Aquatic invertebrates living on two Water plants in a tropical man-made Lake. Hydrobiologia 32: 449–485.

Troubat, J. J., 1975. Un échantillonnage de crevettes d'eau douce en milieu d'herbier. Cah. ORSTOM Sér. Hydrobiol. 9: 291–294.

10. Fish communities of Lake Chad and associated rivers and floodplains

Vincent Bénech, Jean-René Durand and Jacques Quensière

The concept of fish community as used in this chapter must not be confused with the concept of ichthyocenosis from which it is fundamentally different. We call 'fish species community' the possibly biased image given by the sampling of a group of fishes able to be caught in a particular environment at a given time. This definition only names a description — in the meaning of Legendre and Legendre, 1979 — of an ecological entity, but not this entity itself. This remark will be justified by the large complexity of the whole fluvio-lacustrine Chad system in the following discussion.

As described above for other groups, the study of the composition and distribution of fish communities requires the consideration of large spatial, seasonal and between year variability. However the study of fish communities includes certain aspects that increase the difficulties of description and interpretation. The mobility of fishes introduces some additional difficulties, both those of proper sampling and those due to the often considerable migratory activities. The first point sometimes results in biases including some systematic deformations of images of multispecies structures given by the samples. This bias depends on the proper selection of the fishing gear and/or fishing techniques (cf. Section 1), on their use and also on specific behaviours which are themselves varied. According to the richness of data gathered and the aims, we will attempt to minimize this sampling bias by comparing for the same period some groups of data obtained according to different methods. On the contrary some changes will be characterized by comparing some series of surveys of the most repetitive sampling methods.

The migratory movements do not allow the construction of a valid interpretation only from the data on lake communities, and the integration of data on the connecting environments communities, the rivers and flood zones is necessary. Thus in addition to Lake Chad, the study area includes the delta region, the lower reaches of the Shari and the Logone above 10°50'N latitude, the flood plain of north Cameroon and the two temporary rivers that connect the plain to the general system: the El Beïd and the Logomatia (Fig. 1). This is

J.-P. Carmouze et al. (eds.) Lake Chad
© *1983, Dr W. Junk Publishers, The Hague/Boston/Lancaster*
ISBN 978-94-009-7268-1

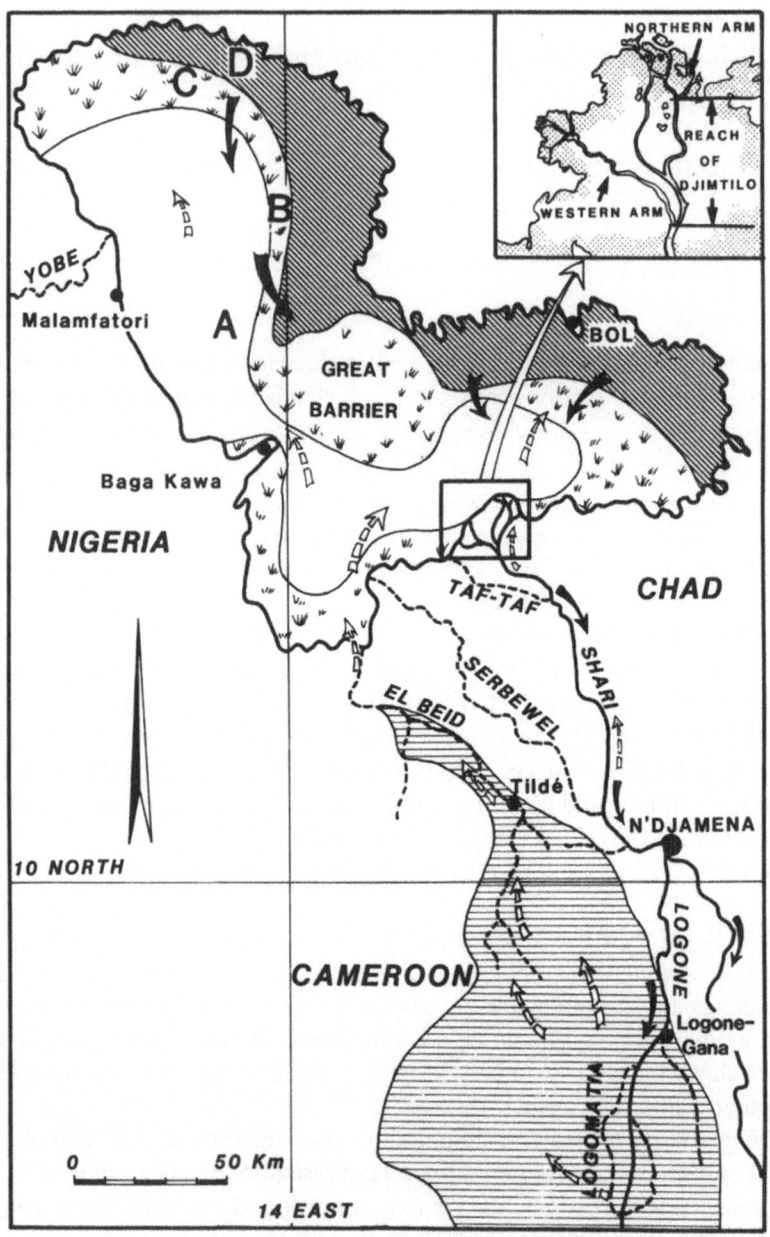

Fig. 1 Map of the study area. Arrows show the migration pattern of adults (black arrows), and juveniles (white arrows) of a typical migratory fish (*Alestes baremoze*). A, B, C, D = sampling sites in the north basin.

[\\\\\\]= Archipelago; [   ] = Open Water; [▼☀☁] = Reed Islands; [▤▤▤] = Flooded areas.

294

the hydrographic system almost corresponding to what is called 'the conventional basin', by the Lake Chad Basin Commission (LCBC).

The hydrology of these environments explains a large part of the spatio-temporal variability found and the description here constantly refers to Chapter 2 of this book. This is so for the average scheme of water supplies and circulation then corresponding to a well-developed lake and to some high or average floods as well as for the years of drought (since 1972) which only saw frequent mediocre floods and a spectacularly reducing lake.

Only an initial general outline is presented here because not all the results obtained from 1966 to 1978 have yet been analyzed. The in-depth study now in progress could not be included in the reduced space of this chapter where after a presentation of methods and data the main results will be discussed under two headings corresponding to the two periods of the lake: the 'Normal Chad' (as defined by Tilho, 1928) from 1965 to 1972 and the change during the Sahel drought from 1972 to 1977.

10.1 Collection and treatment of data

The multi-species approach would be incomplete if it had been based only on the local fisheries. The latter did not include all the permanent and temporary aquatic environments, they could not be operated in all seasons and, except for subsidence fisheries, they were oriented towards the capture of fishes of medium or large size. It was thus necessary to carry out some experimental sampling using several traditional techniques as well as new fishing gear for the Chad basin.

Several sampling methods, from the Kotoko basket trap to the electric trawl were experimented with and used during thirteen years of observation. The required data was mainly obtained from four methods: two foreign — ichthyotoxins and river seines — and two traditional — triangular net of the El Beid and gill nets.

10.1.1 *Fishing with ichthyotoxins*

This method, with possible sources of bias and the results obtained with it, has been described by Loubens (1969, 1970). The poison used contained 5% rotenone and was harmless to man. Its use was justified by the 2 main objectives of the study: to sample environments where other methods could not be used, and to obtain accurate estimations of community structure, fish densities and biomasses in restricted environments.

The poisoning could only be carried out over small areas (one hectare at maximum) in well-sheltered regions. Thus observations were limited to biotopes that are enclosed (subsidence ponds) or specialized (back waters, lake creeks). The open water biotopes had to be sampled by other means.

In spite of these restrictions, precise and acceptable data on biomasses and their composition in the restricted environments could be obtained only by poison fishing. The results obtained by other sampling methods could often be made more explicit by these data. Collected between 1966 and 1968, they essentially concerned a small creek of the southeastern archipelago of the south basin, close to Bol and a secondary arm of the Shari, upstream of Mailao (Fig. 2) during the withdrawal of waters and the first part of the flood.

10.1.2 *Beach seine*

The beach seine is a pulled net capable of catching all the fish encountered above a given size for a given species. Outside the very limited fishing of the Yobé in Nigeria, there were no such local fishing gear. The one we used was made in France and measured 200 × 7 meters with a 40 mm mesh* for the two wings and a 20 mm mesh for the 100 m long central part and for the bag. This net can give excellent results but is not easy to use. It demands a fishing team of 15 to 20 trained men and can only be used in some areas because it is set from a cleared beach and on a clean and firm bottom. In the river, these conditions limited its use to the low water period when the sand beds were exposed and setting was not complicated by currents. In the lake, the use of the seine was limited to the island borders adjacent to the non-muddy bottoms.

It is possible that the total catch underestimated fish biomass to a varying extent because error could result from the escape of fish during setting and pulling-in (above, below and through the net). These escapes were doubtless not the same for all species and all sizes encountered, resulting in some element of bias. Moreover, the gear selection of the beach seine also interfered and essentially corresponds to the passage through the central mesh of 20 mm, as in codend of trawlnet. The mean selection length although not too high (for example it was 150 mm for *Alestes baremoze*) led to a notable underestimation in the abundance of the smaller sized species. From poison fishing data, it is known that 95% of the ichthyomass was composed of fish larger than 100 mm, thus biomass data were less affected than density estimates. It can be deduced (Lauzanne 1977) that the underestimation of the biomass due to gear selection of the seine was about 10%.

The seine was used from 1966 to 1974 essentially in the southeastern archipelago (Bol area, Fig. 2) but also more temporarily in the Shari delta and around Maïlao on the Shari during the subsidence.

---

* For all fishing nets we always give the length of the mesh bar.

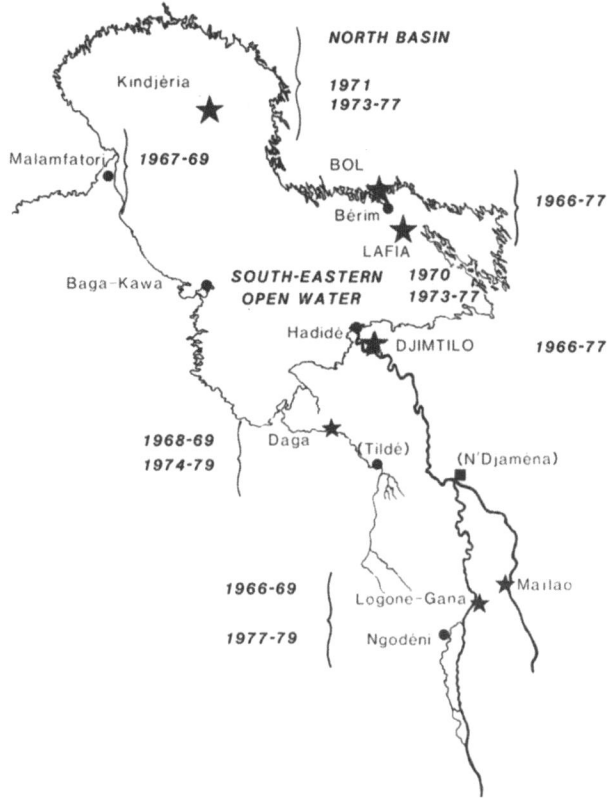

Fig. 2 Sampling sites and periods.
★ = main stations;
● = occasional surveys.

10.1.3 *The triangular nets of the El Beid*

The traditional fisheries of the El Beid were based on the use of triangular nets
of the push net type inside of semi-circular frames built against the permanent
dams found along the river from Tildé up to the lake (Fig. 1). Until 1970 these
nets, whose opening generally measured 4 to 5 m, included some lateral bands
of a medium mesh of 14 mm and a central piece from 6 to 10 mm. After 1974,
the fitting on of the net became slightly different with the entire net except a
frontal band being made of a small mesh (Fig. 3). It has been shown (Durand
1970) that species smaller than 50 mm were rare in the environment and that
the selectivity was negligible for fish between 50 and 200 mm. This size range
included the majority of standard fish lengths in the El Beid where the fisheries
were based entirely on catching 3–6-month old fish leaving the flood plains for

297

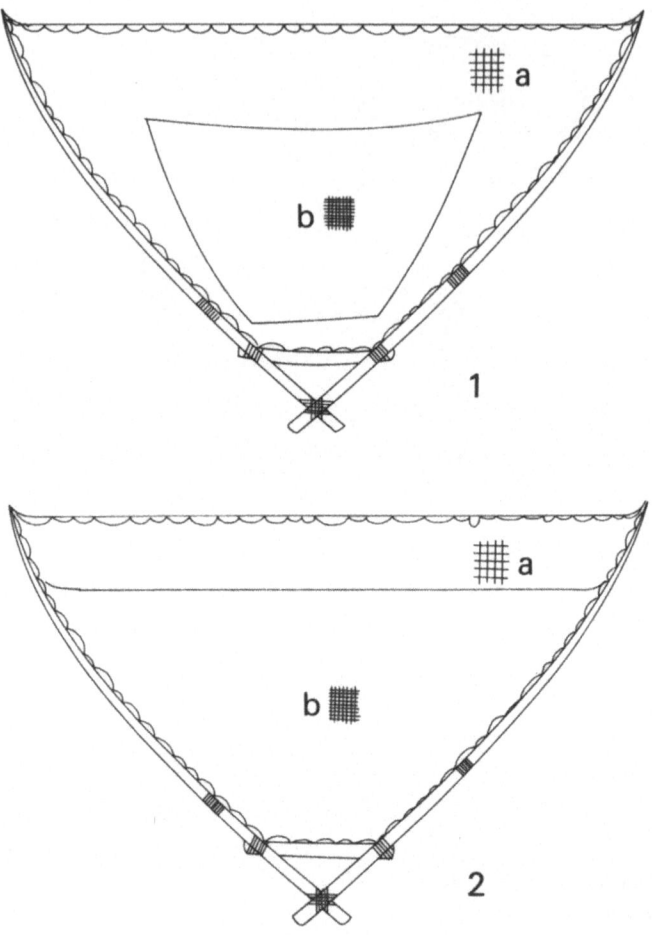

Fig. 3 Characteristics of the 'boulou' (triangular net of the El Beïd).
1 = from 1968 to 1969; 2 = from 1974 to 1979; a = large mesh; b = small mesh.

Lake Chad during the flood withdrawal. Some systematic observations were made in 1968–69 and from 1974 to 1978 at the Daga station.

10.1.4   *Gill nets*

As a sampling method, the gill nets had a serious disadvantage as they did not permit biomass to be calculated and only gave some relative abundance indices of relatively limited size ranges. The size of the fish caught was directly related to the mesh used and this very marked selectivity was the object of several studies. In particular, the efficiency was not independent of the mesh size (Hamley and Régier 1973) and the selectivity curves were frequently far from the normal curves (Bénech 1975; Durand 1978).

Practically however, gill nets had a very considerable advantage because they could be set in all the lake environments and also used as drift nets in rivers outside the period of high water. It also appeared to be a technique familiar to all the fishermen of the region.

Before 1971, two fisheries corresponding to two types of meshes coexisted: large mesh gill nets (GN 80 to 130)* — used in the rivers at the subsidence and in the north basin throughout the year — and average mesh gill nets (GN 25 to 30). The latter mostly caught *Alestes* (*A. baremoze* and *A. dentex*) along with some medium sized species. The large mesh nets caught adults of large species (*Lates, Labeo, Citharinus*, etc.). With the change of the aquatic environment and its exploitation (cf. Chapter 13) the size of mesh decreased in each of the fisheries respectively to 50 and 20 mm.

The absence of small mesh local nets as well as the gap between the medium and large mesh only permitted a limited study of fish populations from specific biotopes to be made. As elsewhere, the local nets were made with fine thread which made them very efficient, though fragile; therefore we chose an experimental homogenous battery including 18 nets. The most frequently used mesh were the following: 10, 12, 14, 15, 16, 18, 20, 22, 25, 30, 35, 40, 50, 60, 80, 110, 115 and 130 mm. This series was able to catch the complete range of fish sizes present in the environments sampled (with the exception of the smallest). Of course, the use of a battery of nets did not solve the difficulties due to gear selectivity, and the reservations made in the introduction remain. The results must be used with caution because the summation of the catch per unit effort did not provide a complete and accurate picture of the entire community because of gear selection and because vulnerability to the net varied with species, depending on morphology and behaviour. Nevertheless, the use of total or specific catch per unit effort for a given mesh, for spatio-temporal comparisons, remains justified and permitted the analysis of relative abundance.

10.1.5 *Sampling stations*

Figure 2 shows the various stations or the regions sampled as well as the periods of observation between 1966 and 1979. The distribution of observations was very unequal temporally as the initial objectives were oriented towards a bio-ecological study of the chosen species, at relatively modest means for the extent of the programme; on the other hand because of the change in the aquatic

---

* In the following text, the abbreviation 'GNx' denotes gill nets of x mm mesh bar.

environment with the drought, the collection of multi-species data in the lake and in the flooded areas became essential.

Between 1966 and 1970, the main concern was to acquire biological data. Five main stations were visited more or less regularly: Bol in the Southeastern Archipelago, Djimtilo in the Shari delta, 17 km from Lake Chad, Mailao and Logone-Gana, situated respectively on the Shari and the Logone about 60 km upstream of the confluence of N'Djamena and Daga on the El Beid in 1968–69 (Fig. 2).

The years 1971–72 constituted a second period in the collection of data. After collecting the biological data the river stations were abandoned, but a permanent team was installed for fishing at Djimtilo and in the Southeastern Archipelago at Lafia to provide continuous sampling for a fairly long period from 1971 to 1973. In 1972, the acceleration of the contraction of the lake led to an interruption of the activities of the permanent team. However, the southeastern Archipelago (Bol-Berim) and the Shari delta (Hadidé) as well as the Southeastern Open Water continued to be regularly visited. The third phase (1973–79) was characterized by the examination of the north basin during the drought, in particular at Kindjéria from 1973 to 1975, and by the revival of fishing in the El Beid at Daga from 1974 to 1979 (Fig. 2). Between 1976 and 1977 a permanent team fished at Hadidé, to compare the migratory passages in the period of the 'Lesser Chad' to those observed before the drought at the end of 'Normal Chad' (1971–72).

In addition to these base data, some surveys were made along the Cameroon border of the south basin during 1968–69, in the Southeastern Open Water during 1970 as well as at the confluence of the N'Djamena, in the flooded zones and their annexes (in particular the Logomatia from 1977 to 1979) and in the entire north basin in February 1971. The western border of the north basin was studied by H. Hopson from 1967 to 1969, from the Nigerian research station of Malamfatori, and complemented the observations made by A. J. Hopson (1964, 1968) (Fig. 2).

10.1.6   *Treatment of data*

The choice of statistical methods was related to the aims pursued, the available information and the general progress in the analytical methods in use by the ecologists. According to the goals pursued, one or another type of treatment could be preferred. It is possible to distinguish biological studies which essentially rely on parametric statistical methods and some adjustments to established models such as growth curves etc.

The study of communities increasingly calls upon descriptive statistics such as hierarchical classification and factorial analysis used from tables of distances, whether euclidian like the χ^2 or non-euclidean such as Kendall's index.

300

The available information was highly variable. In fact, although some samples (by number and weight) were collected very early for each of the species fished, their periodicity and reliability was linked to the aims of the observers. Thus the data collected between 1966 and 1970 were often lacking in the study of communities and the treatments were limited to the catch per unit effort by weight, more appropriate to the study of production. For the following period, however, only the c.p.u.e. in number was used thus the treatments could not be standardized for the whole period of study and did not lead to the same degree of investigation.

The change of development of statistical methods, due mainly to the popularity and increase of descriptive statistical methods, also had a role in the heterogeneity of treatments. The first studies were done essentially by hand from simple calculation methods whereas the more recent studies were based on computers and all the calculations and comparison facilities they provide.

10.2 The communities during 'Normal Chad': 1966–1971

10.2.1 *The lake*

The data for the comparison of multi-species samples were mostly obtained with the help of experimental batteries of gill nets. Thirty distinct stations were visited occasionally between 1966 and 1971 and are shown on the map in Fig. 4. Some complementary observations were made in the Southeastern Archipelago using beach seines (1966–70) and rotenone poisoning (1965–67). The distribution of the principal species will be described by separately reviewing the two basins, then by examining the total data on the richness of the populations.

10.2.1.1 *Distribution of principal species.* The degree of dissimilarity between two zones depended upon species common to both and species peculiar to each of them. The general description made here does not consider rare or accidental species. Only the species that were regularly counted in our catches were considered as they constituted most of the biomass.

For convenience, we have divided the lake into two parts, south and north on each side of the 'Great Barrier' (Fig. 1) and we have considered the major natural zones (archipelagoes, open water). Since more data were obtained from the south basin, we have carried out an initial sorting of species for this zone and then attempted to see how this was represented in the north basin.

10.2.1.1.1 *South basin.* To locate the main factors responsible for community structure in the south basin we shall first consider the results of a correspondence analysis on the results of a series of catches made in the south basin in 1964 and 1965 by the 'Centre Technique Forestier Tropical' (Durand 1973a).

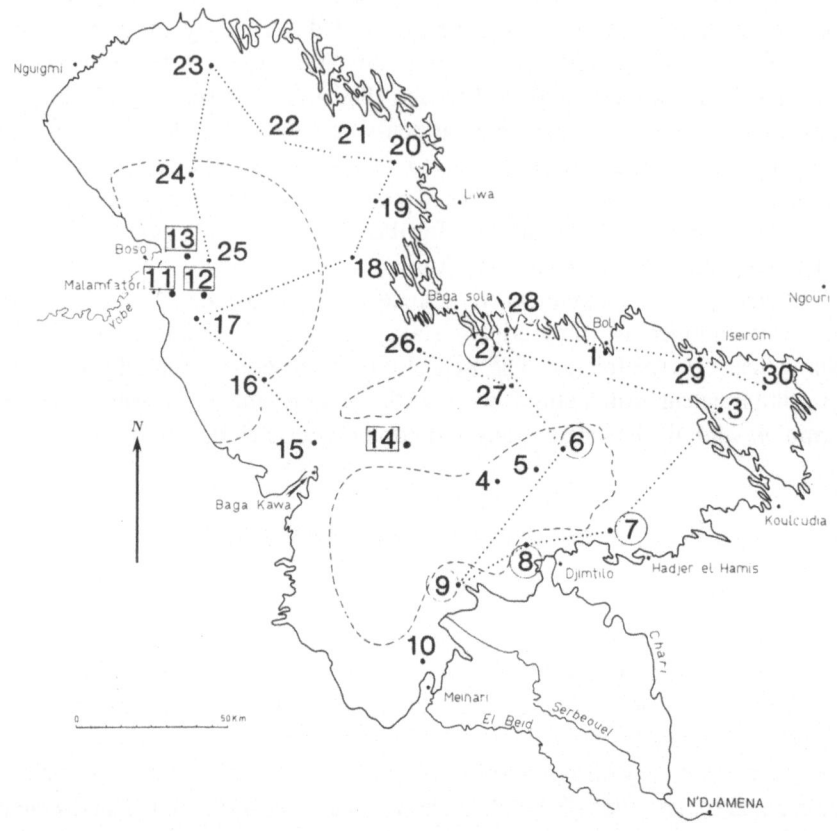

Fig. 4 Locations of the 30 fishing stations. The stations sampled by the Fishing Department of Nigeria and by the CTFT in Chad are respectively indicated with a square and a circle. The dotted line links up the homologous stations: 2, 3, 6 to 9 (CTFT, 1965) 15 to 30 (ORSTOM, January 1971), 26 to 30 (ORSTOM, May 1971).

From November 1964 to December 1965 some periodic catches were made at stations 2, 3, 6, 7, 8 and 9 (Fig. 4). The sorting was not carried out to species but to categories as follows: the Mormyridae, *Hydrocynus*, *Alestes*, *Distichodus*, the Schilbeidae, *Synodontis*, *Labeo* and *Polypterus*.

The percentage of inertia selected for the first three axes: 48.7, 20.3, and 15.8% were important and the position of points in the plane of the axes 1 and 2 was sufficiently explicit (Fig. 5). The ordination of surveys according to axis 1 was in close correlation with the relative abundance of *Alestes*; axis 2 separated the surveys according to the relative importance of *Hydrocynus* on the one hand and of the Schilbeidae on the other. Thus there was a triangular classification: *Alestes*/*Hydrocynus*/Schilbeidae.

The classification of samples led to a grouping of stations in pairs: stations 2

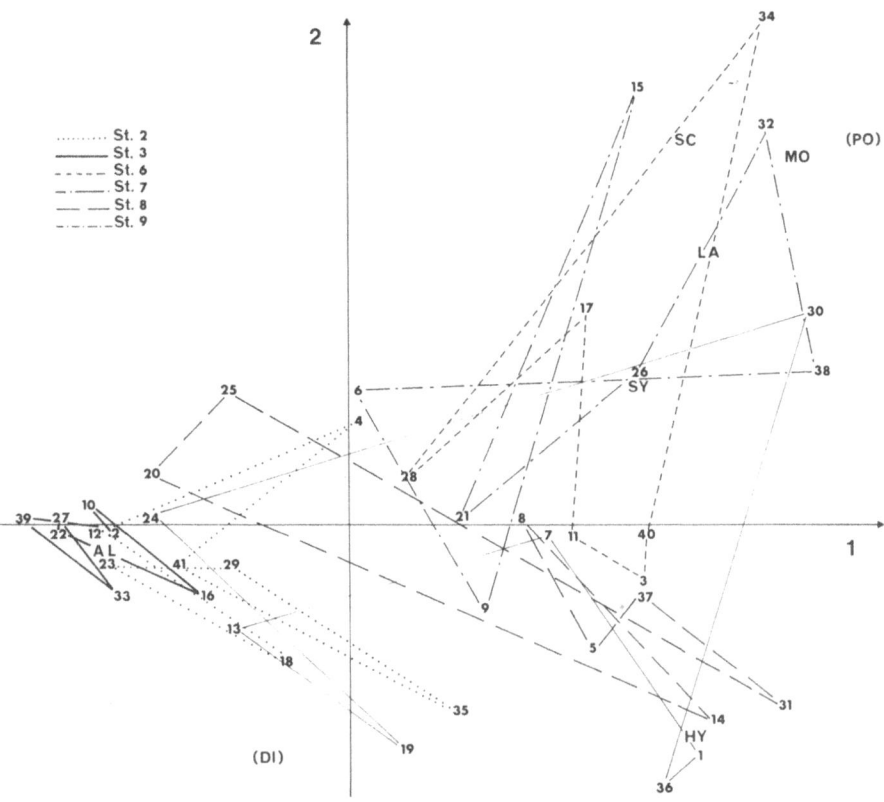

Fig. 5 Correspondence analysis on species and sampled sites (GN 30 in 1965; stations 2, 3 and 6 to 9). Projection in the plane of axes 1 and 2.
AL = *Alestes*; (DI) = *Distichodus* (rare); HY = *Hydrocynus*; LA = *Labeo*; MO = *Mormyridae*; (PO) = *Polypterus* (rare); SC = *Schilbeidae*; SY = *Synodontis*.

and 3 where the preponderance of *Alestes* was well marked (43 to 99% of total catch); stations 6 and 9 where *Alestes* was never dominant and where communities fluctuated through the year. Stations 7 and 8 were composite with the Schilbeidae and not very important and *Hydrocynus* dominant or, sometimes, *Alestes*.

Two points emerged from this analysis: the pairs of stations defined above corresponded to regions of the lake: archipelago (2 and 3), reed islands and open water (6 and 9), deltaic limits (7 and 8). One or two characteristic species were associated with each of these pairs, respectively, *Alestes* (probably *A. baremoze*) for the archipelago, Schilbeidae for the reed islands and open water, *Alestes* and *Hydrocynus* (certainly *H. forskalii*) near the delta.

It must be emphasized that seasonal variations were particularly notable and

303

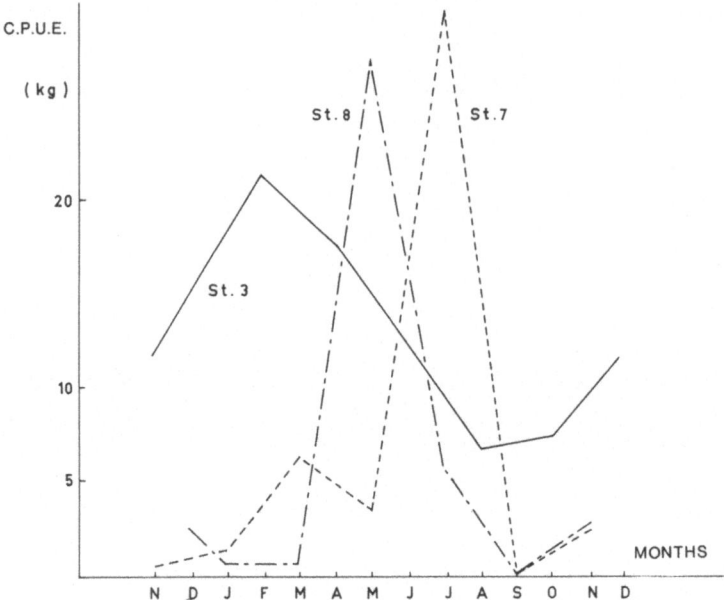

Fig. 6 Seasonal variations in c.p.u.e (GN 30; kg 100 m^{-2} night^{-1}) of *Alestes baremoze* in Shari delta (station 7 and 8) and in the Southeastern Archipelago (station 3).

that every definition of zonations must consider yearly fluctuations. The *Alestes* example was instructive in this respect (Fig. 6). The migration of these fish takes them towards the fluvial system at the time of reproduction and this explains the predominance of *Alestes* in May and July at stations 7 and 8 which were characterized by the permanent presence of *Hydrocynus*.

A. Communities and environments

To specify the relationships between communities and the major types of lacustrine environment presented in the preceding analysis, it is preferable to return to the experimental data that we collected between 1966 and 1971 in the south basin of the lake.

The communities of this zone were described by three series of data from: the Cameroon coast in the south of the basin during 1968 (station 10); Southeastern Open Water in 1970 (station 4 and 5); Southeastern Archipelago at Bol from 1966 to 1969 (station 1). So as not to attach illusory significance to the very heterogeneous data (obtained during various annual cycles and at different seasons), the data on c.p.u.e. were regrouped to provide a table of the species present in each of these zones with total abundance. We have therefore gathered in Table 1 the values found (from 7 to 14 GN following the zones) in

304

Table 1 Catch per unit effort (decagrams 100 m^{-2} night^{-1}) for the main species caught in the south basin between 1966 and 1970. S = GN 10 to 16; M = GN 20 to 40; L = GN 50 to 100. South shore = station 10 in 1968 (cf. Fig. 4); open water = stations 4 and 5 in 1970; archipelago = station 1 from 1966 to 1969. The thick frames outline the group of species characteristic of a zone. Hatching represents the species that are missing or not characteristic.

		South shore			Open water			Archipelago		
		S	M	L	S	M	L	S	M	L
Ubiquitous species	Hydrocynus forskalii	11	99	4	16	117	23	4	56	50
	Eutropius niloticus	21	38		29	69	3	95	46	*
	Lates niloticus	3	21	81	*	1	116	*	14	57
	Hydrocynus brevis	*	10	23	*	5	35		2	16
	Synodontis schall	2	9	*	3	10	16		13	15
	Labeo senegalensis	1	5	14	11	2	16	1	2	13
	Distichodus rostratus	3	22	51	13	4	26	3	8	35
	Pollimyrus isidori	9	*		3			5		
	Schilbe uranoscopus	10	56	4	2	78	32	1	17	1
Characteristic species	Labeo coubie		*		*	*	28			
	Citharinus distichodoïdes			*		*	24			
	Synodontis clarias		*		4	8	5			
	Bagrus bayad		*			1	17			**
	Citharinus citharus	*	*	*	*	*	15			**
	Hyperopisus bebe		12	3		4	12	*	2	28
	Hemisynodontis membranaceus		*		2	4	43			**
	Brachysynodontis batensoda	*	*		2	5	4	3	45	2
	Marcusenius cyprinoïdes	*	19		*	*			5	*
	Petrocephalus bane	1	7		*	*		*	16	
	Alestes dentex	2	24	1	1	4		1	12	1
	Alestes baremoze	2	68		11	30	*	25	153	1
	Heterotis niloticus		*	69			*			**
	Synodontis frontosus	1	14	6		*		1	22	33
	Polypterus bichir	3	63	46			*		5	42
	Chrysichthys auratus	*	8	*		*		*	6	
	Ichthyborus besse	9	*							
	Siluranodon auritus	6								
	Polypterus senegalus	1	17							
	Tetraodon fahaka		2	46			*			
	Mean c.p.u.e.	95	540	423	102	353	458	137	437	312
	N	4281	6088	400	9381	5904	946	8101	11 931	515
	f	47	104	70	64	92	80	43	111	34

* corresponds to catch per unit effort ⩽ 5 g m^{-2} night^{-1}.

** large sized species common in the archipelago and not represented due to the absence of large mesh (only GN 50).

305

three groups*: small mesh GN (10 to 16), medium mesh GN (20 to 40), large mesh GN (50 to 100). These three categories corresponded fairly well to some characteristic size groups: the small mesh caught the adults of small sized species and some young of mostly medium or large sized species, the medium meshes, the adults of medium size (standard length 25 to 40 cm), and the large mesh, the adults of large sized species. Each of the columns of Table 1 was obtained by calculating the arithmetic mean of c.p.u.e. from nets in each category. The total number of fish caught (N) and the corresponding fishing effort (f) at the base of the columns provide a comparison of abundance. Some species that were not caught by the gill nets are not mentioned in Table 1 but taken into consideration from observations obtained by other fishing gear.

The distinction of three sub-regions within the south basin established by factorial analysis was confirmed by the distribution of species established from their abundance and/or their scarcity (their complete absence) in each of the major zones in this region of the lake (Table 1): the archipelago, the open water and the southern portion of the south basin.

B. Ubiquitous and characteristic species

(a) Ubiquitous species

Three species were very abundant everywhere: *Hydrocynus forskalii*, *Eutropius niloticus* and *Lates niloticus*. Four other common ones were associated with these: *Hydrocynus brevis*, *Synodontis schall*, *Labeo senegalensis* and *Distichodus rostratus*.

The ecophases of these species were not equally represented in all the environments: the young *Lates* were found especially along the southern coast (st. 7, 8 and 9); the young *H. forskalii* were less frequent in the archipelago, whereas the young *H. brevis* were not common anywhere. The young of *D. rostratus* and *Labeo senegalensis* were more particularly caught in the open water in October, (st. 5) probably an index of migratory activity.

The case of *Eutropius niloticus* is probably more complex since some individuals of all ages were caught at all stations. There appeared to be a particular abundance of small sized individuals in the region of the reed islands with a c.p.u.e. higher than 4 kg in May 1966 north of station 6. This distribution probably uncovers two distinct groups differing in size at the first reproduction and egg-laying season.

Three of the most common ubiquitous species (the two *Hydrocynus* and *Lates*) were strictly ichthyophagous and the stomach contents of these predators from all regions often contained *Micralestes acutidens* (Lauzanne 1977). This very small sized species (45 mm standard length) was thus also ubiquitous.

---

* Detail of species c.p.u.e. are found in Carmouze et al. (1972). These c.p.u.e. themselves were calculated from previously published raw data (Durand et al. 1972).

Photo 14   An electrified front trawl, experimented in Lake Chad.

Photo 15   Boats used by fishermen: dugout canoes and papyrus canoes.

307

(b) Characteristic species

South of the lake. Three species were caught in this region exclusively: *Ichthyborus besse, Siluranodon auritus* and *Polypterus senegalus. Tetraodon fahaka* was also common in this region along with young *Schilbe uranoscopus* and *Hyperopisus bebe.* However the adults of this last species were very rare in this region, with three species being found only occasionally: *Bagrus bayad, Citharinus citharus* and *Hemisydodontis membranaceus.*

The open water. During the period of 'Normal Chad' there did not appear to be any species that were exclusively associated with the open water; however three species appeared to frequent this zone: *Labeo coubie, Citharinus distichodoides* and *Synodontis clarias.* The importance of the large sized *Hemisynodontis membranaceus* is shown. Although a small sized species, *Alestes dageti* had been found occasionally in other regions; it should be noted here that this zooplanktophagous Characidae was especially abundant in the open water as well as the stomach contents of predators (Lauzanne 1977). In contrast, the species absent from the open water were more numerous: all the *Sarotherodon* species (*S. niloticus, S. aureus, S. galilaeus*) and *T. zillii* which were common in all the regions of the archipelago and vegetation banks (caught by beach seines and poisoning); and another Cichlid of small size, *Haplochromis bloyeti,* dominant in all the vegetation banks of the south basin and in the stomachs of *Lates* (Lauzanne 1977).

Eight species were common south of the lake and in the archipelago, and accidental or rare in the open water (Table 1): *Marcusenius cyprinoides, Petrocephalus bane, Alestes dentex* and *Heterotis niloticus* (only adults of these four species were present); *Synodontis frontosus* and *Polypterus bichir* (young present especially in the south); *Chrysichthys auratus* and *Alestes macrolepidotus.*

Alestes baremoze must also be associated with this group because this zooplanktophagous species was very abundant especially in the zones of reed-islands and the archipelago. Its presence in the open water (Table 1) corresponded to seasonal migrations.

Other species absent from the open water were small sized species only found in the sheltered zones and the vegetation banks. This was particularly true of *Barbus* (especially *B. pleuropholis, B. callipterus* and *B. leonensis*) very common in the archipelago (Loubens 1969) and in the stomachs of predators (mainly *Hydrocynus forskalii*) (Lauzanne 1977).

Archipelago. No species were exclusive to this region. However the zooplanktophage *Brachysynodontis batensoda* appeared to develop particularly in the Southeastern Archipelago.

Few species were abundant in the southern part of the lake and the open water, and absent or rare in the archipelago. In spite of its significant importance in the archipelago, *Schilbe uranoscopus* can be associated with this group because of its particular abundance in the open water and southern part of the lake; the young were found especially near the Cameroonian shore.

308

10.2.1.1.2 *North basin.* Two series of data can be used here: some surveys of experimental fishing carried out initially in all environments of the north basin in January 1971 (stations 15 to 25, Fig. 4); the analysis of the unloading done by fishermen at Malamfatori and Baga-Kawa from 1968 to 1971 for the 90 to 100 GN set in the open water and in the border of reed islands (Durand 1973b).

Three species that were ubiquitous and abundant in the south of the lake, *Lates niloticus, Hydrocynus forskalii* and *Eutropius niloticus* were again found in the north but with some differences in distribution. *H. forskalii* and *E. niloticus* were caught in notable quantities at stations 15, 16 and 24 in the region of reed islands. *Lates* was the only large sized fish caught in all the lacustrine environments. Three of the four other ubiquitous species were found in the north: *Distichodus rostratus* and *Labeo senegalensis*, at least in the open water, and *Synodontis schall* everywhere. The presence of *D. rostratus* is explained by the seasonal migratory pathways (Hopson 1968). Alestes dageti was also common in all fishing environments at all the stations in January 1971, and was an important component of stomach contents of *Lates* in the open water (Hopson 1968; stations 11 and 12).

The large mesh nets of the north basin fisheries (90 and 100 GN) caught the large individuals of a dozen species in the open water and in the region of reed islands. The precise origin of the fish cannot be determined from the data obtained. The unloading, at Malamfatori and Baga-Kawa respectively (stations 13 and 14 of Fig. 4), had a very similar composition (Table 2). Four species clearly dominated all the catches: *Lates niloticus, Heterotis niloticus, Citharinus citharus* and *C. distichodoides* always represented more than 90% of the total catch. *Distichodus rostratus, Bagrus bayad*, and *Labeo* (especially *L. coubie*) were next, while *Hemisynodontis membranaceus* was much less common than in the open water of the south basin.

From the January 1971 surveys and these relatively few data, two groups of species could be distinguished: species characteristic of the southern region found in the north with an analogous distribution and those which did not have a distribution homologous with those of the south. Thus *Alestes baremoze* had exactly the same distribution as in the south basin at the same time of the year. It was absent from the open water and the adults were found in the zones of reed islands, while the young and adults were present in the Northeastern archipelago (st. 18 to 23). The densities were exceptionally high: c.p.u.e. between 5.7 and 9.5 kg (25 to 35 GN) at stations 20 to 22. The *Tilapia* and *Sarotherodon* also appeared to be very abundant in all zones of reed islands and the archipelago along with *Heterotis niloticus*. The less common *Synodontis frontosus* appeared to be caught everywhere in these sheltered zones.

Several species appeared to have a different distribution in the north. *Hemisynodontis membranaceus* has already been shown to be less abundant. Similarly, *Brachysynodontis batensoda* was found in the south of the northern region, without being very abundant, and practically absent elsewhere. *Schilbe*

Table 2 Mean annual species composition (%) for the unloading at Baga-Kawa and Malamfatori (GN 90 to 100). f indicates the total fishing effort ($100 \text{ m}^{-2} \text{ night}^{-1}$)

Species	Malamfatori		Baga-Kawa	
	1968	1969	1968	1970
Lates niloticus	38.1	48.2	36.5	36.5
Heterotis niloticus	18.8	14.1	24.6	14.4
Citharinus spp	29.1	22.1	23.8	30.8
Distichodus rostratus	3.7	6.3	5.3	6.6
Labeo spp.	5.1	3.3	2.9	3.1
Bagrus bayad	0.6	0.7	1.8	3.2
Hemisynodontis membranaceus	0.7	0.1	*	0.7
Others	3.9	5.2	5.1	4.7
Total c.p.u.e.				
($\text{kg } 100 \text{ m}^{-2} \text{ night}^{-1}$)	1.460	1.150	2.400	0.740
f	20 120	38 520	16 580	7800

uranoscopus was abundant in the two reed island stations (15 and 16), rare in the open water (st. 17 and 25) and totally absent from stations 18 to 24. This was also true of several Mormyrid species: *Pollimyrus isidori* abundant at stations 15 and 16, absent from station 18 to 25, *Petrocephalus bane*, *Marcusenius cyprinoides*, *Hyperopisus bebe*.

10.2.1.1.3 *The medium sized species (GN 30) in the whole lake.* In order to cross-check and complete this report we have attempted a more total and systematic analysis with the help of multifactorial analysis (Durand 1973a). This analysis was carried out on a systematic list of 32 species found in 42 surveys carried out between 1966 and 1971 at station 1, 4, 5 and 15 to 30 with GN 30 (Fig. 4).

Because of the great heterogeneity of the surveys and a higher species richness, the direct interpretation of results was less easy than in the preceding case. The first two axes alone corresponded to 23.9 and 20.5% of the total inertia extracted and a third (13.0%) must also be considered.

In the plane of the first two axes and of axes 1 and 3 the Southeastern Archipelago and the Northeastern Archipelago were close; the characteristic species was *Alestes baremoze*, *A. dentex* being secondary. In contrast, three other series of samples, reed islands and Northern Open Waters, southern coast of the lake (st. 10), Southeastern Open Waters (st. 4 and 5) were close and in the plane of axes 1 and 2, the projection in the plane of axes 1 and 3 separated the

310

southern coast from two others. *Schilbe uranoscopus* and to a lesser degree, *Eutropius niloticus* were characteristic species of the Northern and Southern Open Water and *Polypterus bichir* was characteristic of the southern coast.

The interpretation involved several distinct sources of variation.

1. The zones of the lake differed, particularly the archipelagos and the open water zones. The individuality of the southern coast was shown (st. 10) by the presence of *Polypterus bichir* but also by some members of the fluvial fauna which were only found here, and were of major importance (*Mormyrus hasselquisti, Auchenoglanis biscutatus, Synodontis nigrita* ...).

2. Seasonal variations were also evident in the Southeastern Archipelago (st. 1): from 1966 to 1969 the surveys of the first six months form a homogenous group where *A. baremoze* dominated. In contrast, those of the second half of the year were mixed and gave the impression of reflecting some unstable situation, when a variety of species (*Synodontis frontosus, Brachysynodontis batensoda* ...) could be dominant.

3. The annual variations could be considerable here too and we will return to this aspect in the general conclusions. We simply emphasize here that there was some instability even in the absence of major events such as the drought of 1972 and following years. The exploitation of stocks by man could play an important role, as in the scarcity of *Labeo coubie* in the north basin of the lake between 1963 and 1968. But an intrinsic variability could be linked to some successions and long-term replacements such as of *Schilbe mystus* by *S. uranoscupus* (Mok 1975).

10.2.1.2 *Richness of the communities.* The number of species caught in comparable samples was of interest in the study of fish communities. In order to make valid comparisons it was necessary that the fishing gear and its use were identical. Thus the usable data in Lake Chad were obtained by experimental gill nets, particularly from the north of the lake in January 1971 (st. 15 to 25) and from the Southeastern Archipelago in May (st. 26 to 30).

The total number of species caught in one night by the batteries used (10 to 60 GN) varied between 9 and 22 for the north basin of the lake and between 17 and 33 for the Southeastern Archipelago (Fig. 7). The considerable variations were verified for each mesh as well as for the whole group. In the Southeastern Archipelago, stations 26, 27 and 28 were comparably rich with about 30 species whereas more to the east, only 23 species were found at station 29 and only 17 at station 30 or almost less than half the number at station 26. This last was a station of reed islands whereas the five others were poverty of the eastern zone. On the other hand, conductivity values (Fig. 7) increased, passing east, from 120 to 180 μS cm^{-1} at stations 26 to 28 to 240 (st. 29) then 450 μS (st. 30). It thus appeared that the species paucity increased from west to east on a par with the increase of conductivity.

In the north of the lake the three southern stations were clearly richer than all

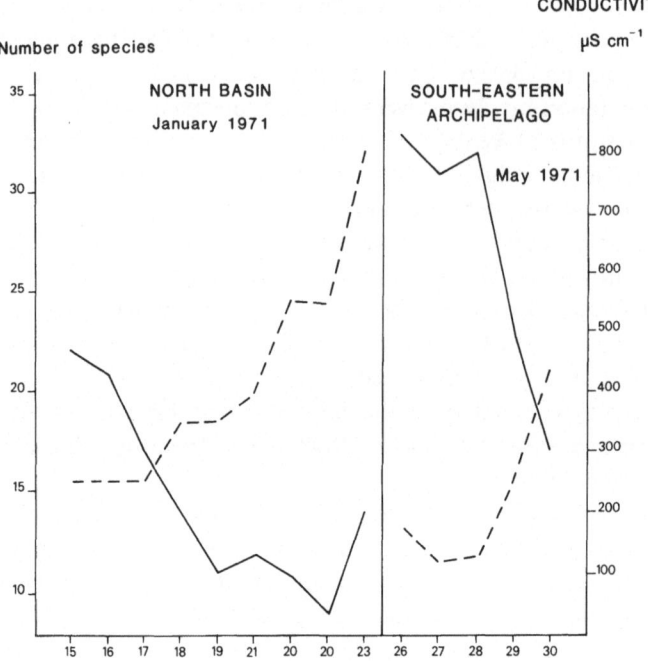

Fig. 7 Number of species caught in the same set of gill nets (GN 10 to 130; continuous line) and for the same fishing effort in different stations ordered according to increasing conductivity (dashed line).

the others, with an average of more than 20 species being caught (not all the small mesh GN were fished at station 17) whereas the other stations were poor or very poor (14 to 9 species). The richest stations were those where the conductivity showed the smallest increase and the stations of the Northeastern Archipelago were the poorest. Several species of Mormyridae were not found in the north of the lake (above 3 to 400 µS cm^{-1}) to the same extent as in the Southeastern Archipelago, independent of the zone type.

The richness of the north could not be directly compared with the Southeastern Archipelago of the lake as the fishing effort was only half that in the latter region. The difference was thus probably underestimated and the Southeastern Archipelago was probably richer than the entire north basin. For the same zone the difference remains considerable between the north and the south, with 17 to 31 species found in the archipelago compared with 9 to 12 in the northeast: 34 species in the Southeastern Reed Islands and 21–22 in those of the north.

In other stations of the south basin where systematic sorting was complete, the classification as a function of conductivity (low in all cases; less than 100 µS cm^{-1} at stations 4, 5, 10 and about 150 at station 1) could not be made. As a

312

whole, the Southeastern Archipelago was richer than the Southeastern Open Water, but the southern coast (st. 10) with a total of 62 species was by far the richest in the entire lake.

10.2.2 *The rivers and their temporary annexes*
Some data are available for three distinct zones: the region of the Shari delta, the riverine environments and the temporarily flooded environments.

10.2.2.1 *The deltaic zone.* The delta region consisted of two main branches of the Shari rejoining near Djimtilo, the village chosen for the initial experimental fishing (Fig. 2). It was not possible to poison fish in this zone because there was no zone of calm waters. However, it was possible to use the beach seine at low water but this could be done only after 1971, and thus the only data obtained were from gill nets. The nets were also subdivided into three main categories: small, medium and large mesh.

(a) Small sized mesh

Gill nets (10 to 16 GN) were used from March 1971 to December 1972 and the corresponding mean c.p.u.e. regrouped by period from 3 to 4 months are gathered in Table 3. The catch by weight was always low. Initially this corresponded to less efficient fishing (relatively thick thread for the small mesh nets set close to the banks) but was also probably related to the scarcity of fish vulnerable to the 10–16 mm mesh. Whatever it was, these fisheries provided useful qualitative results, at least until July/August 1972 as the major perturbation of the very low flood of 1972 commenced in August/September.

Of the 43 species caught at least once over these 16 months, 18 are recorded in Table 3. It only considers the normal catches per mesh size and not the fish whose size is shown without comparison with the mesh of the net. For example, *Schilbe* and *Synodontis* which were caught by their thorny spines, were eliminated after being caught.

Five species appeared to be common in all the surveys: *Schilbe uranoscopus*, *Eutropius niloticus*, two *Polypterus* species (*P. senegalus* and *P. bichir*) and *Hydrocynus forskalii*. Two other species, *Petrocephalus bane* and *Brachysynodontis batensoda* showed significant variations in abundance.

The transitory character of the delta zone was conclusively shown by the presence of *Alestes dageti*, the only endemic species of the lake and *Alestes nurse* and *Ichthyborus*, riverine species caught only in the southernmost region of the south basin of the lake (cf. Section 2.1.1.1).

(b) Medium sized mesh

Two series of data exist for these nets: those obtained when the experimental fishing was with GN 30 drift nets in 1966 and 1967 and those gathered for the

Table 3 Mean catch per unit effort (g 100 m^{-2} night^{-1}) for the small mesh fixed gill nets (GN 10 to 16) at Djimtilo (Shari delta) between April 1971 and November 1972.

Species	1971		1972			T	%
	4–7	10–12	1–3	6–8	9–11		
Marcusenius cyprinoïdes[a]	*	4	*	*	3	1	0.6
Petrocephalus bane	2	3	10	28	2	9	5.2
Pollimyrus isidori	1	1	3	12	*	3	1.7
Hydrocynus forskalii	11	4	7	12	9	9	5.2
Alestes baremoze	12	4	4	3	*	5	2.9
Alestes macrolepidotus	*		1	10	3	3	1.7
Alestes nurse	4	5	10	5	2	5	2.9
Alestes dageti	20	*	1	3		5	2.9
Ichthyborus besse	1	2	5	5	1	3	1.7
Distichodus rostratus	2	8	2	8	1	4	2.3
Chrysichthys auratus	3	1	7	16	2	6	3.5
Schilbe uranoscopus	27	27	25	34	7	24	14.0
Eutropius niloticus	20	40	15	16	18	22	12.7
Brachysynodontis batensoda	1	4		6	74	17	9.8
Synodontis schall	1	1	6	9	5	4	2.3
Lates niloticus	2	16	9	4		6	3.5
Polypterus senegalus	1	4	23	41	12	16	9.2
Polypterus bichir		16	4	20	46	17	9.8
Others	2	21	23	9	13	14	8.1
total c.p.u.e.	110	161	155	241	198	173	100.0
f (nights 100 m^{-2})	222	72	128	139	152		713

[a] The systematic order adopted by Blache (1964) is followed for the species list of Tables 3, 4 and 5.

local fisheries (drift and fixed GN 25 to 32) between July 1971 and August 1972 (Loubens 1973). The corresponding total results are shown in Table 4 for the species representing at least 1% of the c.p.u.e. by weight in one of the data series considered.

During 1966–67, the pelagic community of the delta rivers was dominated by two species which represented together nearly 90% of the catches: *Alestes baremoze* and *Hydrocynus forskalii*. *Alestes dentex* appeared secondarily as well as the two common Schilbeidae, *Eutropius niloticus* and *Schilbe uranoscopus*. The 1971–72 data confirmed the predominance of *Alestes baremoze*; however *H. forskalii* was less abundant whereas *A. dentex* and *S. uranoscopus* appeared more abundant.

Table 4 Relative composition by weight (%) of catches in 1967–1968 (experimental GN 30) and in 1971–1972 for the delta fisheries (after Loubens 1973).

Species	1966–1967 (Exp.) GN 30 Drifting	1971–1972 (Local fisheries) GN 25–32 Drifting	Fixed	GN 50–120 Drifting	Lines
Hyperopisus bebe	*	1.6	0.2	0.7	18.1
Mormyrus rume			0.1	2.4	21.5
Hydrocynus forskalii	28.6	2.6	1.3		
Hydrocynus brevis	0.4	0.3		4.2	
Alestes dentex	5.8	14.2	5.3		
Alestes baremoze	59.5	63.3	50.7		
Citharinus citharus				14.2	
Citharinus latus				5.7	
Distichodus rostratus				23.6	
Labeo senegalensis		0.2	0.8	7.3	
Bagrus bayad				0.7	8.5
Schilbe uranoscopus	2.0	14.0	25.9		3.2
Eutropius niloticus	2.7	0.8	0.9		
Brachysynodontis batensoda	0.1	0.9	11.1		3.2
Hemisynodontis membranaceus		0.2	0.1	36.7	1.7
Synodontis schall		0.3	1.6		31.5
Polypterus bichir	0.1	0.3	1.8	2.2	7.0
Others	0.8	1.3	0.2	2.3	5.3
Mean c.p.u.e.[a]	3.63	3.96	4.86	4.08	8.04
Total f (surveys)[a]	40	786	194	219	84

[a] c.p.u.e. in kg hour^{-1} or night 100 m^{-2} of GN; in kg 1000 hooks^{-1} day^{-1} for the lines. Effort in hours (drifting GN), nights (fixed GN) or days of fishing (lines).

The view of the community obtained from local nets used as fixed nets (only partial because they were only set February to June 1972) was similar to that described above (Table 4): always more than 50% of *Alestes baremoze* and then a notable abundance of *Schilbe uranoscopus*. The same accompanying species appeared the only difference being the 11% of *Brachysynodontis batensoda* in the fixed GN.

The mean c.p.u.e. values showed some very important seasonal variations with the maximum for the medium mesh fixed nets occuring at the end of the withdrawal and at the beginning of the flood (June to August). The comparison

315

of values at 5-year intervals shows a very probable lowering of fishing yield. The parallel trials of experimental GN 30 and local nets of equivalent mesh showed that the latter, made with much finer thread, caught on average, 2 to 3 times more than the experimental nets. It is thus likely that the local nets caught between 7 and 11 kg 100 m^{-2} night^{-1} in 1966–67 as compared with the mean c.p.u.e. of about 4 kg in 1971–72.

(c) Large sized mesh

The only data evaluating the importance of large sized fish are those of Loubens (1973) and of the non-baited multiple hook lines (Table 4). The large mesh drift GN (50–120) caught three species in particular, in the following order; *Hemisynodontis membranaceus*, *Distichodus rostratus*, *Citharinus citharus*, followed by *Labeo senegalensis*, *Citharinus latus* and *Hydrocynus brevis*. The comparison with the lines was instructive because it illustrated the problems of sampling: the three species caught most often with hooks were rare (*Hyperopisus bebe*, *Mormyrus rume*) or absent (*Synodontis schall*) in the 50 to 120 mm gill nets (Table 4). The behaviour of non-baited hooks, the random catches and the differences in composition also partly reflected the heterogeneity of the community as the species caught by the lines were characteristically benthic that is linked to the research on nutrition.

In this example there were thus two complementary fisheries. The catches in the delta fisheries would probably have provided a clear picture of the communities in the reaches containing exploited stages (Loubens 1973).

10.2.2.2 *The rivers.* The Mailao station on the Shari, and the Logone-Gana on the Logone were both situated far upstream of the confluence (Fig. 2) and were only visited at the beginning of the surveys (especially 1966–67). As in the delta, the fixed and drift gill nets supplied the basis of the experimental samples in the absence of active local fisheries. There was also some fishing by beach seines during low water (April to August) and some poisoning in a secondary arm of the Shari.

The main results of the experimental fishing practised in the Shari at Mailao are shown in Table 5. Only the most common species (at least 1% of the c.p.u.e. by weight of one series) were retained. The catches from nets of similar mesh were combined when the sampling effort was insufficient.

In the drift GN 30 a very marked predominance of *Alestes baremoze* was found forming nearly 60% of the total weight caught, a value comparable with that obtained in the delta in 1966–67 and 1971–72. The next two most common species were Hydrocynus forskalii (15.7%) and *Eutropius niloticus* (14.6%). Only three other species exceed 1% of the c.p.u.e.: *Schilbe uranoscopus*, *Alestes dentex* and *Petrocephalus bane*. The presence of *Synodontis nigrita*, *Polypterus endlicheri* and *Alestes nurse* was confirmed by the fixed nets and appeared to be

316

characteristic of the river surveys, as these species were generally not found in the delta.

If all the fixed nets of medium mesh are considered, the total view was fundamentally the same as the preceding one; however there was the lesser importance of *A. baremoze* (26.4%) and greater diversity: 16 species compared with 6 in the drift GN 30 which approached or exceeded 1% of the total c.p.u.e. Among the latter, *Hyperopisus bebe*, *Marcusenius cyprinoides* and *Brachysynodontis batensoda* appeared (confirming that *B. batensoda* normally zooplanktophagous can adopt a very different trophic behaviour in its feeding when close to the bottom, Lauzanne 1977). The importance of *Polypterus bichir* (10% of c.p.u.e. from fixed GN 20 to 40) and the presence of *Synodontis sorex*, confirmed by fishing at GN 50, should also be noted.

The largest mesh nets and the seine also provided some complementary information. The three species of *Citharinus* dominated the catches in the GN 60 to 100 (60%) and in the beach seine (41%). The presence of *C. latus* which is classed among the characteristic river species has already been noted in the delta. We than found the two *Labeo* and the two *Distichodus* species: *L. coubie* and *D. brevipinnis* which were as common as *L. senegalensis* and *D. rostrarus* which were normally dominant. *Lates niloticus* was also common as well as *Synodontis* cf. *schall* (22% of c.p.u.e. in the GN 50). For the first time *Campylomormyrus tamandua* was caught and in notable quantities: 4.7% of c.p.u.e. of GN 50. Finally the abundance of *Heterotis niloticus* and of *Sarotherodon galilaeus* in the seine fishing also illustrated the inefficiency of the gill nets for some species.

The low yields from the small sized mesh in the delta fishing corresponded initially to a limited efficiency of the fishing method and to a lesser importance by weight of small sized fish. The diversity was low and three species dominated the catches: *Eutropius niloticus* and *Hydrocynus forskalii*, confirming the general importance of these two species and also a new species *Alestes nurse*. Among the other species, the presence of *Synodontis nigrita* in the river was confirmed whereas *Petrocephalus bovei* was comparatively new to the lake and the delta.

Not all the results collected on the Logone contradict the description for the Shari and only two points are worth mentioning: the seasonal increase in diversity of catches in the small mesh GN is probably an index of the return of some river species from the flooded zones in subsidence, a phenomenon which was confirmed by the samples from the El Beid (cf. *infra*). Moreover, large adults of *Alestes dentex* which were already very common in the Shari (16% of mean c.p.u.e. from 35/40 GN) were important seasonally. The rush of these fish in the Logone corresponded to the beginning of the draining of the flooded zones of the right bank and also indicated the composition of the traditional main fishing of Logone-Gana (cf. Chapter 13). *A dentex* generally contributed more than 3/4 of the total catches during the first week of fishing or 5 to 10

Table 5 Relative composition by weight (%) of experimental catches in the Shari at Mailao in 1966–1967 (Fi = fixed; Dr = drifting).

Species	Fi GN 10–15	Fi GN 20–25/30	Dr GN 30	Fi GN 35/40	Fi GN 50	Fi GN 60–80–100	Beach seine
Heterotis niloticus							8.6
Hyperopisus bebe		0.2		7.9	2.6		3.2
Campylomormyrus tamandua		0.1			4.7		0.1
Marcusenius cyprinoïdes		3.1	*	0.3			0.1
Petrocephalus bovei	2.4						
Petrocephalus bane	2.4	5.0	1.7	0.5			0.2
Hydrocynus forskalii	19.5	13.4	15.7	5.7	4.9		1.6
Hydrocynus brevis		0.2		0.1	1.0		2.4
Alestes dentex		4.5	3.3	16.0	0.6		5.8
Alestes baremoze	7.8	26.4	58.7	9.5			17.1
Alestes nurse	18.2	0.1	*				*
Citharinus citharus		0.2		0.4	0.7	13.1	4.6
Citharinus latus				3.3		3.1	7.9
Citharinus distichodoïdes						43.6	28.8
Distichodus rostratus	*	0.1			3.3	9.6	0.2
Distichodus brevipinnis				0.3	1.2	6.6	0.3
Labeo senegalensis		1.2	*		8.7	4.7	0.4
Labeo coubie		0.6			3.6	7.1	
Chrysichthys auratus	7.3	2.1	0.2	0.3			*
Auchenoglanis spp.				0.7	2.8	0.8	*
Schilbe uranoscopus	3.6	5.3	3.5	9.2	1.5		1.6
Eutropius niloticus	34.0	13.8	14.6	8.1	1.0		1.7
Brachysynodontis batensoda	3.6	0.8	0.2	15.7	13.3	0.8	0.3
Hemisynodontis membranaceus			*	1.6		2.9	1.2

Synodontis nigrita	2.4	0.9	*		0.8		0.3
Synodontis sorex		0.6	*	3.9	1.1		5.4
Synodontis cf. *schall*		0.8	0.2	22.0	2.0	1.0	
Lates niloticus	*	2.9	0.7	12.1		5.8	
Sarotherodon galilaeus	*						5.3
Polypterus senegalus		1.9	*		0.1		0.1
Polypterus bichir		11.8		5.3	10.4		0.2
Polypterus endlicheri		3.8	0.6		2.8		0.5
c.p.u.e.[a]	0.14	1.9	7.4	2.1	1.5	2.0	100.7
f[a]	27	30	31	23	16	17	50

[a] Cf. Table 4 for the GN. For the c.p.u.e. in kg ha^{-1} and effort in number of hauls.

times more than *A. baremoze*. This was inversely proportional to the numbers generally found in the lake and the rivers.

The rotenone poisoning of an intermittent arm of the Shari at Mailao confirmed the importance by weight of *Lates niloticus, Heterotis niloticus, Sarotherodon galilaeus, Alestes baremoze, A. dentex* and *A. nurse* and all the *Polypterus* species as well as *Mormyrus rume, Synodontis eupterus, Bagrus bayad, Tilapia zillii* and *Sarotherodon niloticus*. Among the small sized fish present in high numbers, there were some adults of small species where *Barbus* dominated: *Barbus pleuropholis, B. punctitaeniatus, B. anema, B. lawrae, B. macrops. Micralestes acutidens* was also generally abundant (whereas *Petersius intermedius* was found in large numbers in a pond dependent on the Logone). Several young were also caught in the intermittent arm at the subsidence: *Synodontis nigrita* and *S. eupterus*, the three *Sarotherodon* and *Tilapia* already cited, the three *Alestes* and *H. forskalii*.

Finally, particular mention must be made of the fish inhabiting oyster beds (*Aetheria elliptica*) in the lower bed of the Shari. The species composition was novel, consisting of the young of medium or large sized species which found a particularly favourable environment there (shelter, food). These young belonged to some species caught in other places near Mailao: in the river (*Synodontis* cf. *schall, Labeo coubie, Chrysichthys auratus*), or in the intermittent arm (*Synodontis eupterus*); to some rare species from other samples (*Mormyrops deliciosus, Malapterurus electricus, Synodontis courteti, Bagrus docmac*) and finally to some species never found elsewhere in our samples *Petrocephalus simus, Nannocharax fasciatus* and *Synodontis filamentosus* (Loubens 1969).

10.2.2.3 *Temporary aquatic environments.* The association of floods of an almost tropical type with a well-marked annual flood, and the very flat plains and the very low river slopes led to some overflowing and flooding of adjacent plains (cf. Chapter 2). This phenomenon which was common in the Sahelo-Sudanian zones, was particularly pronounced in the Chad basin and in its northern part, which is of interest here. In this zone, the lesser Logone was characterized at the Logomatia by some lateral outflows which flooded the north Cameroon plain of about 5000 km² in an average year. The drainage of the flood zone was carried out partially by the Logomatia and principally by the El Beid which flowed into the southernmost part of the south basin of Lake Chad (Fig. 1).

We only have some fragmentary information on the Logomatia prior to 1971. However, some data exist on the flood plain (poisoning of ponds during the dry season) and the El Beid was followed continuously during the 1968–69 flood.

10.2.2.3.1 *The North Cameroon floodplain.* The sampling efforts during the flood period were hindered by considerable growths of grasses (*Echinochloa*) which prevented boat use. There was also a lack of fishing techniques appropriate for such an environment, where the water was generally shallow and transparent with low fish density. Without considering the indirect information supplied by the study of outflows (cf. section on El Beid) we must, therefore, be content with some limited observations on the residual ponds which remains in the depressions of the north Cameroon plain after the El Beid water flowed out completely or before it was flooded by the first rains.

The extremely heterogenous nature of these water bodies (area, depth, vegetation, distance from outflows) combined to create all types of biotopes from the small rapidly drying temporary pond to the large deep pond serving as a refuge until the following rainy season. Yet some species wre constantly found: *Brienomyrus niger, Barbus gourmansis, Neolebias unifasciatus, Synodontis nigrita, Clarias* spp., *Aplocheilichthys* spp., *Epiplatys senegalensis, Polypterus senegalus* and *P. bichir, Sarotherodon* spp. and *Tilapia zillii.* The sedentary species may very grossly define the ubiquitous community of these ponds (Durand 1971).

Some other species were caught occasionally without being abundant: *Alestes baremoze* and *A. dentex, Distichodus rostratus* and *D. brevipinnis, Labeo senegalensis* and *L. coubie, Brachysynodontis batensoda, Synodontis cf. schall. Schilbe mystus* and *Siluranodon auritus* were caught in abundance only in a northern pond not far from the El Beid. We will see that these two species migrated slowly in the El Beid and behaved as trapped individuals which could not migrate before the end of the overflowing period.

Finally, the scarcity or absence of species whose young were very common in the El Beid fishery must be noted in these ponds, particularly *Alestes nurse, Hyperopisus bebe* and *Marcusenius cyprinoïdes.* This is surely an indication of migration in these species.

10.2.2.3.2 *The fishes of the El Beid.* The traditional fishing of the El Beid was carried out from permanent dams constructed perpendicular to the lower bed and extending into the upper bed. These dams were found all along the length of the river, from Tildé until the lake (cf. Section 1.3) with a very high mean density of 270 dams counted from aerial observations in January 1969 (Durand 1970). The fishermen stood facing upstream, immersed to hips, in adjacent bush frames extending the dam into the flood upper bed. They used a previously described triangular net, with a 4 to 5 meter opening (cf. Section 1.3). Various experiments showed that no experimental technique gave results that were efficient and less selective than the traditional nets which were therefore retained for the study. The sampling plan adapted and used at Daga (Fig. 2) has already been described elsewhere (Durand 1970).

If the catches from triangular nets are added to data from poison fishing carried out near the sampling station, the total number of species collected increased to 74, of which the 47 most common ones were kept for a quantitative analysis. Most of the species belonged to a distinct size range of small fish and the study of corresponding ages showed that they were first year juveniles which contributed about 95% in number and weight to the fishery.

Over the course of the flood withdrawal, generally in the month of January, the fishing yields followed a significant change with a very rapid lowering followed after 15 days by a very distinct rise. This phenomenon was accompanied by an appreciative change in the species composition. A comparison of the December surveys with those of the second fortnight of January confirmed that, considering only the 22 most common species (the two rarest species only represented 1.7% of the total number), the Kendall's rank correlation coefficient only reached 0.039, a value much lower than the 0.05 significance threshold: 0.30 for 22 species. The two groups considered thus had no clear affinity. Based on the physico-chemical data this phenomenon appeared to be related to variations in conductivity which themselves reflected the relative importance of the two water bodies: the first flood due to the precipitation in July and August, and then the overflowing of the Logone (Durand 1970).

Although this explanation appears valid, it should however be pointed out that there was great variation in the behaviour of different species. Some species such as *Sarotherodon niloticus* were found at all times while others such as *Brienomyrus niger* were found at the beginning and end of the overflow. This variation in the behaviour was confirmed by the analysis of correlations between surveys and the analysis of interspecies correlations which shades the scheme presented above concerning the time change in the community (Durand 1971).

The analysis of the rank correlation matrix between the surveys showed that inside the first group, no correlation was found between distant samples. It appeared as if the first group was not stable because while retaining some of the main common features the community constantly evolves and could be arbitrarily divided into two sub-groups. The second group, on the other hand appeared more homogenous with some correlations having tendency to increase with time so that community structure appeared to stabilize at least temporarily.

This structure of the surveys corresponds to some species structures that are also fairly marked. The correlation analysis shows three homogenous basic groups of four species. The 22 other species considered in this analysis were classed by function of their average relation to each of the three groups. Figure 8 shows the three basic elements and the most characteristic species in relation to them.

These three groups were characterized essentially by the chronology of appearance of the species making them up. The first group included *Marcusenius cyprinoides, Hyperopisus bebe, Alestes dentex* and *Labeo senegalensis* whose abundance was maximum with a very distinct mode from mid-November to the end of December. Some accompanying species appeared in November before those of group I: *Alestes baremoze, Polypterus bichir* and secondarily *Hydro-*

cynus brevis and *Lates niloticus.* Other accompanying species of group I appeared much later, *Heterotis niloticus,* a rare species; *Distichodus rostratus* also being present in January and *Sarotherodon aureus.* Finally three species whose relations were less clear but whose maximum abundance was in the first two months were *Pollimyrus isidori, Distichodus brevipinnis* and *Mormyrus rume.*

Whereas group I was clearly separated from the following two, groups II and III were partially related by some intermediate species. The four species defining group II were characteristically very abundant at the end of January: *Sarotherodon galilaeus* and *Brienomyrus niger* which were relatively closer to I than *Barbus* spp. and *Clarias* spp. which were closer to group III. Two accompanying species were also found here: *Sarotherodon niloticus* and *Labeo coubie.*

Group III was the most homogenous with *Ichthyborus besse, Siluranodon auritus, Schilbe uranoscopus* and *Synodontis* cf. *schall* which were caught at the beginning of the overflow, then disappeared to reappear only in February when their maximum abundance occurred. There was one accompanying species: *Synodontis nigrita* (Fig. 8).

No clear relationship with the three groups defined above can be put forward for species that did not appear to have a very precise period of abundance. This was particularly true for *Citharinus citharus,* an isolated species; *Brachysyno-dontis batensoda* and *Hemisynodontis membranaceus* which with the less com-

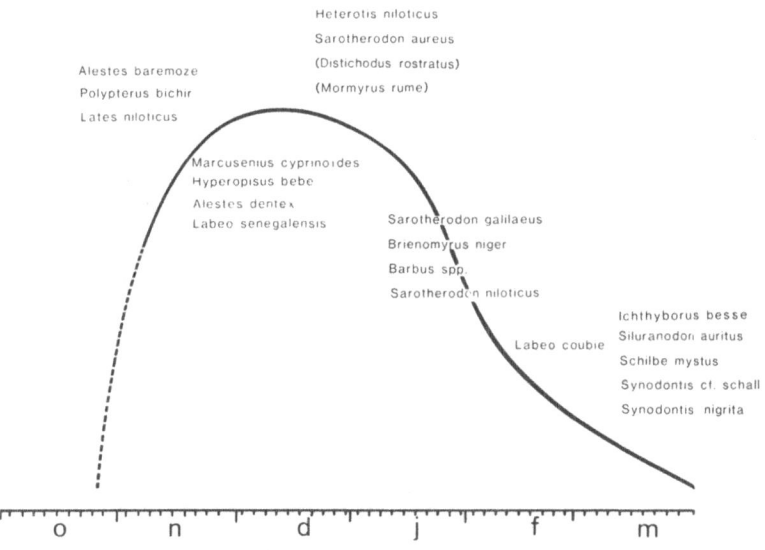

Fig. 8 Groups of migratory species observed at the Daga station (El Beïd) for the different phases of the flood.

mon *Synodontis clarias* made up a small intermediate group; *Alestes nurse*, an isolated and very abundant species which was the peculiar in being common at all times yet had a maximum abundance in the month of February. It would have been interesting to isolate the two species of *Petrocephalus*, *P. bovei* and *P. bane* but these were not separated in the sorting.

In conclusion, the study of species absence may also indirectly help with the attempted description of the communities. For example, we did not find *Xenomystus nigri* of which Blache (1964) said that in some years it could constitute most of the subsidence fisheries in December and January. This is one example of between years variability which will be discussed in the conclusions. A single individual of *Alestes dageti* was caught in four years at Daga, confirming the lacustrine endemicity of this species. Finally we did not find any *Eutropius niloticus* or *Hydrocynus forskalii*, two of the most common species in all the permanent aquatic environments, observations which agree with Daget's (1954) for the Niger.

10.2.3   *Zonation and cycles during 'Normal Chad'*

The general description of the community undertaken here attempted to consider all the aquatic environments of the Lake Chad regions whether they were temporary or permanent, flowing or stagnant. Two major and complementary aspects were examined; the zonation inside a lacustrine system, sensu stricto, during a normal lake phase, and the part played by fluvio-lacustrine or lacustrine migrations. Combining the results led to a classification of species that considered their distribution and their more or less migratory character, both characteristics related to fairly specific ecological characters.

10.2.3.1 *Lacustrine zonation.* The distribution of species in the lake was governed by two main factors: distance from the river system and type of zone, whether open water or archipelago.

In the south basin, the ichthyofauna was most varied along the southern coast, the richness being due to contact with both a permanent (Shari) and a temporary (El Beid) river system. Three species appeared characteristic: *Ichthyborus besse*, *Siluranodon auritus* and *Polypterus senegalus*. Moreover, the young of *Hyperopisus bebe* and *Schilbe uranoscopus* were not found elsewhere in the lake. There did not appear to be any species strictly confined to the open water but on the other hand, some were absent, such as several species of Cichlidae and *Heterotis niloticus* and to a lesser extent. *Alestes baremoze* and *A. dentex*. By contrast, these species were dominant in the Southeastern Archipelago where a clear impoverishment is seen towards the east with the disappearance, for example, of Mormyridae.

In the north basin, with the exception of the southernmost part of the open water which appeared to show a community similar to that of the Southeastern Open Water, some significant differences were found. The differences were related to progressively decreasing species diversity, towards the north east, and to a scarcity of species which were common in the south basin. Thus to the north of an approximate line Malamfatori-Baga Sola (Fig. 9) neither *Schilbe uranoscopus*, nor several species of Mormyridae (*Pollimyrus isidori, Petrocephalus bane, Marcusenius cyprinoides*) occurred. The scarcity of *Brachysynodontis batensoda* and *Hydrocynus brevis* was also to be noted.

10.2.3.2 *Fish migrations.* All the results obtained in the various environments, showed the existence of seasonal variations which corresponded to migrations of usually abundant species. These movements of entire populations were divided into two main types:

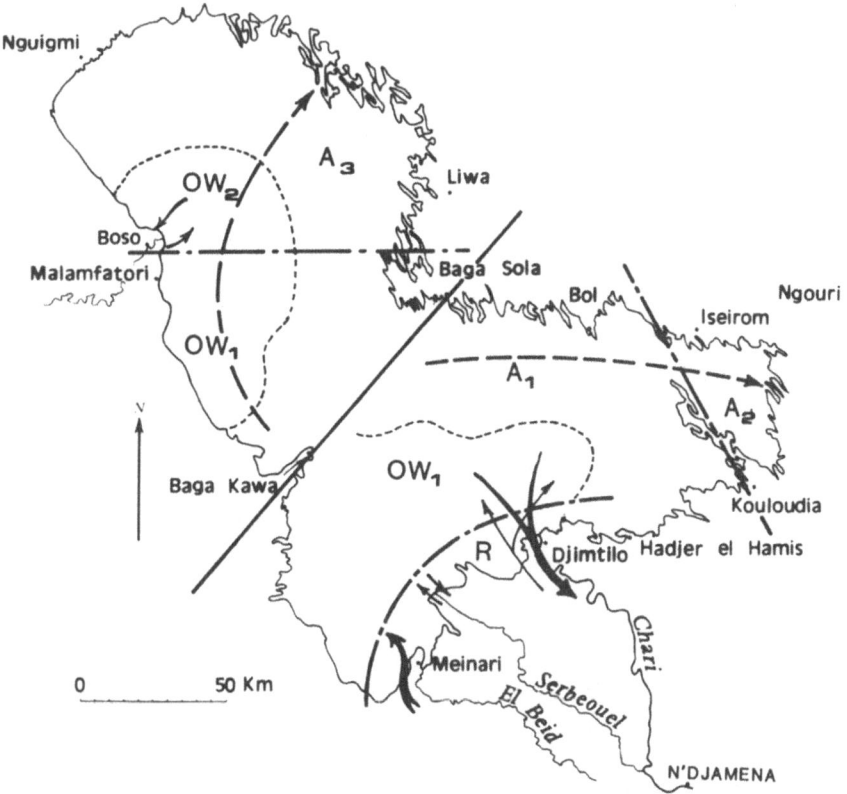

Fig. 9 Diagram of the zonation of the fish. The black arrows correspond to the different migrations observed, and dashed ones correspond to the decreasing number of species in the fish communities.
OW = Open Water; R = Rivers; A = Archipelago.

1. The longitudinal, or 'true', migrations over the course of which the fishes often covered considerable distances between the lake and the rivers or inside the river system. Most of these migrations appeared to be linked to reproduction of which the most common were anadromous migrations at the subsidence or at the beginning of the flood and a reverse migration during the flood;

2. Lateral migrations which corresponded to movements of low amplitude between the rivers and the flooded zones. The latter were invaded at the time of the flood and the fish returned in the lower bed when the flood waters receded at the time of the subsidence. The migratory cycle of *Alestes baremoze* (cf. Fig. 1, Chapter 13) is a good example of the migratory behaviour of several species of the Lake Chad region during the period of 'Normal Chad' (Durand 1978). Two populations of adults lived in the lake and more particularly in the sheltered zones of the reed islands and the archipelago. With a slight shift in time, they both made out widespread reproductive migrations which took reproducing fish to the Logomatia during July and/or August. After spawning, the adults dispersed in the upper layers and the margins of flooded zones, to set out towards the lake at the subsidence. The young fry in the Logomatia were carried away towards the flooded zone of north Cameroon. After spending two to three months in the flood zone, most of them returned to the lake via the El Beid, the direct connections with the river system being interrupted precociously. The young *Alestes baremoze* then remained almost two and a half years in the lake, having undertaken their first reproductive migrations in the rivers.

Due to variations in the migratory cycle, three species groups can be distinguished with varying migratory behaviour.

— *The true migrants.* Only the adults made large-scale longitudinal movements at the time of reproduction. They were *Alestes baremoze, Brachysynodontis batensoda, Distichodus rostratus, Marcusenius cyprinoides* and *Petrocephalus bane.* Three other species should also be noted here: *Hemisynodontis membranaceus, Labeo senegalensis* and *Hydrocynus brevis,* for which the data are not quite complete. *Alestes dentex* was an extreme case because its migrations were greater than those of *A. baremoze.* Although *A. dentex* were very common in Lake Chad, it is possible that they reproduced in the higher reaches of the rivers, since we never saw a single mature individual.

— *The 'mixed' migrants* also make large-scale longitudinal movements but some young did reascend at the same time as the adults, in particular *Schilbe uranoscopus, Synodontis schall, Hyperopisus bebe, Mormyrus rume* and *Eutropius niloticus.* This last species was however distinguished from the preceding ones and those of the first group because no individuals of any age were ever found in the flooded zones.

— *The 'lateral' migrants* were fish living in the rivers and moving towards the extended river bed and flooded zones with the flood and only making short movements into the lower layer. The movements were related to feeding rather than to reproduction and can be divided into two categories. The first were riverine species such as *Alestes nurse* and *Citharinus latus* that were rarely found in the lake. The second were the species present in the lake as well as in the rivers but forming a more or less separate sedentary group. This group included all the Cichlidae, particularly *Tilapia* spp. and *Sarotherodon* spp as well as *Lates niloticus* and *Polypterus bichir*.

A very common predator, *Hydrocynus forskalii* appeared to represent a particular case because it spawned in the lower bed at the time of subsidence. It then appeared to have separate populations: fluvial and fluvio-lacustrine (Srinn 1976). As with *Eutropius niloticus* no representatives of these species were found in the flooded zones or in the El Beid.

10.2.3.3 *Summary classification of the species*. This division of species of the Lake Chad region is partly valid over the Sudano-Sahelian zone. It can be related to the environments inhabited and provides a fairly extensive distribution, which depends upon species behaviour, and various adaptations (trophic plasticity, genetic behaviour, resistance to hypoxia, for example). A detailed classification of the large number of species found (over 100) is in progress but will not be discussed here. It is sufficient to show the variety of types of species distribution.

Few species appeared to be totally ubiquitous in the region considered. Those that were included *Lates niloticus*, *Synodontis schall*, *Labeo senegalensis*, *Distichodus rostratus* and probably *Hemisynodontis membranaceus*, some of whose individuals could be caught in all the lacustrine environments, in the permanent fluvial environments, and in the flooded zones. Although the last three species were migrants, *Lates* was different because of more limited displacements highly motivated by the search for prey. *S. schall* was a mixed migrant (cf. Section 2.3.2)

Several species had a wide distribution which excluded some environments. This characteristic distribution was shown by three very common species: *Hydrocynus forskalii*, *Eutropius niloticus* and *Micralestes acutidens* which frequent all the permanent environments although not a single individual was caught in the flooded zones. or the El Beid. Another type corresponded to species that could be found everywhere except in the northern half of the north basin. They included *Schilbe uranoscopus*, *Brachysynodontis batensoda* and several species of Mormyridae (*Pollimyrus isidori*, *Hyperopisus bebe*, for example). A third important category was represented by species whose distribution excludes the lake open water. This was also a diversified group of quantitative importance since it included, besides several aquatic grass beds species (*Barbus* ...), most of the Cichlidae, and particularly the genera *Tilapia* and *Sarotherodon* as well as Characidae such as *Alestes Baremoze* and *A. dentex*. Some true migratory species and some typically sedentary species were placed together here.

There were some species whose distribution was more limited than that of the preceding species, but were not attached to any particular environment. They included the endemic species of the lake, *Alestes dageti*, and the fluvial species which were never caught in the typical lacustrine environments, namely *Petrocephalus bovei, Ichthyborus besse, Citharinus latus, Siluranodon auritus* and *Polypterus senegalus*. Some species could be placed near this group, having a marked predilection for marshy environments and flooded zones: *Clarias, Ctenopoma, Brienomyrus niger, Synodontis nigrita* and *Barbus gourmansis*. We will see later that most of these species developed some tolerance to the anoxic conditions, which partly explained the change in the lake populations during the drought.

Finally some species had a more specific distribution: *Barilius* was caught on the river sand banks; the characteristic population of *Aetheria* beds in the river: *Petrocephalus simus, Nannocharax fasciatus* and *Synodontis filamentosus* were never been found elsewhere; *Protopterus annectens*, a dipnoid fish, common in all the flooded zones, encysted during the drying up period; *Nothobranchus*, frequent in isolated ponds, laid eggs that hatched after a long time.

10.3 Influence of the drought on the fish communities (1973–1978)

The lake at this time was in the 'Lesser Chad' stage by the definition of Tilho (1928); it started with a severe change in water level caused by the period of drought present throughout the Sahel since 1972. An extreme and lasting modification of the lacustrine hydrology caused serious disturbances in the fish communities as seen in the north basin, the Southeastern Archipelago and Southeastern Open Water. Indirectly they affected the reproductive migrations in the delta zone of the Shari. Moreover, in the aim of considering the rivers influence on changes in lake communities we regularly followed, from 1974, the descent of young fish in the El Beid; they took this outlet of the flooded plains to rejoin the south basin of the lake.

Two phases could be distinguished: that of drying up (1972–1974) and that of 'Lesser Chad' (1975–1977) which only affected the south basin.

10.3.1 *Drying up of the north basin*

Towards the end of the first half of 1973, the drying up of the 'Great Barrier' (Fig. 1) transformed the north basin into a closed basin without inflows other than the low supply of the Yobé because the small floods of 1973 and 1974 did not reach the north of the lake. The mean depth at Kindjéria changed from 3.3 m in September 1973 to 0.5 m in December 1974 (Fig. 10). The transparency was reduced by seven over the same period. The existence of an increasingly abundant seston and phytoplankton involved a large nightly variation in

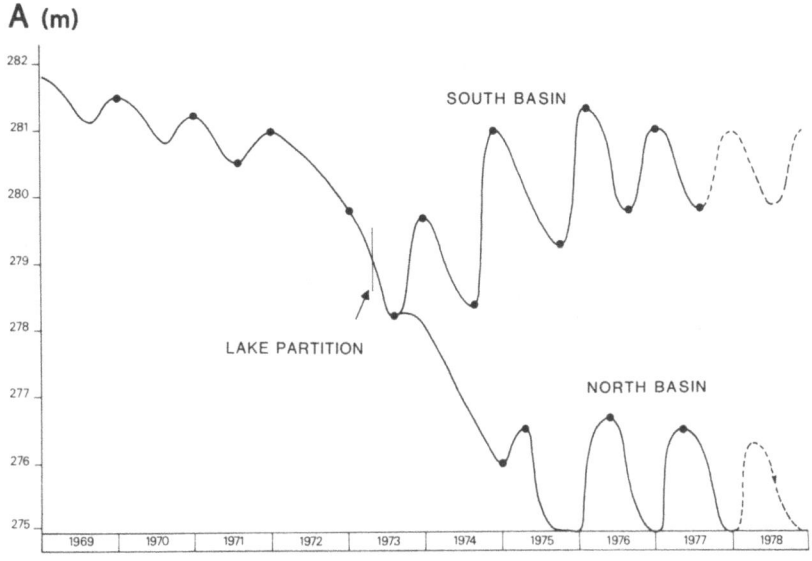

A (m)

282

281

280

279

278

277

276

275

SOUTH BASIN

LAKE PARTITION

NORTH BASIN

1969 | 1970 | 1971 | 1972 | 1973 | 1974 | 1975 | 1976 | 1977 | 1978

Fig. 10 Between year variation in water level (altitude, above sea level) for the lake, and for each of the two basins after August 1973 (lake partition).

oxygen concentration from 0 to 20 mg l^{-1} in 12 hours at 10 cm under the surface (Bénech et al. 1976). The conductivity increased from 900 to 3500 µS cm^{-1} between March and December 1974.

An important barrier of marshy vegetation developed on the exposed ground of the 'Great Barrier' and hindered the outflow of flood waters towards the north in 1974–75. The losses by evaporation were not compensated for and the entire north basin dried completely in November 1975.

At the time of its isolation from the rest of the lake, the fish stocks of the north basin as well as their distribution were practically the same as in 1971, the year of reference for the ichthyology of the northern part of the lake before the drought. No migrations or mass movements were observed towards the south basin before the division of the lake. On the contrary it appeared likely that some stocks in the south basin like *Alestes baremoze* were sheltered in the north to take advantage of a more stable environment (Durand 1978). The concentration of fish then attracted almost all the professional fishermen previously distributed over the delta regions and the southern border of the lake. The development of fishing activity caused a strong increase mortality due to fishing. Also, the degradation of ecological characteristics throughout the basin increased the natural mortality. These two phenomena together contributed to the establishment of a very high total mortality for almost all the species present during the period of 'Normal Chad'. Mortalities were a principal factor in the

329

change in the populations between the isolation of the basin and the second half of the year 1974.

This change was expressed by the progressive disappearance of species like *Heterotis niloticus, Hydrocynus brevis, Citharinus citharus, Tetraodon fahaka, Pollimyrus isidori, Mormyrus rume.* Some mass mortalities following tornadoes were also observed (Bénech et al. 1976). When the water level fell below a critical threshold of about two meters the sediment could be perturbed under the effect of very violent winds. Then there was an explosive development of phytoplankton, a lowering of the transparency and the oxygen content. The fish died directly due to an oxygen deficiency or due to the mechanical action of suspended particles on the branchial tissues. This phenomenon was repeated several times in different places during the rainy season. In September 1974, on the windward banks of several islands, some fringes 0.5 to 1 meter wide and several hundred meters long of dead fish could be seen.

Whatever sampling point was chosen, at the end of 1974, the influence of the drought and the turbulence favored the same group of species, namely *Polypterus*, the *Mochocidae* and *Sarotherodon*, although they reacted very differently to the changes in their environments.

— The Mochocidae (*S. schall* and *B. batensoda*) did not reproduce, and although they suffered considerable mortality, their relative resistance to the reduction in water area included an increase in their density — more so than for the other species (Fig. 11).

— *Polypterus*, represented especially by *P. senegalus* could respire oxygen from air and thus suffered less from anoxia. Due to a lack of data on their reproduction (the gill nets catch few young *Polypterus*) we can only note their growing importance over the course of this period.

— In spite of the considerable losses suffered during the storms the populations of *Sarotherodon* were maintained and proliferated because of their high reproductive capacities. The young became dominant from June 1974 before the growth following the rainy season as shown by the change in the catches (Fig. 12).

During the drought, the increase of *Sarotherodon galilaeus* was much less perceptible than that of *S. niloticus* and especially of *S. aureus* whose development was considerable. The differential success in adapting to the conditions imposed by the drought, was related to the oxygen requirements of the three species. The needs of young *S. aureus* were 30% of *S. galilaeus* while *S. niloticus* was intermediate (Welcomme 1964).

If the same groups of species were favored by the drought, the changes in the fish communities were not identical over the whole basin. The spatial variations became evident in the results of a correspondence analysis which deals with a table regrouping the catch per unit effort from three nets (GN 15, 30 and 50) for the main species caught in four stations of the north basin between 1971 and the end of 1974 (stations A, B, C, D; Fig. 1).

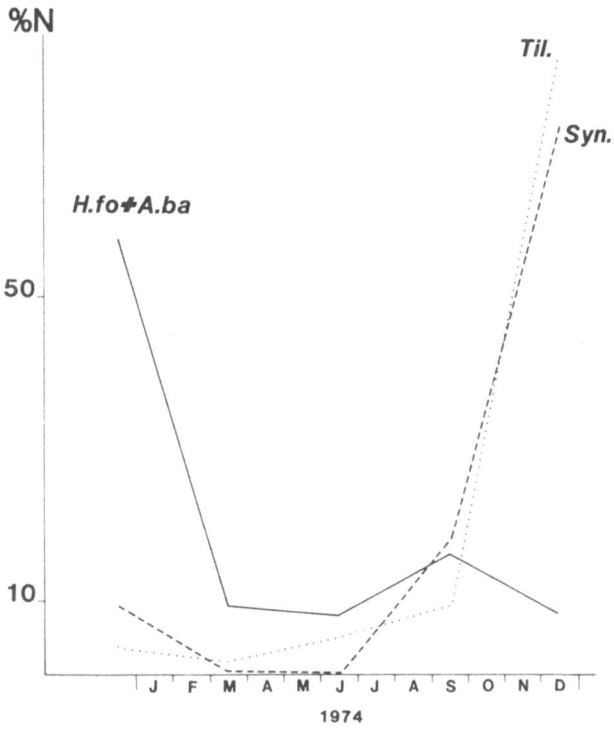

Fig. 11 Variation in the catch per unit effort for a few species at Kindjéria (*N* is the sum of individuals in groups of species or genera considered during the period described).

H.fo = *Hydrocynus forskalii*; A.ba = *Alestes baremoze*; Til = *Tilapia* species and *Sarotherodon* species; Syn = *Synodontis* species.

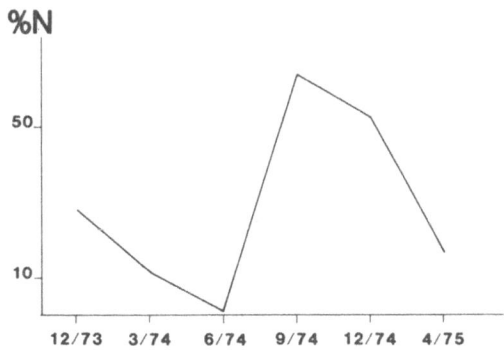

Fig. 12 Experimental fishings at Kindjéria (north basin, GN 11, 15, 35, 40). The number of individuals of young *Tilapia* and *Sarotherodon*, caught in both GN 11 and 15 is expressed as a percentage of the total number of individuals of the same species caught in the whole gill net set.

— In 1971, before the period of drought, few young were found at Station A which was located in the region of the Northern Open Water (Fig. 1). The catches were then primarily made up of large *Hyperopisus bebe*, *Lates niloticus* and *Synodontis frontosus* as well as several *Eutropius*. The community changed fairly little until the beginning of 1974, when the populations flowing back from the archipelago enriched the diversity and the species richness of catches made at the station which was slightly deeper than the eastern border of the lake. *Synodontis* spp. were then dominant, while the importance of *Tilapia-Sarotherodon* increased although it was still not predominant.

— Stations D and C were located in the Northern Archipelago at the northern extremity of the lake and at the limit of the reed islands respectively. In 1971, they had comparable communities composed principally of young *Hydrocynus forskalii*, *Eutropius niloticus*, *Alestes baremoze* and large *Lates niloticus* and *Sarotherodon galilaeus*. As in station A, these communities changed in 1973 and all retained their main characteristics. Over the course of 1974, the disappearance of *Alestes*, *Lates* then of *Hydrocynus* and Eutropius was seen in the northern part (station D). The same phenomenon occurred at station C but with a slight delay. All this happened as first *Alestes* and *Lates* and then *Hydrocynus* and *Eutropius* moved back towards the center of the basin abandoning the place to *Tilapia* which made up most of the fish community until the drying up of these stations towards the end of 1974.

— Station B was located in the deepest region of the lake. As in the other stations the fish community at point B was not modified between 1971 and 1973. By contrast, there was a large decrease in *Alestes* between March 1973 and March 1974 probably because of the considerable fishing effort to which it was subjected throughout the basin. Similarly the apparent development of large *Schilbe* and *Synodontis schall* was noted. The change was regular and the December 1973 samples showed one stage, but after March 1974 the change took place in a different manner and appeared more and more diversified. A rise in abundance was noted for *Eutropius*, *Alestes baremoze* and *A. dentex* of large size. The large *B. batensoda*, usually rare in the region, appeared as well as the large *Sarotherodon aureus* and *S. niloticus*. Some large *Lates* were even caught during 1974. It was thus evident that these fish came from shallower drying up regions of the lake.

In spite of this apparent recovery, the fish community of station B showed the beginning of its final stage from November–December 1974. The presence of *Brienomyrus niger*, the slow development of *Polypterus* stocks, the appearance of a notable number of *Clarias* as well as the multiplication of *Sarotherodon* were the signs of the installation of a marshy fish community. It developed in 1975 until the drying up of the station and it recovered as a result of the brief inundations caused by the floods. After 1975 and the total drying up of the basin, the northern part of the lake was partially flooded each year (Chapter 2). It was then a marshy region like some flooded areas of a particular type colonized by a community with a high productivity mainly based on *Clarias*.

10.3.2 *Southeastern Archipelago*

The Southeastern Archipelago was isolated in April 1973 during the drying up of the reed island region (Chapter 2). The water level then declined very rapidly and the archipelago region would probably have followed the changes found in the north basin if the following flood wave had not reached Bol. It was filtered during passage through the thick band of semi-aquatic vegetation which had developed on the dried ground in April. After 1974, the archipelago was no longer isolated from the open water by a dry ridge but the plant formations, which were installed there (Fotius and Lemoalle 1976) formed a definite obstacle to the free circulation of fish between these two regions.

The study of catches from a gill net battery expressed as presence–absence (Table 6) distinguished two stages in the change of the Southeastern Archipelago communities. The first (1973–74) drying period during which the archipelago was isolated for part of the year corresponded to the lowest water level (Fig. 14). This isolation no longer existed over the second period (1975–77) during which the 'Lesser Chad' was established.

10.3.2.1 *The drying phase (1973–1974)*. The first half of 1973 was still a period of high water. However an increase in the number of fish caught was then observed because the reduction in the water area (90%) caused a concentration of fish and increased the vulnerability of species that were previously caught to a lesser extent by gill nets.

From July to September 1973 at the time of the lowest water, before the arrival of the flood, some species disappeared: *Hydrocynus forskalii, Citharinus citharus, Hemisynodontis membranaceus, Lates niloticus, Alestes dentex* and *Labeo senegalensis*. The species appearing with the reduction in water area also disappeared except *Bagrus bayad* which persisted until November 1973. There may have been a possible emigration during the flood. This fish community change must have been related to some mass mortalities following storms observed in the 1972 rainy season. The sudden decrease of the species number always coincided with the period of storms in June and July (Fig. 14).

The 1974 floods submerged a large number of decomposing plants and caused an oxygen deficit which persisted for at least three months. The appearance of dissolved free CO_2 accompanied the reflooding. This event marked the disappearance of most of the common species from the high water community which had survived the conditions of low water. The Mochocidae which represented 45% of the catches a month before, disappeared in a few days (Fig. 13), while the number of species changed from 23 in September to 8 in October, reaching a minimum of 5 in December (Fig. 14). Only the species

Table 6 Evolution of the species composition of catches by gill nets in the southern archipelago of Lake Chad from 1973 to 1977. (I) Species with supplementary respiratory organs.

Species	1973								1974								1975				1976			1977			
	F	M	A	J	J	O	N	D	J	M	A	M	J	S	O	D	F	Mi	At	D	M	A	At	O	J	F	Mi
Mormyrus rume		■	■	■	■																						
Mormyrops deliciosus			■	■	■																						
Hippopotamyrus harringtoni				■																							
Hydrocynus brevis		■	■	■	■																						
Alestes macrolepidotus		■	■	■	■																						
Labeo coubie									■	■																	
Bagrus bayad		■	■	■	■		■		■	■	■																
Chrysichthys auratus		■	■	■	■				■	■	■																
Hydrocynus forskalii		■	■	■	■				■	■	■																
Citharinus citharus												■															
Hemisynodontis membranaceus																											
Lates niloticus		■	■	■	■		■		■	■	■	■		■													
Hyperopisus bebe		■	■	■	■		■		■	■	■	■	■														
Marcusenius cyprinoides		■	■	■	■		■		■	■	■	■	■														
Petrocephalus bane		■	■	■	■		■		■	■	■	■		■													
Pollimyrus isidori		■	■	■	■		■		■	■	■	■		■													
Labeo senegalensis		■	■	■	■		■		■	■	■	■		■													
Eutropius niloticus		■	■	■	■		■		■	■	■	■		■													
Synodontis clarias		■	■	■	■		■		■	■	■	■		■													
Schilbe uranoscopus		■	■	■	■		■		■	■	■	■		■	■												
Brachysynodontis batensoda		■	■	■	■		■		■	■	■	■		■					■		■		■				
Synodontis frontosus		■	■	■	■		■		■	■	■	■		■				■	■	■	■						
Synodontis schall-gambiensis		■	■	■	■		■		■	■	■	■		■		■	■	■	■	■	■	■	■				
Alestes dentex		■	■	■	■		■		■	■	■	■		■	■	■	■	■	■	■	■	■	■	■	■	■	■
Alestes baremoze		■	■	■	■		■		■	■	■	■		■	■	■	■	■	■	■	■	■	■	■	■	■	■

Alestes nurse-dageti
Distichodus rostratus
Auchenoglanis spp.
Schilbe mystus
Siluranodon auritus
Sarotherodon niloticus
Sarotherodon aureus
Sarotherodon galilaeus
Tilapia zillii
Polypterus senegalus
Gymnarchus niloticus
Brienomyrus niger
(1) Clarias lazera and C. anguillaris
Heterotis niloticus
Polypterus bichir
Polypterus endlicheri

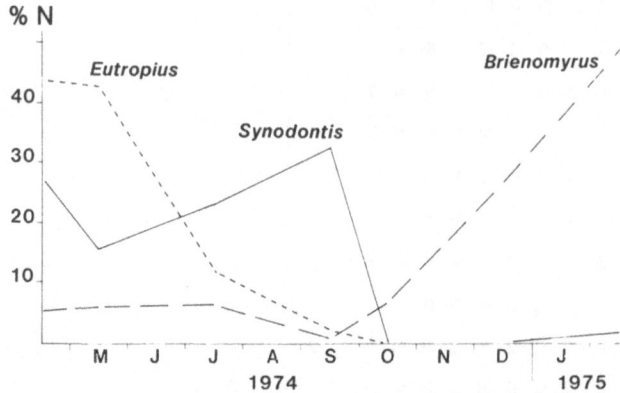

Fig. 13 Variation in the catch per unit effort (in number) for a few species at Bol-Berim (N is the sum of monthly c.p.u.e. for each genus during the period described).

possessing organs of aerial respiration persisted: *Polypterus senegalus*, *Clarias* spp. and *Brienomyrus niger* as well as some individuals of other species: *Alestes nurse-dageti*, *Distichodus rostratus*, *Brachysynodontis batensoda* and *Sarotherodon niloticus* whose traces were found in the October or December, 1974 catches (Table 6).

The two types of severe mortalities recorded at the time of storms or the arrival of flood water were due to hypoxic conditions provoked at first by a dispersion in the water mass of reducing sedimentary compounds and then by the arrival of a considerable mass of anoxic water. The species which disappeared in 1973, endured the hypoxic conditions with much difficulty as shown experimentally for *Alestes baremoze* and *Labeo senegalensis*. Inversely, the species whose few individuals persisted in the very hypoxic water of the 1974 flood, experimentally showed a very good resistance to hypoxia: *Distichodus rostratus* and especially *Sarotherodon niloticus* and *Brachysynodontis batensoda* (Bénech and Lek 1981). A diffuse mortality shown by experimental fishing was superimposed onto the severe and apparent mortality. The normal conditions of existence no longer being assured, the mortality rate rose. The pelagic species of high water disappeared first, then the predators: *Lates niloticus*, *Hydrocynus forskalii*, *H. brevis* and the plankton feeders: *Alestes dentex*, *Hemisynodontis membranaceus*. In this last category *Brachysynodontis batensoda* succeeded in surviving partly due to a more adaptable diet; in 1973 the stomach of these animals contained a considerable proportion of detritus (Im 1977), a starvation diet which halted growth in length (Bénech 1975).

Until the end of 1974, the eventual exchanges with the rest of the lake did not play an important role in the changes in the archipelago fish community because it was isolated at first by the drying of the reed islands zone, then by a cordon of plants several kilometers deep. All the species which abounded in the

336

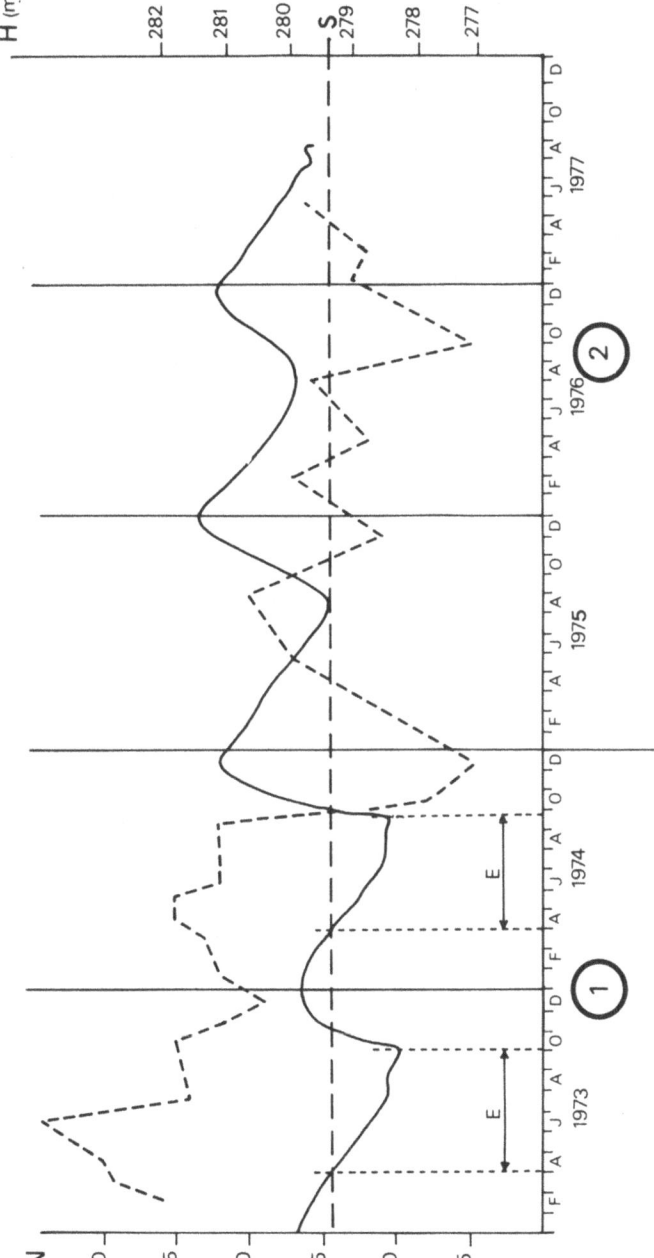

Fig. 14 Changes in the water level (H = continuous line) and in the number of species (N = dashed line) of fish caught by a set of gill nets at **Bol** in the Southeastern Archipelago.
S = level at which the reed islands zone is above water; E = period at which the reed islands zone is above water; 1,2 = first and second period (see explanation in the text).

337

new environment had already been reported in the region, but the 1973 flood may have favored the immigration of some species (*Siluranodon auritus, Brienomyrus niger, Clarias* spp. and *Heterotis niloticus*) who appeared simultaneously.

10.3.2.2 *The 'Lesser Chad'*. During 'Lesser Chad', two factors prevented the resuspension of the archipelago sediments: the very high water level and the development of vegetation belts which limited the action of winds on the open water zones. The only disturbance of the environment was the arrival of hypoxic flood water which eliminated the species sensitive to an oxygen deficit. The reappearance of the latter must be attributed to migrations from the Southeastern Open Water. This colonization was facilitated by the appearance, during 1975, of a navigable channel between Bol and the Shari delta. These recolonizations were by species that developed in the archipelago after the transformation of the environment in 1973: *Sarotherodon* spp., *Schilbe mystus, Siluranodon auritus, Distichodus rostratus, Alestes nurse-dageti* as well as two pelagic species from 'Normal Chad': *Alestes baremoze* and *A. dentex*.

The composition of the characteristic fish community of 'Lesser Chad', consisting of about 15 species (Table 6), appeared well established in spite of a particular seasonal disturbance of the community. All the correlation coefficients (Pearson's Φ coefficient) showed some affinities between all the samples of 1975, 1976 and 1977 except August 1975 and October 1976 which corresponded respectively to the maximum and the mininum of species caught during this period. March showed few affinities with the rest and was the last sample in which three species from 'Normal Chad' were found prior to their disappearance (*Schilbe uranoscopus, Brachysynodontis batensoda*, and *Synodontis frontosus*).

From 1975 the fish community of the archipelago henceforth had less than 20 species whereas 34 were represented in the 1973 catches. With the new appearance of the lake, a seasonal change took form, such that the species richness of the archipelago decreased drastically at the time of the arrival of the flood then increased until the end of the following low water. The arrival of the 1975 flood certainly produced the same effects as that of 1974 (Fig. 14) but the decrease in the number of species caught was less marked in 1975 because of the sampling being much later that year. This was confirmed by the 1976 data as the sampling carried out at the beginning of the flood showed that, as in 1974, the surviving species were those which could use oxygen from the air.

10.3.3 *The zone of contact with the river system*

10.3.3.1 *The Southeastern Open Water*. In comparison to other lake environments the south basin exhibited two original features beginning in the second half of 1973:

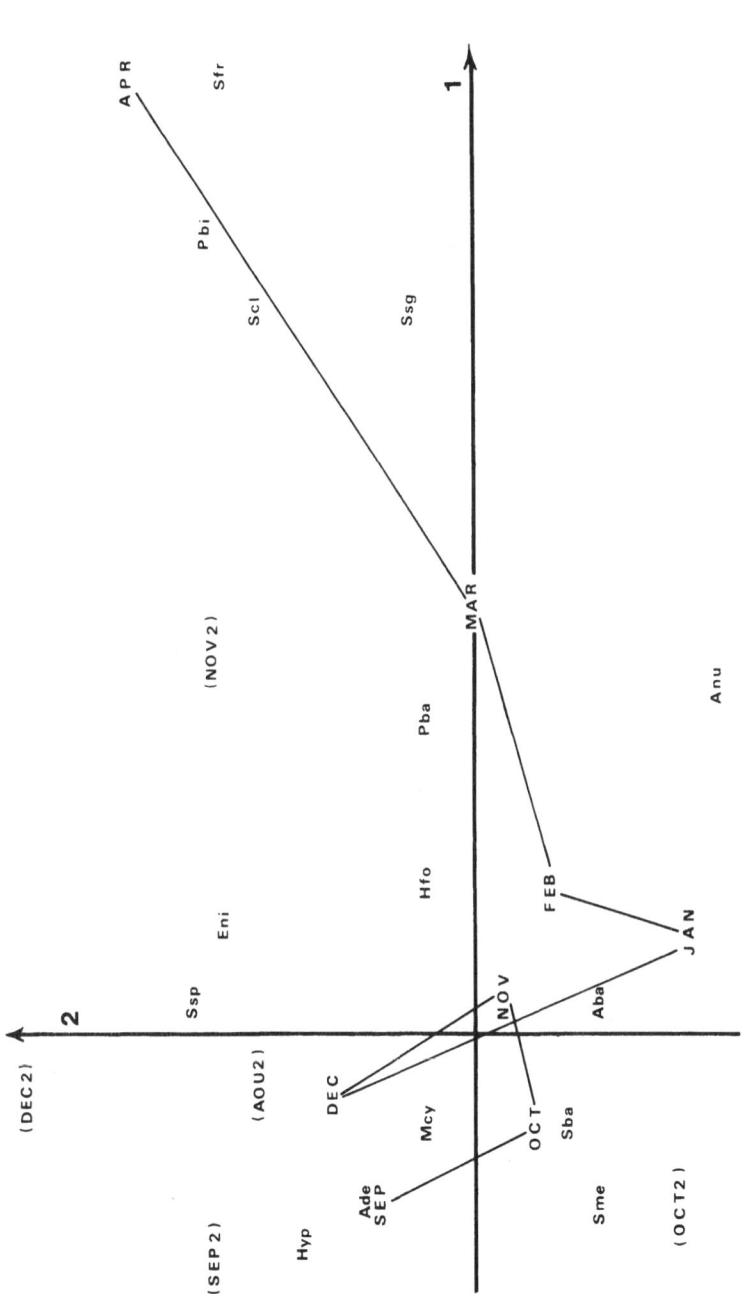

Fig. 15 Changes in the professional fishing catches in the Shari delta between September 1972 and April 1973 shown by the first plan of projections (axes 1 and 2) of a correspondence analysis of the fishing surveys data. Monthly samples subsequently observed (August 1973 to December 1973) did not correspond to a professional fishing. They have been replaced on the plan but do not play in the definition of axes. For abbreviations, see Table 12.
Aba = projection of species;
DEC = projection of monthly sample (first period);
(DEC) = supplementary monthly sample (second period).

339

— a constant connection with the river system and the fish stocks it contained. It should be recalled here, that there were no endemic species of the lake which could be considered as an extension of the river system (Daget 1967), at least in relation to the fish fauna of the basin;

— a change towards a new appearance of the lake characterized by an extension of open water surrounded by a thick vegetation belt unfavourable to sheltering the fish species sensitive to hypoxia.

The change of the fish community of this part of the lake after the drought of 1972–73 must be interpreted in view of these characteristics. As in the South-eastern Archipelago this change included two periods towards the end of 1974.

(a) The subsidence of the lake (1973–1974)

From a hydrological point of view this period corresponded to a considerable reduction in water areas. These conditions also caused an increase in mortality that was less important than in the Southeastern Archipelago or in the north basin due to four reasons:

— at the time of its separation from the northern basin, the southern part of the lake enclosed a majority of open water species and fish stocks less important than those of the other lake regions. The reducing water volume thus led to lower fish concentrations;

— the connections with the Shari being maintained in 1973, some species migrated towards the river system where they found temporary refuge;

— the nature of the sediments also played an important role here as the bottom of the basin was mostly composed of sand and pseudo-sand. There was no notable increase in turbidity with the simultaneous reduction of dissolved oxygen and proliferation of the phytoplankton;

— the region of the Southeastern Open Water was not subjected to the considerable fishing effort seen at the same time in other regions of the lake, especially in the north basin.

From March 1973, the fish community sampled close to the island of Kalom was dominated by *Brachysynodontis batensoda*, mostly adults caught by GN 30 and GN 35. There was also an abundance of *Pollimyrus isidori* and *Eutropius niloticus*. *Alestes baremoze*, *A. dentex*, *Hydrocynus forskalii*, *Chrysichthys auratus* and other characteristic species of the archipelago fish communities were rare. But this rarity was attributable as much to their low abundance in this region at the time of its isolation as to an increase in mortality due to the drought. The proof lies in these species becoming more abundant from 1974 although the conditions of the lake did not improve.

(b) The 'Lesser Chad'

After 1975, the zone of open water increased and arms of open water appeared in the vegetation which was progressively broken down by the maintenance of a relatively high water level in the basin. At the same time the floodplains in the river system recovered a normal flooding, a favourable factor in restoration of migratory stocks (cf. Section 3.4).

Although the information provided here sometimes contained gaps such as the recent changes in the south basin it appeared that a new type of fish community was being established. Thus Shannon's diversity index calculated from relative frequencies of the 16 most abundant species in the annual samples taken between 1973 and 1977, indicated a recovery of the species richness from 1975 (Table 7). This experimental fishing, done close to the island of Kalom also indicated a rejuvenation and a progressive return to equilibrium of the age structure of some populations such as those of *Hydrocynus forskalii*, *Synodontis clarias* or *Labeo senegalensis* despite their being deeply affected by the drought (Table 8).

10.3.3.2 *The Shari delta*. The samples taken in this region of the Shari provided information on the availability of the breeders present in the lake for species which made anadromous spawning migrations. To illustrate the disturbances suffered by the lake populations during the period of 'Normal Chad' and the building up of a fish community of 'Lesser Chad', we will also consider two sets of data collected over the drying up phase in 1972–73 and after the drought in 1976–77.

(a) Drying period (1972–1973)

The data considered here were collected between September 1972 and December 1973 at the time of observations on the delta fisheries (cf. Chapter 13, Section 5). The tendency for change in the sampled fish community appears clearly on the first factorial plane of a correspondence analysis (Fig. 15) on the numerical catch per unit effort of the 15 most abundant species (97.81)% of the total catches) (Quensière 1976).

Data for the months September to December 1972 showed a fairly loose group of species that were dominant during this period: *Alestes baremoze, A. dentex, Marcusenius cyprinoides, Brachysynodontis batensoda*; they were joined

Table 7 Diversity and equitability index of samples taken close to the island of Kalom in the southeastern open waters.

Year	I	I/Imax
1973	1.420	0.384
1974	1.044	0.267
1975	0.920	0.230
1976	1.331	0.341
1977	1.888	0.510

Table 8 Importance of three species in different mesh size gill nets as % (in number of individuals) of the total catch in the Open waters of Lake Chad (East of Kofia).

Species		GN	11	15	20	22	25	30	35	40
Hydrocynus	*forskalii*	1973	0	12.0	6.0	50.0	24.4	7.5	0	0
		1974	7.0	2.6	15.7	15.7	18.9	24.1	14.5	1.7
		1975	2.6	1.8	8.3	8.3	13.0	35.8	18.4	11.9
		1976	5.8	7.7	12.1	6.8	11.1	38.7	15.4	2.4
		1977	4.6	2.3	24.1	9.2	19.5	23.0	16.1	1.2
Labeo	*senegalensis*	1973	0	0	0	0	0	100	0	0
		1974	0	0	1.2	2.1	6.6	12.0	46.3	31.8
		1975	0	0	16.3	28.6	22.5	20.4	2.0	10.2
		1976	0	0.4	1.9	5.4	18.1	53.5	19.3	1.5
Synodontis	*clarias*	1973	0	0	3.0	6.8	9.8	51.8	37.2	3
		1974	1.6	5.2	27.8	16.4	34.6	8.9	4.2	1.3
		1975	4.5	16.0	25.0	22.4	4.5	13.5	7.1	7.1
		1976	6.2	7.9	36.4	40.4	1.1	4.0	3.4	0.6

by *Hemisynodontis membranaceus* in October and November. The month of January was clearly differentiated from the preceding ones by a greater abundance of *Alestes baremoze* which formed 78.1% of the catch. A very clear shift was then seen from a community of dominant *A. dentex* and *A. baremoze* towards a community of dominant *Synodontis*: *Synodontis clarias*, *S. frontosus*, *Polypterus bichir*, *S. schall* and *S. gambiensis*. This phenomenon was not linked to the development of these species but simply to the disappearance of *A. dentex* and *A. baremoze* whose abundance masked it during the preceding months. This is shown clearly by the change in yields from small mesh gill nets (from 25 to 35 mm) that are shown in Fig. 16.

A comparison with yields taken in the preceding months (Loubens 1973) showed a progressive slump in catch per unit effort as a consequence of the drought. In 1972 the exceptionally low flood appeared a month later, but the spawning places and particularly the Yaéré were not flooded.

The chemical water composition before entering into the lake was notably different from other years (Lemoalle 1974) and probably caused some changes in migratory behaviour. Finally the disturbances caused by the lower level of the lake itself (Bénech et al. 1976) caused fish to move towards the north basin which had more stable conditions than the south basin (Durand 1978). This reduction of yield, which was linked to the increasingly difficult fishing conditions in the delta region, incited the fishermen to look for more favorable zones. The month of May 1973 marked the suspension of all professional

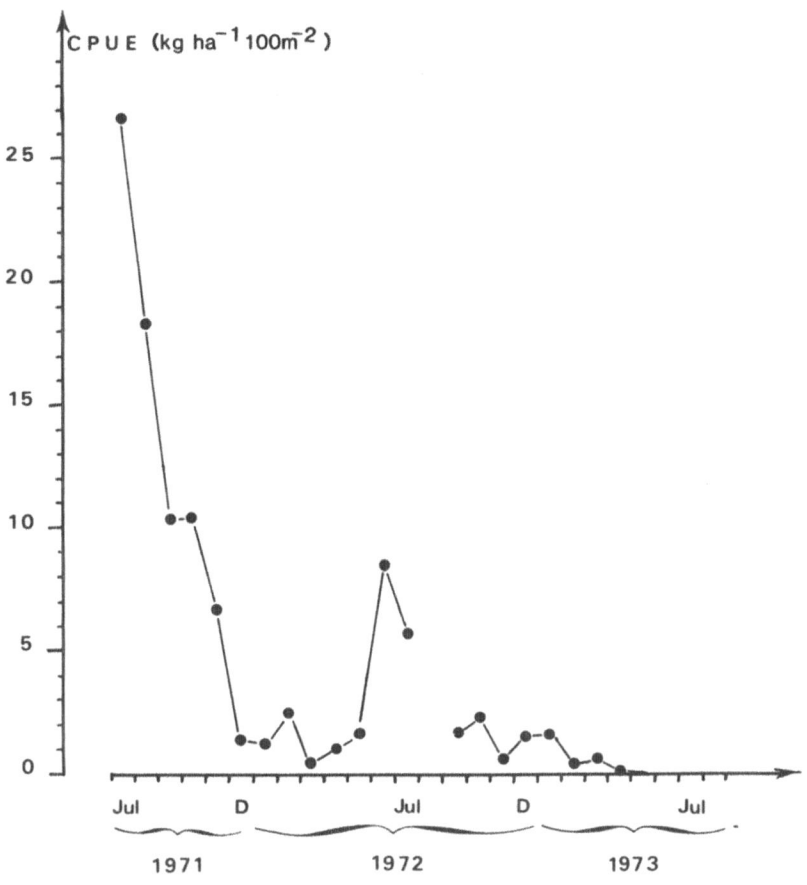

Fig. 16 Changes in the catch per unit effort for traditional fisheries with small mesh drift net in the Shari delta from July 1971 to May 1973.

fishing activity in the delta (Chapter 13); *Alestes baremoze* and *A. dentex* completely disappeared from samples after October. The species diversity thus varied strongly from one month to another due to the mass migrations of Mochocidae: *B. batensoda* in October and *H. membranaceus* in November.

The comparison of the fish communities observed in 1972–73 with those of 1971–72 showed some significant differences between the two periods. *Distichodus rostratus, Citharinus citharus, Labeo senegalensis, Hydrocynus brevis, Bagrus bayad* and *C. latus* no longer occurred among the 15 most abundant species and were replaced by the formerly rarer species such as *Marcusenius cyprinoides, Synodontis frontosus, S. clarias* and *Petrocephalus bane*. The second major difference between the two series of samples was the progressive disappearance of *A. baremoze* and *A. dentex* beginning in the middle of 1973.

343

The change of catches in the delta fisheries during 1972–73 thus signified the profound disorders then present throughout the lacustrine fish communities. It showed the poverty of stocks sheltered by the south basin at the time of the fragmentation of the lake and the selective pressure the hitherto dominant species were subjected to.

(b) 'Lesser Chad' (1976–1977)

The almost continuous observations made throughout the 1976–77 hydrological cycle showed the orientations of the new lacustrine fish community which became established as the lake remained reduced ('Lesser Chad'). The data considered here correspond to the catches made for each of the 40 weekly samples by a battery of 20 gill nets (GN 10 to GN 100). With a weekly average fishing effort of 2000 units (100 m^2 night^{-1}) over the entire period considered, 72 species were caught, of which the 20 most abundant (Table 9), represented

Table 9 Abundance expressed in total number of individuals caught and in percentage of total catches for 20 species represented in the delta fisheries during 1976–77 season.

Species	Number of fish caught	% of total catches
Pollimyrus isidori	25 426	43.78
Schilbe mystus	4370	7.52
Siluranodon auritus	4021	6.92
Petrocephalus bovei	3974	6.84
Alestes nurse	2720	4.68
Brachysynodontis batensoda	2378	4.09
Eutropius niloticus	1690	2.90
Ichthyborus besse	1605	2.76
Petrocephalus bane	1511	2.60
Brienomyrus niger	1108	1.91
Marcusenius cyprinoïdes	993	1.70
Hydrocynus forskalii	809	1.39
Schilbe uranoscopus	796	1.37
Chrysichthys auratus	710	1.22
Alestes baremoze	585	1.01
Distichodus rostratus	585	1.00
Hyperopisus bebe	528	0.91
Polypterus bichir	491	0.85
Synodontis schall	414	0.71
Synodontis clarias	349	0.60
Total	55 058	94.79

94.8% of 58 082 individuals sampled. The results from all the batteries were not directly comparable with the above descriptions. Whereas the observations of Loubens (1973) and Quensière (1976) were for the period exploited by the delta fisheries, the present observations were for experimental fishing using a battery of much more diversified nets. A battery comparable with that used by the fishermen before 1973 (from GN 25 to GN 100) would only have caught 11.4% of the fish now sampled. A comparison of this part of the catch with the above homologous data showed three types of change according to the species in question (Table 10).

— The first type of change was that of species which were abundant during the period of 'Normal Chad' and became rare after 1973. This group included *Citharinus citharus, C. latus, Mormyrus rume, Bagrus bayad* and *Alestes dentex* which were part of the 17 most abundant species in 1971 and 1972 but had become a negligible part of the catch. Although more abundant than the species cited previously, *Hydrocynus brevis, Hemisynodontis membranaceus* and *Labeo senegalensis* also lost their passing importance. *Alestes baremoze* occupied a separate place in this group because although it was rare in the catches of a battery comparable to that of the 1971–73 fishermen it was still present in notable quantities in the catches of the experimental battery. The reason for this

Table 10 Rank of species classed by abundance in 1971–1972, 1972–1973 and 1976–1977. Shari delta fisheries.

Species	1971–1972	1972–1973	1976–1977
Alestes baremoze	1	1	>30
Schilbe mystus	2	5	2
Alestes dentex	3	3	>30
Brachysynodontis batensoda	4	4	1
Hemisynodontis membranaceus	5	2	20
Synodontis schall	6	12	14
Hyperopisus bebe	7	6	16
Hydrocynus forskalii	8	7	4
Distichodus rostratus	9	9	12
Mormyrus rume	10	13	>30
Polypterus bichir	11	10	3
Citharinus citharus	12	27	>30
Labeo senegalensis	13	16	18
Eutropius niloticus	14	11	9
Hydrocynus brevis	15	20	19
Bagrus bayad	16	>30	>30
Citharinus latus	17	29	>30

was that the average size of the migrants had decreased considerably because of the lowering of the age of sexual maturation (Durand 1978).

— The second type of change was that of species described as less abundant in 1971–72 and which become predominant later on such as *Synodontis clarias*, *Schilbe uranoscopus*, *Distichodus brevipinnis*, *Chrysichthys auratus* and *Marcusenius cyprinoïdes* which become abundant from 1972–73. This was also the case of some formerly less abundant predators such as *Lates niloticus* or some species rarely reported in the delta fisheries like *Gymnarchus niloticus*. The success of the latter resulted from the establishment and maintainance of large areas of the lake that are poorly oxygenated and/or unstable. Finally, *Alestes nurse* and *Brienomyrus niger* which were very abundant in the fisheries of the entire experimental battery, remained among the 17 species mostly caught by the reduced battery (mesh bar: 25 mm and over) in spite of their small size.

— The third type of change corresponded to species that were abundant before the drought as well as in the catches of 1976–77. This may be because they did not suffer a recession like *B. batensoda* or because after a decrease in 1972–73 they appeared to reconstitute their stocks like *Synodontis schall*, *Hyperopisus bebe* or even *Hydrocynus forskalii*, *Polypterus bichir* and *Eutropius niloticus* whose importance in 1976–77 was higher than that of 1971–72.

If all the catches made by the complete experimental battery are now considered it is noted that some small sized species that were rare or accidental during the period of 'Normal Chad' became abundant in 1976–77. They were *Pollimyrus isidori*, *Siluranodon auritus*, *Ichthyborus besse*, *Brienomyrus niger* ... Some others became notably more frequent than before: *Petrocephalus bane*, *P. bovei* and the already cited *Alestes nurse*.

To compare the behaviour of migrants sampled in 1976–77 with those described by Loubens in 1971–72, we extracted the variability in seasonal abundance of 19 species that were most abundant between April 1976 and April 1977 (complete experimental battery). To do this, although the periodicity of sampling was not perfect we smoothed the curve of abundance of each species by the method or repetitive moving averages. From the hierarchical ascendant classification obtained (Fig. 17), two groups of species could be distinguished. The first group included those whose maximum abundance was located around low water and it was composed of three groups:

— the first included *Ichthyborus besse* and *Schilbe mystus* whose maximum abundance was at the end of the withdrawal. If the migratory behaviour of *S. mystus* was probable that of *I. besse* was less evident;
— the second group included *Siluranodon auritus*, *Alestes nurse* and *Chrysichthys auratus* whose migratory behaviour was not certain;
— the third group included *Pollimyrus isidori*, *Synodontis clarias*, *Hydrocynus forskalii*, *Alestes baremoze* and *Eutropius niloticus*. The presence of these species was almost constant throughout the year with a more marked abundance at the low water between April and July.

346

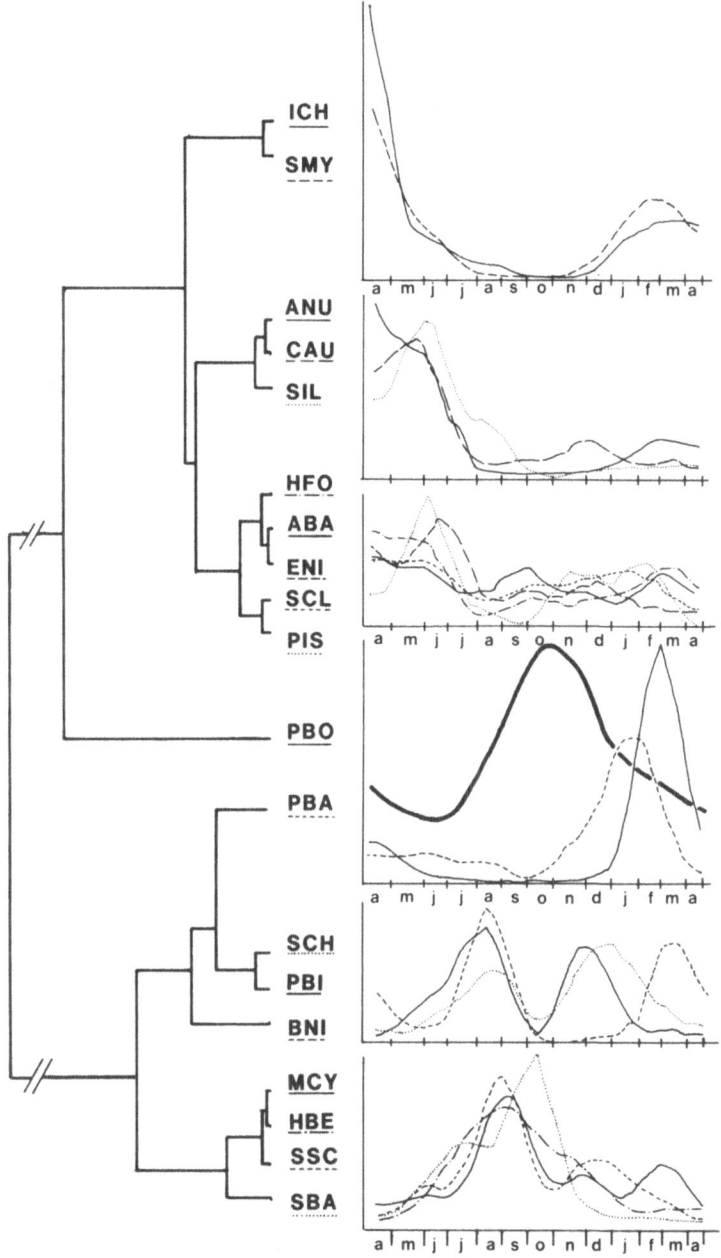

Fig. 17 Smoothed profiles of the catches of the 20 most abundant species in the Shari delta from April 1976 to April 1977 (experimental set GN 10, 11, 12, 13, 14, 15, 16, 18, 20, 22, 25, 30, 35, 40, 50, 60, 70, 80, 90, 100). Species are gathered according to the results of a hierarchical ascending classification (to the left). For abbreviations, see Table 12. The thickest line shows the flood of the Shari river.

A. baremoze was an exception with three periods of very high abundance in May–June, September–October and in February–March corresponding fairly well to the observations of Loubens. The migratory period of *H. forskalii* appeared to be shifted slightly because it was found abundantly from January to May whereas Loubens limited its migration from November to March. There was even less agreement for *Eutropius niloticus* which showed two periods of abundance. The first from April to June almost corresponded to the 1971–72 descriptions and the second from December to February corresponded to the greatest scarcity of species.

The second group of species distinguished by the hierarchic classification included some species whose maximum abundance coincided with the flood period. It was composed of two groups:

— the first included *Synodontis schall, Hyperopisus bebe, Marcusenius cyprino-ides* and *B. batensoda* characterized by a unimodal migrating tendency. The modes were located during the period of flood between July and October, corresponding for *S. schall* and *H. bebe* to the descriptions before the drought. By contrast, *B. batensoda*, 95% of whose catches occurred in May–June 1972 appeared to migrate towards the month of September–October from 1973;

— the second group, *Polypterus bichir, Schilbe uranoscopus* and *Brienomyrus niger* were characterized by some bimodal tendencies. The first mode were contemporary with the preceding group, whereas the second were located at the beginning or end of the subsidence.

Schilbe uranoscopus migration increased considerably in the periods indicated previously, but with an inversion of the magnitude of the passages, with the migration from November to January being more important than that from June to September. *P. bichir* made two migrations, shown by the analysis of data from 1971–72 and 1972–73. Finally, Petrocephalus bovei and *Petrocephalus bane* must be mentioned. Their resemblance is greater although it does not agree with the classification. Their high abundance at the end of the 1977 subsidence was not equivalent to the previous year but appeared to correspond to the catch variations of 1971–72.

10.3.4 *Surveys of the changes in recruitment*

The Yaére-El Beid complex (Fig. 1) played an important role in the stocking of Lake Chad. The overflow of the Logone in the Greater Yaéré of north Cameroon emptied mainly towards the lake by a natural drain: the El Beid (Bénech et al. 1982). The traditional fisheries of this river exploited the juvenile populations which followed the water movement. The composition of catches was an index of the renewing of the lake stocks. In order to understand well the restructuration of lake fish communities, we followed the composition of catches of the El Beid fisheries from the year of the recovery of normal floods (1974).

10.3.4.1 *General survey of catches.* A total presentation of results of four fishing seasons (1974–75, 1975–76, 1976–77, 1977–78) was provided by the projection on two factorial planes (from a correspondence analysis) of species and samples regrouped as a function of the fishing season (Fig. 18). The study of the projections on the factorial planes at first showed a marked opposition on the first axis between two groups of species. One was closely related to the high water period and the other to the period of withdrawal and low water. This distinction was found to be the same for all the years of observation, though 1974–75, the first year after the drought showed a smaller distance between the two groups. Axis 2, defined by the analysis, permitted a better separation between the different periods of subsidence and low water. Axis 3, on the contrary, discriminated between the different periods of high water. The yearly differences suggested by the analysis, corresponded more to the dissimilarity of importance of different floods than to a chronological succession after the drought. The samples of the years of average flood (1975–76 and 1976–77) regrouped in the same way the samples from the years of low flood (1974–75 and 1977–78), as well for the withdrawal periods (axis 2) as for the high waters periods (axis 3). The structure shown by this analysis of results was therefore essentially connected to the hydrology of the El Beid.

10.3.4.2 *Influence of the drought.* All the fish caught in the El Beid could be divided into two groups. The first was made up of less mobile species whose breeding adults made some small transverse migrations at the time of the flood. Most of these species were capable of maintaining themselves in the residual ponds of the plain between two floods. The second group consisted of species whose adults made longitudinal lake–river migrations to spawn close to the Yaéré. These species which were more mobile than the first group, tend to colonize environments that were more favourable to them, more quickly.

These differences can explain the evolution of catches of two small Mormyridae with a short cycle (reproduction at one year) and a similar diet: the migratory *Pollimyrus isidori* and the sedentary *Petrocephalus bovei*. In the catches of the El Beid these two species showed considerable development after the drought of 1972–73. The slowness of the recolonization from the lateral migrations of the river stock explains the shift from the abundance of *P. bovei* by comparison with *P. isidori*.

The same shift was seen for *Brienomyrus niger*, one of the main inhabitants of the Yaéré ponds. Some sedentary species however had a greater capacity for rapid recolonization, e.g. Cichlidae and especially *Sarotherodon niloticus*. The latter has a remarkable resistance to difficult environmental conditions due to its multiple spawning habit and the care given to the young, and can thus respond very quickly to favorable environmental conditions. This particular adaptation allowed them, at the time of the flood of 1974–75, to multiply

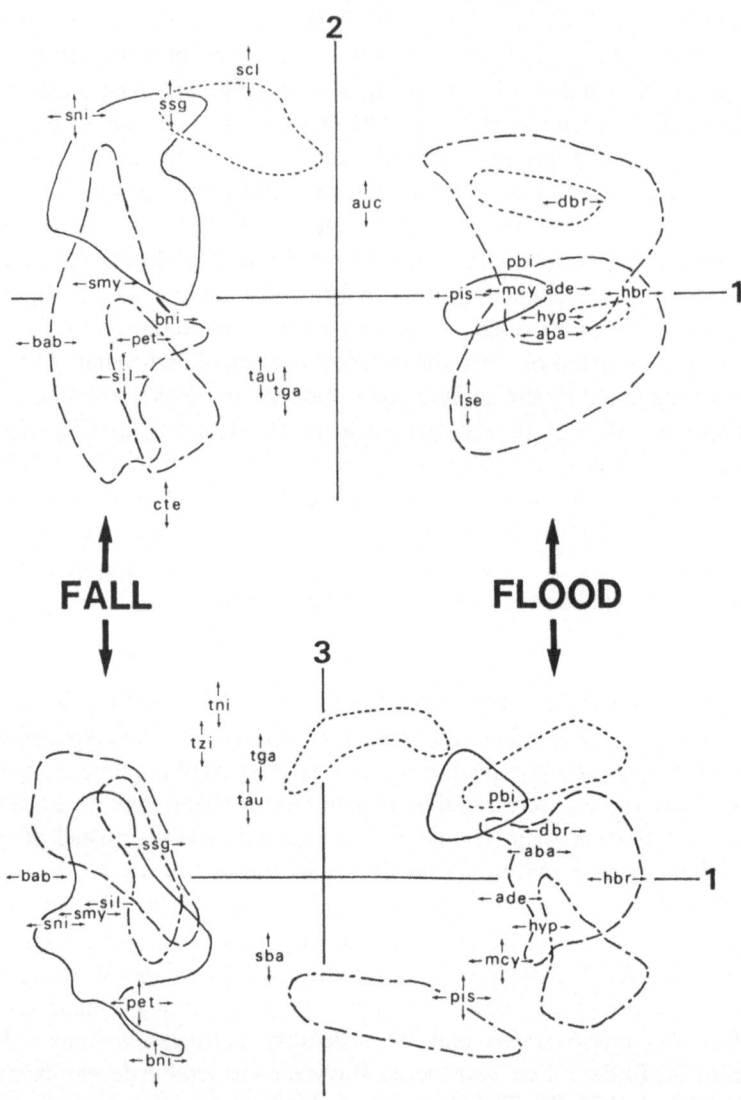

Fig. 18 Correspondence analysis made on catches of the Daga dam (El Beïd river; 1974–1978) coded into four equiprobable classes of abundance. Seasonal variations appear on axis 1 and interannual variations on axis 3 (flood period) and axis 2 (fall period).

1974/75   1975/76   1976/77   1977/78

↑
  tga   Main contribution to vertical axis
↓

←sni→Main contribution to horizontal axis
For abbreviations, see Table 12.

rapidly in the Yaéré thus occupying the place left vacant by the populations that normally colonized this biotope.

Among the large sized migratory species, different types of change could be noted according to the species which can more or less bear the perturbations caused by the drought. Three examples will illustrate this diversity: *Schilbe mystus*, *Citharinus citharus* and *Brachysynodontis batensoda*.

The catches of *Schilbe mystus* declined from 9.4% in 1968–69 (Durand 1970) to 2.6% in 1974–75 and were maintained during the following years at about 1% (Table 11). The lacustrine stocks of *Schilbe* suffered mortalities resulting from the lowering of the water level. Moreover, reproducing only in the rivers, two year classes were absent (1972 and 1973) corresponding to the low floods. In 1974–75 reproduction was still assured from the 1971 age class, then three years old, and spawned just before the dry period. From 1976, the reproduction corresponded to absent year classes. It is the explanation to the higher percentage of young found in 1974–75 than in the following years.

The influence of the drought was more marked for *Citharinus citharus*. It is probable that the lake stocks of this species were almost destroyed. The change in the percentage of young in the catches shows that the restoration of stocks was very slow (Table 11).

In contrast the *Brachysynodontis batensoda* stock recovered rapidly if the change in the El Beid catches was related to the abundance of mature lake fish. This species resists the drought quite well, probably because of its hardiness which was great as seen in the north of the lake as well as in the Southeastern Archipelago. It may also have been due to some of the adults being sheltered in the fluvial system from the end of 1973 as shown by some observations made in the delta region (Quensière 1976).

In the context of 'Lesser Chad', the recruitments showed an unstable species composition. The species that suffered least from the drought or which had a potential for rapid recovery of their stock became important in turn. This was the result of a complex interspecific competition. *Sarotherodon niloticus* benefited from the absence of competitors in 1974–75 and similarly *Marcusenius cyprinoides* in 1975–76 and *B. batensoda* in 1977–78. Thus there was a

Table 11 Number of individuals of three species caught and their % composition in total catches during four fishing seasons at the Daga station (El Beid).

Species	1974–1975		1975–1976		1976–1977		1977–1978	
	N	% N	N	% N	N	% N	N	% N
Schilbe spp.	922	2.6	1348	1.2	1435	1.2	1063	1.0
Citharinus citharus	0	0	12	0.01	40	0.03	38	0.04
Brachysynodontis batensoda	984	2.8	2323	2.1	2790	2.3	10 678	10.4

successive random appearance of abundant species. In fact, since 1974–75 the species richness has increased and the relative abundance of various species has tended to progressively stabilize over the successive years.

10.3.5 *Conclusions*

After a slow retraction period between 1965 and 1971 the exceptionally low floods (on the scale of scientific observations of the basin) of the years 1972–73 and 1973–74 precipitated the hydrological change in the lake. In the space of two years it declined from a 'Normal Chad' to a 'Lesser Chad'. From an ichthyological point of view this evolution included two successive phases. The first corresponded to the drying up with a severe change in the Chad basin, from a stable lacustrine appearance ('Normal Chad') to an unstable marshy appearance. The second corresponded to the move towards the new equilibrium of 'Lesser Chad', different from the first, and with lacustrine and marshy characteristics at the same time. The observations made in the fluvial and lacustrine environments over the drying period showed that the reduction in lake water area acted upon fish communities three basic ways:

— the decrease in water volume caused a concentration of the fish present. It increased inter and intraspecific competition as well as vulnerability to the fishing gear;

— the increasing shallowness of the water layer allowed resuspension of sediments by wave action which caused some mass mortalities by affecting the branchial tissue or by fixing the dissolved oxygen (reducing compounds in the muds). However, the solubility of nutritive substances caused the explosive development of phytoplankton which led to an increase in the daily variability of the oxygen content and dissolved free CO_2;

— the drying up of the shallows at first caused isolation of the parts of the lake richest in fish (the north basin and the archipelago) through the cutting off of river supplies. Abundant marshy vegetation then developed which according to its thickness prevented the supply of flood water to isolated areas and the eventual escape of fish which became trapped there (north basin) or caused organic pollution of water through which they pass (Southeastern Archipelago). Thus some truly anoxic conditions were created that were toxic and increase the mortalities.

Moreover, the absence of water from the upper layers and floodable plains, which normally provided spawning places and/or shelter for the young, prevented the renewing of stocks.

The natural selection operating on the fish communities during the drying up period favoured the development of 'marshy' species endowed with adaptations of diet, reproduction and respiration that allowed them to survive in an unstable environment, at the expense of 'lacustrine' species that are generally migrators with strict preferences. This selection operated both on the lacustrine stocks and their renewal:

352

— on the stocks, by provoking an increase in natural mortality and indirectly by mortality through fishing;

— on the renewal of stocks, by isolating the spawning places as well as by opposing their filling by water. The development of the marshy fraction of the fish community was thus not only linked to the adaptations of these fractions but also to the absence of competition of the other part of the fish community which extinguishes in the lake.

During the second period, the 'Lesser Chad', which is the installation of a new lacustrine appearance, a 'marshy' fish community is maintained which developed over the previous stage. The 'Lacustrine' stocks are restored by the recovery of a normal hydrological balance well marked floods, flooding of spawning places ... and the diversification of lake environments. Here zones of open water and zones of vegetation banks coexist with some important mass residuals of marshy vegetation mostly composed of ambatches (*Aeschynomenes elaphoxylon*).

The study of the drought and its consequences on the lake brings out a major characteristic of the Chadian fish fauna which is the constant coexistence of two groups of species. One is more particularly adapted to low waters and the other to high waters (which we have qualified by 'marshy' and 'lacustrine', respectively). The replacement of the dominance of one group by another is very rapid when a change occurs in the lacustrine environment. As no species are restricted only to the lake, the reconstitution of stocks is always possible from river fish communities.

Table 12 List of the species with their codes and initials used in the figures.

Code	Abbreviation	Name
1	XNI	*Xenomystus nigri*
2	HET	*Heterotis niloticus*
3	HYP	*Hyperopisus bebe*
4	MHA	*Mormyrus hasselquisti*
5	MRU	*Mormyrus rume*
6	MDE	*Mormyrops deliciosus*
9	BNI	*Brienomyrus niger*
11	MCY	*Marcusenius cyprinoïdes*
12	PBO	*Petrocephalus bovei*
13	PBA	*Petrocephalus bane*
14	PIS	*Pollimyrus isidori*
15	GYM	*Gymnarchus niloticus*
19	HBR	*Hydrocynus brevis*
21	ADE	*Alestes dentex*
22	ABA	*Alestes baremoze*

Table 12 (continued).

Code	Abbreviation	Name
23	AMA	*Alestes macrolepidotus*
25	ANU	*Alestes nurse*
26	MAC	*Micralestes acutidens*
28	ICH	*Ichthyborus besse*
29	CIC	*Citharinus citharus*
33	DRO	*Distichodus rostratus*
34	DBR	*Distichodus brevipinnis*
37	LSE	*Labeo senegalensis*
38	LCO	*Labeo coubie*
39	BAR	*Barilius niloticus*
41	BBA	*Bagrus bayad*
42	CAU	*Chrysichthys auratus*
⎰44	ABI	*Auchenoglanis biscutatus*
⎱45	AOC	*Auchenoglanis occidentalis*
⎡46	CAN	*Clarias anguillaris*
⎨47	CLA	*Clarias lazera*
⎣48	CAL	*Clarias albipunctatus*
49		*Heterobranchus* spp
50	SCH	*Schilbe uranoscopus*
51	SMY	*Schilbe mystus*
52	ENI	*Eutropius niloticus*
53	SIL	*Siluranodon auritus*
55	SBA	*Brachysynodontis batensoda*
56	SME	*Hemisynodontis membranaceus*
57	SCL	*Synodontis clarias*
60	SNI	*Synodontis nigrita*
66	SSG	*Synodontis schall-gambiensis*
67	MAL	*Malapterurus electricus*
68	LAT	*Lates niloticus*
70	TNI	*Sarotherodon niloticus*
71	TAU	*Sarotherodon aureus*
72	TGA	*Sarotherodon galilaeus*
73	TZI	*Tilapia zillii*
⎰75	CMU	*Ctenopoma muriei*
⎱76	CPE	*Ctenopoma petherici*
79	PSE	*Polypterus senegalus*
80	PBI	*Polypterus bichir*
82	PAN	*Protopterus annectens*
83	AND	*Alestes nurse-dageti*

Table 12 (continued).

Code	Abbreviation	Name
84	TSP	*Tilapia and Sarotherodon* spp.
85	CSP	*Clarias* spp.
86	PET	*Petrocephalus* spp.
92	BAB	*Barbus* spp.
93	MOC	*Mochocus brevis*
94	EPI	*Epiplatys* spp.
96	—	*Mochocus niloticus*
97	—	*Aplocheilichtys* spp.
98	—	*Aplocheilichtys gambiensis*
99	—	*Hydrocynus* spp.
Regroupments		
25	ANU	*Alestes nurse*
44	AUC	*Auchenoglanis* spp.
75	CTE	*Ctenopoma* spp.
86	PET	*Petrocephalus* spp.
85	CSP	*Clarias* spp.

References

Bénech, V., 1975. Croissance, mortalité et production de *Brachysynodontis batensoda* (Pisces, Mochocidae) dans l'Archipel sud-est du lac Tchad. Cah. ORSTOM, Sér. Hydrobiol. 8: 23–33.

Bénech, V. and Lek, S., 1981, Résistance à l'hypoxie et observations écologiques pour seize espèces de poissons du Tchad. Rev. Hydrobiol. trop. 14: 153–168.

Bénech, V., Lemoalle, J. and Quensière, J., 1976. Mortalités de poissons et conditions de milieu dans le lac Tchad au cours d'une période de sècheresse. Cah. ORSTOM, Sér. Hydrobiol. 10: 119–130.

Bénech, V., Quensière, J. and Vidy, G., 1982. Hydrologie et physico-chimie des eaux de la plaine d'inondation du Nord-Cameroun. Cah. ORSTOM, Sér. Hydrol. 19 (1): 15–35.

Blache, J., 1964. Les poissons du bassin du Tchad et du bassin adjacent de Mayo Kebbi. Mém ORSTOM, 4, 483 pp.

Carmouze, J. P., Dejoux, C., Durand, J.R., Gras, R., Iltis, A., Lauzanne, L., Lemoalle, J., Lévêque, C., Loubens, G. and Saint Jean, L., 1972. Grandes zones écologiques du lac Tchad. Cah. ORSTOM, Sér. Hydrobiol. 6: 103–169.

Daget, J., 1954. Les poissons du Niger supérieur. Mém. IFAN 36: 391 pp.

Daget, J., 1967. Introduction à l'étude hydrobiologique du lac Tchad. C.r. Soc. Biogéogr. 380: 6–10.

Durand, J. R., 1970. Les peuplements ichtyologiques de l'El Beid — Première note — Présentation du milieu et résultats généraux. Cah. ORSTOM, Sér. Hydrobiol. 4: 3–26.

Durand, J. R., 1971. Les peuplements ichtyologiques de l'El Beid. 2ème note. Variations inter et intraspécifiques. Cah. ORSTOM, Sér. Hydrobiol. 5: 147–159.

Durand, J. R., 1973a. Application de l'analyse des correspondances à l'étude de certains peuplements ichtyologiques du lac Tchad. Cah. ORSTOM Sér. Hydrobiol. 7: 55 – 62.

Durand, J. R., 1973b. Note sur l'évolution des prises par unité d'effort dans le lac Tchad. Cah. ORSTOM Sér. Hydrobiol. 7: 195–207.

Durand, J. R., 1978. Biologie et dynamique des populations d'*Alestes baremoze* (Pisces, Characidae) du bassin tchadien. Trav. Doc. ORSTOM, No. 98: 332 pp.

Durand, J. R., Franc, J. and Loubens, G., 1972. Résultats des pêches aux filets maillants et à la senne (1966–70). ORSTOM N'Djaména, 97 pp., mimeo.

Fotius, G. and Lemoalle, J., 1976. Reconnaissance de l'évolution de la végétation du lac Tchad entre janvier 1974 et janvier 1976. Rapport de mission. ORSTOM N'Djaména, 13 pp., mimeo.

Hamley, J. M. and Regier, H. A., 1973. Direct estimates of gill net selectivity to walleye (*Stizostedion vitreum vitreum*). J. Fish. Res. Bd Can. 30: 817-830.

Hopson, A. J., 1968. The gill net fisheries of Lake Chad. Federal Fisheries Service, Maïduguri, 64 pp.

Im, B. H., 1977. Etude de l'alimentation de quelques espèces de *Synodontis* (Poissons, Mochocidae) du Tchad. Thèse doct. spéc., Univ. Paul Sabatier de Toulouse, 150 pp., mimeo.

Lauzanne, L., 1977. Aspects qualitatifs et quantitatifs de l'alimentation des poissons du Tchad. Thèse Univ. Paris VI/ORSTOM, 284 pp., mimeo.

Legendre, L. and Legendre, P., 1979. Ecologie numérique. 1 — Le traitement multiple des donnèes écologiques, 195 pp. 2 — La structure des données écologiques, 247 pp. Masson, Paris et les Presses de l'Université du Québec.

Lemoalle, J., 1975. Bilan des apports en fer au lac Tchad. Cah. ORSTOM Sér. Hydrobiol. 8: 35–40.

Loubens, G., 1969. Etude de certains peuplements ichtyologiques par des pêches au poison. 1ère note. Cah. ORSTOM Sér. Hydrobiol. 3: 45–73.

Loubens, G., 1970. Etude de certains peuplements ichtyologiques par des pêches au poison. 2ème note. Cah. ORSTOM Sér. Hydrobiol. 4: 45–61.

Loubens, G., 1973. Production de la pêche et peuplements ichtyologiques d'un bief du delta du Chari. Cah. ORSTOM Sér. Hydrobiol. 7: 209–233.

Mok, M., 1975. Biométrie et biologie des *Schilbe* (Pisces: Siluriformes du bassin tchadien). 2ème partie. Cah. ORSTOM Sér. Hydrobiol. 9: 33–60.

Quensière, J., 1976. Influence de la sècheresse sur les pêcheries du delta du Chari (1971–1973). Cah. ORSTOM Sér. Hydrobiol. 10: 3–18.

Srinn, K. Y., 1976, Biologie d'*Hydrocynus forskalii* (Pisces: Characidae) du bassin tchadien. Thèse Spec. Univ. Toulouse, 126 pp., mimeo.

Tilho, J., 1928. Variations et disparition possible du Tchad. Ann. Géogr. 37: 238–260.

III. The balance of a lacustrine ecosystem during 'Normal Chad' and a period of drought

11. Phytoplankton production

Jacques Lemoalle

Two aspects of the photosynthetic activity (gross production) of the phytoplankton in Lake Chad were considered. Firstly an attempt was made to determine the photosynthesis-depth curve and the daily gross production. Secondly, an attempt was made to interpret variations which occurred with changing environmental conditions (Lemoalle 1973, 1979).

Changes in biomass and phytoplankton were thoroughly studied at Bol, a station in the Southeastern Archipelago. Environmental variations which were observed throughout the lake during the study period and which resulted from fluctuations in the water level all occurred at this station. The results of the study therefore permitted interpretation to be made of observed changes in different regions of the lake where measurements were made less frequently.

Although Landsat satellite data could be used under certain conditions to extrapolate the discrete *in situ* measurements to the entire water body, the production estimates for the different regions were necessarily approximate and valid only for certain periods. While the main drawbacks were the size, the heterogeneity and the variability of the lake, the considerable changes in environmental conditions during the period of study (1968–1976) required major consideration for understanding the interactions between the different parameters.

11.1 Methods

The principles of the methods are briefly mentioned here, the details and discussion having already been published (Lemoalle 1973, 1979).

The chlorophyll concentration, with no correction for the decomposition products, was estimated after grinding dried filters in 90% cold acetone. A coefficient of 11.9 (Talling and Driver 1963) was applied to the difference of the optical densities which were measured at 665 and 750 nm.

Phytoplankton activity was measured by the oxygen method with incubation in bottles either *in situ* or in an incubator in the laboratory.

J.-P. Carmouze et al. (eds.) Lake Chad
© *1983, Dr W. Junk Publishers, The Hague/Boston/Lancaster*
ISBN 978-94-009-7268-1

Photosynthetic activity (gross production) was designated by A(mg $O_2m^{-3}h^{-1}$). Generally, the hourly rate per unit area ΣA (mg $O_2m^{-2}h^{-1}$) was measured around midday. Successive incubations were used to evaluate the daily production.

The optimum activity which corresponds to the maximum value of A on the photosynthesis-depth curve was designated by A_{opt}. The irradiance in the incubator was determined experimentally so that the photosynthetic activity was equal to A_{opt} measured *in situ*.

11.2 Photosynthesis parameters

The phytoplankton was considered here as a photosynthetic system distributed in a body of water. We examined the shape of the photosynthesis-depth curves in relation to water transparency, phytoplankton concentration and temperature, as well as the relationships between instantaneous gross production and daily production. The experimental determination of these relationships was used to reveal the main factors involved in the photosynthetic activity of the phytoplankton and to provide a set of equations allowing production estimates to be made from the available field measurements.

11.2.1 *Shape of the photosynthesis-depth curves*

The photosynthesis-depth curves determined *in situ* around midday had a simple shape resulting from the environmental characteristics and from the experimental techniques. Due to the homogeneous vertical distribution of the phytoplankton, to the turbulence and the absence of stratification in the euphotic zone, a surface sample could be used for incubation in the whole profiles.

The results obtained at Bol in August 1972 and August 1975 are shown in Fig. 1 as an example of the curves obtained in water containing clay suspensions (1972) and water containing turbidity of organic origin (1975). The surface inhibition was considerable on all days with fine weather. A fairly violent tornado on the evening of 7 August, 1972 decreased the water transparency, SD, and induced the resuspension of deposited but viable phytoplankton. This phenomenon emphasizes the increasing importance of the bottom turbulence with decreasing depth (see Bénech et al. 1976).

Compared with Secchi disc transparency, SD, the optimum production occurred at a greater depth when turbidity was due to clay suspensions (1972) than when organic matter was the major cause of light absorption (1975). This difference was only partly due to the modification in the above-described relations (Chapter 3) between SD and K, the vertical light attenuation coefficient (mean over the 400–700 nm spectrum). The phytoplankton also

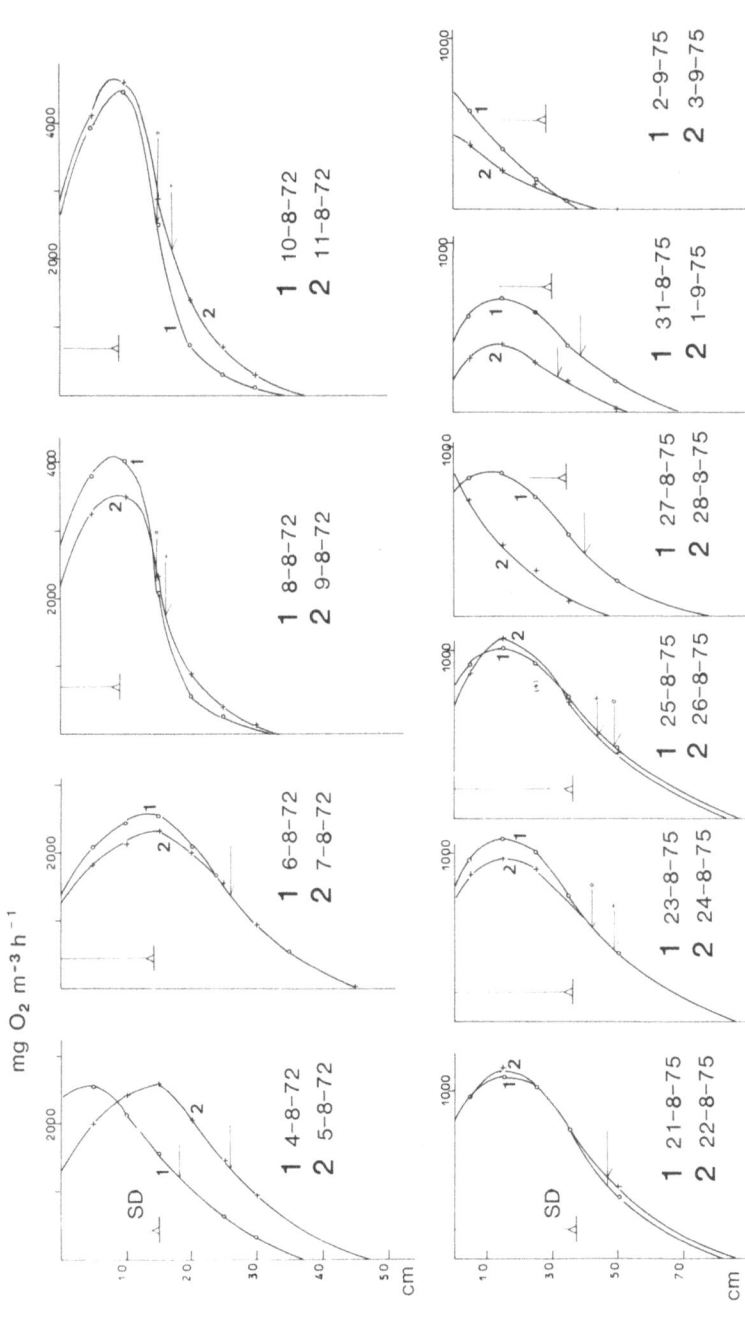

Fig. 1 Examples of photosynthesis-depth curves in Bol during August 1972 (water with suspended clay) and in August 1975 (water with organic matter). Water transparencies are indicated by the Secchi and Z_i by an arrow on each curve. Three hours incubation around midday.

359

responded differently to the light energy, probably because of the different spectral qualities of the types of water.

Given ΣA the integral of the mean hourly production in the water column (mg O_2 m^{-2} h^{-1}), we can define the depth Z_i so that

$$A_{opt}. Z_i = \Sigma A$$

Z_i is a function of water transparency and of the response of phytoplankton to irradiance. In most cases, water transparency in Lake Chad was measured by the Secchi disc. Z_i was thus determined as a function of SD for the different types of water: with clay suspensions from 'Normal Chad' at Bol in 1968–70 (Fig. 2a) or in the whole southern basin in 1970–71 (Fig. 2b), and with organic turbidity (Fig. 2c) of phytoplankton-rich water (Fig. 2d).

The photosynthesis-depth curves obtained during fine weather have been used to compute the following relationships with the 95% confidence limits of Z_i for the range of water transparencies observed, expressed in meters:

—water with clay suspensions

$$Z_i = 2.07 \text{ (SD)} \pm 0.15 \qquad\qquad\qquad n = 142$$

— water with organic turbidity

$$Z_i = 1.20 \text{ (SD)} \pm 0.13 \qquad\qquad\qquad n = 28$$

— phytoplankton-rich water

$$Z_i = 1.55 \text{(SD)} \qquad\qquad\qquad \text{(fitted by eye)}$$

Using these equations the production per unit area, ΣA, can be estimated when A_{opt} and SD are known for a given type of water.

To compare the results obtained in Lake Chad with other primary production models, the experimental value of Z_i can be compared with $Z_{0.5} A_{opt}$, the depth at which the photosynthetic activity is equal to 0.5 A_{opt}. Practically, this depth was fairly well determined since it occurred in the part of the curve where the gradient dA/dz was steepest. It should also be mentioned that this last depth correspond to $Z_{0.5\ 1k}$ from Talling's model (1957). For the Lake Chad results, Z_i and $Z_{0.5\ A\ opt}$, have been estimated directly from the photosynthesis-depth curve. The mean value of the ratio of $Z_{0.5\ A\ opt}$ to Z_i for 170 profiles from different types of water was one (probability greater than 95%): $Z_i = Z_{0.5\ A\ opt}$.

Therefore, the characteristics of the photosynthesis-depth curves in Lake Chad fit Talling's model. The average irradiance I_K as defined by Talling averaged 8.8 J cm^{-2} h^{-1} in water with clay suspensions and 15.9 in water with organic turbidity.

11.2.2 Relationship between biomass(B) and optimal activity(A_{opt})

The optimal activity A_{opt} was dependent upon numerous factors, the main one being the chlorophyll concentration B (mg m^{-3}). Its variations in time and

360

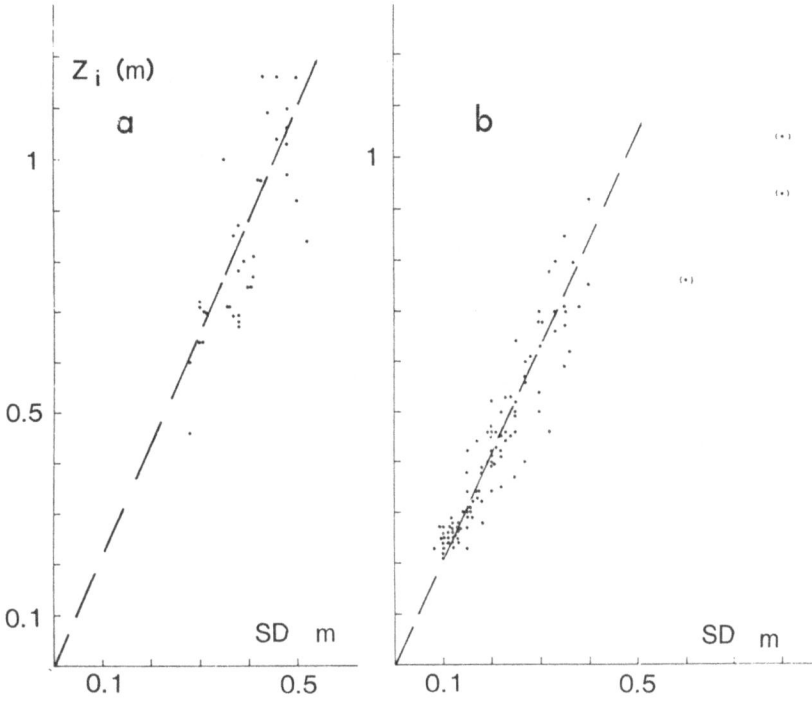

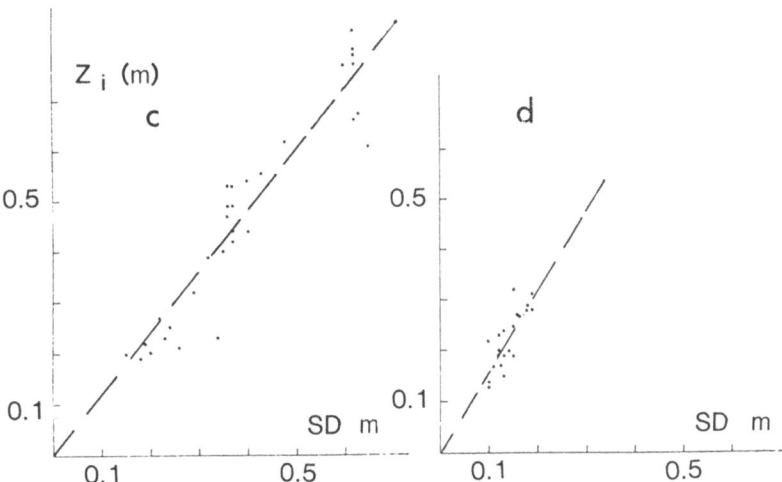

Fig. 2 Z_i as a function of SD, at Bol in 1968–70 (a); in the southern basin in 1970–71 (b), where the points (●) indicate waters with very few inorganic particles; during the period of 'Lesser Chad' with macrophytes at Bol in 1973–76 (c); and during the concentration phase (d).

space were particularly important in Chad, the representative values observed ranging from 4 to 3500 mg Chl a.m^{-3}.

The results obtained, either *in situ* or in the incubator led to the formulation of the regression equation:

$$\log A_{opt} = 1.22 + 1.044 \log B \tag{1}$$

with a correlation coefficient $r=0.97$ for $n=298$ pairs of measurements. This equation represents the mean relation between the two parameters. The correlation coefficient indicates that 95% of the variance of A_{opt} resulted from variations in B. In the range $10 < B < 1000$ mg m^{-3}, the degree of uncertainty about the determination of log A_{opt} from an isolated measurement of B is ± 0.275 (probability 95%). The large variability of B thus concealed the effect of the other parameters which can influence optimal activity and especially the effect of temperature as will be seen later on.

Relation (1) indicates that the specific optimal activity, $\varphi_{opt} = A_{opt}/B$ increased with increasing B, the most frequent mean values ranging from

$$\varphi_{opt} = 18.4 \text{ for } B = 10 \text{ mg Chl } a.m^{-3}$$

and

$$\varphi_{opt} = 22.5 \text{ for } B = 1000 \text{ mg Chl } a. m^{-3}$$

These results are in agreement with the values 20 to 25 which are generally accepted for tropical lakes (Talling 1965b; Talling et al. 1973; Ganf 1975). Moreover, most of the high phytoplanktonic biomass in Lake Chad was observed during the hot season, and relation (1) can be affected by the effect of the temperature on φ_{opt}.

11.2.3 *The effect of temperature*

The effect of temperature on specific optimal activity φ_{opt} was evaluated by the value of the mean $Q_{10} \left(= \exp \dfrac{10}{\Delta T} \Delta (\log A) \right)$ over the range of temperature defined by the difference ΔT.

The mean value of φ_{opt} *in situ* was calculated at Bol (1968–1970) between 11.00 and 15.00 hours in two distinct temperature ranges, from 20 to 23°C and from 29 to 32°C.

Mean T: 20.4°C: mean $\varphi_{opt} = 20.4$ mg O_2 (mg Chl. a)$^{-1}$.h^{-1}

Mean T: 30.0°C: mean $\varphi_{opt} = 23.4$ mg O_2 (mg Chl. a)$^{-1}$.h^{-1}

If the laws of chemical kinetics are applied:

$Q_{10} = 1.15$ between 20 and 30°C

This low coefficient involved the *in situ* phytoplankton which can be generally considered as adapted to their environment.

The results of two surveys conducted in the southern basin in December 1970 and June 1971 gave significantly different specific optimal activity:

December 1970: mean $\varphi_{opt} = 18.82$ n = 30 $_m = 0.78$
June 1971: mean $\varphi_{opt} = 23.07$ n = 33 $\sigma_m = 0.78$

For mean temperatures of 20.0 and 30.5°C, the calculated coefficient was

$Q_{10} = 1.21$ between 20 and 30°C.

These results were applicable to the rather stable conditions in 'Normal Chad' and showed only the apparent value of Q_{10}, since changes in the environment other than the temperature could also occur in different seasons.

Other measurements of the temperature coefficients were made in an incubator, where bottles from the same sample were incubated at different temperatures. Unlike the previous measurements, the phytoplankton was then subjected to thermal shocks (reaching 10° within 15 minutes), while, the other environmental conditions remained similar.

It was then observed that φ_{opt} reached an optimum between 28 and 33°C, beyond which the activity then decreased quickly. Between 20 and 30°C, the different samples showed different behaviour regarding the Q_{10} as well as the value of φ_{opt}. While high values of Q_{10} occurred during the lake flood at Bol in October 1973 (from 2.8 to 4.7), all the other measurements had an average of 1.4.

The Q_{10} values observed in Chad were generally lower than most published values. The adaptation of the phytoplankton to the environment have resulted in the decrease of Q_{10}. Indeed, the mean temperature in the lake was close to the optimum temperature ranging between 28 and 32°C. Therefore, the curve $\log \varphi_{opt} = f(T)$ was close to its maximum and had a shallower slope than in the case of lower temperatures.

Furthermore, it was difficult to estimate the respective roles played by B and T in the variability of A_{opt}, since both variables were most often concomitant in their change ('Normal Chad'). When such was not the case (for instance during the 'Lesser Chad' at Bol), the use of the temperature in step-wise regressions did not improve the determination of A_{opt} and the residual variance must therefore be attributed to the variability of the other environmental conditions.

11.2.4 Daily production

In situ measurements of $\Sigma\Sigma A$, the daily gross production were made through successive incubations, the duration of which was dependent on phytoplankton concentration. The mean relation between $\Sigma\Sigma A$ and the mean hourly production around midday was determined with a probability of 95% for standard weather conditions (28 measurements).

$$\Sigma\Sigma A / \Sigma A = 9.1 \pm 0.3$$

Given the low variability of the day length at the latitude of Lake Chad, this ratio can be considered as constant throughout the year and compares with other determinations in tropical regions (Talling 1965; Ganf 1975).

11.2.5 *Model of photosynthetic activity*

Using the experimental relationships, production in various stations in the lake could be estimated from the measured variables (equations given in Table 1). In the case of statistical relations, two confidence limits (probability of 95%) were calculated: the confidence limit of the regression line (or of a mean value) determined from all the measurements, and the confidence limit of the determination of a single value of the function, the value of the variable being given. In this second case, we observed that the confidence limits did not vary significantly within the range studied.

11.3 Evaluation of production in different regions of the lake

It is difficult to generalise about the whole of the lake with the discrete measurements which were made. However, during each survey we tried to cover the entire range of variation in water transparency and chlorophyll. Therefore, the values given here may be considered as a range of estimations for photosynthetic activity of the phytoplankton in the main regions of the lake.

11.3.1 *Production at Bol and in the southern basin*

The seasonal variations in activity at Bol are given in Fig. 3 for three different years which can be considered as representative of the different periods of change in the lake in this region. The year 1969 represented the period of the low 'Normal Chad' (1968–72) with variations related to the seasonal fluctuations in water level, water transparency and temperature. The hourly production (measured *in situ*) ranged from 0.3 to 0.8 g O_2 m^{-2} h^{-1} with an annual average of 4.2 g O_2 m^{-2} day^{-1}.

The year 1973 was typical of the phytoplankton-rich water phase, along with a considerable and large flood in October. The available data indicate a maximum or 1.1 g O_2 m^{-2} h^{-1} during the warm season with biomass up to 600 mg Chl$a$ m^{-3} and a minimum of 0.2 g O_2 m^{-2} h^{-1} after the flood. The annual average for daily activity, based on the curve of the monthly averages (Fig. 3) was 7.4 g O_2 m^{-2} day^{-1}.

Although higher values were observed in 1974 (up to 2 g O_2.m^{-2}.h^{-1}), the longest period of water circulation resulted in a similar mean activity during 1973 and 1974.

Table 1 Observed relationship between the production parameters.

Equation	Type of water	Statistical error on the relationship	Statistical error on a single measurement	Range
$\Sigma\Sigma A = 9.1\ \Sigma A_{midday}$	suspended clay standard day	± 0.3	± 1.6	$\Sigma A < 1000$ mg $O_2 m^{-2} h^{-1}$
$\Sigma A = A_{opt}\ Z_i$				
$Z_i = 2.07$ SD	suspended clay concentration	± 0.013	± 0.15	$0.1 < SD < 1$ m
$Z_i = 1.55$ SD		—	—	$SD < 0.4$ m
$Z_i = 1.20$ SD	dissolved organic matter	± 0.02	± 0.13	$0.1 < SD < 1$ m
A, 3 hrs incubation		—	± 0.36	mg O_2 m^{-3} h^{-1}
ΣA, 3 hrs incubation		—	$\pm 0.1\Sigma A$	$200 < \Sigma A < 500$ mg $O_2 m^{-2} h^{-1}$
$\log A_{opt} = 1.22 + 1.044 \log B$	all	—	± 0.275	$10 < B < 1000$ (mgChl. a)m^{-1}
$Q_{10} = 1.2$	in situ			
$\varepsilon_{min} = \dfrac{1.95 + k_1}{SD}$	dissolved organic matter	$k_1 = \pm 0.11$	$k_1 = \pm 0.56$	$0.1 < SD < 1$ m
$\varepsilon_{min} = \dfrac{1.39 + k_2}{SD}$	suspended clay	$k_2 = \pm 0.11$	$k_2 = \pm 0.56$	$0.1 < SD < 1$ m
$E_K\ (= I_K) = 8.8$	suspended clay			J.cm^{-2} h^{-1}
15.9	dissolved organic matter			

365

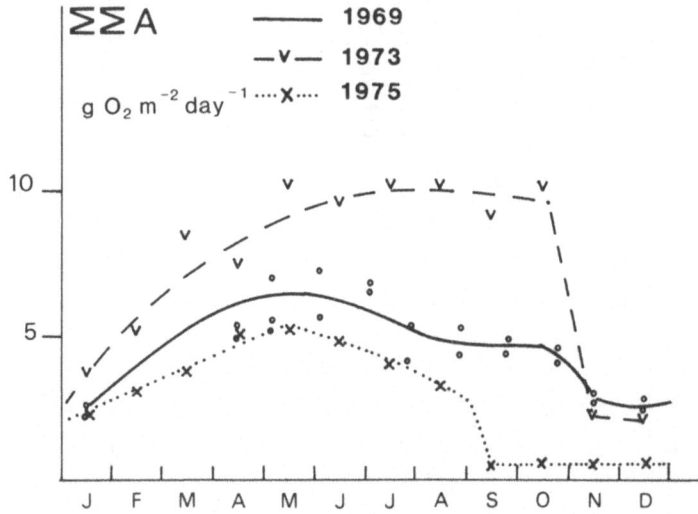

Fig. 3 Seasonal variations in the daily photosynthetic activity (gO_2m^{-2} day^{-1}) of the phytoplankton at Bol, during three years representative of the general changes in this region of the lake.

The return of the water to a 'normal' level was represented by the year 1975 with the action of the macrophytes being considerable when the water moved through the vegetation. Poor utilization of light (I_K higher) resulted in lower production (0.58 g O_2 m^2 h^{-1}) during the warm season. The annual mean production was 2.7 g O_2 m^{-2} day^{-1}.

The general change in B and ΣA over the period 1968–1976 is given in Fig. 4 (semi-logarithmic scale) and emphasizes the variability with time of production related to the environmental conditions.

The results obtained in 1970–71 have been used to estimate the phytoplankton production in the southern basin during the period of 'Normal Chad'. The results of June 1970 and June 1971 were pooled to emphasize the difference between the warm season (low water level) and the cold season (high level) which was influenced by the flood of the Shari. On the whole, 114 measurements (*in situ*) were used to divide the basin into 6 different regions (Fig. 5) (Table 2). This zonation which corresponded roughly to the large natural regions of the basin at that time was related to the zones of water transparency. The production appeared highest in the archipelago itself, and was clearly lower in zone under the influence of the Shari and in the Great Barrier both being transit zones of the flood waters.

During the period of 'Lesser Chad' with the growth of the macrophytes, the open water areas became very small in the archipelago where most of the primary production then resulted from the macrophytes and the epiphytes. The open water in the southern basin maintained a fairly constant area throughout the period 1974–76. The Landsat data were used to evaluate production in this region (Lemoalle 1978):

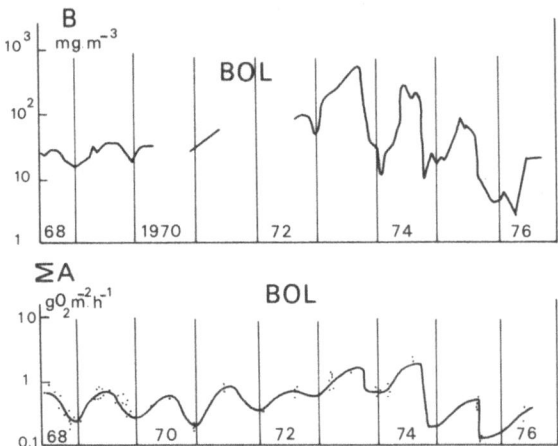

Fig. 4 The change with time of chlorophyll concentration B (mg m^{-3}) and of the production per unit area around midday (gO$_2$ m^{-2} h^{-1}) at Bol.

Fig. 5 Main zones of phytoplankton production in the southern basin during the period of 'Normal Chad' (1970–71).

open waters of the southern basin:
end of June 1975: 3.76 g O$_2$.m^{-2}.d^{-1}
October 8–10th 1975: 0.91 g O$_2$.m^{-2}.d^{-1}

As compared with the results of region 1 in 1970–71, these values show a considerable increase in the photosynthetic activity of the perideltaic region

Table 2 Transparency distribution (SD), optimum activity A_{opt} (mg O_2 m^{-3} h^{-1}) and daily production $\Sigma\Sigma A$ (g O_2 m^{-2} day^{-1}) in the southern basin during 1970–1971. See Fig. 5 for regions.

	June			December		
Region	SD (m)	A_{opt} mg $O_2.m^{-3} h^{-1}$	$\Sigma\Sigma A$ g $O_2.m^{-2}.d^{-1}$	SD (m)	A_{opt} mg $O_2.m^{-3} h^{-1}$	$\Sigma\Sigma A$ g $O_2.m^{-2} d^{-1}$
1	0.20	500	2.3	0.25–0.40	150	0.9
2	0.25–0.30	1000	3.7	0.20	400	1.15
3	0.30–0.35	1200–1500	7.2	0.20	500	1.9
4	0.10–0.15	1200	3.1	0.25–0.40	150	0.9
5[a]	—	—	—	0.12	500	1.6
6	0.10	1000	1.9	0.15	300	0.9

[a]No representative data for June 1970 and June 1971.

after 1973, resulting from higher phytoplankton concentrations and slightly higher water transparencies. This difference can be attributed to the macrophytes, whose main effect was a decrease of the fetch and therefore of the turbulence.

11.3.2 *Production at Kindjéria and in the northern basin*

In order to estimate the production at Kindjéria from chlorophyll and transparency data, various values of the coefficient in the relationship $Z_i = k$ SD were used:

$k = 2.07$ before June 1974 (silty water)
$k = 1.55$ until drying up in October 1975 (phytoplankton-rich water)
$k = 1.20$ during 1976 (organic water)

The change in ΣA at Kindjéria is represented in Fig. 6 as well as the mean values of ΣA during the month of April, in a region of 20 km radius around the station of Kindjéria. These averages are rather close to the results obtained at Kindjéria; this station can be considered as representative of the surrounding region of 1256 km^2.

Given the large and rapid variations in the production at Kindjéria (Fig. 6) and the paucity of measurements, it is difficult to evaluate a mean annual production. Therefore, only the extreme values observed each year and the mean value of ΣA in April as determined on n stations with its standard deviation are given in Table 3.

Supposing that each station from the survey in April 1974 (Fig. 7) was representative of an identical water area, the mean value of the photosynthetic activity in the entire northern basin for 27 stations would have been $\Sigma A = 1.49$ g

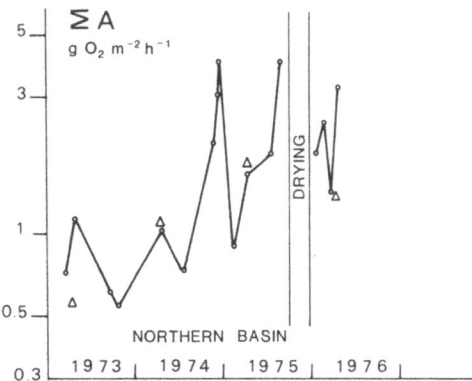

Fig. 6 Changes in hourly midday production in the northern basin at Kindjéria (○) and within a radius of 20 km around Kindjéria (△). Production was estimated from the values of chlorophyll and water transparency.

Table 3 Production in the northern basin, at Kindjéria.

Year	$\Sigma\Sigma A$ $g\ O_2.m^{-2}.day^{-1}$	ΣA $mg\ O_2.m^{-2}\ h^{-1}$	n	σ
1973	4.9–10	570	11	0.22
1974	6.7–36	1150	11	0.33
1975	9.1–36	1730	15	0.77
1976	12.7–30	1310	8	0.86

$O_2.m^{-2}.h^{-1}$ or $13.5\ g\ O_2.m^{-2}\ day^{-1}$ with an area of 6000 km^2, or a total daily production reaching 81 000 tons of oxygen. This result can be compared with the phytoplanktonic biomass which was estimated by Iltis (1977) to be between 167 000 tons and 200 000 tons fresh weight.

11.3.3 *Maximum production values in Lake Chad*

Figures of photosynthetic activity in very productive environments depend upon the method used: the highest values are given by the *in situ* oxygen balance, while the classical method, with incubation in bottles may give estimates that are half of the former in the same environment (Talling et al. 1973; Melack and Kilham 1974).

Four examples of high productions, are given in Table 4. Figure 8 shows daily changes in production measured by successive bottle incubations. The

369

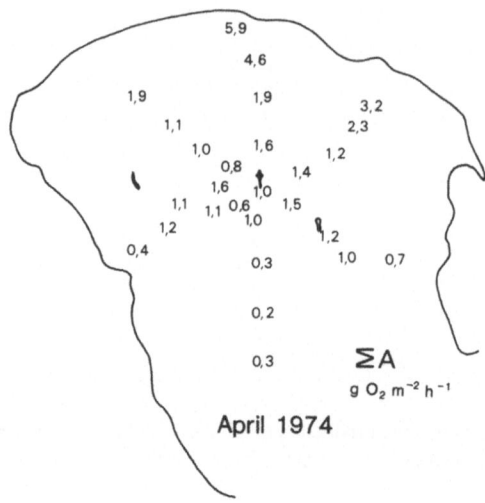

Fig. 7 Distribution of the hourly production around midday ($gO_2m^{-2}\,h^{-1}$) in the northern basin during April 1974.

Table 4 Examples of high photosynthetic activity.

Locality	Date	A_{opt} $g\,O_2\,m^{-3}.h^{-1}$	ΣA_{midday} $g\,O_2\,m^{-2}.h^{-1}$	$\Sigma\Sigma A$ $g\,O_2\,m^{-2}.d^{-1}$	SD m
Baga kiskra	7 March 1974	4.9	1.43	12.2	0.17
Baga kiskra	14 June 1974	10	1.87	16.8	0.12
N'Goudouboul	15 June 1974	28	3.48	31.0	0.10
Bol	28 June 1974	9.8	1.65	14.3	0.18

difference between the morning and afternoon rates was due mainly to the presence of clouds in the afternoon.

Talling et al. (1973) published a detailed analysis of the factors leading to high production in Lakes Kilotes and Aranguadi in Ethiopia. Due to the lack of data on irradiance and chlorophyll concentrations, the same was not possible here. However, it appeared that Lake Chad may approach the highest production values measured in the natural environment. Apart from the four examples mentioned, production higher than $9\ g\ O_2.m^{-2}.d^{-1}$ was measured in the archipelago around Bol as well as in the northern basin.

In these cases, environmental conditions were such that reduced water areas resulted, like isolated ponds, sheltered from the wind and with high water temperature and increased alkalinity caused by concentration of salts.

370

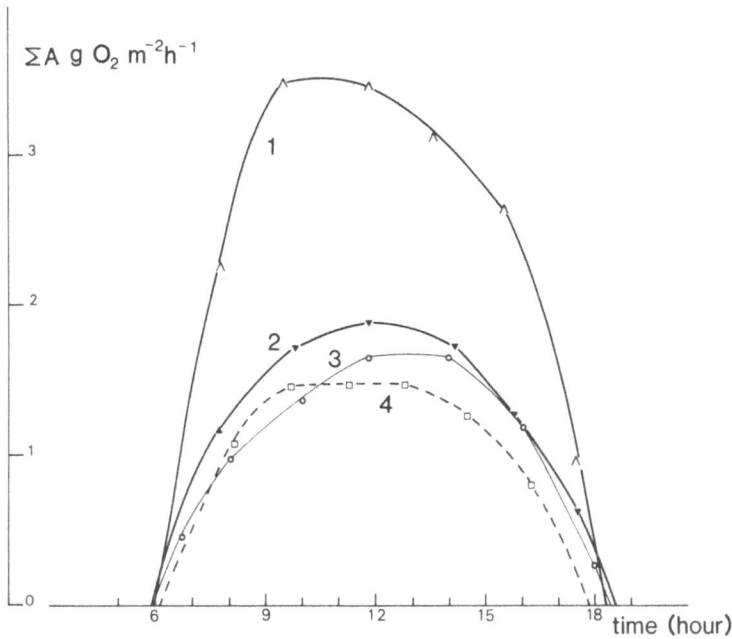

Fig. 8 Changes in ΣA throughout the day with very high production values in the northern basin and at Bol. *In situ* measurements. Each point indicates the mean time of each incubation period. 1 = N'Gouboudoul, June 15th; 2 = Baga Kiskra, June 14th; 3 = Bol, June 28th; 4 = Baga Kiskra, June 14th.

The absence of turbulence due to wind in a shallow environment limits the resuspension of the mineral particles of the sediments, whilst encouraging a vertical circulation at night which is sufficient for a redistribution of the nutritive elements. This situation also leads to the existence of a high value of p which is the percentage of the total light absorption which results from the phytoplankton absorbance.

Moreover, a high alkalinity means a large reserve of inorganic CO_2 and a considerable buffer capacity, both of which lead to high production. Most of the high photosynthetic activities in the tropical natural environment have been measured in rather alkaline waters:

— Lake Mariut, Egypt, 5–6 mé 1^{-1} (Aleem and Samaan 1969);
— Lakes Kilotes and Aranguadi, 51–57 mé 1^{-1} (Talling et al. 1973);
— Alkaline lakes in Kenya and Tanzania, 84–168 mé 1^{-1} (Melack and Kilham 1974).

Only Lake George (Uganda) has high production (12 g O_2 m^{-2} d^{-1}) with rather low alkalinity of 2 mé 1^{-1} (Ganf 1972).

371

11.4 Changes in the phytoplankton in relationship to environmental conditions

The physico-chemical conditions of the environment interact with the three main elements of the primary level: the phytoplankton concentration, the portion of the solar energy effectively used by this phytoplankton (gross production) and the fraction of this energy which may be used by the other trophic levels (net production). From the results obtained in Lake Chad it is possible to show some of the relationships between environmental factors and the phytoplankton (Fig. 9), and the variations in the level Z which strongly influence the other parameters.

In relation to past and present variations in the lake level, the emergent macrophytes are more or less abundant and, as seen in Chapter 5 their action involves four main variables. Amongst these are the chlorophyll concentration B (1) and the water transparency (3) because of their filtration effects on the clay or algal particles in the water. The optical quality of the water is also modified when the inorganic particles are replaced by dissolved organic matter, resulting in a modification of the coefficient k of the relation Z_i: k SD (2). Moreover, macrophytes limit the fetch, damping the short-term variations in water level, and act as a true barrier for the water supply in the northern basin. This is shown by the interaction between the water level and the macrophytes (4). These influences are significant for phytoplankton development and can be used to distinguish two periods in the evolution of the lake: before and after the development of the macrophytes. The general relationships given earlier were valid only for the water areas where the macrophyte had limited influence. When the influence was stronger one has to consider a series of particular cases which were dependent upon the water circulation through the swamps.

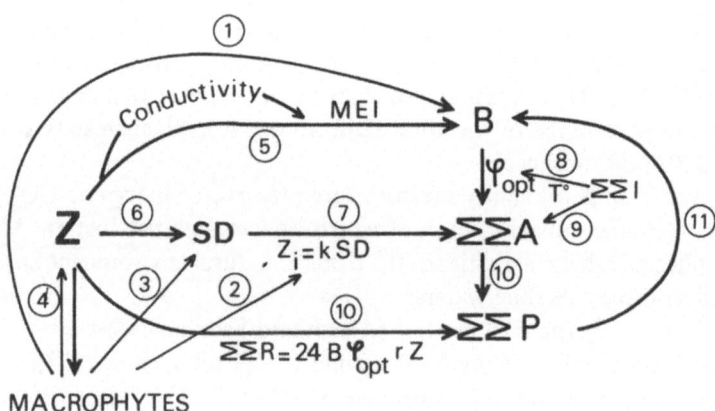

Fig. 9 Diagram of the interrelations between environmental conditions and the parameters of phytoplankton production.

11.4.1 *Phytoplankton concentration*

An increase in the chlorophyll concentration B was observed during a lowering of the water level in different regions of the lake. An increase in the total dissolved solids was generally associated with this lower level as indicated by the conductivity C.

The variations in B (mg Chl. *a.* m^{-3}) are represented in relation to the conductivity C (μS. cm^{-1}, 25°C) on Fig. 11, in the form: log B = f (log C).

In the open waters of the southern basin, the average relationship for the period 1973–76 was expressed by:

$$\log B = 5 \log C - 8.0$$

As the conductivity and the mean level of this region remained fairly constant during the period under consideration, this relationship describes the seasonal variations in a heterogeneous body of water, rather than long-term changes.

In Bol archipelago, chlorophyll increased with conductivity during the concentration phase. After the macrophytes developed, the filtration through the mats of macrophytes disturbed the possible relations between the two parameters (Fig. 10). The B versus C curve at Bol is representative of other stations of the archipelago (Fig. 10) during the concentration period (August 1972–July 1973). The data for the period 1968–1970 which are drawn on the same figure are situated in the continuation of the first cluster of points.

In the northern basin, the measurements have also been divided into two groups: during the concentration phase — from January 1973 to December 1974 — there was a clear relation between chlorophyll and conductivity (Fig. 11). After a first small flood in early 1975, the chlorophyll concentrations remained rather high while the conductivity strongly decreased. The data from Kindjéria during the two years of concentration fit quite well with the data from the whole basin (Fig. 11) and will be used later to represent the phenomena observed in this region.

In addition to our Lake Chad results, the data from the permanent alkaline ponds in Kanem (Iltis 1974) are given in Fig. 12 which summarizes the above mentioned relationships. The upper limit of the chlorophyll concentrations measured in Lake Chad is also shown by a broken line and compared with the limits observed by Talling (1970) in carbonated lakes having alkalinity ranging from 0.5 to 1000 mé l^{-1} (dotted line).

These results lead to several observations:
— during a concentration phase, there was a clear relationship between the conductivity and the chlorophyll concentration in the Bol archipelago and the northern basin. The stations at Bol and Kindjéria were representative of these changes;
— in the open water of the southern basin where there was no real concentra-

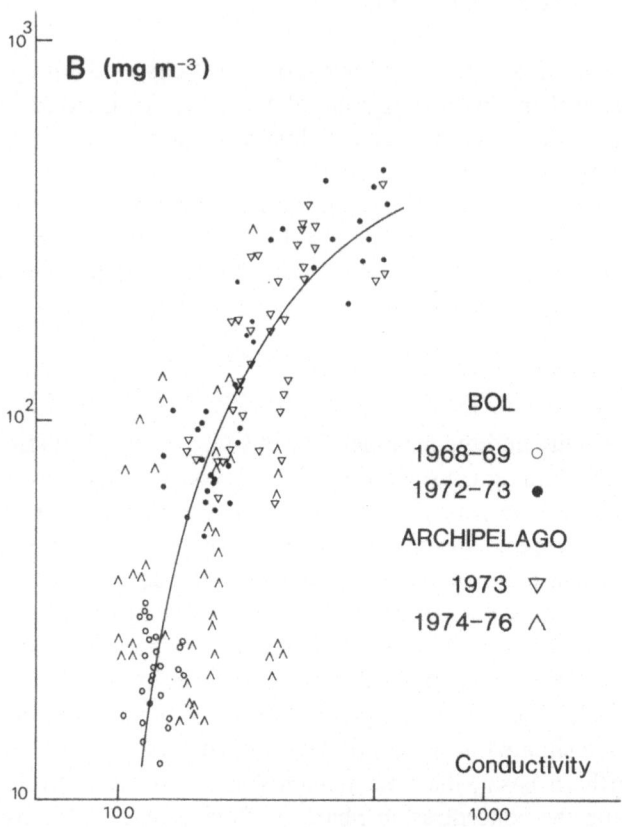

Fig. 10 Relationship between chlorophyll *a* concentration and conductivity at Bol in 1968–69 (○) and in 1972–73 (●) as well as in the archipelago before the development of macrophytes in 1973 (▽) and in the presence of macrophytes in 1974–76 (△).

tion of dissolved solids the seasonal changes in B and C leads to a similar relationship;
— the relationships shown were different for each environment observed;
— an upper limit of B versus C can be determined, which includes all the data from the different regions of the lake (Fig. 12).

The first two points suggest a causal relationship between chlorophyll and conductivity. Actually conductivity itself varied in relation to several other environmental factors.

In the open water of the southern basin, the conductivity varied with the water level and the season: the level was low during the warm season before the occurrence of the flood. After the end of the Shari flood, an increase in temperature, a decrease in the level and in the water transparency (which was

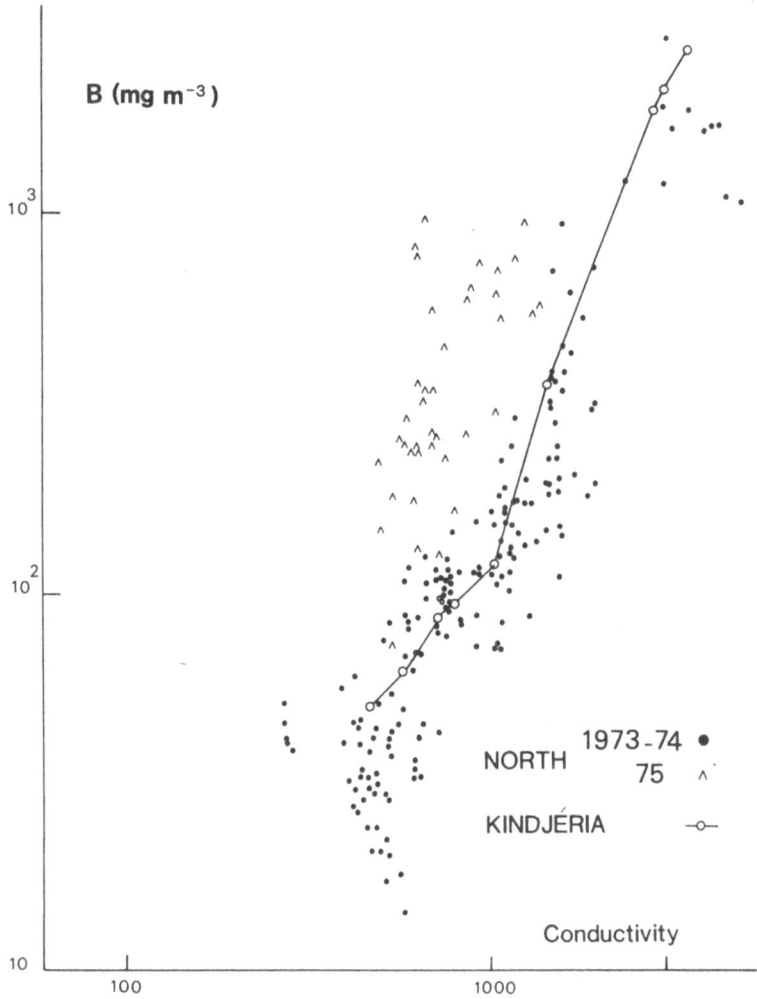

Fig. 11 Relationship between chlorophyll *a* concentration and conductivity in the northern basin. At Kindjéria, the relationship is indicated by the line.

often related to an increase of B in 1975–76) corresponded to the seasonal increase in conductivity.

In the northern basin, the seasonal variations did not appear during the two year drying up phase under consideration. Therefore, the possible influence of temperature may have been overlooked. When the dissolved salt concentration increased, the alkalinity increased and permitted high photosynthetic activity of the high biomass; as the available carbon increased, so did the buffer capacity which reduced the pH fluctuations which resulted from variations in total CO_2.

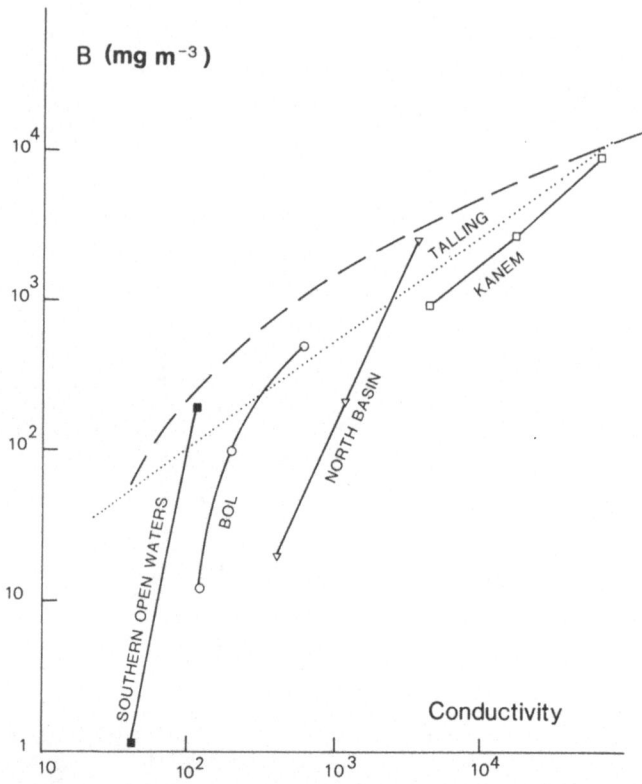

Fig. 12 The change in chlorophyll concentration in relationship of conductivity in the different environments, in the absence of macrophytes. The dotted line indicates the approximate limit of the maximum biomasses observed by Talling in lakes of various alkalinity. The broken line indicates the limit observed in Chad.

Since changes in conductivity and water depth were two related phenomena, a variable including these two parameters may be used: the morpho-edaphic index of Ryder (Rawson 1955; Moyle 1956; Ryder 1972; Henderson et al. 1973). It was applied to the different stations of the lake:

$$MEI = \frac{\text{conductivity } (\mu S \text{ cm}^{-1} \text{ at } 25°C)}{\text{depth (m)}}$$

During the drying up period, before the macrophyte development the regression equations of B with C and MEI were, respectively:

$$\log B = 0.410 + 0.866 \log C \qquad r_1 = 0.705$$
$$\log B = 0.062 + 0.736 \log (MEI) \quad r_2 = 0.80 \qquad n = 244$$

The use of the morpho-edaphic index thus improved the description of the phytoplanktonic concentration (path 5 of Fig. 9).

376

11.4.2 *Photosynthetic activity (gross production)*

If the phytoplankton concentration, B, and the water transparency are the only variable parameters, the production per unit area is proportional to p, the percentage of the light attenuation which results from phytoplankton absorbance. This value is directly related to the amount of phytoplankton per unit area in the euphotic layer.

In the case of Lake Chad where the water transparency SD was measured with a Secchi disc, the following general relationship was proposed (Lemoalle 1979):

$$1/SD = \gamma_W + \gamma_{Fe} + \gamma_B B$$

Then the parameter, p, is expressed as:

$$p = 100 \, \gamma_B . B \, (SD)$$

Using B and SD results, P was calculated for the southern basin in December 1970 and June 1971, for the northern basin in April 1974, and for Bol during the period 1968–70 and from March 1973 to May 1976. For each series, the relationship between ΣA and p are presented in a diagram $\log \Sigma A = f (\log P)$ in Fig. 13, which shows that ΣA was proportional to p.

For the northern basin some values of p were close to 100 and the three highest even exceeded this figure. These high values indicated that the coefficient γ_B used was too high for the phytoplankton considered.

The results obtained in Lake Chad support the above-mentioned theory that production depends upon the proportion of light absorbed by the phytoplankton independently of the absolute concentrations of phytoplankton, mineral particles or dissolved elements.

Concurrently, we observed an increase in photosynthetic activity per unit area at Bol as well as in the northern basin during the reduction in water level. This indicates that, in spite of the increased turbulence at the sediment interface, the increase in phytoplankton concentration resulted in an increase in p; the decrease in SD was proportionally less important than the increase in B.

The photosynthetic yield, E_{tot}, was expressed by the ratio of the assimilated energy (supposing that 1 g O_2 is equal to 3.33 kcal) to the total incident energy over a day. For all the *in situ* measurements in Lake Chad, E_{tot} ranged from 0.04% to 1.80% and was proportional to the percentage, p, which thus appeared as the main factor of variation. The percentage, p, ranged from 3 to 100% according to the stations and to the period of measurement.

The value $E_{tot} = 0.26\%$ which was calculated for Bol during the year 1969 must be considered only as representative of this region during this period. The variability of the lake did not allow extrapolation as was sometimes done elsewhere.

Given the assimilation index φ_{opt} of the phytoplankton in Lake Chad and the

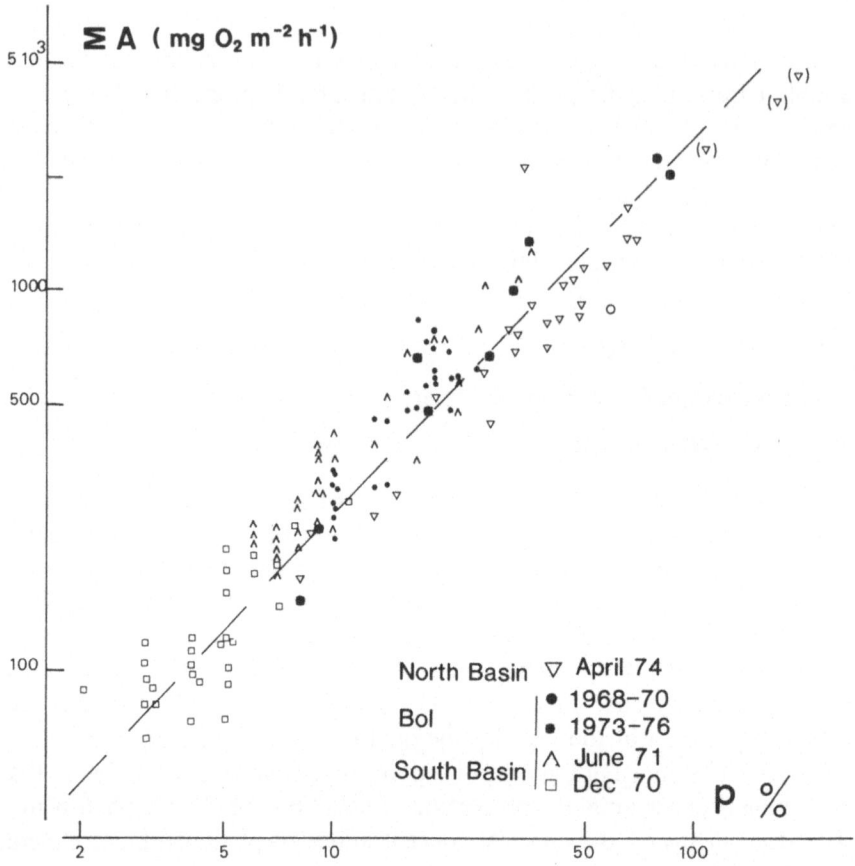

Fig. 13 Relationship between ΣA and p (percentage of the attenuation resulting from the phytoplankton) for various measurements made in Lake Chad (logarithmic scales).

relative constancy of the daily irradiation, gross production depended mainly upon B and SD. However, there was a maximum limit to the phytoplankton concentration beyond which the respiration of the water column becomes higher than the production in the euphotic zone. This boundary condition was defined in relation to the respiration coefficient, r, and to the ratio of the water transparency to the depth, SD/Z (Talling 1970) where $r = R/A_{opt}$, R being the hourly respiration per unit volume. Respiration thus appeared to limit net production and the biomass (path 10 and 11 in Fig. 9).

With the symbols used and the mean conditions for Lake Chad, daily net production ΣΣP is expressed as follows:

$$\Sigma\Sigma P = \Sigma\Sigma A \left(1 - 2.64 \frac{r.Z}{k_1.DS}\right)$$

378

The mean values observed in different parts of the lake imply that r = 0.05 can be considered as a lower limit, while r = 0.9 seems to be the upper limit of the respiration coefficient. Moreover, we noticed that the ratio SD/Z increased when the lake changed from 'Normal Chad' (clay water) to 'Lesser Chad' (organic water) at Bol as well as in the northern basin. Concurrently, I_K of Talling was lower in clay water than in water with dissolved organic matter, as if the phytoplankton adapted to their environment, and the algae likely to use a low irradiance were found when relative water transparency was low and vice versa.

11.5 Comparison with other lakes

The variability of Lake Chad makes it difficult to make a comparison with other more stable environments. It has the same characteristics as Lake Chilwa, as it is shallow, endorheic and unstable. In both lakes, the variations in water level with drying up as the ultimate phase, lead to the same increase in salt and phytoplankton concentration, growth of macrophytes, mortality and recolonization by fishes (Howard-Williams and Lenton 1975, Bénech et al. 1976).

In the previous paragraphs, some temporal variations in the parameters of primary production in the various regions of Lake Chad, have been described. This constitutes a comparison of the lake with itself at different moments of its development. The range of variations observed in gross production (from 1.4 gO_2 m^{-2} day^{-1} in the peri-deltaic region during 1970–71, to 30 g O_2 m^{-2} day^{-1} in the northern archipelago of the northern basin during 1974) is equivalent to that observed by Brylinski and Mann (1973) in their first synthesis of the I.B.P. results from 55 lakes distributed from 0 to 65° latitude.

Since biomass and phytoplankton production resulted from the interaction of numerous environmental conditions, we have compared the main parameters measured in some representative environments.

With the introduction of the morpho-edaphic index the variations in the chlorophyll concentration in relation to the environmental conditions could be described for Lake Chad and other water bodies such as the Kanem lakes and Lake Fitri. These lakes constitute a type of tropical carbonated lake whose suspended solids decrease when conductivity increases.

In a graph of log B = f (log MEI), the points representative of Lake·George (MEI 87, B = 200 mg/m^3) or Lake Mariut in Egypt (MEI50 resulting from alkalinity, a very considerable biomass) (Aleem and Samaan 1969) have much higher biomasses than Lake Chad as compared with their morpho-edaphic index. The same holds true for Loch Leven in Scotland whose mean biomass is, moreover, sensitive to the climatic variations. In comparison with Lake Chad, these three shallow lakes have a common feature: light absorption in the water results mainly from the phytoplankton as shown by the close relationship between B and SD or K (Aleem and Samaan 1969; Bindloss 1974; Ganf 1972).

If we consider the morpho-edaphic index, Lake Chad was characterized by a low biomass during the normal period as compared with the tropical shallow lakes for which data are available. The suspended minerals were responsible for this situation and brought Lake Chad closer to lakes where turbulence is considerable at the bottom level: Lake Balaton (Entz 1964), IJsselmeer (Lijklema 1976) Neusiedlersee (Dokulil 1973) some lakes in Australia (Kirk 1977). As stated by Henderson et al. (1973), who found it necessary to use corrective terms in the MEI when dealing with fish production, it also appears that the MEI needs the same type of correction when applied to phytoplankton biomass.

The results obtained in Chad for the photosynthetic activity per unit area have shown the importance of the percentage of light attenuation resulting from the phytoplankton itself. In this respect, the points representative of shallow lakes varying in latitudes combine with the data from Lake Chad in the graph log $E_{tot} = f$ (log p) of Fig. 14, although the seasonal variations in irradiance and production are much more important under temperate conditions.

The values of φ_{opt} observed in the different environments confirm the distinction made by Talling (1965b) between temperate and tropical lakes. More recently, Lastein and Gargas (1978) showed that, for 18 shallow lakes in Denmark, φ_{opt} and I_K were mainly dependent upon the temperature, the difference between eutrophic and oligotrophic lakes being of lesser importance. If the impact of the temperature on φ_{opt} seems clear cut, the same does not hold true for I_K which depends upon the attenuation per unit of pigments. In Lake Chad, I_K was estimated at 8.8 J cm^{-2} h^{-1} (400–700 nm) in clay water and at 15.9 J cm^{-2} h^{-1} in organic water, which emphasizes the importance of the phytoplankton adaptation.

In shallow lakes, maximum biomass and production depend upon the light climate defined by the ratio of the water transparency to the depth, as well as by the features of the phytoplankton, I_K and K_B. In a discussion about the influence of the morphometry on lake productivity, Richardson (1975) and Horne et al. (1975) put forward opposite arguments on shallow lake productivity. Nutrient cycling is faster in shallow lakes, as a result of the absence of stratification (Richardson), but they can be placed in an unfavourable position by the suspended minerals and a shallow euphotic layer (Horne et al.). Of course, some examples support both views.

The influence of the nutrients and especially of nitrogen and phosphorus was not considered in this study as concentration measurements alone are insufficient, and it would have been necessary to deal with the dynamics of the various forms of these two elements. However, the values of φ_{opt} observed in Lake Chad suggest that these nutrients may not be limiting optimal photosynthetic activity.

We noticed that highest biomass and production rates were usually observed

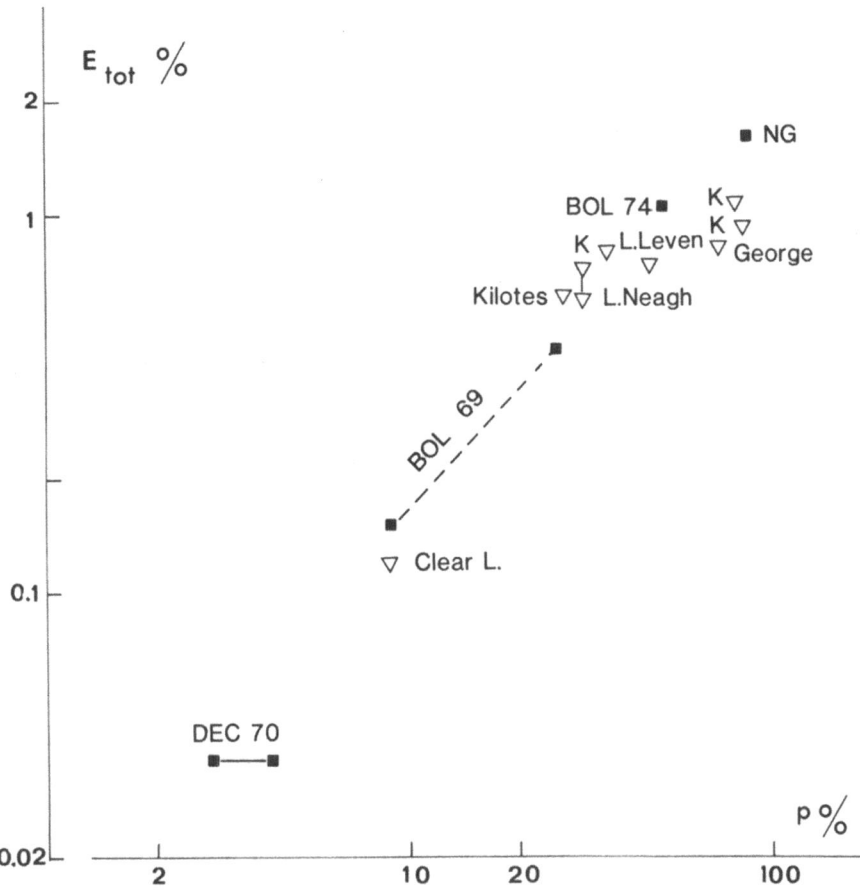

Fig. 14 Relationship between the total photosynthetic yield E_{tot} and the percentage, p, of light attenuation resulting from the phytoplankton. Measurements made in Lake Chad (■) and in other shallow lakes (▽).

when the environmental conditions (SD/Z) were most favourable to positive net production. There was however, an exception, when the filtration effect resulting from the water circulation through the macrophytes limited the phytoplankton concentration. Apart from this case, the photosynthetic yields equalled the highest values observed in other eutrophic environments.

If different parameters of the primary production are considered separately, we may conclude that Lake Chad did not show any particular feature. However, the suspended clay turbidity during the period of 'Normal Chad', and the filtration effect through the macrophyte mats during the period of 'Lesser Chad', largely contributed to its individuality when considering phyto-planktonic biomass and production together.

It thus appears that the biological development of a lake depends upon its physical features when the nutrients are not strictly limiting, and that turbulence is of prime importance in shallow lakes.

11.6 Comments on the use of remote sensing

Landsat satellites which make observations of the earth carry multispectral scanners with four bands: MSS4 from 500 to 600 nm, MSS5 from 600 to 700 nm, MSS6 from 700 to 800 nm and MSS7 from 800 to 1100 nm in the near infra-red. The data from these satellites were used to extrapolate to larger water areas the discrete field measurements made along a single route (Lemoalle 1979).

The water column could be considered as homogeneous, especially in the morning at about 9.30 a.m. when the satellite data were acquired. With a clear sky, the surface radiance, L (energy reflected by the lake surface) was representative of the water column and depends upon the optical properties of the water (Morel and Prieur 1977). If a relationship existed between the MSS radiance in band X and the field data, it was possible to extrapolate this relationship to the whole surface and estimate the distribution of the field parameter from the distribution of radiances.

Generally, there is a simple relation between the radiance L_X, as it is observed by the satellite, and the reciprocal of the water transparency as measured with the Secchi disc, 1/SD. This relationship can be extrapolated if the body of water is composed of only one type of water, that is to say that the relative proportions of the different substances playing a significant part in the light attenuation must remain constant. For instance, the relationship between the transparency of clay or organic waters and the luminance were clearly different.

For a lake of a given type of water, it is therefore possible to estimate the distribution of water transparency. If, moreover, there is a relationship between 1/SD and the chlorophyll concentration B, it is also possible to estimate the distribution of the phytoplankton, and for each picture element the photosynthetic activity which is proportional to the product $B \times (SD)$, may also be computed.

Such conditions are not always met. But, when they are, a synoptic picture of the distribution of the useful parameters over a large area provides information that cannot be obtained by field measurements only, given the size of the lake.

References

Aleem, A. A. and Samaan, A. A., 1969. Productivity of lake Mariut. Egypt. Int. Rev. Ges. Hydrobiol. 54: 313–355 and 491–527.

Bénech, V., Lemoalle, J. and Quensière, J., 1976. Mortalités de poissons et conditions de milieu dans le lac Tchad au cours d'une période de sécheresse. Cah. ORSTOM. Sér. Hydrobiol. 10: 119–130.

Bindloss, M. E., 1974. Primary productivity of phytoplankton in Loch Leven, Kinross. Proc. r. Soc. Edinb., B, 74: 157–181.

Brylinsky, M. and Mann, K. H., 1973. An analysis of factors governing productivity in lakes and reservoirs. Limnol. Oceanogr. 18: 1–14.

Dokulil, M., 1973. Planktonic primary production within the Phragmites community of lake Neusiedlersee (Austria) Pol. Arch. Hydrobiol. 20: 175–180.

Entz, B., 1964. Light conditions of lake Balaton, a shallow lake in Hungary. Verh. int. Ver. Limnol. 15: 260–264.

Ganf, G. G., 1972. The regulation of net primary production in lake George, Uganda, east Africa. In: Productivity problems of freshwaters, pp. 693–708. Z. Kajak and A. Hillbricht-Ilkowska (eds.), Warszawa, Krakow.

Ganf, G. G., 1975. Photosynthetic production and irradiance — photosynthesis relationships of the phytoplankton from a shallow equatorial lake (lake George, Uganda). Oecologia 18: 165–183.

Henderson, F., Ryder, R. A. and Kudhongania, A. W., 1973. Assessing fishery potentials of lakes and reservoirs. J. Fish. Res. Bd Canada. 30: 2000–2009.

Horne, A. J., Newbold, J. D. and Tilzer, M. M., 1975. The productivity, mixing modes, and management of the world's lakes. Limnol. Oceanogr. 20: 663–666.

Howard-Williams, C. and Lenton, G. M., 1975. The role of the littoral zone in the functioning of a shallow tropical lake ecosystem. Freshwat. Biol. 5: 445–459.

Iltis, A., 1974. Le phytoplancton des eaux natronées du Kanem (Tchad). Influence de la teneur en sels dissous sur le peuplement algal. Theses d'État, Univ. Paris VI, 271 pp.

Iltis, A., 1977. Peuplements phytoplanctoniques du lac Tchad. III. Remarques générales. Cah. ORSTOM Sér. Hydrobiol. 11: 189–199.

Kirk, J. T. O., 1977. Use of a quanta meter to measure attenuation and underwater reflectance of photosynthetically active radiation in some inland and coastal southeastern Australian waters. Aust. J. mar. Freshwat. Res. 28: 9–21.

Lastein, E. and Gargas, E., 1978. Relationship between phytoplankton photosynthesis and light, temperature and nutrients in shallow lakes. Verh. int. Ver. Limnol. 20: 678–689.

Lemoalle, J., 1973. L'énergie lumineuse et l'activité photosynthétique du phytoplancton dans le lac Tchad. Cah. ORSTOM Sér. Hydrobiol. 7: 95–116.

Lemoalle, J., 1978. Application des données Landsat à l'estimation de la production du phytoplancton du lac Tchad. Cah. ORSTOM Sér. Hydrobiol. 12: 83–87.

Lemoalle, J., 1979. Biomasse et production phytoplanctoniques du lac Tchad (1968–1976). Relations avec les conditions de milieu. ORSTOM, Paris, 3 II pp.

Lijklema, L., 1977. The role of iron in the exchange of phosphate between water and sediments. In: H. L. Golterman (ed.), Interactions between sediments and freshwater, pp. 313–317. Junk-Pudoc, The Hague.

Melack, J. M. and Kilham, P., 1974. Photosynthetic rates of phytoplankton in east African alkaline, saline lakes. Limnol. Oceanogr. 19: 743–755.

Morel, A. and Prieur, L., 1977. Analysis of variations in ocean color. Limnol. Oceanogr. 22: 709–722.

Moyle, J. B., 1956. Relationships between the chemistry of Minnesota surface waters and wildlife management. J. Wildl. Mgmt. 20: 303–320.

Rawson, D. S., 1955. Morphometry as a dominant factor in the productivity of lakes. Verh. int. Ver. Limnol. 12: 164–175.

Richardson, J. L., 1975. Morphometry and lacustrine productivity. Limnol. Oceanogr. 20: 661–663.

Ryder, R. A., 1972. The limnology and fishes of oligotrophic glacial lakes in north America (about 1800 A.D.) J. Fish. Res. Bd Canada 29: 617–628.

Talling, J. F., 1957. the phytoplankton population as a compound photosynthetic system. New Phytol. 56: 133–149.

Talling, J. F., 1965. Comparative problems of phytoplankton production and photosynthetic productivity in a tropical and a temperate lake. Memorie Ist. Ital. Idrobiol. 18 (suppl.): 399–424.

Talling, J. F., 1970. Generalized and specialized features of phytoplankton as a form of photosynthetic cover. IBP/PP Technical meeting, Trebon, 1969, pp. 431–445. In: Prediction and Measurement of Photosynthetic Productivity.

Talling, J. F., Wood, R. B., Prosser, M. V. and Baxter, R. M., 1973. The upper limit of photosynthetic productivity by phytoplankton: evidence from Ethiopian Soda lakes. Freshwat. Biol. 3: 53–76.

Talling, J. F. and Driver, D., 1963. Some problems in the estimation of chlorophyll a in phytoplankton. Proceedings, Conf. on Primary Productivity Measurement, Marine and Freshwater, Hawaii 1961, U.S. Atomic Energy Comm., pp. 142–146.

12. Secondary production (zooplankton and benthos)

Christian Lévêque and Lucien Saint-Jean

In order to understand the dynamics of an ecosystem, it is necessary to determine the contribution of the different groups to total biomass as well as to evaluate their turnover rates and the amount of organic matter that each produces per unit time. Therefore, the zooplankton and zoobenthos production were studied as part of the research on the productivity of Lake Chad conducted from 1968 to 1973.

For this purpose, the biology, growth rates and the dynamics of the main species present in the environment which represented a considerable part of the biomass were studied. These different parameters allowed the evaluation of the annual mean production of the different groups and therefore of the biological production of the environment.

12.1 Zooplankton

List of symbols used

D_i = duration of a developmental stage (time interval between the moults which limit the stage)

D_e = duration of the embryonic stage (from laying to hatching)

D_n = duration of the nauplius stage (Copepods)

D_c = duration of the copepodid stage

D_j = duration of the juvenile development of Cladocerans from birth to first laying

D_p = duration of the juvenile development of Copepods ($D_p = D_n + D_c$)

D'_e = time interval between two successive layings

N_E, N_N, N_C, N_A, N_M, N_F, N_T = number of eggs, and embryos, of nauplii, copepodids, adults, males, females and the population (total number)

W, B, P = individual weight, biomass and production

W_i = mean weight of an individual of stage i

W_{iA}, W_A, W_{fA} = initial, mean and final weight of the adults

ΔW_i = increase in weight through stage i

J.-P. Carmouze et al. (eds.) Lake Chad
© *1983, Dr W. Junk Publishers, The Hague/Boston/Lancaster*
ISBN 978-94-009-7268-1

Z = age expressed as a unit of biological time. These units are equal to the duration of the embryonic stage in days
b = birth rate

12.1.1 *Biological aspects and population structure of the main species*

12.1.1.1 *Reproduction, recruitment and population structure.* The presence of egg-bearing females and juvenile instars in the populations of planktonic copepods throughout the year showed that reproduction and recruitment were continuous. However some variations existed over the year (Fig. 1a). The number of eggs and nauplii of *Tropodiaptomus* decreased to two clearly defined minimum levels (Fig. 1b) corresponding to the cool season (January–February) and to the rainy season (July–August). These minimum levels did not result from a decrease in fecundity since the number of eggs per female (N_E/N_F) remained more or less constant (Fig. 1b) as did the proportion of adults in the population (Fig. 1d). In fact, the minimum levels observed resulted from a decrease in the abundance of adults during these periods.

A rather similar phenomenon was observed in the Cyclopoids but in the cool season the decrease in numbers was less pronounced in *Thermocyclops* which was rare especially during the rainy season (Fig. 1c). The opposite phenomenon occurred in *Mesocyclops*.

According to the available data, it seems that population structures remained fairly stable throughout the period of 'Normal Chad' while considerable changes and above all a big increase in the fecundity of the Calanoids (Table 7) were observed during the drying up of the lake.

Few results are available on variation in fecundity and population structure of the Cladocera, but it can be assumed that they were low, similar to the variations in the ratios N_J/N_T (Table 1) and N_E/N_A (Fig. 2) of the main species. The observations conducted from March 25th to April 24th, 1968 when the temperature increased by 6°C (Fig. 2) show that the fecundity remained rather constant in *Moina* and *Diaphanosoma* and varied in *Daphnia barbata*, according to temperature (decrease of fecundity with temperature increase).

12.1.1.2 *Duration of embryonic development.* The duration of embryonic development (D_e) of the main planktonic crustacean species was determined by laboratory breeding at 24°C (January 1969) and at 29–30° (August 1968 and 1969, July 1973). For three species it was determined by *in situ* breeding at 18–20°C (*Moina micrura, Thermocyclops neglectus, Mesocyclops leuckarti*) (Gras and Saint-Jean 1969, 1976).

The results obtained are represented by the following sigmoid logistic curve:

$$1/D_e = \frac{K}{1 + e^{b(\hat{\theta} - \theta)}}$$

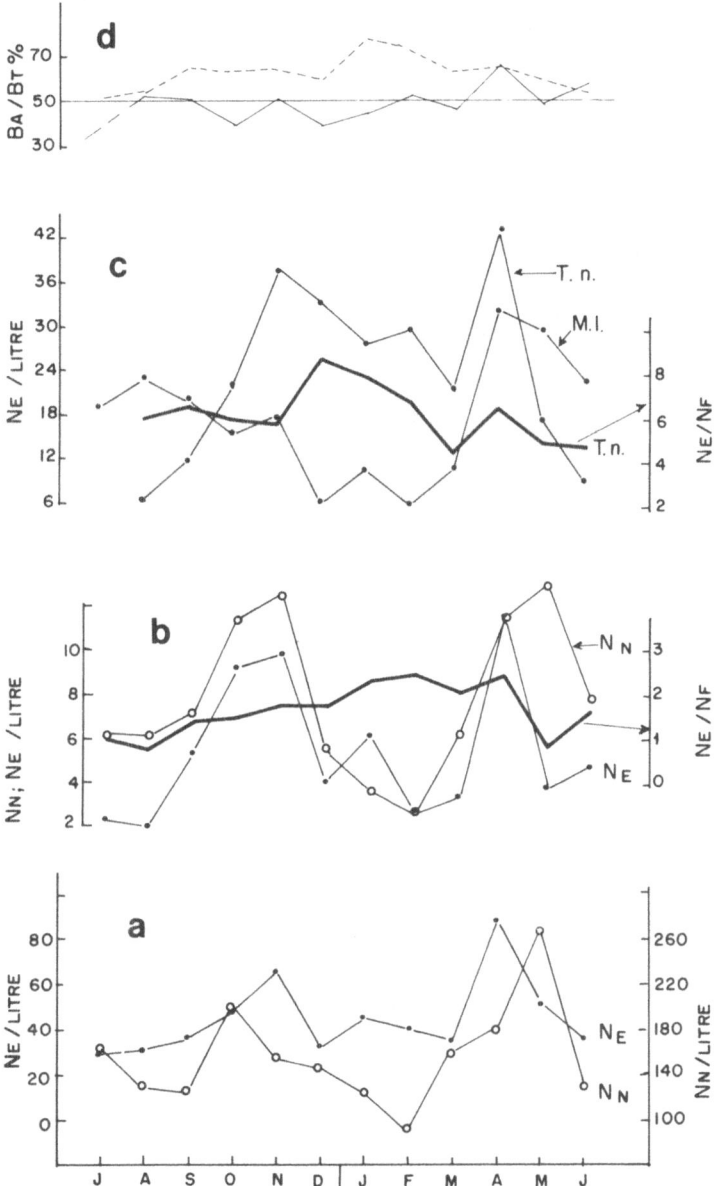

Fig. 1 Variation in the reproduction (number of eggs N_E), fecundity (N_E/N_F), recruitment (number of nauplii N_N) and proportion of adults expressed in terms of weight (B_A/B_T) of Copepods during 1964–1975. (a) Total Copepods, 98% of the nauplii belong to the Cyclopoids on average; (b) Calanoids (*Tropodiaptomus*); (c) Cyclopoids; (d) Cyclopoids (dashes) and Calanoids (lines).

Table 1 Data on zooplankton population structure (percentage of nauplii (N), copepodids (C), adults (A) and young (Y) in relation to the total number) of the main species or groups in the southeastern Archipelago during the period of Normal Chad (1964–1965, 1968) and during the period of 'Lesser Chad' (1973). * = Dominant species; M.m. = *Moina micrura*; D.e. = *Diaphanosoma excisum*; B.l. = *Bosmina longirostris*; D.b. = *Daphnia barbata*; T.i. = *Tropodiaptomus incognitus*; T.g. = *Thermodiaptomus galebi*; T.n. = *Thermocyclops neglectus*; T.i.c. = *Thermodiaptomus incisus circusi*; M.l. = *Mesocyclops leuckarti*.

	CALANOIDS			CYCLOPOIDS			Y(%)					
	N%	C%	A%	N%	C%	A%	Diapt.	Cycl.	M.m	D.e	B.l.	D.b.
1964–65 (annual)	27 (T.i.*+T.g.)	48	25	75 (T.n.+T.i.c.+M.l.)	20	5	75	95	–	–	8	–
1968 (30 days)	58 (T.i.)	26	16	–	–	–	84	–	69	72	56	76
January 73 (15 days)	21 (T.i.+T.g.*)	16	64	62 (T.n.+T.i.c.+M.l.)	24	14	37	86	59	63	64	–
April 73 (15 days)	50 (T.g.)	15	35	81 (T.n.*+M.l.)	11	8	65	92	59	64	–	–

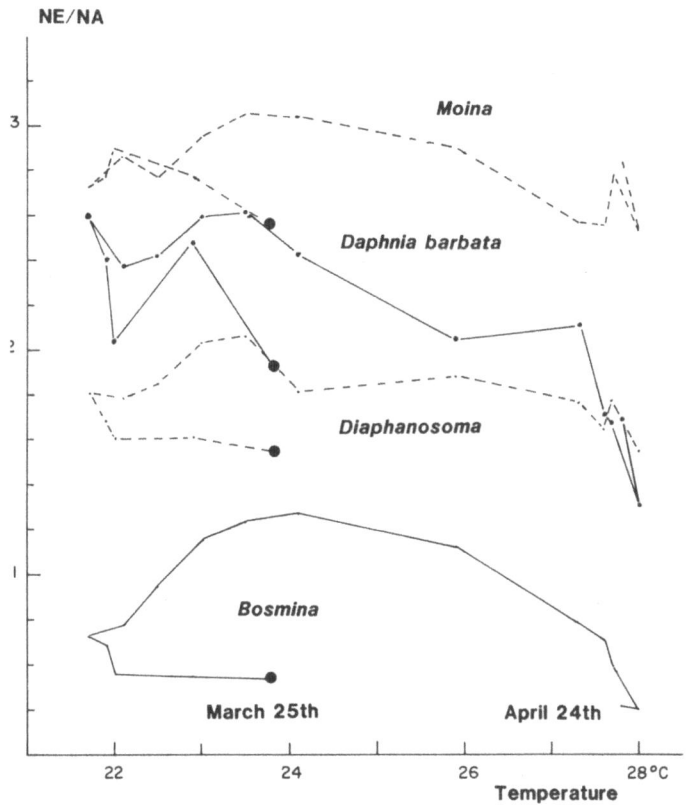

Fig. 2 Variation in the number of eggs per female (N_E/N_A) in relation to the temperature in four species of Cladocera at Melia from March 25th to April 24th, 1968 (Southeastern Archipelago).

It is the model used to describe the variations in the rate of embryonic development ($1/D_e$) in relation to temperature (θ). K is the maximum value $1/D_e$ when temperature increases; $\hat{\theta}$, the optimum temperature of embryonic development is the value of the temperature at the inflection point of the curve; b is a constant (Gras and Saint-Jean 1976).

The median part of the sigmoid curve (Fig. 3) is comparable to a straight line; $1/D_e = 1/S \ (\hat{\theta} - \theta_0)$, where θ_0 is the 'theoretical developmental threshold' and corresponds to the intersection point of the straight line with the *x*-axis and $1/S$ is the slope of the straight line. The temperature range corresponding to this median part whose approximate limits can be fixed at $\hat{\theta} \pm 6°C$, is generally situated in the optimum development zone according to Winberg (1971) and coincides with the annual temperature range of Lake Chad (19 to 30°C) for most species.

The values of the different parameters of the logistic equation and its

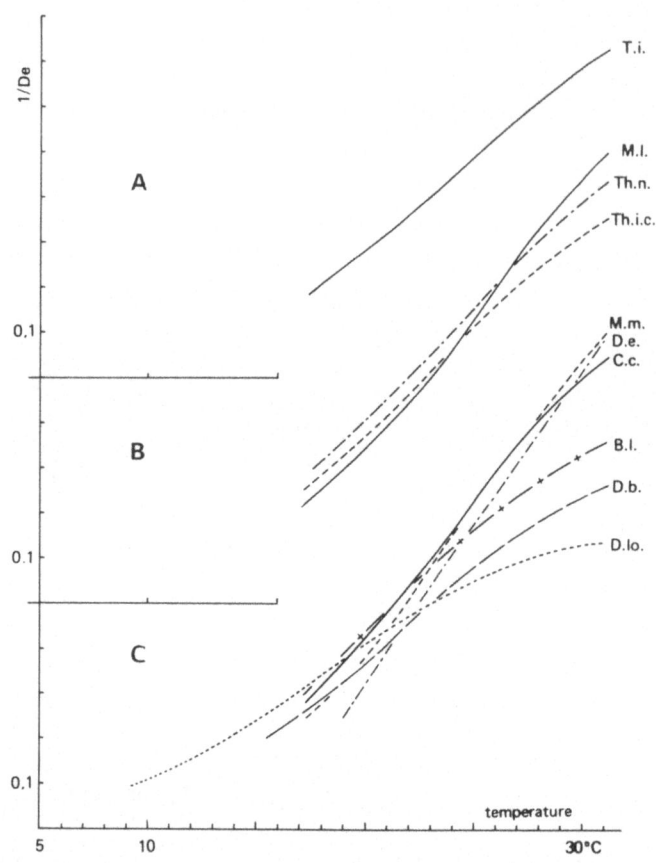

Fig. 3 Embryonic development of the main planktonic crustaceans (logistic curves). T.i. = *Tropodiaptomus incognitus*; M.l. = *Mesocyclops* cf. *leuckarti*; Th.n. = *Thermocyclops neglectus*; Th.i.c. = *Thermodiaptomus incisus circusi*; M.m. = *Moina micrura*; D.e. = *Diaphanosoma excisum*; C.c. = *Ceriodaphnia cornuta*; B.l. = *Bosmina longirostris*; D.b. = *Daphnia barbata*; D.lo. = *Daphnia longispina*.

simplified expression will be found in Table 2 as well as the duration of embryonic development (in hours).

The embryonic development was of similar duration both in the Copepods and the Cladocerans under study; it ranged from 1 to 1.6 days at 30°C, from 1.4 to 2 days at 25°C and from 2.5 to 3 days at 20°C.

The interspecific differences were rather more pronounced at high temperatures. So, at 30°C, three groups could be distinguished:

— species which grew very quickly $(23 < D_e < 25$ h) such as *M. micrura*, *D. excisum*, *C. cornuta* and *M.* cf. *leuckarti*;

— species which grew more slowly $(31 < D_e < 38$ h) such as the three species of *Daphnia*, the two Calanoids and *B. longirostris*;

Table 2 Features of embryonic development of the planktonic crustaceans: parameters of the logistic equation ($\hat{\theta}$, K/2, b) and its linear approximation ($1/D_e = 1/S$ ($\theta - \theta_0$)); duration development (D_e) at different temperatures calculated according to the logistic equation (after Gras and Saint-Jean 1976.)

Species	$1/D_e = 1/S$ ($\theta - \theta_0$) (in days)	$\hat{\theta}$	K/2	b	D_e (hours) 20	25	30°C
B. longirostris	0.0452 (θ–10.5)	21.0	0.475	0.21	56	36	29
D. longispina	0.0309 (θ–6.9)	18.4	0.355	0.19	59	43	37.5
D. barbata	0.0391 (θ–10.4)	22.0	0.455	0.19	65.5	41.5	32.5
D. lumholtzi	0.0328 (θ–9.0)	21.2	0.40	0.18	67	45	36
C. affinis							28.5
C. cornuta	0.058 (θ–12.7)	23.7	0.64	0.20	58	33	24
M. micrura dubia	0.0656 (θ–14.1)	24.6	0.69	0.21	63	33.5	23
D. excisum	0.0685 (θ–15.1)	25.3	0.70	0.21	69.5	35.5	23.5
T. incognitus	0.0406 (θ–13.1)	24.1	0.445	0.20	88	49.5	35
T. incisus circusi	0.0461 (θ–12.0)	23.0	0.505	0.20	67	40	29.5
T. neglectus	0.0476 (θ–11.7)	24.3	0.60	1.175	62.5	38	27.5
M. cf. *Leuckarti*	0.0595 (θ–14.4)	25.4	0.655	0.20	72	38	25.5
T. galebi							33.5

— species with intermediate features such as *T. incisus circusi* and *T. neglectus*. The second species had developmental features similar to those of *Thermocyclops hyalinus consimilis* found at Lake George (Burgis 1971).

Below 20°C, the previous classification was reversed since *B. longirostris* and *D. longispina* had the most rapid growth. These two species whose optimum temperature of embryonic development were lower (Table 2) are cosmopolitan forms that were well represented in the temperate zone and prevailed in Lake Chad during the cool season. On the contrary, the other Cladocera species with a higher optimum developmental temperature and whose optimum zone had shifted towards high temperatures, were more abundant in the warm season (cf. Gras and Saint-Jean 1976).

Moreover, it was observed during experiments that the duration of embryonic development of the Copepods *M. leuckarti* and *T. neglectus* decreased by about 10% in July 1973 at 30°C, compared with August 1968. This difference can be accounted for by the influence of nutritional conditions.

12.1.1.3 *Duration of juvenile development*
— Cladocera

The duration of juvenile development (D_j) was also determined in the laboratory at 29–30°C (August 1968) and at 24°C (January 1969) (Gras and

Saint-Jean 1969, 1978). The mean duration of juvenile development (D_j) of the species studied at 25 and 30°C will be found in Table 3 as well as the duration of the cycle, egg to egg ($D_j + D_e$). It also shows the relative duration of juvenile development ($D_j D_e$) which is the ratio between the duration of juvenile development at the temperature, θ, and the duration of embryonic development at the same temperature. This ratio D_j/D_e did not vary much at 25°C and 30°C.

Roughly the same groups of species as previously identified for embryonic development were found again. The first one was composed of *M. micrura, D. excisum* and *C. cornuta* which had rapid embryonic and juvenile development, the cycle from egg to egg ranging from 73 to 110 h at 25°C, and from 52 to 81 h at 30°C. The second one corresponded to the three species of *Daphnia* whose cycle ranged from 165 to 198 h at 25°C. *B. longirostris* appeared to occupy an intermediate position.

M. micura was characterized by a very rapid juvenile development whose duration barely exceeded the duration of embryonic development ($D_j/D_e = 1.2$). The ratios D_j/D_e which were much higher in *Daphnia* can be compared to those obtained for *Daphnia* species under temperate conditions.

The number of juvenile stages varied with the individuals under laboratory conditions. Generally, it is 2 or 3 in *M. micrura*, 4 in *C. cornuta* and *C. affinis*, 5 in *D. lumholtzi*, 4 to 6 in *D. barbata* and *D. excisum*.

In the natural environment, the evaluation of the number of juvenile stages according to the size structure of the population showed that there were some modifications from 1968 to 1973. So, 3 to 4 stages were observed in *M. micrura* in 1968, 2 to 3 in 1972 and 2 in the Eastern Archipelago in 1973; and 5 to 6

Table 3 Cladocera: mean duration of juvenile development (D_j); relative duration in relation to the duration of embryonic development (D_j/D_e) and duration of the cycle from egg to egg (after Gras and Saint-Jean 1978).

Species	25°C (January 1969)			30°C (August 1968)		
	D_j (hours)	D_j/D_e	Duration of cycle (hours)	D_j (hours)	D_j/D_e	Duration of cycle (hours)
Bosmina longirostris	86.8	2.37	123.8	70.0	2.57	
Daphnia longispina	154.5	3.60	198.5			
Daphnia barbata	132.0	3.14	173.5	102.1	3.26	
Daphnia lumholtzi	120.3	2.63	165.3	104.9	3.01	
Ceriodaphnia affinis					1.97	
Ceriodaphnia cornuta	60.6	1.80	99.6	47.4	1.98	81.4
Moina micrura	40.2	1.20	73.4	28.9	1.26	51.9
Diaphanosoma excisum	75.4	2.13	110.9	54.6	2.32	78.1

stages were observed in *D. excisum* in the same region in 1968, 5 in 1972 and 4 in 1973.

+ Copepods

The results obtained from 4 sets of observations are listed in Table 4. Calanoids and Cyclopoids had a juvenile development (D_p/D_e) identical in duration for a given year: 12.6 to 17.3 in *T. incognitus* and about 18.5 in *T. neglectus* and *M.* cf. *leuckarti* in 1968–69; 7.7 in *T. galebi* and about 6 in *T. neglectus* and *M.* cf. *leuckarti* in 1973. These durations were much longer than those of the Cladocera where D_j/D_e ranged from 1.2 to 3.6. However, it can be pointed out that the values observed in 1973 were much lower than those observed in the Cyclopoids in 1968–69.

The durations of the Copepodid (D_c) and nauplius (D_n) stages were roughly identical in the Cyclopoids $(D_c/D_n \simeq 1)$ but very different in the two Calanoids studied: $D_c/D_n \simeq 4.5$ in *T. incognitus* and 2 in *T. galebi*. The different copepodid stages did not have the same duration in the four species; the first three stages were shorter than the following ones while stage C5 was the longest.

12.1.1.4 *Time interval between two layings*. In Cladocera, the emergence of the young from an egg by incubation often occurs just before the moult and subsequent egg-laying so that the time interval between two layings (D'_e) is roughly equal to the duration of the embryonic development (D_e).

In Lake Chad, the interval between hatching and laying, as estimated in culture, was 20 minutes for *D. excisum*, *C. cornuta*, *C. affinis* and *M. micrura*, 1 h for *D. longispina* and *D. humholtzi*, 3 h for *D. barbata* and 4 h for *B. longirostris*.

In the Copepods, D'_e was much higher than D_e, the mean values of D'_e/D_e ranging from 1.5 to 2.4 in 1968 and 1969 (Table 5) (Gras and Saint-Jean 1976). On the contrary, in 1973, this time interval could be compared to that of the Cladocera (a few hours) probably because of the improved nutritional conditions of the natural environment.

12.1.1.5 *Growth of the Copepods*. Growth expressed as weight was determined for the main species after evaluating the weight, W_i, and the duration, D_i, of the stages at 30°C (Gras and Saint-Jean 1979). The results of the observations conducted on *T. incognitus* in August 1969 and on *T. galebi*, *T. neglectus* and *M. leuckarti* in July 1973 are listed in Table 6 as weight of the different stages (W_i), duration of these stages and daily increase in weight $\Delta W_i/D_i$).

The growth curves corresponding to the above-mentioned experimental data were determined by expressing the duration of embryonic development as a unit of time (or of age) (Fig. 4). In the four species under study, it was estimated that growth was exponential during the nauplius stage. It remained so during the copepodid stage of the Cyclopoids *M.* cf. *leuckarti* and *T. neglectus*, while it assumed a parabolic shape during the copepodid stage of the Calanoids, *T. galebi* and *T. incognitus*.

Table 4 Mean instar duration (days) in Copepods; D_e = embryonic development; D_n = duration of the nauplius stage; D_c = duration of the copepodid stage; D_p = duration of juvenile development ($D_c + D_n$); D_c1 to D_c5 = duration of the five copepodid stages (after Gras and Saint-Jean).

Species	Dates	Temperature	D_e	D_n	D_c1	D_c2	D_c3	D_c4	D_c5	D_c	D_p
Thermocyclops neglectus	01/69	24°	1.71	16.00	←———		11.0	———→	4.5	15.50	31.50
	07/73	30°	1.01	2.93	0.57	0.52	0.51	0.60	0.86	3.06	5.99
Mesocyclops cf. *leuckarti*	01/69	24°	1.77	18.00	←———	9.5	——→	←—	5.5 —→	15.00	33.00
	07/73	30°	0.96	2.66	0.48	0.45	0.48	0.64	0.87	2.92	5.58
Tropodiaptomus incognitus	01/69	24°	2.25	6.50	←———		20.10	———→	10.40	30.50	37.00
	08/68	30°	1.45	3.20	←———		10.00	———→	5.00	15.00	18.20
	08/69	30°	1.47	4.84	2.18	2.45	3.50	5.70	6.80	20.63	25.47
Thermodiaptomus galebi	07/73	30°	1.40	3.45	0.99	0.97	1.08	1.83	2.50	7.37	10.82

Table 5 Time interval between two successive layings (D'_e) in Copepods. Values observed in laboratory during January 1969 and August 1968. The range values are mentioned in brackets (after Gras and Saint-Jean 1976).

Species	1969 January (24–25°C)		1968 August (29–30°C)	
	D'_e (hours)	D'_e/D_e	D'_e (hours)	D'_e/D_e
T. incognitus	99 (63–109)	2	53 (46–77)	1.5
M. cf. *leuckarti*	57 (47–64)	1.5	38 (30–40)	1.5
Th. neglectus	62 (44–86)	1.5	54 (41; 46; 76)	2
Th. incisus circusi	90 (73–109)	2.2	72 (45–95)	2.4

Therefore, the growth equations for the copepodid stages are as follows:

$Cyclopoids$ *M.* cf. *leuckarti*: $W = 0.155\, e^{1.086(Z-2.77)}$

$T.$ *neglectus*: $W = 0.155\, e^{0.866(Z-2.9)}$

Z is the age expressed in units equal to the duration of embryonic development, D_e (in days); $Z = 0$ at birth, $2.77 = D_n/D_e$ of *M. leuckarti* and $2.9 = D_n/D_e$ of *T. neglectus*. The value 0.155 correspond to the initial weight (in µg) of the Copepodids *M. leuckarti* and *T. neglectus*.

$Calanoids$ *T. galebi*: $W = 0.247\,(Z-1.06)^{2.1}$

T. incognitus: $W = 0.249\,(Z-1.5)^{1.3}$

Given a temperature other than 30°C and the same value for D_p/D_e, the daily increases in weight were calculated after evaluating D_i at this temperature. For this evaluation it was assumed that the relative duration, D_i/D_e, of the stages remains constant and knowing D_e from the laws of variation of D_e previously determined in relation to the temperature. Insofar as the D_i/D_e are assumed to be constant, the shape of the curve does not change regardless of temperature (Gras and Saint-Jean 1981).

12.1.1.6 *Conclusions.* The laboratory observations showed a very great acceleration in juvenile development of the Cyclopoids from 1969 to 1973. The ratio D_p/D_e as well as the relative duration of the nauplius (D_n/D_e) and copepodid (D_c/D_e) stages decrease by about three times between the two above-mentioned sets of observations.

In the Calanoids, although the comparison seems to be less convincing because the species studied over the two periods were not the same, it also appears that there was an acceleration of development. The relative duration of the development observed in 1973 ($D_p/D_e = 7.7$) for *Thermodiaptomus* was, in fact, much shorter than the lowest value observed for *Tropodiaptomus* (12.6) in 1968–69. A more justified comparison can be made if the copepodid stages only

Table 6 Planktonic copepods: mean dry weight of an individual ($\overline{W}_i$ in μg) for each stage; absolute duration of these stages (D_i in days) and daily increase in weight ($\Delta W_i/D_i$). E = embryonal stage; N = nauplius stage; C1 to C5 = copepodid stages; Ac = adults in growth; A = adult. Results obtained in August 1969 for *T. incognitus* and in July 1973 for the other species (after Gras and Saint-Jean).

	Tropodiaptomus incognitus			*Thermodiaptomus galebi*			*Thermocyclops neglectus*			*Mesocyclops* cf. *leuckarti*		
	$\overline{W}_i$ (μg)	D_i (days)	$\Delta W_i/D_i$	$\overline{W}_i$ (μg)	D_i (days)	$\Delta W_i/D_i$	$\overline{W}_i$ (μg)	D_i (days)	$\Delta W_i/D_i$	$\overline{W}_i$ (μg)	D_i (days)	$\Delta W_i/D_i$
E	0.20	1.47	0	0.08	1.40		0.015	1.01		0.015	0.96	
N	0.40	4.84	0.09	0.25	3.45	0.15	0.05	2.93	0.05	0.05	2.66	0.05
C1	0.85	2.18	0.23	0.85	0.99	0.59	0.20	0.57	0.17	0.20	0.48	0.23
C2	1.50	2.45	0.34	1.50	0.97	0.99	0.30	0.52	0.27	0.35	0.45	0.40
C3	2.70	3.50	0.38	2.85	1.08	1.32	0.50	0.51	0.42	0.60	0.48	0.67
C4	4.60	5.70	0.43	5.30	1.83	1.81	0.85	0.60	0.67	1.10	0.64	1.28
C5	7.30	6.80	0.48	9.70	2.50	2.55	1.35	0.86	1.27	2.35	0.87	3.07
Ac		4.38	0.52		1.41	3.25						
A	11.00			17.40			2.15			4.55		

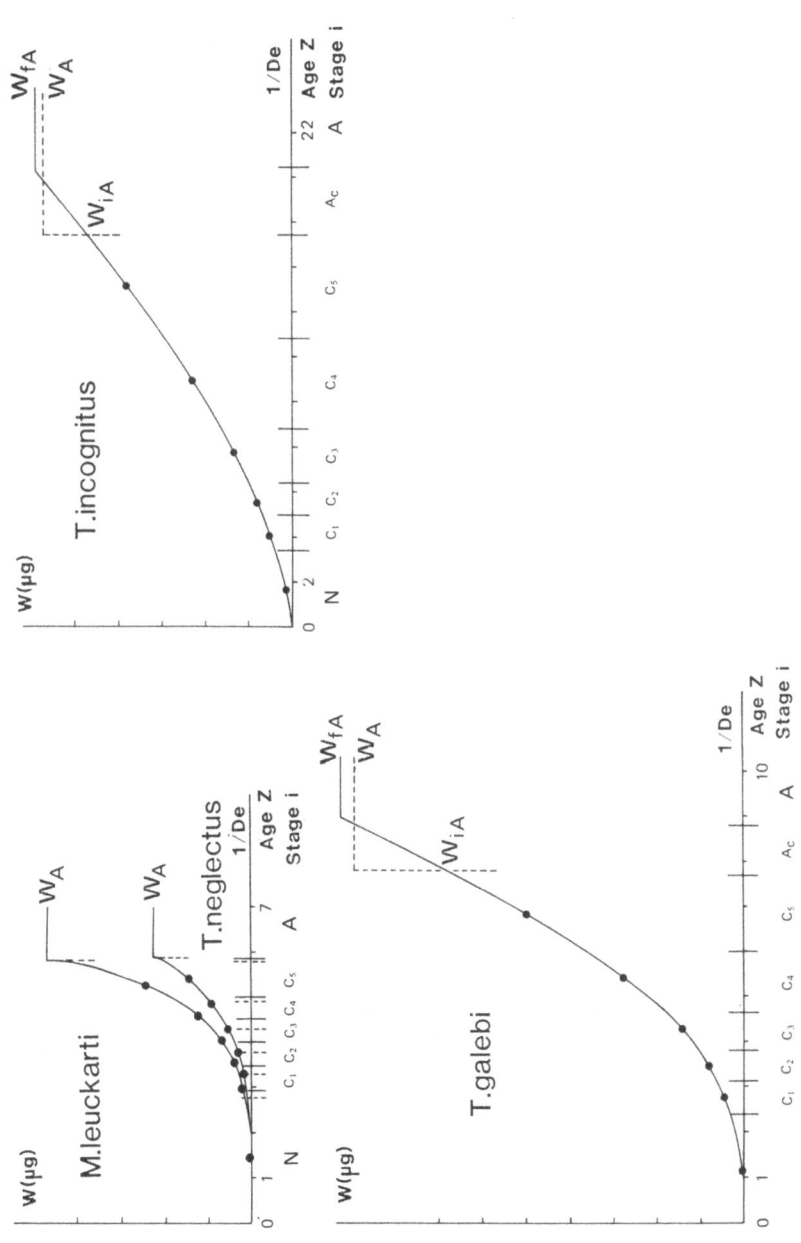

Fig. 4 Growth in dry weight for four species of Copepods. In the range of ages, the unit is equal to $1/D_e$ (D_e, in days), W_A = mean weight of adults; W_{iA} and W_{fA} = initial and final weight of adults; N = nauplii; C_1–C_5 = copepodid stages; A_c = adults in growth; A = adults reaching the end of their growth.

of each species are considered. The value of 5.3 and 10.4 respectively are even more different (Gras and Saint-Jean 1981).

With a simultaneous analysis of the conditions of the natural populations and, above all, a comparison between the populations sampled at Melia in March–April, 1968 and those sampled at Tchongolerom in 1972–1973, significant modifications seem to be revealed which are in keeping with the previous acceleration of development. A reduction in the number of juvenile stages was observed in the two main species of Cladocera, *M. micrura* and *D. excisum*, probably along with a decrease in the duration of juvenile development. So, for *M. micrura* D_j/D_e would move from 1.8 in 1968 to 0.7 in 1973 and for *D. excisum* from 3.2 to 1.6 (Gras and Saint-Jean 1978).

Moreover, there was an increase in clutch size in all the filtering or mixed diet species (Table 7). A reduction in the time interval between layings could also be noted in the Cyclopoids from the laboratory observations of 1973. All these modifications of biological parameters could have resulted from an improvement in nutritional conditions following an increase in the algal biomass per unit volume during the drying up period of the lake (Iltis 1977).

During 1973, from the population point of view, there was an increase in the proportion of adult Copepods as well as the Cladocerans, *M. micrura* and *D. excisum*. The above-mentioned modifications of decrease in D_j/D_e, and increase in the size of layings account for this. This phenomenon may occur along with an increase in the birth rate (Gras and Saint-Jean 1978).

12.1.2   *Evaluation of production*

12.1.2.1   *Methods.* In Cladocera, production was calculated from the formula: $P = bB$ where B is the biomass expressed in mg of dry weight per m^3, and b is the birth rate of the population expressed by the relation:

$$b = \frac{1}{D_e} \ln \left(1 + \frac{N_E}{N_T} \right)$$

D_e being the duration expressed in days of the embryonic development and N_E/N_T the ratio between the number of embryos (N_E) and total free individuals (N_T), found in the sample of the population considered.

In Copepods, the formula is:

$$\sum \frac{\Delta W_i}{D_i} \cdot N_i \quad + \quad \frac{W_E \cdot N_E}{D_e}$$

| Production of the nauplius and copepodid stages | Production of adults in the form of eggs |

398

Table 7 Size of the layings (number of eggs/number of egg-bearing females) in the main species of planktonic crustaceans during the period of Normal Chad (1964–65 and 1968) and during the period before the drying up (1972–73) in the different stations of the eastern archipelago. For the years 1964–65, it is the average obtained in the five stations sampled each month. In 1968, it is the average of 16 samples taken in March–April. In 1972 and 1973, it is the average of 2 to 8 samples (after Gras and Saint-Jean).

Dates	Archipelago 1964–65	Melia 1968	Tchongolerom					Tchongolerom–Melia			
			May 1972	June 1972	August 1972	January 1973	March 1973	June 1972	August 1972	January 1973	March 1973
Tropodiaptomus incognitus	3.8	3.5	6.8	7.0	5.9	9.3	–	8.0	3.2	7.5	–
Thermodiaptomus galebi	9.1	5.0	12.1	11.9	11.8	22.0	27.8	15.2	9.8	16.9	29.1
Thermocyclops neglectus	15.3	–	14.0	16.4	18.0	22.7	23.3	–	13.8	16.2	24.7
Mesocyclops leuckarti	42.0	–	44.9	37.7	37.2	34.4	45.0	–	36.6	47.7	42.4
Moina micrura dubia	–	3.2	4.0	4.0	4.5	6.4	5.7	4.9	3.6	3.9	5.3
Diaphanosoma excisum	–	2.1	2.3	2.6	2.9	3.2	3.1	3.3	2.6	3.0	3.8
Bosmina longirostris	–	1.0	1.9	1.7	–	3.4	–	–	8	3.1	–

ΔW_i is the increase in weight during the stage i with a duration D_i and a number N_i, W_E is the weight of eggs with the number N_E.

12.1.2.2 *Results*. The daily production and the daily production rate $(P/\bar{B})$ were calculated from samples taken over several days and at different times (and temperatures) at three stations situated in the Eastern Archipelago of the lake (Table 8).

It should be noted that the production rate varied considerably with species, the greatest difference being observed in January 1973 when the production rate of *Moina* was 20 (Tchongolerom) or 27 times (Tchongolerom−Melia) higher than that of the Calanoids. The classification of the species according to decreasing production rate is:

1. *Moina* $0.27 < P/\bar{B} < 0.87$
2. *Diaphanosoma* ⎫
      Bosmina   ⎬ $0.16 < P/\bar{B} < 0.50$
      Daphnia   ⎭
3. Cyclopoids $0.10 < P/\bar{B} < 0.27$
4. Calanoids $0.01 < P/\bar{B} < 0.07$

On the other hand, the distribution of production between the different developmental instars was irregular as shown by the biomass distribution and the variations in production rate according to the stages (Table 9). Given their low production rate (ranging from 0.003 to 0.033), the adult Copepods which represented a high percentage of the biomass, thus represented only a small percentage of the production of their populations (20 to 50% for Calanoids and 5 to 10% for Cyclopoids). These distributions, especially in the Calanoids, showed great variability which can be accounted for (Gras and Saint-Jean, in preparation). The analysis of distribution of the five copepodid stages revealed a sharp difference between *Mecocyclops* cf. *leuckarti* and *Thermocyclops neglectus*. The first 2–3 stages contributing a higher percentage of production in *Mesocyclops* (Fig. 5) because of their greater abundance. In order to make them independent of temperature, production and production rate are expressed as D_e, the duration of embryonic development (in the case of monospecific populations) or U, a unit combining the D_e values of species comprising the community, as a function of time.

This expression shows (Fig. 6) that the production rate of the population remained rather similar in the different series of observations ranging from 0.18 (at Tchongolerom in May, 1972) to 0.38 (at Tchongolerom-Melia in March, 1973). The main cause of variation in this rate was the respective abundance of the Calanoids (minimum production rate) and the Cladocera, especially *Moina* (maximum rate).

The production of the crustacean populations was also estimated over a year during the period of high water in the Eastern Archipelago (1964–65) and for

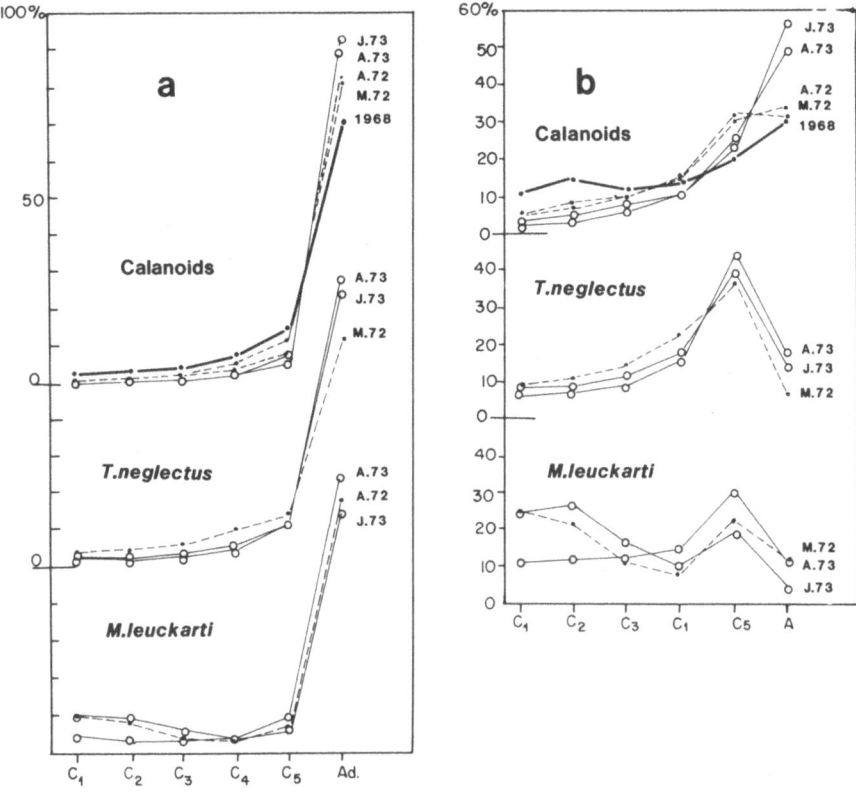

Fig. 5 Distribution of the biomasses (a) and production (b) according to the developmental instars in Copepods during 1968 and in the series of May (M) August (A) 1972 and, January–February (J) and March–April (A) 1973 at Tchongolerom (Eastern Archipelago).

the whole lake in the cold season (February–March 1971), with the production rates observed at Melia in March–April 1968 (Tables 10 and 11).

In the three large groups, the Cladocera, Calanoids and Cyclopoids as well as the total community (Fig. 7), the production rates were naturally at minimum level during the cold season. The variation was regular and rather low in amplitude over the year, the ratio between maximum and minimum values only reaching 2.45 (0.27/0.11). Moreover, spatial distribution was not very heterogeneous if each main region is considered individually like the distribution in biomasses observed in 1971. Production rates were of equal value everywhere, but the production (per m^3) was much lower in the Southern Open Water than in the rest of the lake (Table 11) (as was biomass).

If the annual community production rate observed at a mean temperature of 26.2°C, that is to say 63.7 (Table 10) during 1964–65, is applied to the biomass

Table 8 Mean biomass, B, (in mg dry weight per m³), mean daily production, P_j (in mg dry weight per m³) and daily P/B ratio of zooplankton in three stations of the Eastern Archipelago during different periods. M.m. = *Moina micrura*; D.e. = *Diaphanosoma excisum*; B.l. = *Bosmina longirostris*; D. = *Daphnia*.

Date	Mean temperature (in °C)		Cladocera				Cladocera (total)	Calanoids (total)	Cyclopoids (total)	Total
			M.m	D.e	B.l	D.				
March-April	24.5	B	22.8	18.6	90.0	9.7	141.1	185.5	65.4	392.0
		P_j	10.8	5.3	18.3	2.0	36.4	13.1	7.3	56.8
		P/B	0.47	0.28	0.20	0.21	0.26	0.07	0.11	0.15
May 1972	29.5	B	17.4	30.8	3.4	0	51.6	236.0	77.3	364.9
		P_j	12.0	9.1	1.2		22.3	17.1	16.9	55.4
		P/B	0.69	0.30	0.34		0.43	0.07	0.22	0.15
June 1972	30.4	B	17.9	29.6	0.3	0	47.8	149.4	40.9	238.1
		P_j	16.3	14.7	0.1		31.1	10.5	9.5	51.1
		P/B	0.86	0.50	0.35		0.64	0.07	0.23	0.21
August 1972	28.6	B	28.5	34.3	ε	0	62.8	142.2	16.3	221.3
		P_j	22.9	16.3	–		39.2	9.9	4.0	53.1
		P/B	0.80	0.48	–		0.63	0.07	0.24	0.24
January-February 1973	19.1	B	10.5	4.2	2.0	0	16.7	139.3	43.6	199.6
		P_j	4.2	0.8	0.5		5.5	2.7	4.3	12.5
		P/B	0.40	0.19	0.26		0.33	0.02	0.10	0.06
March-April 1973	25.6	B	6.1	3.3	0	0	9.4	58.2	178.6	246.2
		P_j	5.3	1.3			6.6	3.1	35.7	45.4
		P/B	0.87	0.41			0.70	0.05	0.20	0.18

Melia

Tchongolerom

Tchongolerom-Melia

Period					ε					
August 1972	28.6	B	22.9	30.4		3.6	56.9	99.5	21.7	178.1
		P_j	14.7	12.0		1.2	27.9	6.0	5.6	39.5
		P/B	0.64	0.40		0.33	0.49	0.06	0.26	0.22
January-February 1973	19.1	B	26.7	3.6	5.1	0	35.4	75.6	101.3	212.3
		P_j	7.3	0.6	1.1		9.0	0.8	12.9	22.7
		P/B	0.27	0.16	0.21		0.25	0.01	0.13	0.11
March-April 1973	25.6	B	42.1	10.2	0	0.3	52.6	112.0	245.9	410.5
		P_j	28.3	3.2			31.5	6.7	60.6	98.8
		P/B	0.67	0.31			0.60	0.06	0.25	0.24

Table 9 Distribution of the production and biomass between the nauplii (N), the Copepodids (C) and the adults (A) of Calanoids and Cyclopoids in the populations sampled during 1968, 1972 and 1973 in the stations of the Eastern Archipelago of Lake Chad. The production is expressed as mg m^{-3} per unit of biological time. P_{AE} is the adult production in the form of eggs. In the Cyclopoids, $P_A = P_{AE}$; in the Calanoids, P_{AE} must be added with a term corresponding to a production by increase in body weight.

	Date	Calanoids					Cyclopoids			
		P_N (% P_T) / B_N (% B_T)	P_C (%) / B_C (%)	P_A (%) / B_A (%)	P_{AE} (mg)	P_T (mg) / B_T (mg)	P_N (% P_T) / B_N (% B_T)	P_C (%) / B_C (%)	P_{AE} (%) / B_A (%)	P_T (mg) / B_T (mg)
Melia	March-April 1968 (n=16) — P	29.9	49.5	20.7	2.2	25.9		88.6[a]	11.3[a]	
	— B	8.5	27.8	63.7		185.5		16.3	40.9	
	May 1972 (n=5) — P	13.4	59.6	27.0	3.0	25.5	26.9	67.3	5.8	19.6
	— B	2.2	19.7	78.1		236.0	13.2	31.2	55.6	77.3
	June 1972 (n=2) — P	18.7	49.9	31.4	2.1	15.0	30.3	61.8	7.9	10.4
	— B	2.8	14.4	82.8		149.4	15.7	27.4	56.9	40.9
	August 1972 (n=5) — P	17.8	54.7	27.4	2.1	15.6				
	— B	2.3	18.2	79.5		142.2				
Tchongolerom	January-February 1973 (n=8) — P	2.7	42.7	54.6	4.3	11.1	13.4	75.0	11.6	12.4
	— B	0.3	7.5	92.2		139.3	4.8	23.9	71.3	43.6
	March-April 1973 (n=8) — P	9.5	46.3	44.2	1.6	6.2	33.1	55.7	11.2	55.2
	— B	1.2	10.7	88.1		58.2	12.8	19.4	67.8	178.6

Tchongolerom-Melia									
August	12.2	69.8	18.0	0.7	9.7				3.3
1972	2.1	29.9	68.0		99.5				101.3
(n = 5)									
January–February	18	50.0	32.0	0.4	2.5	18.7	78.0		3.3
1973	0.8	5.7	93.5		75.6	8.6	31.9		59.5
(n = 8)									
March–April	18.5	39.2	42.3	4.2	14.5	36.4	57.2	6.4	97.2
1973	2.9	10.8	86.3		112.0	14.8	24.3	60.9	245.9
(n = 8)									

[a] % of $P_C + P_N$ or $B_C + B_A$.

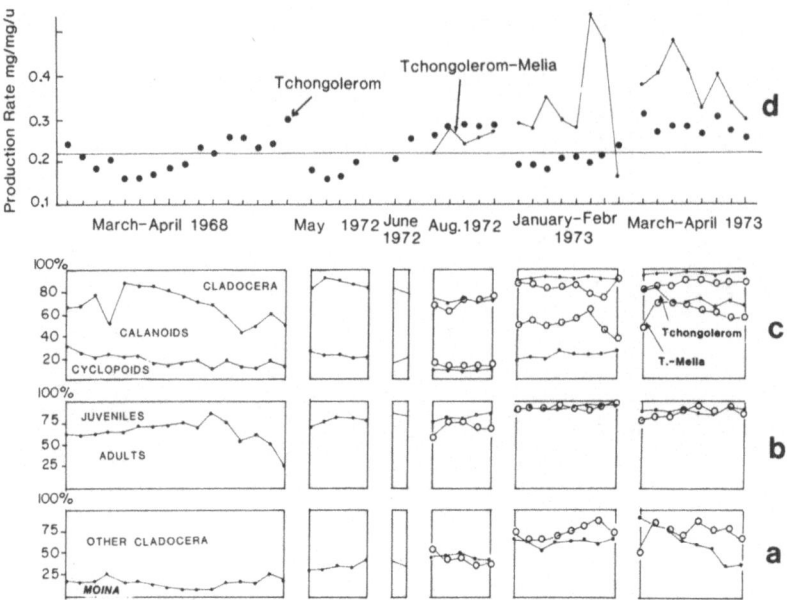

Fig. 6 Variations in the different series of samples during 1968, 1972 and 1973; (a) in the proportion of *Moina* in relation to the other Cladocerans; (b) in the proportions of young and adults expressed in terms of weight in the calanoid populations; (c) in proportions of Cladocerans, Cyclopoids and Calanoids; (d) in the production rate of the population per unit of biological time (unit designated by U which is, in each series, a weighted average of the values of the duration in the embryonic stage of the species belonging to the population).

values observed in 1971, the three major natural regions of the lake had the following production values:
— 567 kg ha^{-1} year^{-1} (dry weight) for the Northern basin;
— 440 kg ha^{-1} year^{-1} for the Southern Archipelago and the Great Barrier;
— 159 kg ha^{-1} year^{-1} for the Southern Open Water.

The annual production can be estimated as 780 000 tons dry weight, or an average of 433 kg ha^{-1} during the period of 'Normal Chad', for the whole lake (about 18 000 km^2).

There are only a few studies of zooplankton production in tropical Africa. Burgis (1971) quotes daily production rates for *Thermocyclops hyalinus* in Lake George ranging from 0.08 to 0.26 according to the method of calculation used. The first value which seems to be retained by the author is considerably lower than that of 0.17 which was calculated for Cyclopoids in Lake Chad during 1964–1965 at a mean temperature of 26.3°C similar to that recorded in Lake George (25 to 26°C). In Lake Turkuna (Ferguson 1975), the production rate N_E/N_T. De for *Tropodiaptomus banforanus* ranging from 0.04 to 0.56 is considerably higher than those calculated for Calanoids in Lake Chad since the latter

406

Table 10 Monthly values of the daily production, biomass and the production rate in the central zone of the Archipelago in the south basin during 1964–1965 (averages of 5 stations).

Months		1964						1965						P annual ($mg\ m^{-3}$)	B annual	P/B annual
		07	08	09	10	11	12	01	02	03	04	05	06			
Moina micrura	P_j	13.9	9.1	10.2	5.3	9.3	4.4	2.4	3.2	6.1	20.4	11.7	2.5	3148.1	16.5	190.8
Diaphanosoma excisum	P_j	5.5	6.4	15.3	10.9	13.1	3.9	1.3	1.7	8.1	11.0	18.7	10.7	3242.4	27.5	117.9
Bosmina longirostris	P_j	5.7	3.4	1.6	6.7	7.2	3.6	5.7	10.5	5.3	11.4	5.4	3.3	2123.1	29.6	71.7
Daphnia	P_j	0.2	1.3	1.5	4.9	5.9	6.2	10.2	16.2	10.2	10.7	1.4	0.4	2101.8	27.8	75.6
Ceriodaphnia	P_j	19.1	15.2	11.8	13.9	4.9	9.9	3.4	1.4	1.1	15.1	11.8	7.3	3494.9	23.1	151.3
Cladocera	B	99.7	84.7	94.1	114.1	137.0	126.5	136.2	182.5	122.2	199.1	130.5	67.5		124.5	
(total)	P_j	44.4	35.4	40.4	41.7	40.4	28.0	23.0	33.0	30.8	68.6	49.0	29.2	14 110.3		113.3
	P/B	0.44	0.43	0.43	0.37	0.29	0.22	0.17	0.18	0.25	0.34	0.38	0.43			
Calanoids	B	101.4	110.8	138.6	220.3	224.1	108.1	83.7	60.6	118.1	238.0	273.2	145.6		151.9	
(total)	P	9.2	8.6	9.7	15.2	13.3	4.3	2.3	2.0	6.4	14.3	17.7	12.2	3504.0		
	P/B	0.09	0.08	0.07	0.07	0.06	0.04	0.03	0.03	0.05	0.06	0.07	0.08			23.1
Cyclopoids	B	41.0	39.2	44.9	55.9	79.6	41.3	58.7	53.6	73.6	77.4	73.4	38.4		56.4	
(total)	P	10.5	7.9	9.2	13.7	10.4	5.8	6.5	6.3	11.9	10.6	14.2	10.7	3580		
	P/B	0.26	0.20	0.21	0.24	0.13	0.14	0.11	0.12	0.16	0.14	0.19	0.28			63.5
Total	B	242.1	234.7	277.6	390.3	440.7	275.9	278.6	296.6	313.9	514.5	477.1	251.5			
production	P	64.1	51.9	59.3	70.6	64.1	38.1	31.8	41.3	49.1	93.5	80.9	52.1	21 194.3	332.8	63.7
mean daily P/B ratio	P/B	0.27	0.21	0.21	0.18	0.15	0.14	0.11	0.14	0.16	0.18	0.17	0.21			
Temperature in °C		29.3	28.4	28.8	28.5	25.6	22.2	20.8	21.8	24.5	27.0	27.6	29.9	mean annual temperature		26.2

Table 11 Mean daily production of the zooplankton (in mg m^{-3}) for each zone (A,B,...) and ratio $\Sigma P_j/\Sigma B$ calculated at a temperature of 22°C in the different regions of the lake sampled in February–March, 1971 (see text); N = nauplii; C = copepodids; A = adults.

	Southern open waters		Northern open waters and Archipelago				Southern Archipelago and reed islands				
	A (10)	B (10)	C (10)	D (10)	E (10)	F (10)	G (10)	H (9)	I (10)	J (5)	K (2)
M. micrura	4.5	1.4	2.4	0.7	0.2	1.5	7.6	1.2	18.5	17.8	9.2
D. excisum	2.0	1.1	1.8	0.1	0.2	1.5	4.4	3.9	16.5	15.1	6.7
B. longirostris	1.0	0.7	8.6	4.6	3.3	11.7	10.8	19.6	9.6	4.0	13.6
Daphnia	1.3	0.3	2.9	15.8	12.6	15.2	4.4	0.1	4.6	3.6	29.3
Ceriodaphnia	0.2	+	0	0	1.0	0	+	0	2.0	5.2	0
N. Cyclopoids	0.8	0.8	1.1	1.0	1.2	1.0	1.5	1.8	2.4	1.7	1.2
C. Cyclopoids	4.5	2.3	4.5	6.0	5.9	4.7	3.7	6.7	8.0	7.2	6.3
A. Cyclopoids	0.1	0.3	0.5	0.4	0.4	0.4	0.5	0.8	0.8	0.4	0.7
N. Calanoids	0.3	0.1	0.2	0.2	0.7	0.3	0.7	1.1	0.6	0.3	1.0
C. Calanoids	1.4	2.2	1.9	0.8	3.5	0.6	1.3	1.5	2.3	3.1	1.5
A. Calanoids	0.2	0.4	0.7	0.4	1.3	0.3	0.7	0.7	0.7	0.8	0.6
Cladocera	9.0	3.5	15.7	21.2	17.3	29.9	27.2	24.8	51.2	45.7	59.0
Cyclopoids	5.4	3.4	6.1	7.4	7.5	6.1	5.7	9.3	11.2	9.3	8.2
Calanoids	1.9	2.7	2.8	1.4	5.5	1.2	2.7	3.3	3.6	4.2	3.1
Total P (mg m^{-3})	16.3	9.6	24.6	30.0	30.3	37.2	35.6	37.4	66.0	59.2	70.3
$\Sigma P/\Sigma B$	0.18	0.10	0.13	0.16	0.11	0.17	0.15	0.14	0.18	0.19	0.18

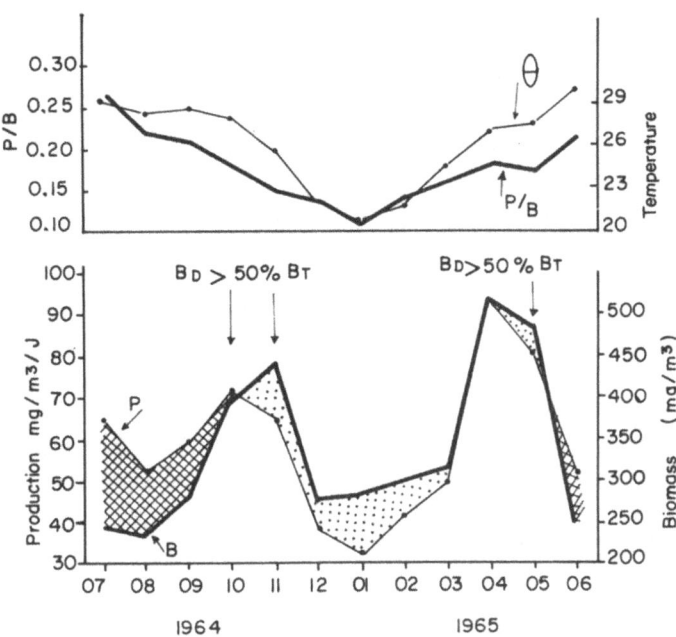

Fig. 7 Variations in the biomass (B), the production (P) and the production rate (P/B) of the planktonic microcrustaceans in the Eastern Archipelago during 1964–1965. B_D = biomass of the Calanoids; B_T = total biomass.

was 0.06 in 1964–1965 and ranged from 0.01 to 0.07 in 1972–1973. These variations are probably due to differences in the populations and the ecological conditions, but they also result from the different methods used.

12.1.2.3 *Between year variability.* Referring to the measurements of production made from 1968 to 1973, an increase in the production rate per unit D_e was observed in the Cladocera from 1968 to March 1973, especially in *Moina* at the Tchongolerom station. In the Calanoids, this rate was stabilized at about a mean value of 0.10. But it increased sharply in the Cyclopoids and Cladocera. In these three groups, there was a simultaneous increase in the fecundity of females (N_E/N_F) and probably a decrease in the duration of juvenile development leading to an increase in production rate, but also an increase in the proportion of adults that decreased this rate at least in the Copepods. The increase in fecundity and the acceleration in development observed in 1973 may have resulted from an improvement in nutritional conditions. As already shown, the consequences of these modifications varied with the populations: increase in the production rate of the Cladocera and Cyclopoids, stability of the Calanoids. It is interesting to note that the increase was greatest in species

409

(*Moina, Diaphanosoma* and Cyclopoids) where the increase in proportion of adult weight was lowest and that there was stability in the Calanoids where both the proportion of adults and this increase were highest.

A number of observations (Gras and Saint-Jean, in preparation) suggest that the previous variations represent a change, following the decrease in the water level preceding the drying up of the lake. This change was part of significant alterations in the composition of the populations and applied only to the region of the Eastern Archipelago. Between the periods of high and low waters the changes occurring in the composition were:

— a big decrease in the biomass of Cladocera that corresponded to the apparent disappearance of the genera *Daphnia* and *Bosmina*; while it was about 130 mg m^{-3} (37% of the total) at high water, it was only 40 mg (16.6%) during 1972–1973;

— during the 1972–1973 period, the substitution of *Tropodiaptomus* prevalent at high water by *Thermodiaptomus* which remained alone in March 1973 followed by the accelerated disappearance of this group just before the drying up. The Calanoids which represented 45% of the biomass on average during the period of high water represented only a quarter of it in March, 1973 and disappeared after the drying up period;

— an increase in the number of Cyclopoids, especially *Thermocyclops*.

The biomass of Cladocera was reduced by the biomass of the genera that disappeared: *Daphnia-Bosmina*. The remaining species, *Moina* and *Diaphanosoma*, approximately maintained their biomass but did not fill the gap left by the other genera. Therefore, at low water, only the most productive species remained and moreover, their production rate increased. The same phenomenon seemed to occur in the Copepods with a decrease in Calanoids and an increase in the number of *Thermocyclops*. On the whole, a relative stability in the production rate of the population was reached as already indicated.

With the above-mentioned changes in community composition, two states of production can be distinguished in this region of the lake. The first was during the period of high water up to 1972, when the Cladocera predominated forming 35 to 40% of the biomass and 2/3 of the production. The second corresponded to the low water when they contributed only about 40% of the production, that is to say a percentage similar to that of the Cyclopoids whose importance increased.

12.2 Benthos

12.2.1 *Biology of the main species and population structure*

The biology of some species and seasonal changes in their population structure must be studied before evaluating the production of the natural populations through classical methods. Observations on the benthic fauna of Lake Chad have referred mainly to the molluscs (Lévêque 1973a), although some data about Chironomids (Dejoux 1976) and Oligochaetes (Dejoux et al. 1969) are also available.

12.2.1.1 *Molluscs.* Quantitative samples were taken over a year from different biotopes of three sites on the lake (Bol, Samia, Baga Kawa) as well as in the delta of the Shari. The size structure of the populations of the main species was determined each time a sample was taken and the dimensions considered were height of shell for Prosobranchia and its length for Lamellibranchia. From the analysis of these demographic measurements some data were obtained on the biology of the species. They were supported by observations from *in situ* cultures in boxes containing previously sifted sediment, covered with a mosquito net and immersed on the bottom. Individuals of a known size were put into these cultures and measured at more or less regular intervals (Lévêque 1971).

Corbicula africana has a seasonal reproductive cycle that was shown by the analysis of its size structure at various stations (Lévêque 1973). However, the juveniles in the population occurred at the beginning (Bol) or the end of the cool season (Samia, Shari delta) according to the sites, and the reproductive season lasted several months. The life-span in culture seemed to range from 1 to 2 years under the usual lacustrine conditions, but it could be longer in certain particular environments. It was thus possible to follow the growth of a cohort in the Shari delta from December, 1967 when it was one year old, up to 1969 when it was three years, with the biggest individuals reaching 20 mm. In 1965, some larger *Corbicula* were collected in a drying backwater situated close to the Shari delta. The factors limiting the life span and growth of *Corbicula* in the lake are not known.

Bellamya unicolor is a viviparous and dioecious species whose juveniles in the distal end of the uterus at birth are about 3 mm. Some biological aspects of this species were studied with *in situ* cultures. It was thus possible to follow growth and specify the age at first reproduction (Fig. 8) which was about two and a half months in the warm season (28–30°C) and three and a half months in the cool season (20°C). The survival curves (Fig. 9) showed that the maximum life span was 15 months, 50% of the individuals disappearing between 5 and 7 months. Reproduction was continuous, but fecundity followed a seasonal cycle with a maximum in the warm season and a minimum in the cool season. The number of births in culture expressed monthly for each adult individual (mixed sexes) ranged from 7 to 1.5 (Fig. 10). In the size structure of the natural populations, both an adult and a juvenile pattern could be observed, clearly distinct throughout the year, suggesting a high death rate in the latter, since recruitment was continuous.

Though little is known about *Cleopatra bulimoides* the demographic struc-

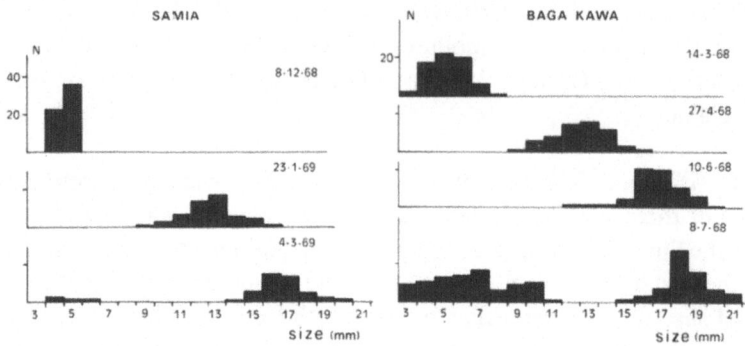

Fig. 8 Bellamya unicolor: changes in size structure of *in situ* cultures showing the growth and occurrence of the first births (Lévêque 1973).

tures showed that reproduction could occur throughout the year with maximum values during periods that varied according to the stations. Since the adult size was reached three or four months after birth in culture it can be assumed by comparison with *B. unicolor* that first reproduction occurs about 3 months after birth. The maximum life-span appeared to range from 1 to 2 years and the largest sizes observed were 14–15 mm.

Melania tuberculata is considered ovoviviparous and parthenogenetic (Morrison 1954). According to its demographic structure, reproduction occurred throughout the year; however, with a major decrease in the cool season. In cultures first reproduction occur about two and a half months after birth and the size ranged from 9 to 10 mm. Under the general conditions of Lake Chad, maximum size ranged from 17 to 18 mm, but, as in the case of *Corbicula*, growth exceeded this limit under certain conditions and individuals from 20 to 30 mm were collected in a few shallow places in the lake.

When analyzing demographic structures, it can be pointed out that the young Prosobranchia, with a few exceptions, were generally not very numerous in relation to adults in the natural populations, although their reproduction was continuous. This anomaly can be accounted for only by considerable mortality after birth resulting partly from the predation exerted by fishes. Lauzanne (1975) compared the size of the molluscs caught in the natural environment with those of the molluscs contained in the stomachs of three malacophagous fishes in the Eastern Open Water: *Synodontis schall*, *Synodontis clarias*, *Hyperopisus bebe*. It appears (Fig. 10) that the three species of fish behaved similarly and their predation was exerted mainly on young individuals. However, the phenomenon was a little less clear in *Corbicula* whose young were consumed less than the young Prosobranchia.

In the Eastern Open Water where these malacophagous fishes represented about 7% of the total weight of the catches in gill nets in 1970, the predation on

412

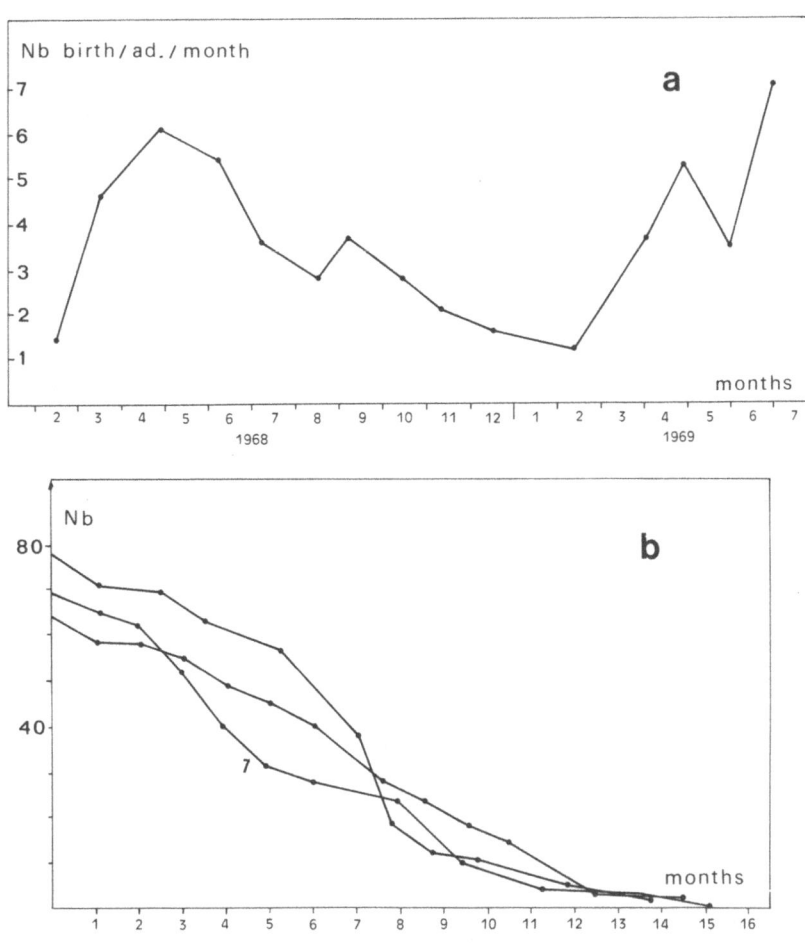

Fig. 9 Bellamya unicolor in *in situ* cultures at Bol; (a) monthly number of births for each adult individual over the year; (b) survival curves for three cohorts (Lévêque 1973).

young molluscs must have been important quantitatively and played a prime role in the dynamics of the snail populations. Of course, this phenomenon occurred in the whole lake and it accounts for the small proportion of young molluscs in the populations. In fact, Lauzanne (1975) considered that *S. schall* consumed about two hundred molluscs a day and *S. clarias* about one hundred.

It was often difficult to evaluate the growth of the molluscs only by studying their size structures, especially as reproduction was continuous or at least staggered in time. Therefore, we referred to the *in situ* cultures to determine the growth curves and the parameters of von Bertalanffy's equation were calculated through the method of instantaneous increments from the results thus obtained (Table 12).

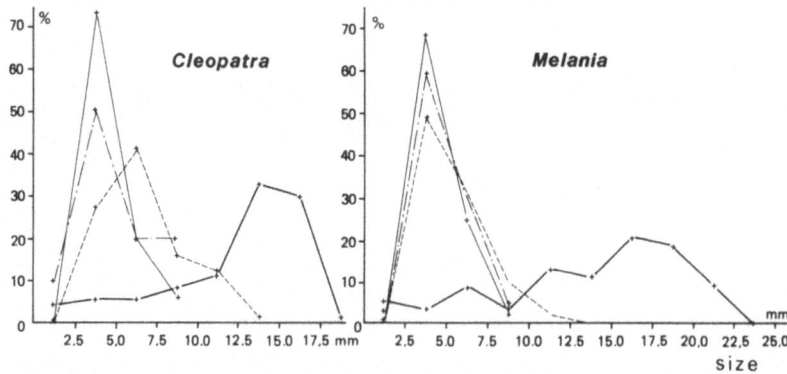

Fig. 10 Distribution of the size classes for two species of benthic molluscs consumed by fishes. The benthic population is represented by thick lines, the individuals present in the stomachs of *Synodontis clarias* by thin lines, *Synodontis schall* by dashed lines and *Hyperopisus bebe* by dashed and dotted lines.

Table 12 Growth of the benthic molluscs in Lake Chad: values of the parameters K and L∞ from Von Bertalanffy's equation for the main species at various stations.

	K	L∞ (mm)
Corbicula africana		
— Bol	0.221	10.0
— Baga Kawa	0.115	14.2
— Shari delta	0.127	13.9
Caelatura aegyptiaca		
— Bol	0.152	26.8
— Baga Kawa	0.114	33.8
Caelatura teretiuscula		
— Bol	0.159	28.1
Mutela rostrata		
— Bol	0.179	59.3
Mutela dubia	0.129	76.4
Cleopatra bulimoides		
— Baga Kawa	0.920	11.6
— Samia	0.624	12.1
Melania tuberculata		
— Baga Kawa	0.307	15.9
Bellamya unicolor		
— Bol	0.969	19.5

414

The influence of local ecological conditions on the growth rate and the resulting maximum sizes, can be seen by comparing the growth curves of the same species from various stations. In particular, this was true for *Caelatura aegyptiaca* which grew more rapidly at Baga Kawa than at Bol. On the other hand, many species grew less rapidly in the cool season probably because of the drop in temperature which was rather marked at that time.

The size–weight relationships calculated for the dry organic weight (without shell), the weight of the shells and the fresh weight including the shell (Lévêque 1973) were used to determine the growth curves for weight and to calculate the increase in weight for the four main species over 15 days. There were two distinct groups:

— *Bellamya* and *Cleopatra* grew very quickly during the first months, after which first reproduction occurred. Growth was then reduced and the maximum size of the oldest individuals was only slightly above that at first reproduction. This maximum size varied a little according to the sites and the ecological conditions, but the growth rate remained the same. In fact, these species have a maximum life span which did not seem to exceed one and a half years.

— In contrast, in *Melania* and *Corbicula*, first reproduction and the death of the individuals occurred before the animal reached its potential maximum size. This explains the absence of an asymptote in the growth curve of *Corbicula*, since it was determined only for a life-span of one and a half year while the species can live and grow for a much longer time. The same held true for *Melania* as shown previously.

There were, therefore, basic differences in production between the two groups. The production resulting from the increase in weight of the *Bellamya-Cleopatra* group was very high in the young individuals and less important in the adults, while in the *Melania-Corbicula* group, production from growth occurred throughout the entire life of the individuals at least when the latter were of the size commonly found in the lake.

12.2.1.2   *Oligochaeta and insects*. The duration of the larval stage (from egg to adult) was determined for two Chironomids. It was 17 days at 26°C and 13 days at 30°C in the laboratory for *Chironomus pulcher* (Dejoux, 1971) and 18 days between 18 and 23°C for *Tanytarsus nigrocinctus* (Fig. 11) under almost natural conditions (Dejoux 1976).

An attempt was made to evaluate the duration of larval growth for a few species of Chironomids based on the rhythm of emergence during a lunar cycle in the cool season in January 1973 (Dejoux 1976). It seems that the majority of these species have a larval stage with a duration of 15 days or a little less. However, these results should be checked in culture.

Oligochaetes reproduce throughout the year, but reached a maximum in the cool season (Dejoux et al. 1969). This group was not studied in detail.

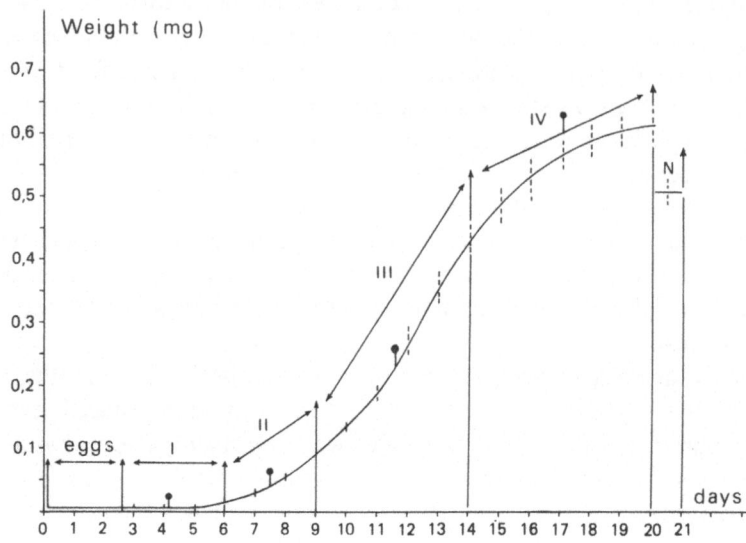

Fig. 11 Growth curve in weight and growth stages of *T. nigrocinctus*.

12.2.2 *Production*

Among the three groups of benthic organisms investigated, the molluscs are the only ones to have been thoroughly studied (Lévêque 1973a). Production was evaluated in various biotopes at three sites situated in different regions of the lake. Here *in situ* cultures and regular sampling allowed the study of growth rates and changes in demographic structures of the populations of the main species.

Since reproduction is continuous or at least evenly distributed in time, it was not possible to distinguish cohorts and to use simple methods of calculating production. Therefore, a method was developed (Lévêque 1973a) based on the evaluation of an instantaneous growth rate of the population (G) at the time of each sample.

The annual P/$\bar{\text{B}}$ ratio was calculated for the populations of each species (Table 13) to be able to apply the results thus obtained to the whole lake and to draw a parallel between the different stations. The P/$\bar{\text{B}}$ values varied for the same species according to the stations. However, they were maintained within rather narrow limits. The highest P/$\bar{\text{B}}$ values were found for *Bellamya* and *Melania* which were the predominant species in the whole lake (Lévêque 1972) according to biomass and numbers.

From a knowledge of the life-span of the species, the results obtained can be compared to the theoretical values of P/$\bar{\text{B}}$, determined in relation to the life-span for populations in equilibrium (Lévêque et al. 1977). The observed values

Table 13 Mean biomasses (in g m^{-2}) and annual P/B̄ ratios for the main species in the stations under study.

Species	Station	Organic matter production		Shell production	
		B̄ in g	P/B̄ annual	B̄ in g	P/B̄ annual
Melania tuberculata	Samia 1	3.5	4.4	28.5	5.0
	Samia 2	0.7	3.0	5.6	3.1
	Baga Kawa 1	0.9	5.3	6.8	5.7
	Baga gawa 3	1.5	4.8	12.3	5.2
Cleopatra bulimoides	Samia 1	3.6	2.0	27.5	2.3
	Samia 2	3.6	1.7	28.0	2.1
	Baga Kawa 1	3.6	2.5	26.3	3.0
	Baga Kawa 3	3.4	3.5	25.5	3.6
	Bol 2 1968	0.8	2.4	5.8	2.8
	1969	0.5	2.8	3.6	3.4
	Bol 3 1968	0.3	2.5	2.2	2.9
	1969	1.8	3.5	13.7	4.1
Bellamya unicolor	Baga Kawa 2	2.8	5.5	18.0	6.0
	Baga Kawa 3	2.1	6.1	13.4	6.5
Corbicula africana	Samia 1	1.1	2.8	34.8	3.0
	Samia 2	0.5	2.4	16.1	2.6
	Baga Kawa 1	1.7	2.3	56.6	2.6
	Baga Kawa 2	3.7	2.9	121.4	3.0
	Baga Kawa 3	1.3	2.1	43.1	2.2
	Bol 3	1.3	2.8	46.3	3.1

of P/B̄ for *Bellamya* (life-span: 1 year; P/B̄ = 5.8) and *Melania* (1.5 years; P/B̄ = 4.4) are close to the theoretic values. The same was true for *Corbicula* (P/B̄ = 2.6) if it is estimated that this last species can reach an average of 2 years. On the other hand, the value observed for *Cleopatra* (P/B̄ = 2.6) is much lower than the theoretical value expected for a life-span ranging from 1 to 1.5 years. This phenomenon could have been caused by a variety of factors including a miscalculation of the life-span, death rate or growth. However, we also noted that the young stages were rather rare in the populations of this species, while the adults, whose production was low, prevailed. This finding is comparable to the observations on the populations of the lake where a decrease was noted from 1968 to 1970 in the density of *Cleopatra* which was replaced by *Melania* in several biotopes in the south basin. Therefore, the population would no longer be in equilibrium and would decline, explaining the rather low P/B̄ values of *Cleopatra*.

When calculating production, we could also show that a rather strong relation existed between the instantaneous growth rate of the population (G), which corresponded to a daily P/B̄, and the average weight of individuals (W̄) (Fig. 12). Such empirical relationships are interesting from a practical point of view insofar as they make it possible to quickly calculate production of sampled populations from the mean weight of the individuals (Lévêque 1973a).

From knowledge of the molluscan stocks in Lake Chad in 1970 and the P/B̄ ratio for the main species, we estimated the annual production of molluscs in Lake Chad (Table 14). It was 279 000 tons of organic matter (dry weight) and 1 883 000 tons for the shells, that is an average of 14.5 g/m²/year and 98 g/m²/year respectively. If the caloric equivalents are used (Table 15), then

Table 14 Mean P/B̄ ratios calculated for the main species of benthic molluscs in Lake Chad and evaluation of the annual production for the whole lake in 1970. The values are expressed in tons (dry weight).

Species	Organic matter		Shell	
	P/B̄	P	P/B̄	P
Melania	4.4	60 852	4.8	513 000
Bellamya	5.8	162 122	6.2	996 500
Cleopatra	2.6	44 996	3	183 200
Corbicula	2.6	4711	2.8	124 000
Caelatura	2.0	6410	2.2	9300
Total		279 091		1 883 500

Table 15 Evaluation of the annual production of benthic molluscs in the various zones of Lake Chad in 1970 (Fig. 11, Chapter 8).

Zones	Area (km²)	P (g m⁻² yr⁻¹)	P (Kcal m⁻² yr⁻¹)
1	3082	0.1	0.4
2	3871	35.3	141
3	1501	24.1	96
4	2133	25.6	102
5	2290	11.4	46
6	2083	3.8	15
7	4259	3.0	12

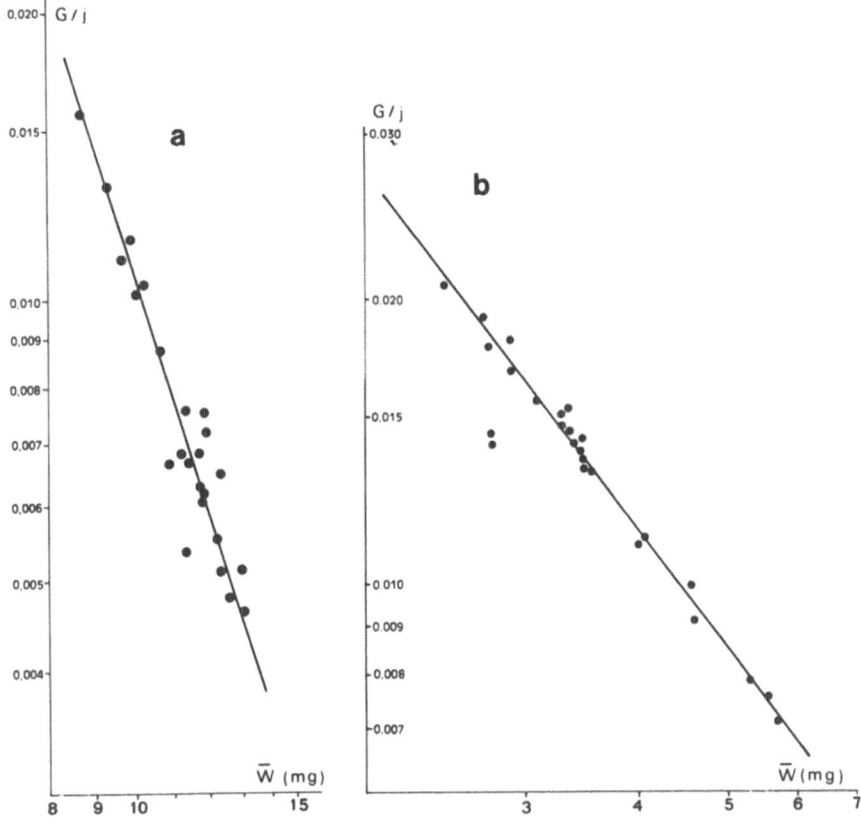

Fig. 12 Relationships between the instantaneous growth rate of the populations (G) and the mean weight of an individual belonging to these populations (W) expressed in mg of the dry body weight; (a) *Cleopatra bulimoides*; (b) *Melania tuberculata* (from Lévêque 1973).

the total production of organic matter was 1116×10^9 Kcal, that is an average of 58 Kcal/m²/year.

Most of the production resulted from the three species of Prosobranchia (Table 14) among which *Bellamya* played the most important role. In other respects, the distribution of this production was irregular in the lake (Table 14), since the zones with the highest production were those which had the highest biomasses such as the Northeastern Archipelago, the Northern Open Water and the Great Barrier.

It is obvious that the above-mentioned estimates are valid only for the year 1970 and they can undergo profound modifications according to the lake level and its water area. However, these observations can be applied to the period of 'Normal Chad' since the biomass estimates made in 1968 were similar to those for the year 1970. The production of shells was considerable and required about

700 000 tons of calcium as the shells contain 37% calcium in the form of aragonite. This was four times the annual mean supply of calcium to the lake or half the dissolved lacustrine stock of this element (Carmouze 1976). Therefore, the molluscs play a significant role in the regulation of calcium in Lake Chad.

Observations on the production of *T. nigrocinctus* (Chironomidae) showed that the average daily $P/\bar{B}$ ratio was 0.24 in the cool season at a station situated in the north basin (Dejoux 1976). However, this result cannot be applied to the total biomass of the benthic insects of which Chironomidae represent only a part.

12.2.3 *Energy balance*

Energy ingested as food can be utilized in different ways: part of it is used to make organic matter, production (P); another part is used for metabolic processes and can be measured by oxygen consumption, respiration (R); finally, a part of it is excreted without being digested (F) and as excretory products (U). The equation, $C = P + R + F + U$, expressed in comparable energy bits (the calorie) represents an energy budget (Klekowski 1970). Assimilation is also defined by $A = P + R$.

For the benthic molluscs of Lake Chad we only determined the assimilation budget, because the evaluation of excretion raised problems that were impossible to solve with the means available.

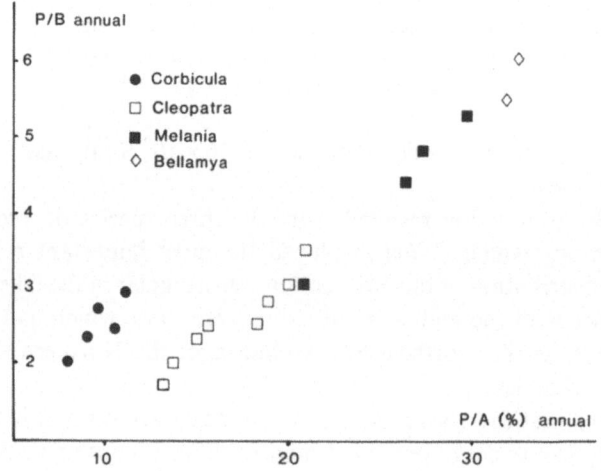

Fig. 13 Relationships between the annual P/B and the annual P/A for the natural populations of the various species of benthic molluscs in Lake Chad. Each point corresponds to a population of the species under studies.

420

Table 16 Measurement of oxygen consumption of the benthic molluscs; values of the constants a and b for the relation $R = aW^b$ and correlation coefficient (r) of the straight line between log R and log W. R is expressed in terms of microliter/individual/hour and W in terms of mg of dry organic weight (D.O.W.) or fresh weight including the shell (F.W.).

Species	Locality	Temp. °C	Date	Number of replicates	Size range (mm)	R = f(D.O.W.) a	b	r	R = f(F.W.) a	b	r
Cleopatra bulimoides	Baga Kawa	21.5	26/1/70	25	7 à 12	1.97	0.768	0.884	1.02	0.531	0.884
	Baga Kawa	29.5	22/10/69	46	8 à 13	2.51	0.790	0.921	1.40	0.527	0.918
Melania tuberculata	Samia	29	25/10/69	38	5 à 13	1.60	0.930	0.980	0.24	0.788	0.980
	Samia	29	26/10/69	41	5 à 14	1.59	0.909	0.987	0.23	0.793	0.986
	Samia	20.5	23/1/70	32	7 à 14	1.25	0.747	0.976	0.22	0.684	0.976
Bellamya unicolor	Baga Kawa	20.5	26/1/70	43	9 à 18	2.39	0.829	0.974	0.44	0.743	0.974
	Baga Kawa	30	19/10/69	45	6 à 17	2.56	0.905	0.979	0.40	0.811	0.979
	Bol	20.5	20/1/70	30	8 à 19	3.16	0.702	0.984	0.46	0.710	0.983
	Bol	32	20/8/69	27	8 à 19	7.30	0.666	0.991	1.18	0.671	0.991
	Bol	26	23/8/69	66	5 à 17	2.17	0.790	0.975	0.32	0.797	0.975
Corbicula africuna	Delta	29.5	28/10/69	43	4 à 13	7.36	0.638	0.997	0.86	0.578	0.991

Table 17 Calorific values for the tissues of the five main species of benthic mollusc in Lake Chad (Lévêque 1973).

Species	Locality	Number of replicates	Calorific value in cal g^{-1} dry weight	% ash	Calorific value in in cal g^{-1} without ash
Bellamya unicolor	Samia	5	4141 ± 65	15.7 ± 1.4	4919 ± 95
Bellamya unicolor	Baga Kawa	3	3982 ± 197	16.7 ± 1.5	4777 ± 319
Cleopatra bulimoides	Samia	3	3989 ± 140	15.4 ± 1	4730 ± 127
Melania tuberculata	Samia	3	3818 ± 232	17.7 ± 3.4	4636 ± 88
Melania tuberculata	Baga Kawa	1	3678	22.6	4751
Corbicula africana	delta du Chari	3	4974 ± 25	12.8 ± 1.6	5707 ± 83
Caelatura aegyptiaca	Baga Kawa	5	3972 ± 149	19.7 ± 2	4949 ± 164

Respiration was measured experimentally in the field during different seasons (Lévêque 1973b) by using individuals which had just been collected and whose physiological condition was *a priori* closer to the natural conditions than those of the individuals maintained in culture. The relationships between the amount of oxygen consumed (in µl/ind/h) and the weight of individuals expressed in fresh weight including the shell (F.W.) or in the dry weight of the organic matter (D.O.W.) were calculated for different size ranges and at different temperatures (Table 16).

The oxycalorific coefficient of 4.86 cal/ml of oxygen (Winberg 1971) was used to convert the volume of oxygen consumed into calories. These results, as well as the measurements of production were used to determine assimilation for the different populations of the main species and to calculate the utilization

Table 18 Production, respiration and assimilation (in Kcal m^{-2} year^{-1}), annual P/B ratio and P/A (%) for the different populations of benthic mollusc that were studied in Lake Chad.

	R (Kcal m^{-2} yr^{-1})	P (Kcal m^{-2} yr^{-1})	A (Kcal m^{-2} yr^{-1})	P/A (%)	P/B annual
Bellamya unicolor					
Baga Kawa station 2	120.9	58.4	179.3	32.5	5.5
Baga Kawa station 3	116.9	55.0	171.9	32.0	6.1
Cleopatra bulimoides					
Samia station 1	183.8	29.2	213.0	13.7	2.0
Samia station 2	163.9	25.6	189.5	13.3	1.7
Baga Kawa station 1	164.1	36.8	200.9	18.3	2.5
Baga Kawa station 3	187.7	49.3	237.0	20.8	3.5
Bol station 3 1968	13.6	2.5	16.1	15.6	2.5
Bol station 3 1969	90.2	23.0	113.2	20.3	3.5
Bol station 2 1968	43.3	7.8	51.1	15.2	2.4
Bol station 2 1969	24.7	5.8	30.5	19.0	2.8
Melania tuberculata					
Samia station 1	172.7	61.8	234.5	26.4	4.4
Samia station 2	30.8	7.9	38.7	20.5	3.0
Baga Kawa station 1	43.4	18.4	61.8	29.8	5.3
Baga Kawa station 3	34.0	12.8	46.8	27.3	4.8
Corbicula africana					
Samia station 1	145.1	16.9	162.0	9.7	2.8
Samia station 2	60.0	6.3	66.3	8.9	2.4
Baga Kawa station 1	202.3	21.3	227.4	11.1	2.3
Baga Kawa station 2	446.7	51.9	498.6	10.4	2.9
Baga Kawa station 3	116.4	11.5	127.9	9.0	2.1

efficiency of energy assimilated for growth (P/A) (Table 18). It seems that there was a rather linear relation between the annual P/$\bar{\text{B}}$ and P/A for the three species of Prosobranchia (Fig. 13). On the contrary, for P/B ratios that were identical in size the P/A ratio was much lower in *Corbicula*. This difference could be accounted for by the fact that the ratio of dry body weight to the shell weight ranged from 1/7 to 1/9 in Prosobranchia, while it was four times lower in *Corbicula* (1/35). Therefore, it would be necessary for this last species to use a greater amount of energy per unit of dry weight than Prosobranchia, in order to form its shell. The P/A values observed in the molluscs of Lake Chad were consistent with most previous results which had already been obtained with other mollusc species (Lévêque 1973b).

References

Burgis, M. J., 1971. The ecology and production of copepods, particularly *Thermocyclops hyalinus*, in the tropical Lake George, Uganda. Freshwat. Biol. 1: 169–192.

Carmouze, J. P., 1976. La régulation hydrogéochimique du lac Tchad. Trav. Doc. ORSTOM, No. 58, 418 pp.

Dejoux, C., 1971. Recherches sur le cycle de développement de *Chironomus pulcher*. Can. Ent. 103: 465–470.

Dejoux, C., 1976. Synécologie des Chironomides du lac Tchad (Diptères, Nématocères). Trav. et Doc. ORSTOM, No. 56, 161 pp.

Dejoux, C., Lauzanne, L. and Lévêque, C., 1969. Evolution qualitative et quantitative de la faune benthique dans la partie est du lac Tchad. Cah. ORSTOM Sér. Hydrobiol. 3: 3–58.

Ferguson, A. J. D., 1975. Invertebrate production in lake Turkana (Rudolf)-Symposium on the Hydrobiology and fisheries of Lake Turkana, 25–29th May, 1975, 28 pp., mimeo.

Gras, R., 1970. Poids individuel, durée de développement et production des différents stades de *Tropodiaptomus incognitus* (Crustacés, Copépodes). Cah. ORSTOM Sér. Hydrobiol. 4: 63–70.

Gras, R. and Saint-Jean, L., 1969. Biologie des crustacés du lac Tchad. 1. Durée de développement embryonnaire et post-embryonnaire: premiers résultats. Cah. ORSTOM Sér. Hydrobiol. 3: 43–60.

Gras, R. and Saint-Jean, L., 1976. Durée du développement embryonnaire chez quelques espèces de Cladocères et de Copépodes du lac Tchad. Cah. ORSTOM Sér. Hydrobiol. 10: 233–254.

Gras, R. and Saint-Jean, L., 1978a. Taux de natalité et relations entre les paramètres d'accroissement et d'abondance d'une population de Cladocères à reproduction par parthénogenèse. Cah. ORSTOM. Sér. Hydrobiol. 12: 19–63.

Gras, R. and Saint-Jean, L., 1978b. Durée et caractéristiques du développement juvénile de quelques Cladocères du lac Tchad. Cah. ORSTOM Sér. Hydrobiol. 12: 119–136.

Gras, R. and Saint-Jean, L., 1981a. Durée de développement juvénile de quelques copépodes planctoniques du lac Tchad. Rev. Hydrobiol. trop. 14: 39–51.

Gras, R. and Saint-Jean, L., 1981b. Croissance en poids de quelques Copépodes planctoniques du lac Tchad. Rev. Hydrobiol. trop. 14: 135–147.

Klekowski, R. Z., 1970. Bioenergetic budgets and their application for estimation of production efficiency. Pol. Arch. Hydrobiol. 17: 55–80.

Lauzanne, L., 1975. La sélection des proies chez trois poissons malacophages du lac Tchad. Cah. ORSTOM Sér. Hydrobiol. 9: 3–7.

Lévêque, C., 1971. Equation de von Bertalanffy et croissance des mollusques benthiques du lac Tchad. Cah. ORSTOM Sér. Hydrobiol. 5: 263–283.

Lévêque, C., 1972. Mollusques benthiques du lac Tchad: écologie, étude des peuplements et estimation des biomasses. Cah. ORSTOM Sér. Hydrobiol. 6(1): 3–45.

Lévêque, C., 1973. Dynamique des peuplements, biologie et estimation de la production des mollusques benthiques du lac Tchad. Cah. ORSTOM Sér. Hydrobiol. 7: 117–147.

Lévêque, C., Durand, J. R. and Ecoutin, J. M., 1977. Relations entre le rapport P/B et la longévité des organismes. Cah. ORSTOM Sér. Hydrobiol. 11: 17–32.

Morrisson, J. P. E., 1954. The relationships of Old and New World Melanians. Proc. U.S. nat. Mus. 103: 357–394.

Winberg, G. G., 1971. Methods for the estimation of production of Aquatic animals. Academic Press, London and New York, 175 pp.

13. The exploitation of fish stocks in the Lake Chad region

Jean-René Durand

Fishing exploits a major resource in the Sahel because the potential yield of continental fisheries is estimated to be several hundred thousand tons, most of which originates from the central Niger delta in Mali and the Lake Chad region. This tonnage represents a similar quantity of protein to that produced by cattle and also has significant export value. Only recently has the value of this resource been recognized and so the wide range of knowledge required for proper management has not been aquired. The development of fisheries in the Lake Chad region can be examined initially from data collected since 1963, mostly on total catches. But the optimization of the fishing industry still requires a considerable amount of biological and socio-economic research.

The fisheries of Lake Chad cannot be considered in isolation as has generally been done for the other disciplines in this book. There is remarkable agreement between the permanent as well as the temporary river and lake environments. The climatic variations associated with the very flat morphology of the center of the Chad basin, cause the seasonal existence of large flood zones, fed by rivers and then drained towards the permanent environments. These flooded zones offer shelter and food to the young of several species and thus they constitute a considerable portion of total lake production.

The life cycle of several commercially important species can be described in three phases: an upstream reproductive migration of adults in the rivers at the beginning of the flood, growth of the fry in the flooded zones and the major river beds and finally a return to the lake where both adults and juveniles remain. The life cycle stages of *Alestes baremoze*, whose biology and dynamics were particularly studied (Section 6) are summarized in Fig. 1 (Chapter 10). They occurred in the fishing zone studied, above 10°50' latitude north which included most of the fishing activities of the Chad basin.

The fishing industry of Cameroon was first described by Monod (1928). This study only deals with the northern course of the Logone and the Shari in the Chad basin. A complete list of traditional techniques and their usage until 1955 was provided by Blache and Miton (1962). Recently, more specialized studies have been done on some fisheries of Lake Chad and the lower Shari and

J.-P. Carmouze et al. (eds.) Lake Chad   425
© *1983, Dr W. Junk Publishers, The Hague/Boston/Lancaster*
ISBN 978-94-009-7268-1

Logone streams: Durand (1970a, b, 1971, 1973, 1980); A. J. Hopson (1968); J. Hopson (1969, 1972); Loubens (1973); Quensière (1976) and Vidy (1982).

Blache and Miton showed that the traditional fishing techniques were well adapted to the many situations encountered in a complex and varied aquatic network. Therefore the very large variety of fishing gear in use resulted from the local differences in hydrography and seasonal rhythms:
— low water fishing during which the best yields are obtained;
— river fishing with drift nets during the flood;
— fishing in flooded zones during high waters;
— exploitation of lateral migrations during subsidence.

No systematic surveys were made which allow a precise comparison to be made 20 years later but it is certain that many traditional techniques are not used anymore. This is the case with 'zemys' fishing practised by the Kotokos at the Logone-Shari confluence and downstream. These collective fisheries have disappeared since the sixties when Monod counted 68 large pirogues in 1925 and Blache and Miton 169 in 1955.

13.1 The fisheries

The recent fishing history of the Lake Chad region has been marked by a combination of two phenomena: the Sahel drought and an accelerated development of fishing activities. Two periods can be distinguished: pre-1971, with little or moderate exploitation and a post-1971 period of drought with an increase in human predation on the aquatic environments.

13.1.1 *Before the lake contraction*

13.1.1.1 *The rivers and their tributaries*. From the delta to the confluence, the Shari is characterized by the absence of adjacent flooded zones. Therefore it is the only possible route for species which migrate from the lake to the rivers and whose migrations are related to the flood patterns upon which fisheries activity depends. This latter was very important and the four major fishing methods were organized around it: fixed gill nets, drifting gill nets of small mesh (28.5 mm in general), nets of large mesh (60 to 70 mm), unbaited multiple hook lines (Loubens 1973).*

Unlike fishing in the lake, the bulk of the catch came from small mesh nets in this area, where most of the *Alestes baremoze* catches were made between 1965 and 1970.

---

* All the mesh sizes cited refer to the edge of the mesh (in mm) measured from knot to knot, being thus half the classical 'stretched' mesh. The abbreviation 'GNx' is used for 'gill net of size x mm'.

Photo 16   View of a traditional fishing on the El Beid River, during the dry season.

— In the rivers, above the Logone and Shari confluence, the seasonal fishing techniques often remained the most diversified. The total yield was still less in the Logone than in the Shari. This was doubtless due to the very extensive flooded zones on the right and left banks of the Logone which did not really have an equivalent in the Shari.

Small or large mesh fixed or drifting gill nets, unbaited multiple hook lines were again found as well as triangular nets (sakamas) used from the pirogues, casting-nets (introduced by the fishermen of the Benoué basin etc. ... The major outflow of the flooded zones on the right bank of the Logone drained the north and south Ba-Illi region (cf. Fig. 22, Chapter 2). When fishing, the Kotokos install a screen dam made from the stems of *Echinochloa*, supported by solid plant beams driven in the river bed.

Stick seines, which are 20 to 23 mm gill nets were found in this region. The nets, about 0.5 m high and 15 to 20 m long are fixed on some small sticks which support the net opening. The fishermen draw several attached sheets that scrape the sandy bottom up to the shore. Large sized fishes like *Tilapia* and *Citharinus* were caught but most of the catch consisted of *Alestes baremoze, A. dentex* and *Hydrocynus forskalii*. These fisheries operated at low waters on the uncovered sand banks and their importance has become reduced over these last few years.

— There were two flooded zones south of the lake: the triangle between Logone

427

and Shari, south of Logone-Gana, dependent on the river system, and 'yaéré' of North Cameroon, whose main exit, the El Beid, directly rejoins Lake Chad. The first of these zones provided the recruitment of the great traditional fishery of Logone-Gana which only caught adult fishes returning to the Logone during their anadromous migration when the subsidence begins. In the yaéré of North Cameroon however, the fishing took place in the subsidence outlets, catching fingerlings aged from about two to five months.

In the region of the Logomatia which drained the southern waters of the North Cameroon floodplain (Fig. 1, Chapter 10), little fish traps built with small spaced wicker blocked many hand-made channels dug in the river bank. These fisheries only caught fingerlings and the yields were relatively low as most of the flood waters passed through the El Beid. This river is barred by many permanent fish dams from Tildé up to the lake (Fig. 1, Chapter 10) and about 270 were counted in January 1969 (Durand 1970a). Nets with triangular frames similar to the sakamas mentioned earlier were used for fishing. Young fishes constituted the bulk of the catch.

13.1.1.2 *The lake.* Before 1972, the south basin was characterized by a general weakness of fisheries, that were practically absent from the southeastern archipelago and the south and southeastern open water (Fig. 1, Chapter 2). Only on the southern border and especially in the Shari delta, was activity notable with two types of gear being used: large mesh nets, as in the north basin, and the non-baited multiple hook lines, simply placed on the bottom or hung horizontally between surface and bottom.

Fishing was more active in the north basin with a large predominance of large mesh nets (GN 60 to 130 with a trend towards GN 90). Limited at first to the zones in contact with the reed islands and the western coast with the northern open water, they slowly expanded throughout the basin, due to the increased use of outboard motors. The northeastern archipelago however was hardly exploited, corresponding to a relative scarcity of large sized fishes.

The bulk of the catch of average sized fishes, of little importance before 1970, was caught by seines on the Yobé river (Fig. 1, Chapter 10). Seasonally, these fisheries used locally made nets, of 20 mm mesh, placed right up-stream of the mouth of the Yobé and used between September and February (J. Hopson 1972). Some gill nets were used along the western coast, near the Yobé. The use of the new boats and outboard motors allowed the extension of this fishery, which was installed in 1969 in the northeastern reed islands with GN 25 to 30 (J. Hopson 1972).

The diversity of fishing activity in the above description shows that the gill nets dominated, and that the fisheries corresponding to the lower bed of the rivers, to the delta and to Lake Chad were the ones that provided practically all the production for the past ten years.

This situation was quite different of that described by Blache and Miton as

Photo 17 General view of a traditional fishery occurring once a year at Logone Gana during the annual drying up of the North Cameroon flood plain.

the gill nets then only appeared as one technique among others and being fairly specialized allowed either the capture of a particular species, or allowed fishing at certain times. It appears however that this technique of fishing is very old since it was cited by Monod (1928) for the Shari delta and Boyd Alexander (1907, in A. J. Hopson 1968) for the mouth of the Yobé. The increase in total fishing effort corresponded to the introduction of nylon for making the gill nets. A. J. Hopson (1968) placed its appearance at about 1958 but Mann (1962) noted that during 1961 most of the nets used on the lake were still locally made by the fishermen. Thus it is possible to the start of gill net fishing from 1961–63 as being due to the introduction of nylon and most especially to the industrial production of net material at accessible prices. According to our estimates, the total fishing effort in the north basin was multiplied by about 50 between 1967 and 1971–72 when in the same time the total production multiplied by 5 (Durand 1973b).

13.1.2 *Changes in fishing during the drought*

The intensification of gill net fishing was accompanied by a progressive displacement of fisheries towards the lake and by their establishment there. In

Photo 18 A fisherman and air-dried fish.

these last years, it supplied most of the fish production of the Chad basin (Durand 1973). To show the change in fishing habits, Blache and Miton can be quoted (1962, p. 19): 'Le lac reste très peu exploité et ne l'est que dans sa bordure sud-ouest et dans la région des ilôts-bancs de la bordure nord-est' whereas the north basin of the lake provided between 50% and 80% of the total production between 1967 and 1975 (Stauch 1977). This change in the fisheries of the northern Chad basin had already begun in the fifties since Blache and Miton had already noted the tendency to 'des implantations de plus en plus denses de groupements d'émigrés venus du sud dans les zones nord' ... It is thus possible to say that the search for fishes and for increased yields pushed the fishermen increasingly closer to Lake Chad, a tendency that was favoured by the introduction of nylon which allowed the generalized use of gill nets. To this must be added the proximity of Nigeria whose protein needs opened up a large market which easily absorbed the increased fish production.

Until 1971, the progressive displacement of fishing areas towards the delta region was accomplished independently of the lowering of the lake. Starting in 1972 and 1973, the in-between years, the catches in the river diminished very noticeably and the fishermen increased in number by the border of the lake or sometimes fished in the lake from the Delta encampment. During 1973, river fishing was practically negligible and the lowering of the lake helped in exploiting the whole lake. This almost depleted the stocks before the drought of the north began in 1974. After being displaced massively towards the north, the fishermen fell back again toward the south basin, starting in 1975. It was the south basin which supplied the essential catches although significant ones were made in the marshy zones of the north basin temporarily flooded each years for several months.

The fishing techniques did not change particularly between 1965 and 1977* and the gill nets still supplied most of the fishing effort in the lake region. The meshes on the other hand changed a lot. There were initially two principal types of nets: (a) gill nets of large mesh (GN 80 to 130) used especially as fixed nets in the rivers at low water and in the north basin throughout the year. These nets caught most of the large sized fishes: *Lates niloticus, Labeo* spp., *Citharinus* spp, *Hydrocynus brevis, Bagrus* spp., *Heterotis niloticus, Hemisynodontis membranaceus* ... (Durand 1973); (b) Salanga gill nets (GN 25 to 30): the salanga represent all the *Alestes baremoze* and *A. dentex* fished in abundance until 1972. Average sized fishes (25 to 30 cm standard length) were taken including *Hydrocynus forskalii, Brachysynodontis batensoda*, various Schilbeidae, Mormyridae. ...

The apparent absence of meshes lower than GN 25 on the one hand and those between GN 40 and 70 on the other guaranteed that most of the fish caught were adults.

Since 1971 the fishing of *Alestes* almost ceased with the disappearance of the stock that earlier supplied 8000 to 10 000 tons of fish a year (cf. Section 6.2). The large meshes were slowly been replaced by smaller meshes: GN 70 then 60 and 50. Observations made during November 1977 in the Shari delta showed that the meshes ranged from 40 to 75 mm with a dominance of GN 55.

The environmental changes have evidently involved some modifications of the fish populations and the major features of this change were given in Chapter 10. It should simply be noted there that the initially common fishes were slowly replaced by better adapted species, particularly tolerant of temporary or permanent deoxygenation of the environment (*Clarias* for example).

13.2 Processing and commercialization

Ideally one would like to assess total fish catches from effort and yield data for the principal fisheries and the most important commercial species. This would avoid sometimes erroneous extrapolations and would permit the actual change in the fisheries to be extensively analyzed. This is impossible in the Lake Chad region, except for some data as those shown in Section 4, and we have to

---

* Seine fishing had developed in the Shari above the confluence after 1975 (Franc and Vidy, personal communication). During May 1977 these nets (GN 14 to 25) only caught small sized fish.

evaluate here the fresh fish production from the statistics of circulation of smoked and/or dried fish.

The consumption of fresh fish occurred immediately around the lake and rivers (in particular by the two large urban centers, N'Djamena and Maïduguri). All of the remaining catch was preserved by smoking/drying or by drying alone. Sun-drying was used for the small sized fishes (for example, fishery of the El Beid) and certain fishes of average size sold commercially under the name 'salanga'. This category corresponded almost solely to *Alestes baremoze* and *A. dentex*, which were always dried whole (Nigerian side of lake), or eviscerated and placed flat (Logone and Shari). for most of the large sized fishes (*Lates niloticus, Heterotis niloticus, Citharinus* spp., *Distichodus rostratus, Bagrus bayad, Hemisynodontis membranaceus, Labeo senegalensis* and *L. coubie, Gymnarchus niloticus, Hydrocynus brevis, Clarotes laticeps, Mormyrus rume* etc. ... in order of approximate importance), the most common processing was marketed under the name 'banda' prepared for the south Nigerian markets.* The fishes were scaled, eviscerated and cut in pieces. Only some of the bony pieces (heads of some species, spines ...) were thrown away. After smoking (or often only superficial burning), the fishes were again dried for one week. After this they were packed and forwarded to the south via Maïduguri.

The major objective of processing was to obtain good quality drying. A well-dried fish is much less attractive to fish eating insects, which were very abundant. Although average losses were ignored, up to 50 or 60% of the initial dry weight could be lost during transport between production and marketing sites. The insufficient drying resulted from a very hasty process or from poor climatic conditions (increased humidity from May to October).

13.2.1 Conversion coefficients

To obtain estimates of actual catches of fresh fish, it was necessary to consider the many factors operating on the dry weight or the smoked weight. This is, among others, an area where the absence of serious studies was particularly felt. Three factors had to be taken into consideration: the auto consumption (feeding of fishermen's families), the weight losses in the processing of fresh fish, the losses caused by ichthyophagous insects during storage and transport.

13.2.1.1 *The banda.* Outside of so-called auto-consumption, a considerable amount of fish is consumed fresh around the lake and in the urban centers. Not

---

* In the following text these two vernacular names will be retained; 'salanga' corresponds to *Alestes* caught with gill nets of average mesh (25 to 30 GN) then dried, and 'banda' includes the many species smoked after being caught with large mesh gill nets (50 to 130 GN but usually 80 to 100 GN).

Photo 19 Dried fish (*Alestes*) commonly called 'Salanga'.

all the rest was converted to banda (e.g. smoked *Gymnarchus* sold separately; *Lates* that reached a certain weight and which were dried and sold on the markets around the lake ...); failing precise data, we suggest that 20% of the large sized fishes were not converted to banda.

The transformation coefficient of fresh fish by smoking/drying is also subject to controversy. It actually varies with species, time of drying and seasonal humidity. Mann (1962) recorded it as 2 and 3 on two smoked fish samples after only one day of drying; Hopson (1964) estimated that 70% of fresh weight was lost at the end of the drying. Here the value 3 will be used considering that it is a minimum value and that a better calculation would be closer to 3.5 or even 4. According to the literature, the weight losses during transport by insect attacks range from 30 to 70%. On average, we estimated that 25% of the banda weight was lost between the end of drying and the inspection at Maïduguri. In total, the combination of these three coefficients, gives a coefficient of conversion between fresh fish and banda of $0.8 \times 0.33 \times 0.75 = 0.2$. In other words, there would have to be about five tons of fresh fish for one ton of banda recorded at Maïduguri.

13.2.1.2 *The salanga.* We started from some direct calculations of conversion coefficients for the salanga. Fish processing in the traditional way, drying, was carried out at the end of March, beginning of April and was very quick. The dry

433

weight/fresh weight ratio was almost identical for the two species: 3.49 for *Alestes dentex* and 3.53 for *Alestes baremoze*, thus 3.5 will be adopted for the salanga.

Theoretically, the auto-consumption (here this term includes all the production that had not appeared in the general circuit of commercialization of dry fish) should also be considered. There was also a lack of quantitative estimates and any precise numbers cannot be projected. It is however certain that the auto-consumption quantity should be considerably less for the salanga than for the banda, because it was a rather expensive product.* Considering the other causes of loss (transport, attack by insects like *Dermestes* between the site of fishing and Maïduguri), we estimate the difference between the fresh production and the commercial production (that was 10% of the fresh production).

From the various corrections one can say that a ton of commercial salanga corresponded to 1110 dry kg on the site of production and thus to a fresh production of 3900 kg of fish (Durand 1978).

13.2.2 The marketing network

The banda and salanga took quite different routes from the production site. The only thorough description that has been made was for the years 1963–64 (Couty and Duran 1968).

The configuration of the lake basin and the enormous demand for fish in the Nigerian market resulted in a relatively simple scheme for moving dried smoked fish (banda) which nearly all travelled through the city of Maïduguri in northern Nigeria (Fig. 1). The trucks loaded with bags of banda arrived in Maïduguri by the route of N'Djamena, which attracted the production of the low river system of the delta and of the south basin, by the route of Baga-Kawa which corresponded to the catches of the north basin. The traffic of 'banda' corresponded mostly to the fishing by large mesh nets (thus the larger fishes) which represented 80 to 90% of the total lake production.

As the salanga were then fished in the rivers, especially in the Logone upstream of the confluence, the total production was sent directly to North Cameroon or southern Chad. With the displacement of the fishing downstream, the route from N'Djamena to Maïduguri became important. However, in general the fish were unloaded in Cameroon in the southern region of the lake or even at Woulgo directly in Nigeria. Thus chadian taxes in N'Djamena could be avoided. From about 1967 the north basin of the lake became a new producing area for salanga which travelled by the route Baga-Kawa/Maïduguri.

---

* Like a number of other Charicidae, the *Alestes* have appreciable seasonal fat reserves that can be converted to oil. On the other hand, the dried meat is used like a seasoning in the sauces for millet and rice.

Cameroon remains the major center of dried fish whether directly by Maltam and Maroua or through northern Nigeria (Fig. 1). No recent data exist to

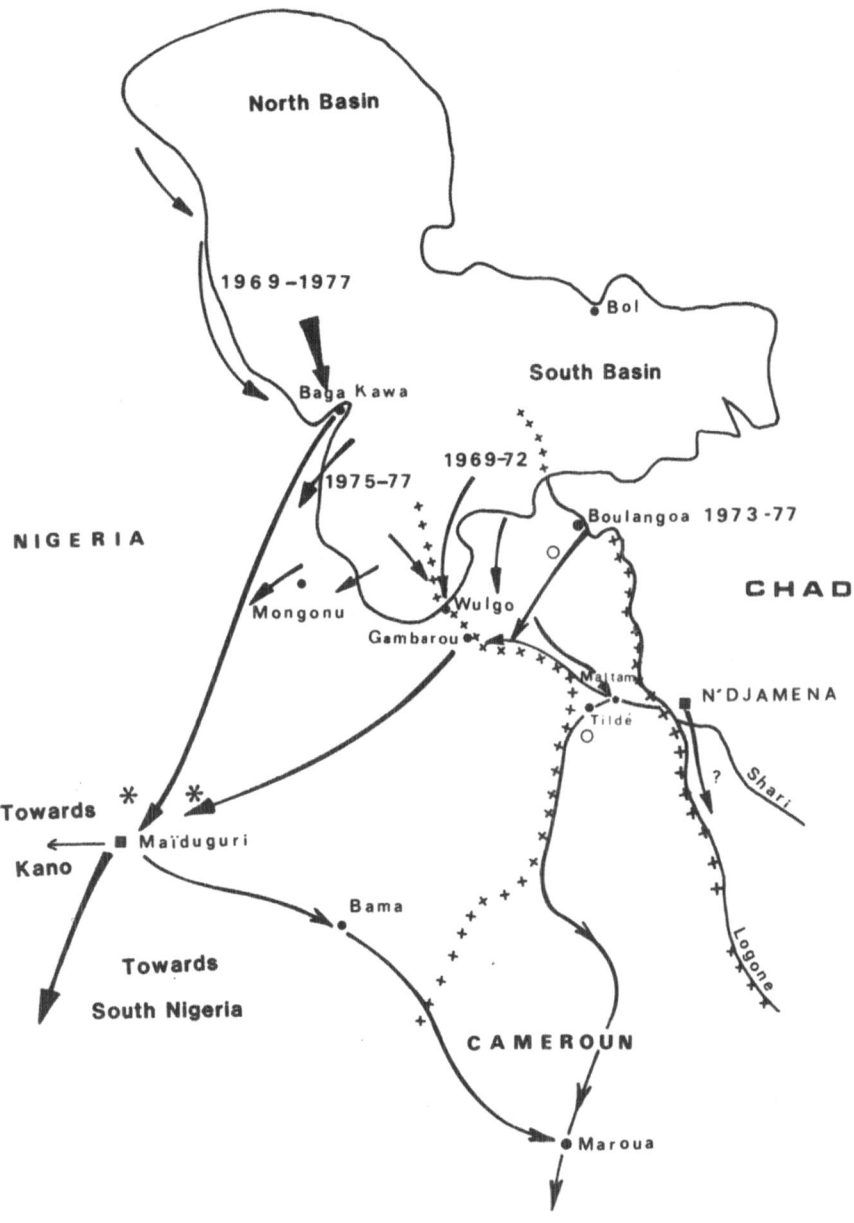

Fig. 1 Main routes for the traffic of the dried fish in the Lake Chad region; xxxx = boundaries; * = existing check-points; o = projected check-points.

435

permit theories to be proposed about the eventual Chad routes. From mid-1975, the salanga trade stopped completely with the depletion of *Alestes* stocks.

There was thus a convergence towards Maïduguri of the two major routes for the fish and an interesting possibility existed to control the commercial quantities entering this city. Beyond, the fish were sent to southern Nigeria via Jos and Kaduna: Lagos, Enugu, Benin-city, Onitsha and Kano, these representing on average the destination of 60 to 80% of the total tonnage (Stauch 1977).

The recent change in the environment involved some modifications in the routes between the fishing grounds and Maïduguri:
— after 1975, the route of Baga-Kawa did not exclusively correspond to the number of fish caught in the north basin. Production there diminished very markedly but was partially taken over by fishing in the Nigerian region of the south basin. The banda rejoined the Baga-Kawa route, especially from the south via Mongonu;
— the development of a very dense population of immersed macrophytes in the south basin forced the fishermen to stay in the open water and the delta. Otherwise, the density of the vegetation was such that the traditional channels of Woulgo and the Cameroon coast could not be taken. This explains the development of the center of Boulangoa that was established since 1973 and by which most of the fish caught in south basin of the lake passed in 1977 (Fig. 1).

13.3 Exploitation of lake stocks: the banda

The major fishing effort in the Lake Chad region was based on large mesh gill nets. They could also be used in the riverine environments (fixed and drifting nets) and in the lake as well as in the main routing ways of banda (Section 2.1.1).

Actually, the analysis that we are going to make here deals particularly with the lake stocks (as shown in the title) because they supplied most of the bulk of total catches since 1969 (between 70 and 95%). In addition, the only useful data on catch per unit effort from the point of view of the exploitation of stocks came from the north basin of the lake.

13.3.1 *Changes in total catches*

The first statistics on the movement of banda allowing the calculation of total catches in the fisheries by large sized gill nets were obtained from 1961 to 1963 (Mann 1962; Hopson 1964). After an interruption of more than five years, the collection of statistics on fishing was resumed in 1969 and continued until 1977 (Stauch 1977). We have described above (Section 2.2) the recent observations

436

made at the entry to Maïduguri; but from 1961 to 1963 the traffic inspection was done at the city exit, thus preventing us from estimating from old data the respective supplies of the rivers and the lake. On the other hand, it is likely that the total tonnages recorded on entry and exit at Maïduguri were comparable because this city was only a transit site of a product bound for southern and western Nigeria. Table 1 shows the half-yearly tonnages of fresh fish and their equivalents to the quantity of banda inspected on the two routes. The corresponding change is shown in Fig. 2.

The total catches constantly increased between 1969 and 1974, at first very quickly, more than 75% between 1970 and 1971 and over 44% between 1971 and 1972, then more slowly between 1972 and 1973 (over 16%) and from 1973 to 1974 (over 15%). The 220 000 tons caught during 1974 represented a maximum that was followed at first by a very rapid initial lowering of catches (41% less between 1974 and 1975), later slowed down (lower than 16% between

Table 1 Half-yearly tonnage of fresh fish corresponding to the tonnage of 'banda' inspected at the entry to Maïduguri: Baga (BG) and N'Djamena (NDJ) roads from July 1969 to June 1977 in thousands of fresh tons (after Stauch 1977).

	BG	NDJ	Total	
1969 II	20.0	8.8		28.8
1970 I	23.6	8.1	31.7	65.5
1970 II	23.2	10.6	33.8	
1971 I	36.0	14.4	50.4	115.0
1971 II	50.3	14.3	64.6	
1972 I	55.3	19.4	74.7	165.7
1972 II	68.1	22.9	91.0	
1973 I	73.1	17.3	90.4	191.5
1973 II	80.5	20.6	101.1	
1974 I	89.7	28.0	117.7	220.0
1974 II	82.9	19.4	102.3	
1975 I	50.7	18.5	69.2	128.9
1975 II	33.8	25.9	59.7	
1976 I	38.6	20.3	58.9	108.2
1976 II	29.9	19.4	49.3	
1977 I	37.2	14.8		52.0

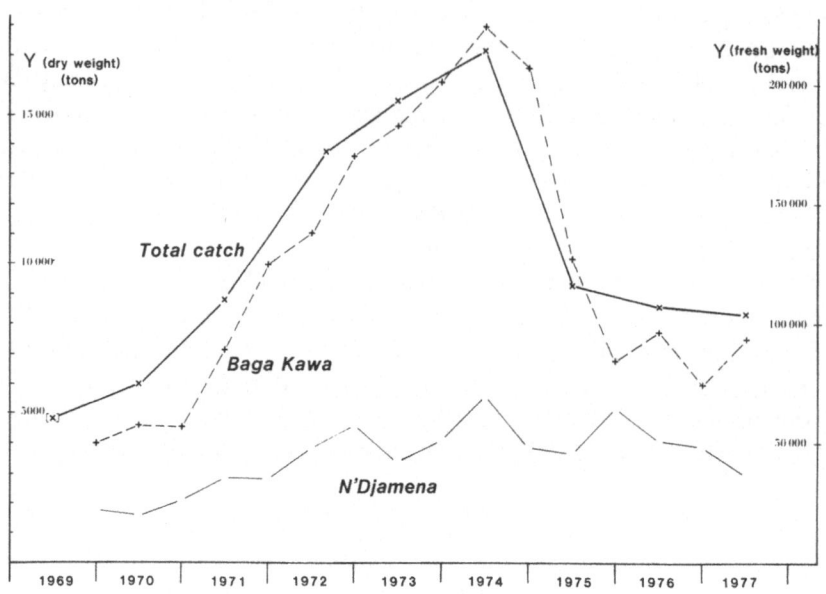

Fig. 2 Half-yearly statistics for the traffic of the Banda (tons); o———o = N'Djamena Road; x---x = Baga Kawa Road; x———x = total fresh weight (after conversion).

1975 and 1976). The partial data of 1977 give the impression of a stabilization of total catches between 100 000 and 110 000 tons. This change was explained by the increased fishing effort coincident with the lowering of the lake and by the increase of drying between 1972 and 1974. The maximum catch of 1973 and 1974 was due to a partial depletion of stocks, whose living space was considerably reduced within a few months. After the transition phase of 1976 when the lake returned to a certain stability, the permanent environment was limited to the south basin only. The total catches of 1976 and 1977 were thus the result of sustained exploitation of stocks but on almost half the aquatic space.

During 1969–70 the fish arriving by the N'Djamena route were caught in the rivers or in the delta; from the beginning of 1972, the proportion of catches in the south basin of the lake became appreciable then predominant. Between 1973 and 1977 the lake is estimated to have supplied 90 to 95% of the total catches.

From 1961 to 1963, the annual tonnage estimated at the exit of Maïduguri stayed the same: 6 to 7000 tons of banda or 30 to 35 000 tons of equivalent fresh fish. It therefore appears that the increase in production was relatively slow between 1963 and 1970 (Fig. 4). It will be seen later that it was however probable that the fishing activities had a remarkable development during the same period (Section 3.3).

13.3.2 The catch per unit effort

Some systematic studies were carried out from 1968 to 1971 on the landings from large mesh gill nets operating in the north basin. The observations were made in particular in two villages of the west coast, Baga-Kawa and Malamfatori, that were already the object of a detailed study (Durand 1973). Some experimental fishing GN 95 was done from 1963 to 1967 in the open water of the north basin of the lake (Hopson 1968).

Table 2 shows the mean catch per unit effort reconstructed for various periods between August 1963 and June 1971. The lowering of yields was very rapid from 1963 to 1966: from 18.3 to 4.6 kg. It was followed by a relatively stable period between 1967 and 1969 then the yields again showed a significant lowering during 1971, the average catch per unit effort for over 16 000 nights of fishing being less than 0.8 kg. Although the direct comparison of experimental catch per unit effort obtained during 1963–67 with that obtained for the local fisheries (1968–71) is not really valid it is necessary to emphasize that the average values were 25 to 30 times less from January to June 1971 than from August to December 1963.

Such a lowering of yield of course affected all the important species (Table 2). There appear to have been two different phases: a general lowering until 1967, when the catch composition had a tendency to simplify with a progressive dominance of *Lates*, a continued reduction in the number of *Citharinus* and *Heterotis* and an almost total disappearance of *Labeo coubie*.

Figure 3 shows two specific types of change: that of *Labeo coubie* which

Table 2 Average catch per unit effort (kg 100 m^{-2} night^{-1}) for the large mesh nets (90 and 95 mm) between August 1963 and June 1971 in the open water of the north basin of Lake Chad. N is the number of effort units on which the observations were made. The main species of fish caught are shown by the abbreviations: LN = *Lates niloticus*; HN = *Heterotis niloticus*; CI = *Citharinus citharus* and *C. distichodoides*; DR = *Distichodus rostratus*; LC = *Labeo coubie* (*: c.p.u.e. lower than 0.05 kg).

	LN	HN	CI	DR	LC	Others	Total	N
1963: August–December	7.4	1.8	0.9	0.4	7.2	0.6	18.3	47
1964: April and June and 1965: August–December	3.0	0.7	1.7	0.3	3.0	*	8.7	94
1966: January–July	1.3	0.5	1.6	0.1	0.8	0.3	4.6	90
1967: January–December	0.4	0.3	0.5	*	0.1	0.1	1.4	183
1968: January–December	0.6	0.3	0.4	0.1	0.1	*	1.5	20 123
1969: January–November	0.6	0.2	0.3	0.1	*	*	1.2	38 522
1971: January–June	0.5	*	0.1	*	*	0.1	0.7	16 467

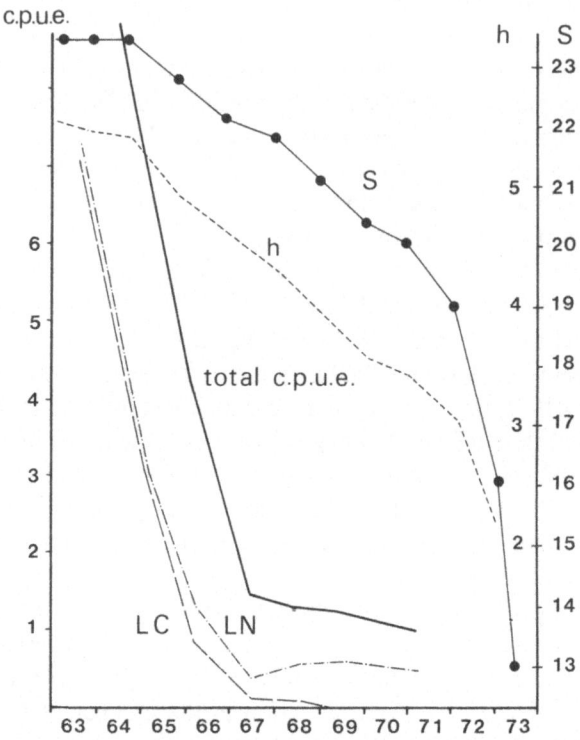

Fig. 3 C.p.u.e. evolution (GN 90 and 100; in kg 100 m^{-2} night^{-1}) between 1963 and 1971 in the Open Water of the Lake Chad northern basin; LC = *Labeo coubie*; LN = *Lates niloticus*; h = water level gauge at Bol (in meters); S = lake area (in 10^3 km^2).

dominated the catches during 1963 and then showed a negligible c.p.u.e. in 1967; that of *Lates niloticus*, which after following a similar change to that of *Labeo*, stabilized in 1966 and has since maintained itself at about 0.5 kg 100 m^{-2} night^{-1}.

Unfortunately no systematic observations of the local catches were made after 1971. Some scattered observations provide qualitative information on the change in the populations (Bénech 1975): *Lates* had become very rare by the end of catches with *Clarias*. Species more or less adapted to the conditions of the particular environments that were frequently hypoxic were present, namely: *Gymnarchus niloticus, Heterotis niloticus, Polypterus senegalus* and *Protopterus annectens* (Bénech et al. 1976; Bénech and Lek 1981).

The recent change in species composition of the catches corresponded of course to the considerable lowering of the lake since 1972; but during the preceding years the very rapid lowering of the catch per unit effort between 1963 and 1968 and the modification of relative abundance before 1971 are

440

explained on the one hand by the slow but already considerable change in the environment and changes in the fishing effort on the other hand.

Figure 3 also shows the height of the maximum annual lake level between 1963 and 1973 and the corresponding area of water (after Carmouze 1976). The morphology of the lake basin intervenes below the 4 m level which corresponds to a lake area of about 20 000 km^2. For the period of our study, after passing a maximum from 1962 to 1965, the lake area decreased regularly and the rate of lowering did not really accelerate as it did after 1971. Between 1963 and 1967, the change was not considerable (23 500 to 22 000 km^2) whereas the catch per unit effort dropped from about 15 to 1.5 kg 100 m^{-2}. Therefore it was the change in the fishing effort which explained the change which was particularly spectacular for *Labeo coubie*: its stocks were almost unexploited during 1963 but disappeared 4 years later (Fig. 3). Such a phenomenon must be related to the resilience of a species that was not under human predation before.

13.3.3 *The fishing effort*

The increase of total fishing effort has been shown to correspond to the introduction of nylon to make the gill nets. Thus the development of gill net fishing can be placed from about 1961–1963, a period corresponding to the first observations on catch per unit effort. No estimates of total fishing effort have been made since. However, based on a few observations, the total fishing effort can be evaluated from the accompanying data on total catches and the total catches per unit effort. This change can be estimated for the north basin of the lake between 1963 and 1971. Supposing that the c.p.u.e. of large mesh nets (GN 90 to 100) were well represented throughout the fisheries where the banda originated and also that the later experimental observations (1963–1967) can be put on the same basis as those obtained from the fisheries. We see that these assumptions are all the more acceptable because the extent of the phenomena was well marked.

Table 3 summarizes the most characteristic values for the total catches (in tons of fresh fish) and the catch per unit effort (in kg 100m^{-2} night^{-1} in the north basin of Lake Chad between 1963 and 1972. These values were transferred onto Fig. 4 with the estimates of the total effort obtained; the latter, shown as f, thus represented an hypothetical number of fishing nights for gill nets of 90/100 mm mesh. The total catches for 1967 were calculated by supposing a regular growth between 1963 and 1967. Other data (Durand 1973) however suggest that the c.p.u.e. during 1971 was higher than the value shown here, hence the hypothetical values, 1 and 1.5 kg. Finally, to cover the absence of c.p.u.e. data for 1972, they may be considered to have increased, given the lowering of lake volume, and are estimated at 1.5 and 2.0 kg.

The fishing effort multiplied by 20 between 1963 and 1969 (Fig. 4) and its

Table 3 Assessment of the total fishing effort (f) for the
fisheries of the north basin of Lake Chad between 1963
and 1972. the catch per unit effort (kg 100 m^{-2}) data are
those of GN 90 to 100 and T is the annual tonnage of
fresh fish corresponding to smoke-dried fish inspected on
the Baga-Kawa route at the entrance of Maïduguri. The
values in parentheses are hypothetical.

Year	T (tons)	c.p.u.e.	f (10^{-7})
1963	30 000	18.3	0.17
1967	(35 000)	1.4	2.50
1969	40 000	1.2	3.33
1970	46 800	(1.0)	4.68
1971	86 300	0.7	12.33
		(1.0)	8.63
		(1.5)	5.75
1972	123 400	(1.0)	12.34
		(1.5)	8.23
		(2.0)	6.17

annual increase during this period would have been 5×10^6. The latter
essentially corresponds to the very rapid lowering of c.p.u.e. whereas the total
catches increased only by about 30% in six years. Between 1969 and 1970, the
c.p.u.e. stabilized at about 1 kg 100m^{-2} and f increased by 10^7. During 1971, f
increased again by 4×10^7 for the same value of c.p.u.e., and, supposing that the
c.p.u.e. increased during 1971 and 1972, f stabilized between 6 and 8×10^7.

From the foregoing, it can be concluded that *between 1963 and 1972 the total
fishing effort was multiplied by about 40 in the fisheries of the north basin of Lake
Chad while at the same time, the total catches quadrupled.*

Starting in 1972, the environment changed profoundly, causing higher
vulnerabilities, and the use of the catch per unit effort data is no longer valid.
The maximum catches of 1973 and 1974 corresponded to the depletion of the
stocks and one may simply say that the apparent stability of the total catches of
1975 to 1977 actually concealed a very important species variability corresponding to a succession imposed by a very rapid change of environmental
conditions.

In conclusion, two major phenomena played a role in the changes in the
fisheries between 1960 and 1977. At first, there was a very pronounced
development of human predation on the stocks and this intensification of
fishing will doubtless continue with the introduction of new methods (nylon,

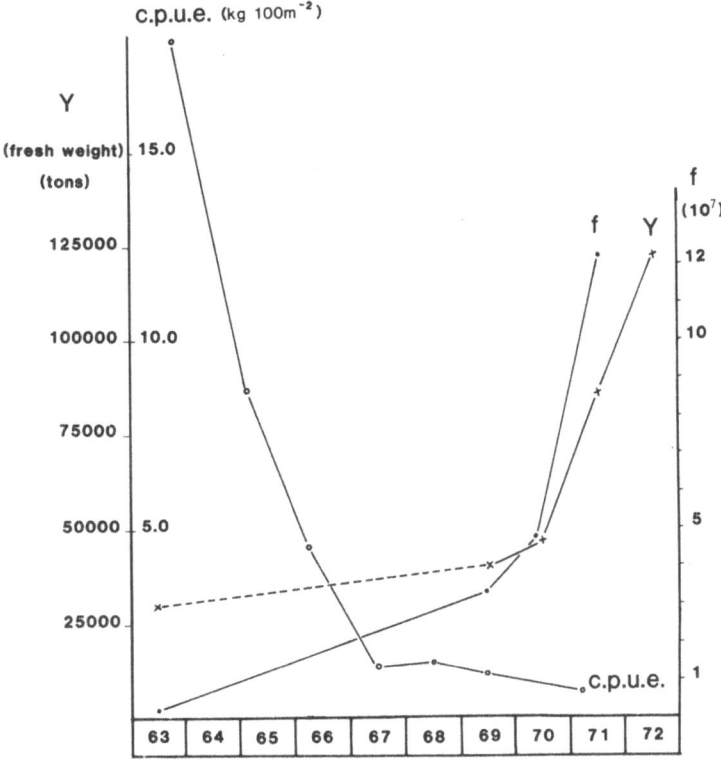

Fig. 4 Data for large mesh gill nets in Lake Chad northern basin (GN 90 and 100): yearly tonnage (Y), c.p.u.e. (kg 100 m^{-2} night^{-1}) and corresponding fishing effort (f in night × 100 m^2); cf. Table 3.

outboard motors). The second factor was the fall in lake level with the drought which at first accentuated the increase in fishing effort (1973–1974) and then reduced the available stocks (since 1975).

The prospects for fisheries can be examined by considering these two factors: (a) large variability of the environment involving parallel variations of potential catches; (b) increased level of fishing effort.

The latter will take place in the present small lake, and also if there is a future return to the maximum extension of the large Lake Chad, a point that will be developed later in our conclusions.

13.4 Riverine fisheries: the Shari delta

The riverine fisheries were very active before 1970 on the Shari between the lake and the confluence and on the mid-reach of the Logone and the Shari up-river from N'Djamena. There was then a progressive movement of fishermen towards the lake, with a tendency to look for increased catches by going down the rivers and concentrating progressively on the delta. This movement reached its maximum by 1971–72, a time when the rivers above N'Djamena were not fished any more, and when the fishermen still did not dare to really venture into the lake, fearing very difficult navigation conditions. At this time, the study of deltaïc fisheries was undertaken by Loubens (1973) and extended later by Quensière (1976). These observations lasted two and a half years and corresponded to the end of the normal period (1971–72) and the beginning of the drought (1972–73).

The Shari is connected to the lake by a delta with two branches of similar importance; the area studied extended to 12 km on the northern arm, with a width of 150 to 450 m (Fig. 1, Chapter 10). In this zone, 275 pirogues operated three types of fishing gear, resulting in four types of fishing: gill nets of small mesh used as fixing or drifting, drift nets of large mesh and nonbaited multiple hook lines.

The principal period of activity depended on the hydrological cycle and on the fish migrations. Small mesh drifting GN have a considerably reduced activity in very high waters (October) and during low waters (March to May). Fixed GN of small mesh were used during the lowering (February to July); drifting GN of large mesh during high waters from August to December and finally lines were used during low waters from April to July.

There was no concentration of fishing means nor of fish landings and the fishing enterprise was a lone fisherman, owner of his pirogue and sometimes supported by a paid helper. The fishermen, of very diverse ethnic origins work spread out along the length of the reach, and throughout it the physical, biological (migrations) and human factors show a great spatio-temporal variability. Statistically, this is an extreme method of fishing as it is entirely individual, the only relatively favourable factor being the sufficiently raised level of fishing effort in the study area. This type of fishing has rarely been studied and the need to acquire quantitative information necessitated the adjustment of adapted methods, based on direct observations and on the collaboration of fishermen.

Loubens and Franc (1972) and Loubens (1973) give a detailed description of the methods used to determine fishing efforts, catch per unit effort and total catches for each fishing method. The fishing effort was determined by counting the pirogues while fishing at different times of the day for the drift nets. As the average area of the nets was known, it was thus possible to extrapolate to 24 hours to obtain the daily effort characterizing the study period. The effort was also obtained by counting fishing gear and the pirogues for the fixed nets and the lines. The results obtained between July 1971 and December 1973 are indicated in Table 4.

Three major characteristics are immediately apparent on examination of this

444

Table 4 Daily fishing efforts in a reach of the Shari delta between July 1971 and December 1973 (after Loubens 1973, and Quensière 1976).

Month		Drift nets $(100 \text{ m}^{-2} \text{ h}^{-1})$		Stationary nets $(100 \text{ m}^{-2} \text{ day}^{-1})$	Lines $(1000 \text{ hook day}^{-1})$
		Small mesh	Large mesh		
1971	J	1180		30	340
	A	720	80		340
	S	470	470		
	O	220	800		
	N	1290	350		
	D	1640	30		
1972	J	1320			
	F	2750		90	
	M	390		500	
	A	210		370	50
	M	610		370	170
	J	1310		500	170
	J	1250		60	170
	S	2040			
	O	1760	440		
	N	670	300		
	D	1110			
1973	J	1370			
	F	1230			
	M	200			
	A	90			
	A	290			
	S	100			
	O		100		
	N		100		
	D		300		

table: the general predominance of drift nets of small mesh; a marked seasonal variation in fishing effort for each fishing method considered and a highly marked between year change during the observation period.

For the annual cycle 1971–72, the catch per unit effort showed obvious variations. Concerning the small mesh GN (Fig. 5), starting from high values

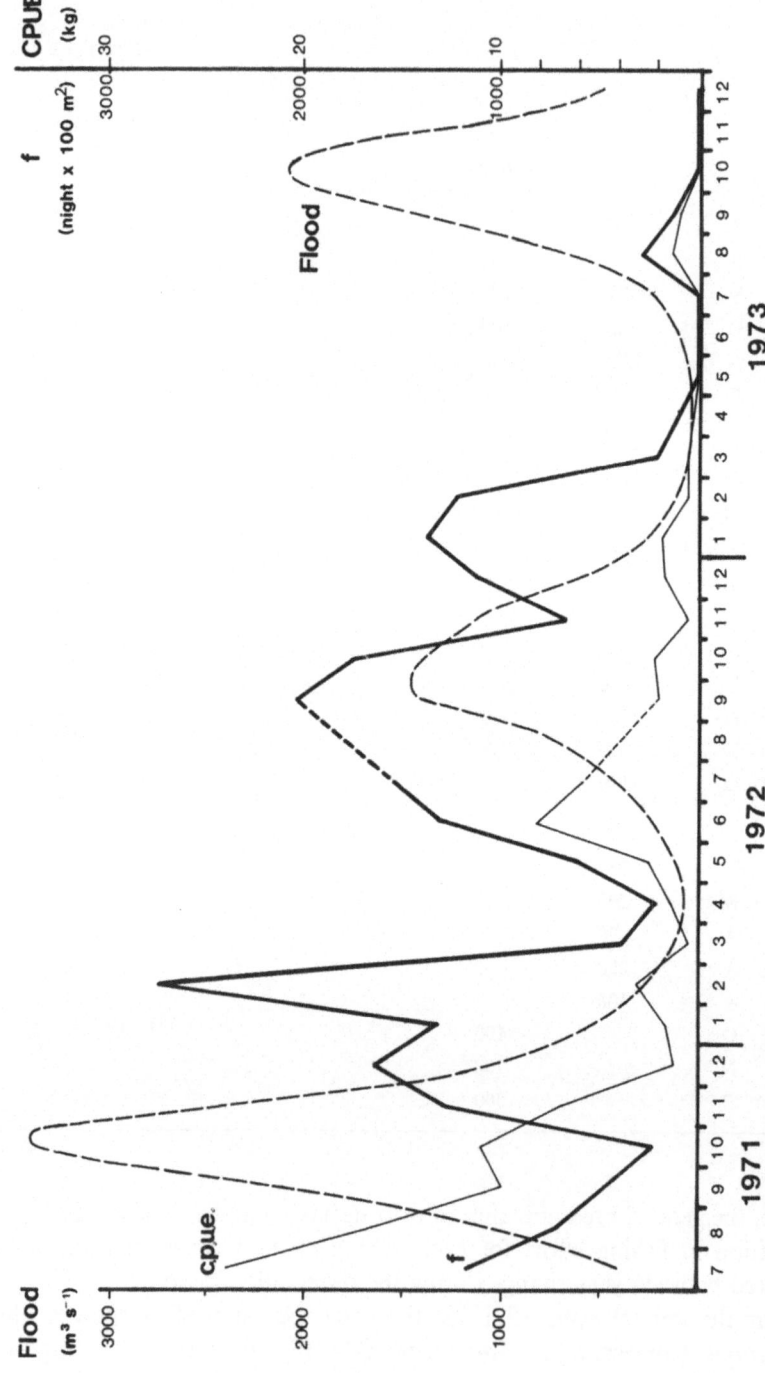

Fig. 5 Daily total fishing effort and corresponding c.p.u.e. between July 1971 and December 1973 for small mesh drift nets in a Shari delta reach (after Loubens 1973 and Quensière 1976); dashed line = flow of the Shari river at N'Djamena.

during July and August 1971, the average c.p.u.e. progressively dropped until December. It then stayed low until May, then increased during the flood of 1972 with a maximum in June. The c.p.u.e. showed an appreciative lowering during July 1972 in relation to homologous values of 1971 (Fig. 5). This lowering was followed by a collapse in 1973 as the average c.p.u.e. from February to September was about $0.5 \, \text{kg} \, 100 \, \text{m}^{-2} \, \text{hour}^{-1}$, that is, 10 to 15 times less than the yields obtained during 1971 at the same time. The large mesh drift nets were the only gear in use at the beginning of October 1973.

The interpretation of concomitant variations of c.p.u.e. and fishing effort for small mesh GN involves the fishermen, the fishes and environmental change both for the river and the lake. Figure 5 puts together the data on the mean c.p.u.e., the daily efforts and the Shari flood for the entire period studied. The small mesh drift GN activity was governed by fluvial hydrology for the period corresponding to a normal flood (July 1971 to June 1972): the effort dropped to very low values during the flood (normal during 1971), culminated over the course of the subsidence with a new minimum when the lowering was the most marked. All this was independent of the catch per unit effort although this varied considerably over the course of this period. It reflected the difficulty of operating during the excess current in October, and with the water and current shortage during the subsidence. This interpretation was confirmed during the following annual cycle: due to the exceptional low value of the 1972 flood volume, the fishing efforts remained very high during September and October 1972; in November, the time of maximum flood, they were still higher than in the preceding year, although the c.p.u.e. stayed very low. The first third of 1973 was similar to 1972 but from May and up to July the fishing completely stopped, the low waters being very marked and the c.p.u.e. remained at a very low level. It was this low c.p.u.e. which explains, that although the placing of drifting nets would not have suffered the pronounced lowering, they disappeared completely by 1973 while they were still very active from March to June 1972 (Table 4).

During 1973, the positioning of the fishing effort was progressively modified with a gathering of all the fishermen in the downstream part of the reach where they expected to have the best catches. This expectation was not achieved and the fishermen finally learned that in the isolated north basin of Lake Chad it was possible to work easily and with raised yields, so they left the delta and went there to try their luck (Quensière 1976). This explains why in August the fishing was not done by professionals but mostly by the farmers for whom the fishing was complementary and who therefore did not look for optimal exploitation.

Total catches were the sum of the concomitant variations of yields and efforts and they became almost negligible during the last six months of 1973. Table 5 summarizes the results between July 1971 and December 1973 for the four fishing methods. The fixed nets and the multiple hook lines were not used

Table 5 Monthly and annual total catches (tons) for the reach of the Shari delta studied; sdr = small mesh drift nets; sfi = small mesh fixed nets; ldr = large mesh drift nets; mhl = non-baited multiple hook lines (*corresponds to total monthly catches of less than five tons).

Month	1971					1972					1973		
	sdr	sfi	ldr	mhl	Tot	sdr	sfi	ldr	mhl	Tot	sdr	sfi	Tot
January						70				70	70		70
February						170	20			190	10		10
March						50	60			110	*		*
April						10	50		10	70	*		*
May						30	310		60	400			*
June						170	250		50	470			*
July						270	30		20	320			*
August	450		10	60	520						10		10
September	200		80		280	110				110	*		*
October	110		100		210	130		70		200		10	10
November	230		40		270	10		10		20		30	30
December	150		10		160	60				60		20	20
Total	1140		240	60	1440	1080	720	80	140	2020	90	60	150

during 1973. Over the course of the first annual cycle, from August 1971 to July 1972, total catches amounted to 3000 tons and were about eight tons hectare^{-1} for the reach studied. Considering that the reach did not represent the entire north branch of the delta and by granting to the west branch this same importance as the north branch, about 8000 tons would have been caught in the delta (Loubens 1973). The apparently very high density does not correspond to a sedentary biomass but mostly to a permanent replacement of migrant populations rejoining the lake or the river system following the hydrological cycle. Thus the catches of the delta must be integrated inside the large ensemble where the migrations occur, as was attempted in the previous section.

During the second annual cycle, from August 1972 to July 1973, total catches collapsed to 600 tons, five times less than during the course of the preceding cycle. This catastrophic change then continued, as shown by the comparison of September to December data: 920 tons were taken in 1971, 390 in 1972 and only 60 in 1973.

Actually the fishermen did not leave the deltaïc fishing grounds during the period of very low floods of 1972 and 1973. Indeed although this created some unfavourable conditions during the 1973 low waters, conditions during the maxima of 1972 and 1973 were better than during an average year. The lake was the refuge of fluvial stocks between two anadromous reproductive migrations and it was this lowering which encouraged them to move in step and invade the lake. This change was unavoidable because the displacement towards the lake had already begun several years earlier with the rapid increase of fishing effort and the concomitant lowering of yields.

13.5 Fishing and flooded areas

In spite of their temporary nature, the flooded zones were colonized very rapidly by several species that represented a significant biomass. The wealth of this environment involved the development of traditional fisheries, very hardy activities that also showed an obvious seasonal character.

The two large flooded zones dealt with here both depended on the Logone: the flooded zone of Ba Illi (cf. Fig. 22, Chapter 2) between the left bank of the Shari and the right bank of the Logone and the flooded plain of North Cameroon.* These two zones differed in their morphology and in their connection to the permanent hydrological system.

The flooding of the plain between Logone and Shari originated upstream

---

* The Yobé, a poorly known intermittent river will not be considered here. Although its water supplies to the lake are low, it is important because its basin reaches 85 000 km^2 and extends westwards in Nigeria as far as the Jos plateau.

from several outlets on the right bank of the Logone and from the major outlet of the 'Grande Courant' that originated downstream of Lai. The flood waters were drained by the N'Gourkoula which joins the Logone a little upstream from Logone-Gana and other various communications probably existed between the plain and the Logone in several places downstream of the last best known outlets. the morphology of the plain led to a relatively varied but temporary aquatic environment containing depressions, channels, flat vegetation zones and regions of raised cover. Several openings on the permanent river environment and the easy internal communications explain why this zone was mostly open and affected the fish populations found there.

The large flooded plain of North Cameroon was very flat and interrupted only by hillocks on which the Kotoko villages are placed and where the only trees were found. The plain was exclusively occupied by grasses (*Echinochloa*) in water depth of 0.70 to 1.00 m. The major origin of the flood was the Logomatia which was the only important tributary on the left bank. The drainage of the flooded plain occurred partly through the Logomatia which served as an outlet, but also through several temporary drains which rejoined the El Beid flowing into the southernmost region of Lake Chad (Fig. 1, Chapter 10).

During the flood, the dispersion of fish made them difficult to catch and the flooded environments were often poorly accessible, thus concentrating all the fishing activities on the tributaries during the subsidence. Three fishing zones should be noted in particular: the large dam upstream of Logone-Gana, the fisheries of Logone-Gana, the fishing weirs of the El Beid. The first two were good examples of the diversification and complementarity of the traditional fishing techniques used in the exploitation of these aquatic environments; on the contrary, in the El Beid fisheries, a single technique was used for the bulk of the catch.

13.5.1 *The traditional fishery of Logone-Gana*

The main outflow of the flooded zones of the right bank of the Logone, upstream of Logone-Gana called Koulambou or N'Gourkoula drained the regions of the North and South Ba-Illi plains (Fig. 22, Chapter 2). When the Kotoko fishing head of Logone-Gana had decided that fishing could begin, generally at the end of December, depending on the degree of subsidence, the river was completely blocked by a series of small spaced screens that made the passage of fishes trying to reach the Logone at the beginning of the subsidence difficult. Several fishing techniques were then used: when the fishes reached the dam they could turn back upstream, but then they were likely to be caught by the large Kotoko basket traps whose opening is turned downstream and which occupied the whole river bed (Fig. 6A). Some species tried to jump the barrier (particularly *Hydrocynus*) and often fell into the pirogues downstream. The

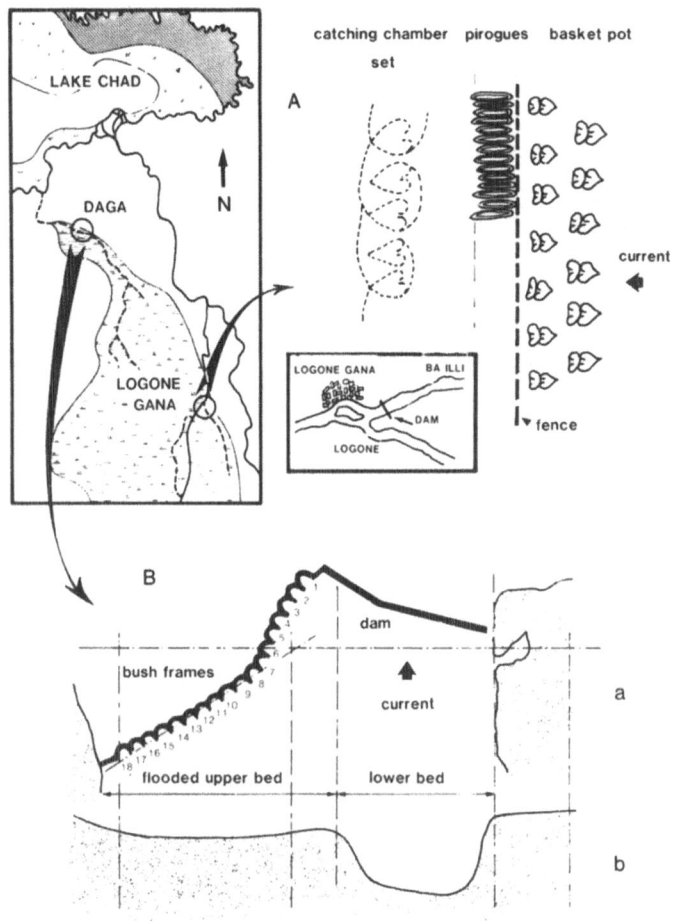

Fig. 6 Two examples of the traditional fisheries on the outlets of the flooded areas; (A) the great fishing of Logone Gana (cf. text); (B) a fishing dam in the El Beid river; (a) plan; (b) cross section; 1 to 18 = numbers of the fishing chambers.

fishes avoiding the basket traps and the pirogues met the large Kim and Kabalaï capture chambers set up downstream of the first dam right across the river. Last but not least, numerous gillnets of average mesh size were set during the first week.

The main feature of these subsidence fishings was that they dealt with adult fishes only. *Alestes* constituted the bulk of the first day's catches and the dominant *Alestes dentex* was 10 to 20 times more abundant than *A. baremoze* which was generally very abundant in all the other environments (cf. Chapter

10). *Alestes* was then replaced by still water species which were carried away at the end of the fall.

The basket traps were set only for about a week and then removed; the capture chambers, however, remained on for one or two months and the residual waters of the reach were still exploited by fixed gill nets, non-baited multiple hook lines etc. ... Total catches could be appreciable about 250 tons were caught in 1969 during the first three days which generally gave the highest catches. The variability of yields from one year to the next must have been particularly high because of the underlying natural variability, linked to the yearly hydrology, and the human estimates which sometimes led the fishing head to authorize very late the placing of the dams, when the first migrations had already begun.

13.5.2 *The Logomatia fisheries*

The seasonal fishing activities in the Logomatia region were followed for two consecutive years, 1977–79 by Vidy (1982). Comparisons with the data from before the drought were therefore not possible and the results of these two years cannot be considered in the general context.

Subsidence fishing developed from the beginning of October in the down-stream end of the Logomatia, a temporary river which well represented an arm of the Logone, an outflow zone through the flood plain of North Cameroon and a zone of returning water from the flood plain to the Logone. Five types of fishing gears, corresponding to eight different ways of fishing were found there: gill nets; triangular nets or 'sakamas' similar to those of the El Beid; little catch weirs on man-made channels; unbaited multiple hook lines and dams.

The total calculated production of the reach was similar in 1977–78 and in 1978–79: about 300 tons, mostly comprising the young of a few months as for the El Beid fish. It is interesting to note that this homogeneity of total results hid the considerable differences between the catch per unit effort from one year to the next as the very weak flood of 1977 involved an increased vulnerability of fishes and very high yields. An analogous study of the 'Kafue flats' (Welcomme 1979) implied that 'a weak flood affected the stock not only by limiting the recruitment but also by augmenting the efficiency of fishing towards the young fish' (Vidy 1982).

The satellite observations of 1975 and 1978 lead to an estimation of 60 000 hectares for the area drained by the Logomatia during similar floods. This represented about 15% of the maximum flooded area for the entire Yaéré (400 000 ha). Given the very low population density in this region, we can estimate that the fish production on the Logomatia was well represented by what was caught. We thus arrive at some yields between 5 and 7 kg ha^{-1} for five months flooding. It is true that these very low figures refer to the exploitation of stocks that did not always regain their past importance (e.g. *Alestes baremoze*, cf. Section 6.2) after the drought. It is however possible that the low yields were partly due to the difficulties of exploiting a flood plain where communications were a problem.

All the same, the 300 tons caught during the four months of traditional fishing activity in the Logomatia were quantitatively significant because it was practically the only commercial resource in this region. One needs to question whether this is sensible as the general economy of fishing might suffer from catches of young fish of two to five months. It would be theoretically more rational to take individuals that have completed their potential growth, but the young fish from the Logomatia and Logone could be food for ichthyophages (*Hydrocynus, Lates*...) which have a tendency to gather in the rivers bordering the subsidence outflows. We will return to this problem with the El Beid fisheries.

13.5.3 *The El Beid fisheries*

The El Beid only flowed for five to eight months per year, when the waters of the North Cameroon plain arrived: through the El Beid, they flowed out towards Lake Chad where they reached the southernmost region of the south basin. The river bed was well marked, from 50 to 60 m wide and from April to July only consisted of a string of muddy ponds. The supply of the El Beid was dependent on local precipitation which corresponded to the first flows of August and September and of the Logone flood which included the maximum flood of El Beid in December. The river then overflowed and the subsidence lasted until the end of March.

The floods of the El Beid were evidently highly dependent on the Logone floods, the inundation of the North Cameroon plain being all the more important as the flood of the Logone was strong and prolonged above the level from which it flowed out towards the lower plain. Thus the irregularity of the floods was particularly high, the extremes in the last 25 years corresponding to a mean annual volume estimated from 87.5 m^3 s^{-1} (1953–54) to $0 m^3$ s^{-1} (1972–73 and 1973–74); the between year mean could be estimated at about 40 m^3 s^{-1} corresponding to an annual volume of 1.3. thousand million m^3, or only about 4% of the Shari supplies to the lake.

The importance of the El Beid for the fish stocks of the Lake Chad region was due to its connecting role between the floodplain, where the young of several species gathered and obtained shelter and food, and Lake Chad. It was by far the easiest way to leave the flooded plain at fall as the water passed mainly to the north: 85% of the flooded zone could be considered to be drained via the El Beid and only 15% via the Logomatia (Bénech et al. 1982).

The El Beid fisheries were based on the presence of fish dams associated with the use of triangular nets, the 'Boulous'. Aerial observations showed that these permanent dams were distributed throughout the length of the river from Tildé to the lake (Fig. 1, Chapter 10). In January 1969 there were about 270 (Durand 1970b). During 1968–69 fishing was active on 165 dams while in total 200 could be used normally. The average density for the whole was almost 2.0 km^{-1} and for active dams alone 1.2 km^{-1}, i.e. one dam every 800 m. The distribution was actually unequal as two thirds were found in the lower part of the course of the El Beid. The maximum density of the dams was found about 100 km away from the flooded zones where only 150 m separated them from each other. The development of these structures corresponded to old and steady fishing activities.

The dams were made of interlaced thorny branches secured very tightly onto trunks sunk into the river bed. They were renovated every year and they diverted the fishes to the minor bed where the dams were extended by some semi-circular joining frames, with their opening turned upstream (Fig. 6B). Following the morphology of the bed, the number of fishing places was highly variable: from 4 to 28, the average being 15.2 per dam. Nets were always placed in very shallow water, a meter at maximum and at the time of the subsidence, the frames in the major bed were progressively emptied and the fish reached the lower bed.

The catches almost exclusively included individuals under one year, on average 3 to 5 months (Durand 1970a). Table 6 shows the relative importance by number and weight of 18 major species caught over the whole 1968–69 fishing season, as well as their average length. The comparison with the maximum sizes observed for the Chad basin (LMO) showed that they were mostly juveniles, much smaller than the sexually mature fish.

Total catches during the fishing season 1968–69 were estimated at 7.7 tons for the studied dam. The extrapolation of these figures to 150 active dams in 1968–69 gave a total figure of 1200 tons. By its location and number of fishing frames (18) the chosen dam could be considered as representative of the average dam of the El Beid. On the other hand, it appeared that the year of observation was lower than a normal year because there was a certain correspondence between the total yield and the importance of the flooding, thus of the flood. The 1968–69 flood was low with an annual volume lower than half of the average of the last 25 years; the low yields obtained at the time of a mediocre flood gave rise to a diminishing fishing activity and an emphasized decrease in total catches in comparison to a year of normal flood. It can thus be concluded that the total catches of an average year should clearly exceed the 1200 tons given above.

Considering the importance of catches made in the El Beid of 3 to 5 months old juveniles, one can wonder if this practice is not harmful to the total adult catches of the same species in the lake or in the rivers. We were able to examine this for *Alestes baremoze*, the only sufficiently well-known species. The young

454

Table 6 Relative importance by weight and number of the main species caught by triangular nets over the fishing season 1968–69. $\bar{L}$ is the average length in the catches and LMO the maximum length observed in the Chad basin.

	W_T (%)	N_T (%)	$\bar{L}$ (mm)	LMO (mm)
Sarotherodon niloticus	17.4	6.6	97	395
Hyperopisus bebe	12.8	8.3	162	508
Marcusenius cyprinoïdes	11.1	9.5	125	330
Alestes nurse	9.5	9.0	97	218
Sarotherodon galilaeus	9.0	5.5	93	410
Sarotherodon aureus	6.7	5.0	98	260
Brienomyrus niger	4.4	12.7	69	130
Alestes dentex	3.3	2.8	127	410
Distichodus rostratus	3.2	1.1	37	625
Polypterus bichir	3.1	0.2	340	660
Clarias spp.	2.8	4.4	90	(890)
Schilbe uranoscopus	2.7	11.1	74	340
Alestes baremoze	2.1	3.8	91	330
Labeo senegalensis	1.6	1.3	108	550
Citharinus citharus	1.3	0.4	133	580
Siluranodon auritus	1.2	7.5	60	123
Mormyrus rume	1.1	0.4	194	870
Distichodus brevipinnis	1.0	0.1	188	588
Total	94.3	89.7		

weigh between 7.5 and 10 g at four months, the average age reached when they are caught in the El Beid. The weight of 100 of these juveniles has been compared to the yield they would represent in the riverine fisheries for various combinations of the instantaneous coefficients of mortality. F varies from 0.05 to 1.5 and M is 0.3 or 0.5. Figure 7 summarizes the results obtained for each sex separately as there is a difference for growth in weight.

In most of these cases, the gain would be appreciable since the ratio of the ponderal yields of these 100 recruits would range between 2 and 6, supposing that — as it has been verified otherwise — the sex ration is close to 50%.

We tried to estimate the consequences on the total catches by supposing that the dynamic parameters of *A. baremoze* stocks represented those of an average stock among those of Lake Chad. The total catches were estimated at 1200 tons for the 1968–69 fishing season, thus corresponding to a rather poor year. A probable range for the entire catch of the juveniles of the region might be from 1000 to 3000 tons. Table 7 shows the results obtained for each of these two figures in the various hypotheses of mortality. It is seen that the increase in total

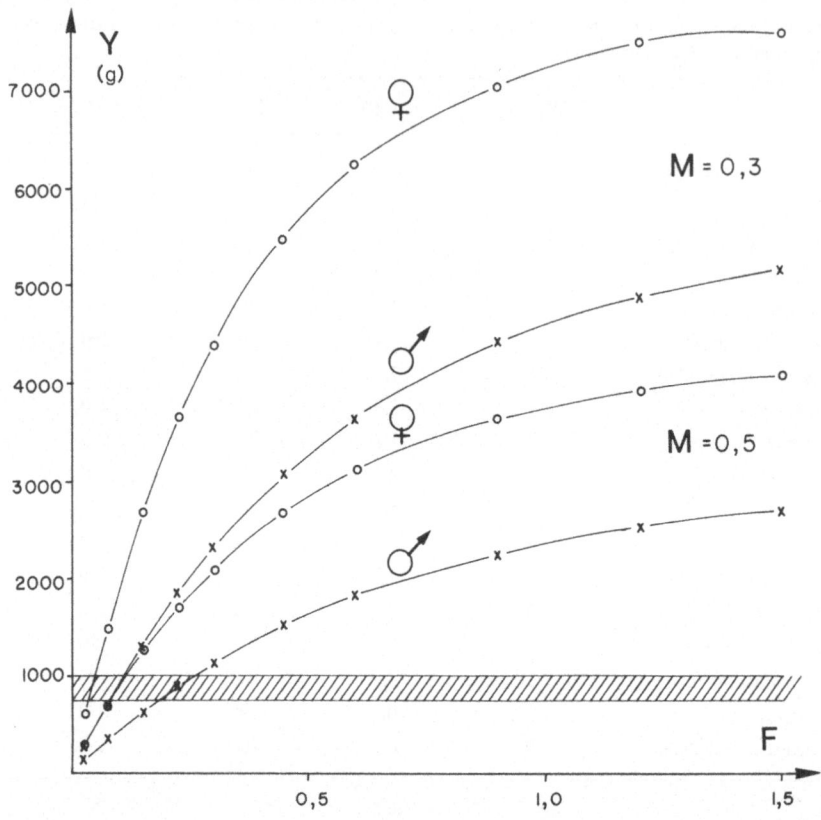

Fig. 7 Comparison of the yield (for 100 four-months old recruits) in the riverine fishery (males: $t_c = 3$ years; females; $t_c = 2.75$ years) with the weight of these 100 young *Alestes* caught during the fall fishing (▨).

Table 7 Potential catches (tons) in the fluvio-lacustrine fisheries if there was no fishing of juveniles in the El Beid (cf. text). M and F are the instantaeous coefficients of mortality.

El Beid catches	1000		3000	
M	0.3	0.5	0.3	0.5
F ⎧ 0.4	4600	2200	13 800	6600
0.7	6000	3000	18 000	9000
1.0	6700	3500	20 100	10 500
1.5	7000	4000	21 000	12 000

catches should be between 5000 and 15 000 tons. Of course, it must not be inferred that all the juvenile catches have to be stopped because one ignores the extent to which the rivers and the lake would have reached this large number of supplementary recruits without increased competition and increased natural mortalities.

The temporary aquatic environments thus develop traditional fishing activities that are characterized by their very seasonal nature and their occurrence mostly on the outlets. The latter are chosen because fishes in search for permanent rivers or laker at the subsidence concentrate and get through them, as well for adults (Logone-Gana) as for juveniles (Logomatia and El Beid).

The flooded zones had many beneficial effects in the general economy of fishing. Above all there was an increase in the general productivity because of the considerable extension of the water area, where new sources of food developed very rapidly with the flooding (phytoperiphyton, zooperiphyton, insects ...). They also offered shelter for the young, as is particularly clear for plains such as North Cameroon's. The density of vegetation cover was such that the fishes of large size could not penetrate it easily and the predators were relatively rare. The type of connection with the permanent environment also played a role because it was probable that the natural predation rate on the young was much higher in the rivers than in the lake where the relative dispersion of prey and the presence of some shelter must have diminished the risks that they encountered. Thus this was another positive aspect of the flooded plain of North Cameroon.

Nevertheless it cannot be denied that the predation exerted by man was important in the El Beid. It was not obvious, despite the considerable fishing system in use, that removal was very high. If the total catches in the flooded area during 1968–69 are reduced to 350 000 hectares, a yield of 4 kg ha^{-1} is obtained, similar to the results of Vidy (personal communication) for the Logomatia. A recent global estimate for the entire flooded plain of North Cameroon, taking into account every kind of gear, led to a 14 kg ha^{-1} figure (Quensière, personal communication). The latter represented the actual total production which remained far lower than the figures given by Welcomme (1979) who gives a range of 40 to 60 kg ha^{-1} year^{-1} for the mean sustained production of flooded tropical zones.

13.6 The exploitation of a fluvio-lacustrine stock: *Alestes baremoze*

This species was chosen for the study of population dynamics and exploitation of stocks for several reasons, in particular its importance in the Chad basin as well as in all the large basins of the Sudano–Sahelian region. The potential yield of stocks for the whole region must be between 30 and 50 000 tons in a normal climatic context and the construction of large reservoirs in the region should be favourable to the development of zooplanktophagous lacustrine stocks. The results obtained can be extrapolated to other stocks similar to this species and showing analogous characteristics. On the other hand this species benefits from a particular commercialization route that allows the calculation of fishing quantity to be obtained in some conditions with some precision. Finally, most of its biology is known, providing the essential data for demographic studies. After a brief review of major biological features, the evolution of stocks and catches since 1969 will be discussed, followed by some aspects of exploitation.

13.6.1 *Main biological characteristics*

The biology of this species and its behaviour in the Chad basin are strongly influenced by the existence of the lake and the migratory cycle originates from Lake Chad. This is so both for the adults which leave the lake for upstream reproduction only and for the immature fish which stay inside the lake until the time at first reproduction, that is to say at the end of the third year when Lake Chad has its normal extension. On the other hand, if the permanent lacustrine character weakens with the fall of the lake, they return to the biology observed in the western river basins of the Sahel where *Alestes baremoze* reproduce mostly at two years. There cannot be more than one spawning a year due to the short reproductive season and the necessary maturation times. Above 130 grams — corresponding to an age of about 30 months — there is a linear relationship between the fecundity (in thousands of eggs) and the females weight. The fecundity is high and corresponds to 230 000 eggs kg^{-1}. Associated with the percentage of mature females at each age, this relationship permits the study of the stock fecundity as a function of the exploitation rate (Section 6.3.3).

The abundance of *Alestes* is partly explained by the fitting of the biological cycle to the hydrological cycle as they spawn mostly by the border of the flooded zones at the time of the Logone overflowing. The latter carries the young fry lower down inside the flooded plain where they find rich and varied food and are protected from predation by the density of aquatic and semi-aquatic macrophytes. The catadromous migrations of many fish via the El Beid allow them to rejoin the lake without being exposed to an intense predation pressure when descending the river (Fig. 1). The lacustrine environment is well suited to the zooplanktophagous *Alestes* which finds shelter in the archipelago, the vegetation and the shallows of the east and the north.

The influence of a permanent lacustrine environment is also apparent in the case of growth. The factor controlling growth in length is certainly temperature, the variations are similar in river and lake environments but the possibility of permanent zooplankton food in the lake allows *Alestes* to acquire a very high condition during the period of growth check in the cold season while the rivers

458

in subsidence do not have any substitute food. This results in better total growth than in the Ivory Coast and Senegal. These results were confirmed by observations made in the eastern region of the southern Sahelian zone (lakes Albert and Rudolph) where growth was clearly more rapide than in Chad and accomplished mostly during residence in the lacustrine environment (Hopson J. 1975). The zooplankton diet there allowed the best food yields and, as the annual temperature variations were very small, there was no growth check.

The study of growth in length confirmed that there is a clear difference between sexes, and for the same age *Alestes* the weight is 25 to 35% greater in large adult females than in males. From four months growth fits the von Bertalanffy model well (Fig. 8). In the very young fish growth is clearly more rapid because it occurs from September to November in the flooded zones where the presence of an abundant zooperiphyton is particularly favourable. Starting from the age of four months, the two following equations are thus retained:

$$\male \quad L_t = 251.80 \left[1 - e^{-0.058\,(t+2.91)}\right]$$
$$\female \quad L_t = 292.00 \left[1 - e^{-0.043\,(t+3.82)}\right]$$

with monthly values of k, as time is given in months.

The marked condition cycle, with an annual difference of about 30% for adults, is characteristic of several Sudano-Sahelian migratory species for which the phase of river displacement is generally a phase of scarcity. These changes in weight for length directly affect the production. From the equations given

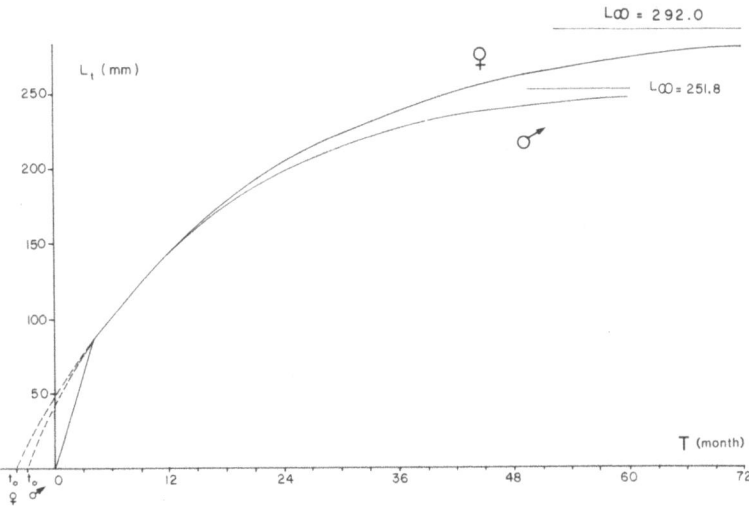

Fig. 8 Length growth curve for *Alestes baremoze* males and females in the Lake Chad basin after Von Bertalanffy's equation. Between 0 and 4 months, the best fit is the straight line $L_t = 20.785\,t$, (L in mm; t in months).

459

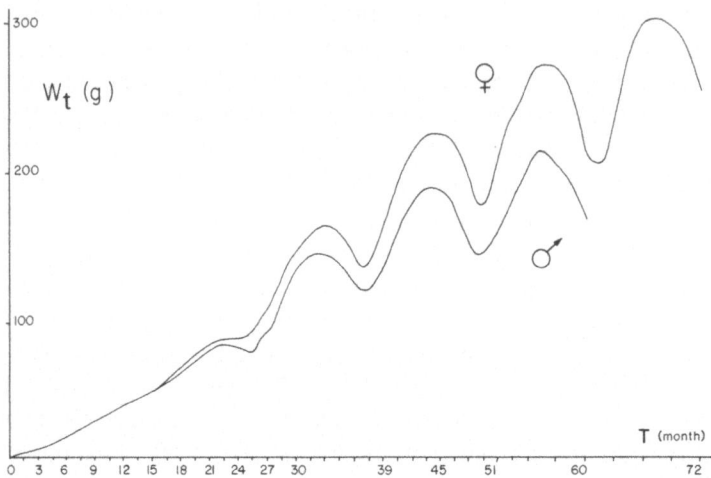

Fig. 9 Weight growth curves taking condition factor variations into account for *Alestes baremoze* males and females in the Lake Chad basin.

above describing the growth in length the average growth by weight curves can be obtained (Fig. 9), by considering the existence of different allometric coefficients for the sexes, the immatures and the adults and by integrating the seasonal variations of condition. Estimates of average weight as a function of age have been used in the production models.

13.6.2 *Changes in the A. baremoze stock*

The configuration of the lake basin and the enormous demand for fish in the Nigerian market resulted in the relatively simple scheme of traffic for dried fish given in Section 2.2. It should be noted merely that from the fishing statistics collected from 1969 to 1977 (Stauch 1977) and from some corrections and extrapolations of results, it has been possible to reconstruct the total quantity of *Alestes baremoze* (fresh weight) corresponding to the quantity checked on the commercialization routes. These estimated annual catches are shown in Table 8 and on Fig. 10.

From 1969 to 1971 and to a lesser extent 1972, most of the catches were made in the river system and the delta region (graph A on Fig. 10). From 1972, the proportion of lake catches increased rapidly: 13.7% in 1972 to 60.3% in 1974. In the river system, however, the decrease was continuous from 1970 to 1976: from more than 8000 tons to an apparent disappearance. On the whole during the period 1975–77, an average of 50 to 100 times less *Alestes baremoze* were fished than during a normal year.

Table 8 Estimated annual catches (tons) of *Alestes baremoze* between 1969 and 1976 based on controls at Maïduguri, after corrections; A = lower river system and delta (1969–72) and south basin (1973–76); B = north basin.

	A	B	Total
1969	6120	240	6360
1970	8070	210	8280
1971	4435	535	4970
1972	4380	695	5075
1973	2345	1425	3770
1974	1310	1990	3300
1975	*	145	145
1976–77	*	*	*

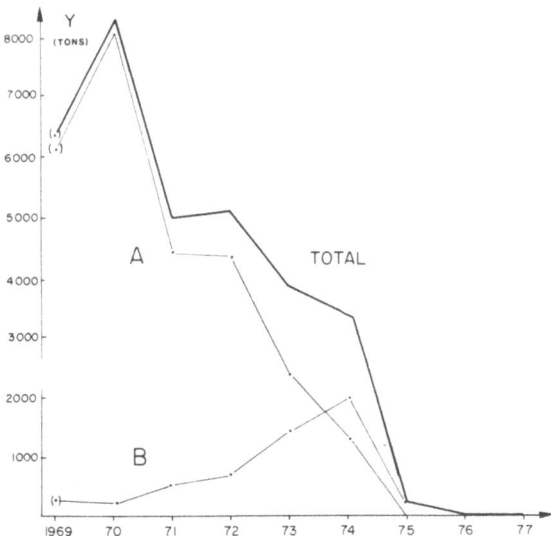

Fig. 10 Estimated yearly catches (Y) for *Alestes baremoze* between 1969 and 1977 from Maïduguri. Check-points data, after correction; (A) river, delta and south basin; (B) north basin.

To outline the analysis of the general change in stocks, it is necessary to know the change in the catch per unit effort (c.p.u.e.), failing that of the effort itself which remains the most difficult to understand and assess in the individual artisanal fisheries. A satisfactory description can be obtained from the experi-

461

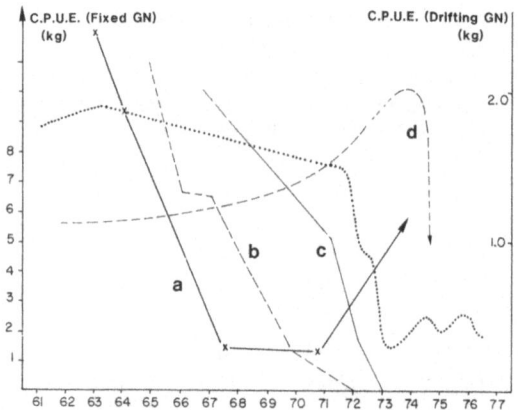

Fig. 11 Schematic evolution of the c.p.u.e.; ----=experimental fishing; —— local fishing; (a) large mesh GN, north basin; (b) GN 30, drift nets, riverine fishing; (c) GN 28.5, drift nets, Shari delta; (d) GN 25–30, fixed nets, north basin. The dashed line shows the Lake Chad water level evolution.

mental data especially in as much as the observed phenomena have been particularly marked (Fig. 11). As it is impossible to attach the same importance to experimental and to local c.p.u.e. we have simply indicated the tendencies as they appear to be, based on the partial data obtained between 1965 and 1975. In the fluvial system the catch per unit effort from drift nets was clearly much higher during 1966–67 in the river upstream of N'Djamena as well as in the delta. These catches corresponded to a little exploited stock. There was then an appreciable reduction in yields, in a ratio of about 10 to 1 until 1972. The exact speed of this lowering is not known but the c.p.u.e. had already been lowered considerably by 1971 and it is thus possible that this change had been fairly continuous since 1966–67. The experimental data of the southern archipelago showed a similar but earlier phenomenon, as the catches were already low for the adults in 1969–70. While the c.p.u.e. in 1965 (CTFT) reached 15 kg, they were only 1 to 2 kg 100 m^{-2} night^{-1} during 1969 (Fig. 12). On the contrary, the density of *Alestes* in the north basin was higher in 1971 and later increased in 1973 and 1974 due to the division of the lake and the rapid reduction of the remaining water volume.

During normal times in the Chad system, the north basin is the most marked lacustrine environment and thus a refuge for a large part of the various fish stocks. The use of large mesh nets developed much earlier there than fishing with average size gill nets. This explains the slower reduction in the catch per unit effort for *Alestes baremoze* (Fig. 11). Between 1966 and 1971, although there was no direct influence of the early fall of the lake on the catches of *Alestes*, their behaviour was affected as during this period adults of the southeastern archipelago disappeared. It is probable that they went to the north basin where they found some very deep waters and where the zooplankton

462

remained considerable. Even if this phenomenon could have played a certain role, the change in the fishing effort in the river system was preponderant between 1966 and 1971.* Four to five years later, we recognize the same scheme as for the large mesh gill net fisheries in the north basin (cf. Section 3.3).

From the beginning of 1971, the change in the lake environment was felt much more clearly. The catches per unit effort as well as the average sizes decreased rapidly in the delta fisheries during 1972, before the catastrophic effects of this very low flood could be recognized. The degradation of environmental conditions in the south of the lake resulted in a gathering of *Alestes* in the north basin and an apparent absence of the 1972 migration, well before the interruption of the communications between the two basins. In 1973 there was practically no more fishing in the delta and the fishermen gathered in the north basin which was isolated: the concentration of fishes led to an increase of catches during 1973 and 1974, then to a disappearance of stocks in 1975. From 1973 to 1975, fishing did not play any role as the fishes were anyway in a critical situation that had to end with a natural collapse. The change in total catches is in agreement with all these hypotheses (Fig. 12).

In conclusion, the fishing effort, corresponding to traditional techniques until 1960–65, though significant, has not been considerable. Through the introduction of nylon and industrial gill nets the total effort was considerably increased and became too high between 1970 and 1972. The shift of the Sahel drought led to the disappearance of stocks in 1975. The reconstitution of *Alestes* stocks, again allowing considerable catches, depends on the total fishing effort level and on the way it would be distributed.

13.6.3 *Dynamics and exploitation*

Three complementary aspects of the dynamics of *A. baremoze* populations will be examined: the mortalities, the yields per recruit and the fecundity of the stock.

13.6.3.1 *Instantaneous mortality rates.* To approach the problem of production and yields per recruit separate estimates of instantaneous rates M and F are needed. No direct calculation can be made for *Alestes baremoze* in Lake Chad. However, there is a correlation between the type of growth observed and species longevity: the former is generally slower as the latter increases. In a study of the relationship between growth rate and natural mortality Beverton and Holt (1959) showed that there was a more or less linear relationship between M

---

* This change in total effort intervened at first through its total increase, but also through the progressive concentration of the fishing in the delta area.

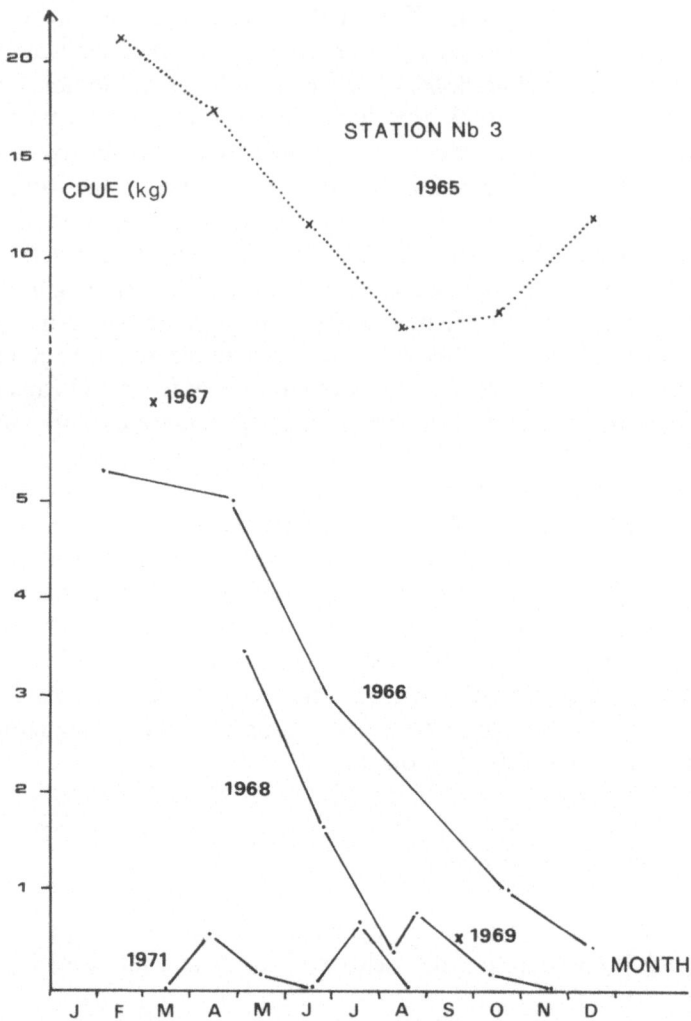

Fig. 12 C.p.u.e. for *Alestes baremoze* in the Southeastern Archipelago (GN 30; kg 100 m^{-2} night^{-1}); c.p.u.e. always equals 0 from January 1972. The dashed line shows the evolution of the c.p.u.e. at station 3 (CTFT; *Alestes baremoze* and *A. dentex* gathered).

(instantaneous rate of natural mortality) and k (Bertalanffy coefficient) for some well known marine fish families.

Studies on estimates of M for tropical fishes of continental Africa are rare if not non-existent. In a study of *Pseudotolithus elongatus*, a species with a longer life span than *Alestes*, Le Guen (1971) found values of M ranging between 0.162

(M 0/1) and 0.439 (M 4/5). Similarly, Fontana (1979) recorded values of M between 0.2 and 0.6 for Congolese demersal fishes.

Alestes baremoze has a short life-span (five years for males, six years for females). Moreover, the variability of environmental conditions in the Sahelo-Sudanian zone (periodic drying of flooded zones, severe deoxygenation and frequently increased temperatures in the shallow water) must involve a very high natural mortality. Here M is taken as 0.3 to 0.5, corresponding to the first phase of the study with a well developed lake environment. With the drying of the lake, the so-called natural mortality would have increased.

From estimates of Z and M, the major traits of the changes in mortality due to fishing F can be inferred. From 1966 to 1970, F must have been low (about 0.4) and the exploitation moderate (Fig. 13). The exploitation intensified slowly, leading to F value of 0.6 to 0.8 during 1971–72. From the end of 1972 until December 1974, when the stocks disappeared, Z again increased maybe up to 1.70.

These two phases of increased mortality due to fishing are not at all similar:

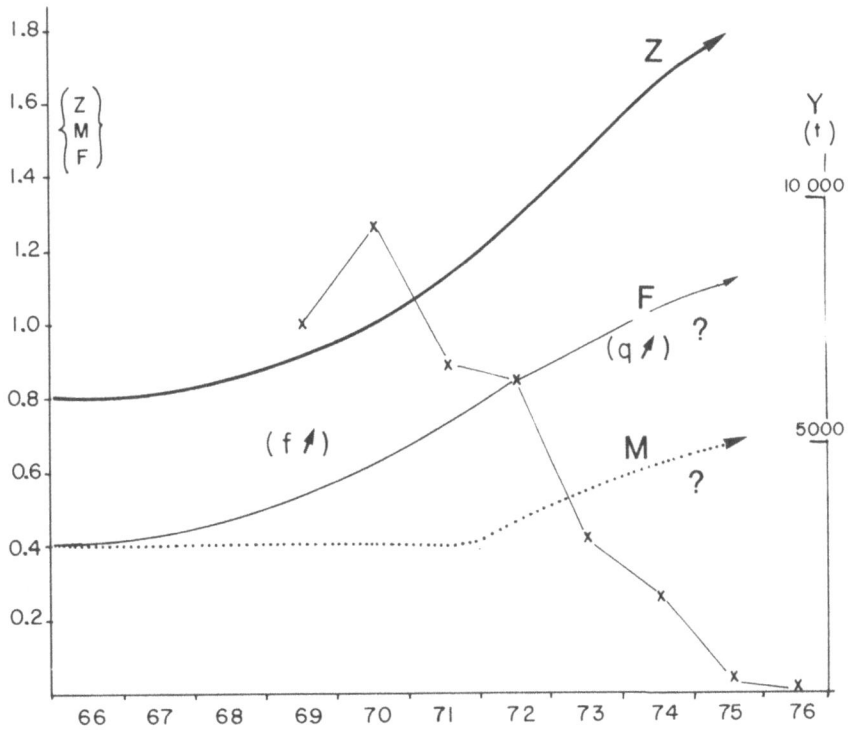

Fig. 13 Probable schematic evolution for instantaneous mortality coefficient of *Alestes baremoze*. The thin line corresponds to the yearly total catches.

465

in the first phase the increase in fishing effort, f, as well as its concentration in the delta zone probably led to the increase in F. In the last phase just preceding the disappearance of the stock, there was not necessarily an increase in total fishing effort. It was limited by the increasingly reduced distribution area of *A. baremoze* and also by increased vulnerability of fishes concentrated in the environments where the fishing was much easier, thus q, which represented the mortality per unit of fishing effort in the expression $F = q \times f$, increased. There was also a parallel increase in M over the course of the drought.

13.6.3.2 *Yields per recruit.* Beverton and Holt's model was chosen to describe the variations in yield per recruit as a function of the mortality coefficients M and F on one hand, and of the age at first catch, t_c, on the other hand. Four cases have been considered, for males and females separately, for two likely values of M, 0.3 and 0.5, in the scale of hypothetical values going from 0.25 to 4.75 years for t_c and from 0.1 to 0.5 for F. Figure 14 shows the types of results obtained for males with $M = 0.5$; the drawn curve corresponds to maximum yields per recruit for a given F and the black part corresponds to the situation prevailing from 1966 to 1971.

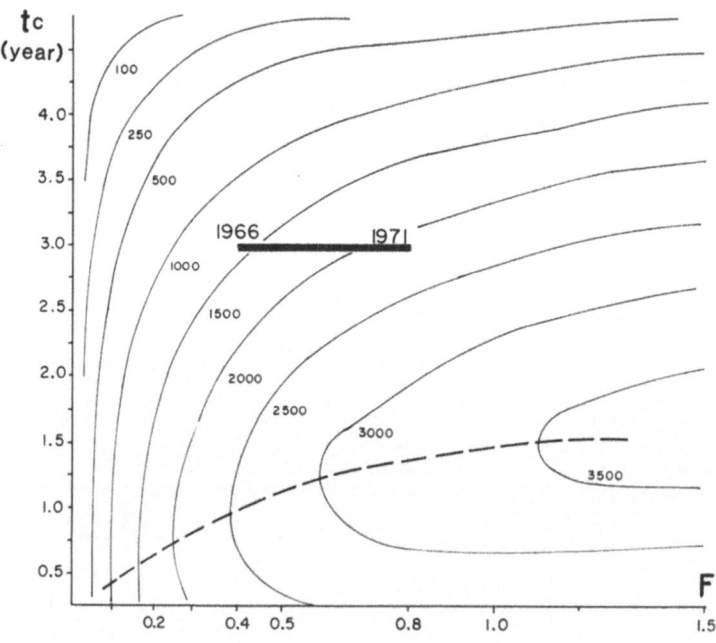

Fig. 14 Contours of equilibrium yield per recruit (Beverton and Holt) for the males of *Alestes baremoze* in the case of $M = 0.5$. The dashed line corresponds to a Y maximum for a given F. The black strengthened segment corresponds to the situation observed between 1966 and 1971.

Between 1966 and 1971, the fishing was mostly fluvial and only caught adults during spawning migrations from the lake. These adults were mostly in their third year and they reproduced at the age of three years. They began to invade the river system during the first quarter and thus the average age at recruitment could be estimated to be about 2.5 years for the two sexes, as the fishes enter the fishing zone as soon as they leave the lake. This average age at recruitment corresponds respectively to 215 mm for males and 224 mm for females.

The river fishing was then entirely based on the use of gill nets. The average length of *A. baremoze* caught by GN 28.5 was about 240 mm and the recruitment sizes were thus clearly lower. The sizes corresponding to the average age at first catch must be higher than those reached at 30 months, which is why the estimated average ages at first catch were respectively chosen as 2.75 years for males and 3.0 for females. Therefore, to have an estimate of the total yield for the species, it is necessary to add the yields corresponding to the t_c chosen above (Table 9), supposing that fishing mortality and recruitment are identical for both sexes. The contribution of females is always higher and represents more than two thirds of total yields as long as F values stay low and still 60% for F = 1.5.

Over the course of the period considered here the total fishing effort probably increased progressively and F doubled, passing from about 0.4 to 0.8. The corresponding change in yields per recruit is shown in Fig. 14 for the males.

It is certain that the stock was clearly underexploited in 1966 at the beginning of our observations and the increase of the mortality due to fishing was completely positive while leading to a considerable increase of yields: for M 0.3, Y increased 48% in males, 32% in females when F went down from 0.4 to 0.8;

Table 9 Estimation of total yield, Y (in grams for 200 recruits of 4 months, 100 males and 100 females) as a function of M and t_c.

F	M = 0.3			M = 0.5		
	Y♂(t_c = 3)	Y♀(t_c = 2.75)	Y(♂ + ♀)	Y♂(t_c = 3)	Y♀(t_c = 2.75)	Y♂ + ♀
0.03	294	638	932	142	293	435
0.08	705	1493	2198	341	690	961
0.15	1318	2686	4004	640	1257	1897
0.23	1851	3642	5493	904	1724	2628
0.30	2315	4409	6724	1135	2111	3246
0.45	3071	5523	8594	1520	2701	4221
0.60	3646	6251	9897	1821	3117	4938
0.90	4422	7052	11 474	2244	3635	5879
1.20	4882	7419	12 301	2514	3926	6440
1.50	5162	7599	12 761	2691	4103	6794

for M 0.5 there was an increase of 52% in males and 42% in females (Table 9). Beyond this a new increase in fishing effort would have much less effect. A new doubling, from 0.8 to 1.5 gives no more than a gain of 15 to 20% for the males and about 30% for the females. Although considerable in absolute value, this increase in catches would have implied, all else equal, a doubling of fishing effort and thus a very clear decrease of catch per unit effort.

It is thus more likely to be a decrease in the age at the first catch that could lead to a new improvement of yields. Figure 15 shows the so-called eumetric fishing curves as they are obtained when writing out the values of t_c corresponding to a maximum yield for a given value of F. It is seen that the various hypothetical ages at first catch are placed mostly above the eumetric curve. For males, the optimum age at first catch would be placed at 1.5 (F about 0.9); for females, the best value of t_c would be near 2.35 (M 0.3) and 1.75 (M 0.5). It must be noted however that the corresponding gain of yield to this level of fishing effort is not very high: from 5 to 20% according to sexes and M values.

Until now we supposed that F remained constant and therefore that the fishing effort was regularly distributed throughout the year. In reality, however,

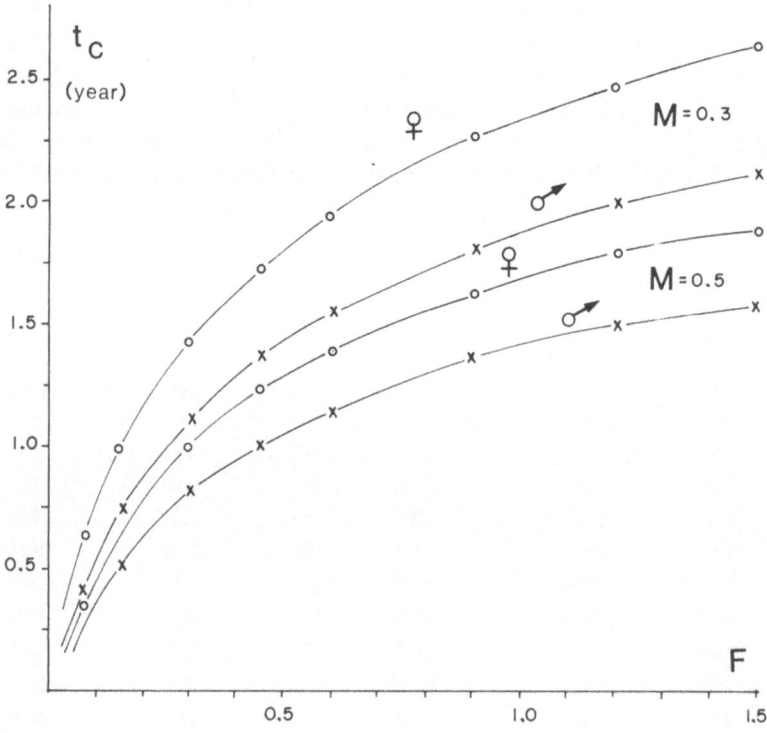

Fig. 15 Eumetric fishing curves for *Alestes baremoze*.

468

the fishing activity was rhythmically related to the hydrological cycle of the rivers. This cycle was accentuated in the Sahelo-Sudanian zone, where during the two extremes, low and high waters, there was a very clear reduction in fishing activity. The fish dispersed during high water with the increase in current speed, and, during low water, some techniques (drifting GN) could not be used and the fish took refuge in the lake (cf. Section 4).

To our knowlege, there have been very few attempts at direct evaluation of fishing effort for continental artisanal fisheries and none within tropical Africa prior to the study of Loubens (1973) in a reach of the Shari delta. this study has been summarized in Section 4.

In order to see if the annual modulation of fishing effort, thus of F, introduced an appreciable difference, we recalculated the yields with the help of the Ricker model. For males, when $M = 0.3$, it was supposed that the quarterly values of F would have corresponded to variations of f found by Loubens (Section 4). The comparisons made for constant F and F changing quarterly showed that there was a slight increase in yields due to the fact that the most active fishing of the first quarter caught fish in optimum condition.

To know if the systematic catch of individuals in a better condition had a considerable influence on the total catches, we used the Ricker model by simulating a concentrated fishing effort during the quarter when the *Alestes* have the best condition, supposing that the fishing was closed during the nine other months. If the quantitative aspects only are considered, the gain is noticeable by comparison to the model at uniform F: from 8 to 15% for the F/t_c couples chosen for 1966–1970.

13.6.3.3 *Stock fecundity.* The study of stock fecundity is based on the notion of theoretical fecundity per recruit introduced by Le Guen (1971). From the relationship between weight and fecundity ($\varphi = 0.345\ W - 25$ where φ, the fecundity, is given in million of eggs and W in grams) and taking into account the proportion of ripe females of a given age we obtain the relationship between female fecundity and age; we admit here that the average number of spawnings for the entire female stock is equal to one, this should slightly overestimate the absolute fecundity since a part of the lacustrine females, especially in the context of a greater Lake Chad, did not reproduce each year.

The theoretical fecundity is the sum of individual fecundities at a given time so that $\varphi = \Sigma N_t \varphi_t$. It thus depends on the structure of the stock and of fishing. Expressed as a function of virgin stock theoretical fecundity, it allows, for a given combination of F and t_c, the estimation of the relative level of stock reproductive potential (Garcia 1977). The various fecundities for various combinations of t_c, F and M have been calculated here (Table 10). The results are given for an initial total number of 10 one year females during the period 1966–70.

The two values circled in Table 10, 2798 when $M = 0.3$ and 1504 for $M = 0.5$

Table 10 Fecundity of *Alestes baremoze* stock as a function of F and t_c (years) for M = 0.3 and 0.5 ——: fecundity of females from 1966 to 1970 (in millions of eggs for 10 females of one year).

F \ t_c	1.25	1.75	2.25	2.75	3.25	4.25	5.25
M = 0.3							
0.04	2458	2545	2596	2646	2689	2759	2798
0.2	1508	1751	1932	2120	2298	2587	2760
0.4	848	1144	1390	1665	1940	2412	2718
0.8	305	555	824	1152	1497	2161	2646
1.6	60	197	403	638	1105	1886	2540
2.0	31	135	300	400	700	1600	2300
M = 0.5							
0.04	1338	1386	1407	1433	1455	1487	1504
0.2	848	985	1085	1187	1279	1415	1489
0.4	498	672	803	970	1116	1344	1473
0.8	194	353	514	710	908	1246	1446
1.6	70	139	280	488	714	1120	1405
2.0	19	99	229	434	665	1085	1391

practically represent the fecundity of virgin stock as they refer to theoretical values of the age at first catch, 5.25 years, which is close to the maximum age reached, and to some extremely low values of fishing mortalities: F annual = 0.04. Rather than build the curves of general isofecundity similar to those obtained for the yields per recruit, the relative fecundity for $t_c = 2.75$ and $t_c = 1.75$ have been shown here (Fig. 16). When the age at first catch is 2.75 years — the estimated mean value in the fluvial fisheries — the relative fecundity of the stock is still 35% for M = 0.3 (and 42% for M = 0.5) when F = 1.0. Thus it may be concluded that the traditional exploitation which does not catch females before 30 or 36 months protects the reproductive potential of the stock even if the exploitation is quite heavy (F = 1.0). If a new type of exploitation were to be introduced, with females caught before sexual maturity ($t_c = 1.75$), the relative fecundity of the stock would remain appreciable: 20% for Z = 1.5 (M = 0.5; F = 1.0) (Fig. 16).

In conclusion, three results can be considered in the forecasting of optimal exploitation of *Alestes baremoze* stocks: (a) the suppression of fishing dams on the El Beid could involve an average annual gain of some hundreds of tons in the fluvio-lacustrine fisheries; (b) the concentration of fishing effort during the months when the *Alestes* have the best condition would lead to another gain; (c) the stock fecundity is not, *a priori*, a limiting factor.

Obviously, (a) and (b) cannot be practiced due to strong socio-economic constraints.

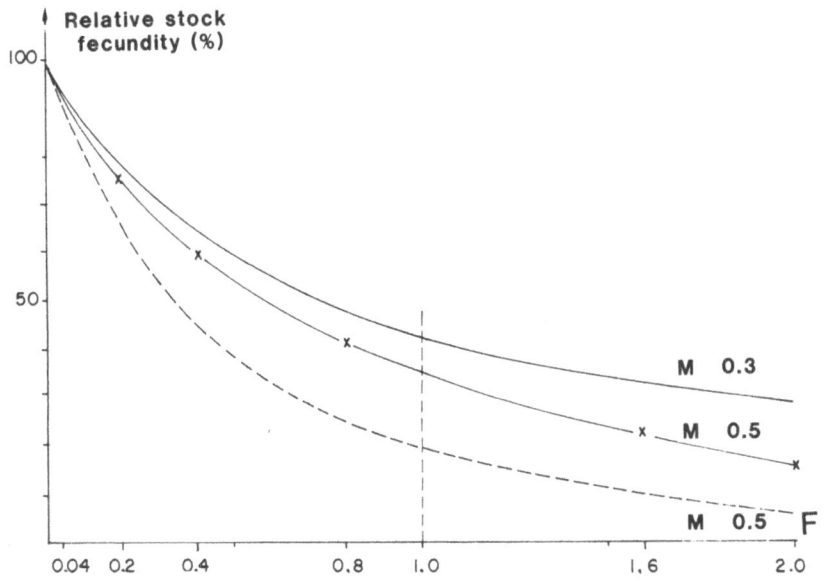

Fig. 16 Relative fecundities for the stock of *Alestes baremoze* (%); —— = t_c = 2.75 years; - - - -
= 1.75 years.

13.7 Rational exploitation and prospects of fisheries

In comparing the total catches obtained for the *Alestes baremoze* stock and for
all the large sized fish, between 1969 and 1977 some spectacular differences can
be seen (Fig. 17). Between 1970 and 1974, the *Alestes* catches were reduced by
three fifths, although during the same period the production of large mesh
fisheries tripled from 65 000 to 220 000 tons. Some analogous tendencies appear
between 1974 and 1975 with a very rapid reduction of catches in the second
year. However, some fundamental differences in the reaction of stocks were
noted later when the *A. baremoze* practically disappeared during 1976 and 1977
and the total catches of average or large sized fishes stabilized around 100 000
tons (Fig. 17).

The explanations on the preceding pages considered first the nature of the
stocks compared, monospecific and multispecies; in the latter species replaced
each other in a succession due to the changing of the lake environment. Some
intervening human factors also partially affected the changes in catches as seen
in the underexploited multispecies stocks between 1969 and 1971 when the
catches of *A. baremoze* in the rivers were not far from the optimum. However, it
is clearly evident that the primary factor remained the change in the environment which was effective in two ways:

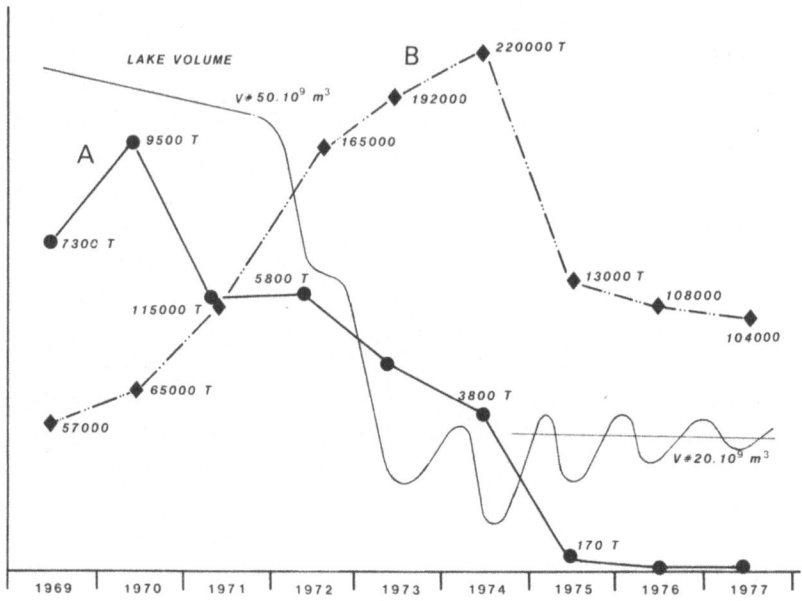

Fig. 17 Compared changes in total catches of salanga (*Alestes*, curve A) and banda (large size fish, curve B).

— on the *annual recruitment*, by the importance of the river flood and of some temporary flooded environments which were directly dependant on it;
— on production, through the development of permanent aquatic environments which conditioned the general productivity of the ecosystem.

Any rational exploitation, or, in a more realistic way, any attempts bringing a contribution to it, will depend on the various foreseen patterns of environmental evolution. The following will be examined in this view.

13.7.1 *Possible trends of A. baremoze exploitation*

For the dynamics study and for the estimates of yield per recruit (Section 6.3) we have limited ourselves to the period characterized by a well-developed Lake Chad and a unique fluvial fishery. From 1971–72, an intermediate period began with a concentration of fishing effort in the deltaic region and the change in catch per unit effort which fell in 1972. For some months very small mesh nets appeared (about GN 25) indicating a lowering of the age at first catch. This was a very transitory stage because there was an almost immediate transfer of fishing effort towards the south basin of the lake during 1973, the north basin in 1974 and a rapid depletion of stocks that would have been killed by the drought

472

in any case. Contrary to several species whose stocks were either maintained or developed after 1973, *A. baremoze* was very vulnerable to the deteriorating environmental conditions.

Although the period 1972–77 barely resembled the classic equilibrium exploitation, it is instructive to examine all the hydrological factors that contribute to the good condition of fish stocks in general and of *Alestes baremoze* in particular in the Lake Chad region and to the resilience of stocks subject to severe climatic risks.

Two hydrological parameters were paired:

— the Logone flood on which the flooding of North Cameroon and the flood of the El Beid depended (cf. Chapter 2);

— the total volume of the lake which itself was a function of yearly supply balances by the Shari. The latter affects the total level of stocks through all the susceptible lacustrine factors that play a role such as shelter for the young, available space, food etc. ...

It should be emphasized that these two aspects were practically independent because, apart from exceptional periods, the supplies of Logone to the lake change scarcely from one year to the next and played a relatively minor role in the lake level.

The beneficial effects of the Logone flood can be shown in two ways: (Bénech and Quensière, personal communication).

— *growth of juveniles* (Fig. 18). There was a positive correlation between the overflowing of the Logone and the average weight of some species from El Beid, the one migratory, the other sedentary. First growth is thus stronger since the juveniles stayed longer in the flooded plain. The fishing yield showed the same tendencies (Fig. 18). The 1974–75 fishing season is placed separately because it corresponded to the first flooding after two consecutive years of drought and the flood accumulated over these two years (livestock, vegetation cover ...) may have led to an exceptional productivity;

— *recruitment of migratory species*. There was a satisfactory relationship between the overflow volume of the Logone and total catch per unit effort for a number of migratory species (Fig. 18). For 1974–75, the catch per unit effort did not follow the relationship for the same reasons as above. The data did not permit further species analysis because for the years of observation the effect of the variability of the Logone overflow may have been concealed by the consequences of the drought on the reproductive potential. However, that a connection existed between the importance (extent) of the Logone flooding and recruitment of migratory species is highly probable. Moreover this does not exclude the Shari from being able to play a role, probably less important, with the notable extension of its major bed during high waters.

The reduction of *Alestes* stocks between 1972 and 1977 was expressed by an apparent disappearance of adults in the fishing statistics but also by a lack of recruitment during 1972, 1973 and probably 1974. We saw that during 1972

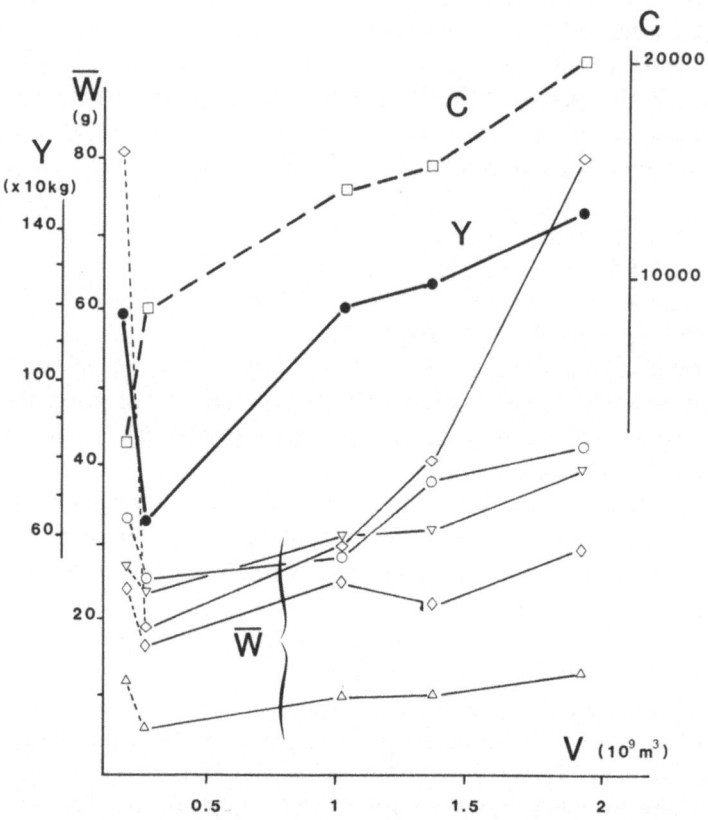

Fig. 18 Relationships between the flood volume of the Logone river (V), and the specific mean individual weights ($\overline{w}$), the global yield (Y), the c.p.u.e. in number for some migratory species (C) in the El Beid river.

and 1973 the weak floods of the Logone scarcely included any overflowing into usually flooded plains and that the El Beid did not flow at all. The young spawned during September 1972 and 1973, which were moreover fewer that in a normal year, thus could not rejoin the permanent hydrographic regime. During 1974 the lake was at its minimum and in spite of an almost normal river flood the recruitment did not appear to be considerable. On the other hand, during November 1977, *Alestes baremoze* again began to be caught in the delta fisheries. The nets used, GN 15 to 20, were catching 180 to 200 mm *Alestes*, two-year old fish during September 1975 when the presence of older *Alestes* appeared much more sporadic.

We may come to the conclusion that the recruitment of *Alestes* was practically stopped for three years, from 1972 to 1974, and it was only in 1975 that the spawns enabled a rebuilding of stock. This partial reconstitution in these conditions can only be explained by their very high fecundity: it was enough that a small group of spawners survived to be able to reproduce in 1975; it confirms that the fecundity should not be a limiting factor in any of the possible patterns.

Besides these disturbances, linked to a regression or an extension of Lake Chad, two stable situations appear to be plausible:
— existence of a large lake of 18 000 to 24 000 km² corresponding to some normally occurring rains and floods;
— Lake Chad reduced to a single south basin as a permanent environment, the annual floods being much less on the average than in the first case and not sufficient to fill in the north basin.

The first situation corresponds to that from 1966 to 1971 and the second prevailed since 1973. In the last case, the south basin was not homologous to typical lacustrine environments because as a whole it was much more influenced by the Shari flood (Chapter 2).

This rather fluvio-lacustrine character had two major consequences:
— sexual maturity occurred earlier and the first reproduction took place at an average of two years. The fishing in the river system would then operate from an age at the first catch up to 1.75 years, which has been seen to correspond to the best exploitation;
— on the other hand, the sustained yield per unit lake area for such a fishery would be lower than that of the large lake because the higher annual instability would make it less productive.

For the exploitations of stocks corresponding to a well-developed Lake Chad, and by referring to an average situation, particularly for recruitment, two situations should be considered to find the best combinations of t_c and f — thus F — for an optimal exploitation.

1. The habits of fishermen are not radically modified and they are reluctant to set *Alestes* nets in the waters of a lake regaining ground. The situation of the years 1966–1970 again occurs with the movement of migrating adults which direct the fishing so it only develops in the rivers and the delta. The age at the first catch is then a little lower than three years and cannot be reduced. The use of 25 mm mesh nets would improve the yields somewhat, in particular for the males.* Moreover, the optimum fishing effort would correspond to instantaneous coefficients of mortality due to fishing of about 0.8 to 1.0. In these conditions it must be possible to obtain an average yield a little higher than that obtained during the four years for which we have some observations: an

---

* Recall that in all the cases, the best exploitation would be led with 25 to 28 mm mesh when the actual legal minimum size is 35 mm resulting in an underexploited stock where the catches would only deal with the largest sized females.

average of 7500 tons from 1969 to 1972. An increase of 20% leading to annual catches of about 9000 tons would appear to be reasonable.

2. Lake fishing is kept up in spite of the raising of lake waters, giving a combination of two fisheries. The fluvial fisheries would continue to be exploited seasonally at the time of migration of the large adults. In the lake, a moderate fishing could take place, with 22 to 25 GN, on 18 to 30 juveniles. We saw that the age at first catch should be lowered in order to obtain an optimal exploitation. It is not obvious that fishing on adults would be of great benefit because they are more dispersed than the juveniles who have a tendency to gather in certain regions of the archipelago particularly in the northeast of the lake. The combination of limited fishing efforts in the lake and the river would lead to an increase in yield due to the lowered age at first catch. The gain, depending on the sex and level of F, would average from 15 to 20% by comparison to the previous one and thus the balanced total catches reach 10 to 11 000 tons.

13.7.2 General prospects of lake exploitation

It is not possible to analyze the involvement of multispecies stocks in the varying exploitation and environmental circumstances, as has been done above for a single species. Even if the specific data on biology and demography had been obtained, it would be still very difficult to build up something serious since interspecific relations intervene. However, it is possible to consider the total productivity of the lake through total fishing yields.

13.7.2.1 *The yields*. For flat waters like those of Lake Chad, a direct relationship can immediately be noticed between the area occupied by the waters — and the incident solar radiation — and the fish production through the various trophic chains. Thus the total catches of the fisheries depend on the total extent of the lake water and it is interesting to estimate the average annual yields per hectare from 1969 to 1977.

The total estimates must be affected by some factors such as the local production of the flooded zones and the rivers, the state of the lake (in particular the importance of the vegetation and of different types of plants that are more or less favourable to aquatic life). Nevertheless, by comparing total production and areas of basins and lake over the course of the period 1969–1977 (Fig. 19) some yields per hectare are obtained that could be calculated more precisely in further studies.

The figures used for estimations are gathered in Table 11 and yields obtained for the entire lake and the north and south basins considered separately, are shown in Fig. 20. To distinguish the production of the two basins it is necessary to arbitrarily assign, starting from 1975, a major part of the tonnages counted on the Baga-Kawa road to the south basin.

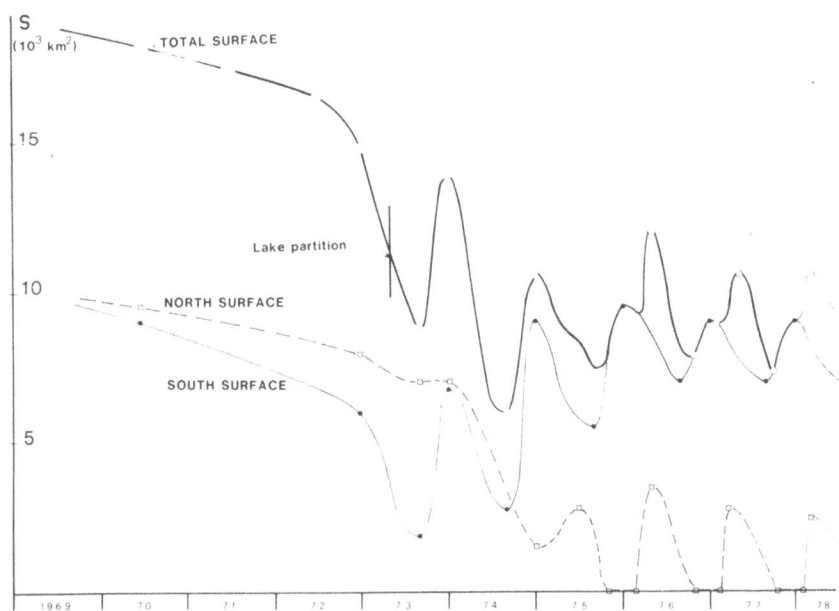

Fig. 19 Water areas approximate variations for Lake Chad between 1969 and 1978.

Table 11 Approximate calculation of water areas (S_s = south basin; S_n = north basin; S_t = total area in km²) and corresponding average annual yields (kg ha⁻¹) between 1969 and 1977.

Year	Water areas (km²)			Yields (kg ha⁻¹)		
	S_s	S_n	S_T	Y_s	Y_n	Y_T
1969	9500	9500	19 000	26.4[a]	42.5	34.4
1970	9100	9200	18 300	30.8[a]	51.2	41.0
1971	8500	9000	17 500	40.8	96.6	69.5
1972	8000	8600	16 600	59.9	144.4	103.7
1973	(4000)	7500	11 500	98.3	206.9	166.2
1974	(5000)	(4300)	9300	105.4	406.5	244.6
1975	6800	(1700)	8500	127.8[b]	248.8[b]	152.0
1976	7600	(1200)	8800	112.1[b]	191.7[b]	123.0
1977	8000	(1000)	9000	106.0[b]	186.0[b]	115.5

[a] In 1969 and 1970 the values of lacustrine yields were overestimated because the riverine fisheries were still notable. The mean values in parentheses correspond to the years 1973 and 1974 for the south basin and since 1974 for the north basin where seasonal fluctuations have been particularly important.

[b] It was supposed that a growing part of fish checked on the Baga-Kawa route had been fished in the south basin.

477

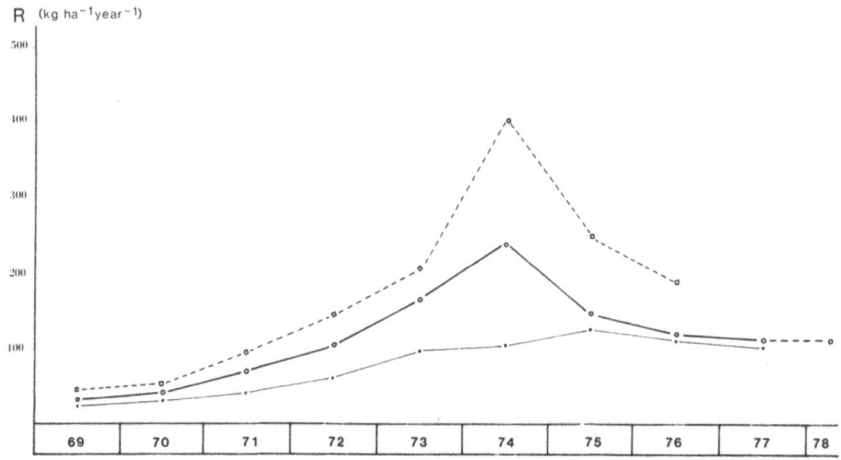

Fig. 20 Yield estimates (R, kg ha^{-1} year^{-1}) between 1969 and 1978 for the north basin (- - - -), the south basin (thin line) and the whole Lake Chad (thick line).

It should be emphasized here that the foreseeable increase in yield was particularly significant for the entire lake area; it increased from 34 kg ha^{-1} in 1969 to 104 in 1972 and 245 in 1974. After a severe decrease in 1975 (142 kg ha^{-1}) it appeared to stabilize at about 100 to 120 kg ha^{-1}.

The yields were always higher in the north after 1969, corresponding to a relative underexploitation of the southern lake. The yields per hectare regularly increased during four years: 51 kg in 1970, 97 in 1971, 144 in 1972 and 207 in 1973. The depletion of the north basin in 1974 corresponded to a yield of 407 kg ha^{-1} and from 1975 to 1977 the yields stayed very high, about 250 to 180 kg ha^{-1} year^{-1}.

The change in yields in the south basin was quite different. By considering that a part of the fluvial fishery accounted for catches in the statistics of the N'Djamena route from 1967 to 1972 and that the Baga-Kawa route also corresponded to the south basin starting from 1975, the yields per hectare can be estimated to increase from 10/15 during 1969–70 to 100 in 1973 (Table 11). After 1973 the yields appeared to stabilize at about 105 in 1977 and did not show the fluctuations recorded in the north basin. A similar value was found for the entire lake in 1972 when it was reflooded and did not appear to give way to an excessive exploitation.

Thus, 100 to 120 kg ha^{-1} seems to be an average yield which is possible to maintain under the adopted exploitation conditions. It does not mean that, in the frame of rational stock exploitation, it would be impossible to increase the yields in Lake Chad fisheries.

478

13.7.2.2 *Prospects of lake fisheries.* The level of total catches that can be achieved each year in Lake Chad depends first on its extension and then on the pattern of exploitation. No forecasts can be made on an eventual rise of the lake that would reflood the north basin, but it is evident that the latter would have increased the total catches to about 180 000 tons. Unfortunately, one cannot be optimistic in the short term due to the high evaporation and the actual deficit of the north basin: two to three exeedingly strong floods in succession are needed to bring about an average lake, still far from the extension of 'Greater Lake Chad' in the years 1960–65. This unfortunately improbable scenario means that the lake may not recover its past extension in the next few years especially as the recent floods have often been very low. Thus the total volume flowing during the course of the hydrological cycle year 1979–80 has been similar to the two low floods of 1972–73 and 1973–74.

It is thus likely that the extension of lake waters in the next few years will stay the same as over the recent period. Since 1975 the average area of the lake was seen to be close to 9000 km^2 with a permanent lake reduced only to the south basin, about 8000 km^2. A stable system appears there: on the one hand, a reduced south basin is unlikely since it would mean a new worsening of the Shari supplies; on the other hand, the excess waters in average years form bogs in the shallows of the north basin and evaporate before the following flood appears.*

According to the yields obtained, the annual total catches should thus stabilize at a level of about 90 to 100 000 tons, provided that the exploitation stays balanced, both for the gill nets and for the total fishing effort level. As already seen, the case of multispecies stocks is too complex as they are constituted of biologically varied species that may necessitate distinct exploitation patterns for each species. Nevertheless, some reasonable 'mean' modifications could be adopted. Scattered observations in the Shari delta during November 1977 showed that the catches were very diversified with nearly twenty species caught, especially *Tilapia* spp., *Hemisynodontis membranaceus, Heterotis niloticus, Hydrocynus* spp., *Citharinus* spp. The common occurrence of probably immature individuals in the catches indicates that the mesh size was already too small, and that it would have been necessary to prevent the large mesh from getting below 60 to 70 mm.

Before the environmental changes influenced the fisheries, the effect of the fishing effort was already considerable and an excessive rate of exploitation was reached by 1971 for *Alestes baremoze*. A precise analysis for the total fisheries of the Lake Chad region will not be possible until specialized studies to estimate

---

* These are particular temporary aquatic environments which have not been considered among the usually flooded zones of the lake Chad region. It is nevertheless certain that the productivity of these bog zones is very high (*Clarias* spp., for example).

the major fisheries parameters (specific catch per unit effort, size distribution with gear, age of fish caught, fishing effort ...) have been carried out. If the estimates of total catches from the road-checks start again and are reinforced, then it would be possible to understand the change in stocks and appropriate measures could be taken to insure the rational management of resources which are still important in spite of a very unfavourable climatic context.

At the present time, it is difficult to go any further in stock management. Later, a predictive model could probably be built for particularly well-known stocks. For migratory species with a fluvio-lacustrine cycle, it should be possible to find recruitment indices built from hydrological indices combining various characteristics (floods, flooded zones ...) one or two years before, but the total multispecies yield (here a high value of about 100 kg ha^{-1} which will be discussed further in Chapter 17) appears to be linked only to the extension of permanent waters, essentially the Lake Chad area. The forecast can solely be made from one year to the following due to the effect of each annual flood in relation to the variability of supplies that cannot be modelled as they depend on the climate of the entire African continent.

References

Bénech, V., 1975. Effets de la sècheresse sur les peuplements de poissons dans le lac Tchad et le Delta du Chari. Notes techniques ORSTOM N'Djaména, 9, 14 pp., mimeo.

Bénech, V. and Lek, S., 1981. Résistance à l'hypoxie et observations écologiques pour seize espèces de poissons du Tchad. Rev. Hydrobiol. trop. 14: 153–168.

Bénech, V. Lemoalle, J. and Quensière, J., 1976. Mortalités de poissons et conditions de milieu dans le lac Tchad au cours d'une période de sècheresse. Cah. ORSTOM Sér. Hydrobiol. 10: 119–130.

Bénech, V., Quensière, J. and Vidy, G., 1982. Hydrologie et physico-chimie des eaux de la plaine d'inondation du Nord-Cameroun. Cah. ORSTOM. Sér. Hydrol. 19, in press.

Beverton, R. J. H. and Holt, S. J., 1959. A review of the life-spans of fish in nature, and their relation to growth and other physiological characteristics. In: Wolstenholme, G. E. W. and Connor, M. O. (eds.), Ciba Foundation and Colloquia on Ageing, London, Churchill, 5, 142: 180 pp.

Blache, J. and Miton, F., 1962. Première contribution à la connaissance de la pêche dans le bassin hydrographique Logone–Chari–Lac Tchad. Mém. ORSTOM 4, 143 pp.

Carmouze, J. P., 1976. La régulation hydrochimique du lac Tchad. Trav. Doc. ORSTOM, No. 58, 418 pp.

Couty, P. and Duran, P., 1968. Le commerce du poisson au Tchad. Mém. ORSTOM, 5, 252 pp.

Durand, J. R., 1970a. Les peuplements ichtyologiques de l'El Beid — Première note — Présentation du milieu et résultats généraux. Cah. ORSTOM Sér. Hydrobiol. 4: 3–26.

Durand, J. R., 1970b. Les peuplements ichtyologiques de l'El Beid. Observations sur les rendements. ORSTOM Fort-Lamy, 14 pp., mimeo.

Durand, J. R., 1971. Les peuplements ichtyologiques de l'El Beid. 2ème note. Variations inter et intraspécifiques. Cah. ORSTOM Sér. Hydrobiol. 5: 147–159.

Durand, J. R., 1973. Note sur l'évolution des prises par unité d'effort dans le lac Tchad. Cah. ORSTOM Sér. Hydrobiol. 7: 195–207.

Durand, J. R., 1978. Biologie et dynamique des populations d'*Alestes baremoze* (Pisces, Characidae) du Bassin Tchadien. Trav. Doc. ORSTOM, No. 98, 332 pp.

Durand, J. R., 1980. Evolution des captures totales (1962–1977) et devenir des pêcheries de la région du lac Tchad. Cah. ORSTOM Sér. Hydrobiol. 13: 93–111.

Fontana, A., 1979. Etude du stock démersal côtier congolais. Biologie et dynamique des principales espèces exploitées. Propositions d'aménagement de la pêcherie. Thèse, Université Pierre et Marie Curie Paris VI, 300 pp.

Garcia, S., 1977. Biologie et dynamique des populations de crevettes roses (*Penaeus duorarum notialis* PEREZ-FARFANTE, 1967) en Côte d'Ivoire. Trav. Doc. ORSTOM, No. 79, 275 pp.

Hopson, A. J., 1964. Annual report (1963) *Federal Fisheries Service*, Lake Chad Research Station, Malamfatori, Lagos, 34 pp.

Hopson, A. J., 1968. The gillnet fisheries of Lake Chad. Federal Fisheries Service, Maïduguri, 64 pp.

Hopson, J., 1969. A preliminary study on the biology of *Alestes baremoze* in the Malamfatori area. Ann. Rept. Malamfatori, 1966–67: 51–83.

Hopson, J., 1972. Breeding and growth in two populations of *Alestes baremoze* (Joannis) (Pisces: Characidae) from the northern basin of Lake Chad. Overseas Res. Publ., 20, 50 pp.

Hopson, J., 1975. Preliminary observations on the biology of *Alestes baremoze* (Joannis) in Lake Rudolf. Symp. Hydrobiol. and Fish. Lake Rudolf, 25 pp., mimeo.

Le Guen, J. C., 1971. Dynamique des populations de *Pseudotolithus (Fonticulus) elongatus* (Bowd, 1825). Poissons Sciaenidae. Cah. ORSTOM Sér. Océanogr.: 3–84.

Loubens, G., 1973. Production de la pêche et peuplements ichtyologiques d'un bief du delta du Chari. Cah. ORSTOM Sér. Hydrobiol. 7: 209–233.

Loubens, G. and Franc, J., 1972. Etude méthodologique pour la récolte de statistiques de pêche basée sur l'observation de pêcheries d'un bief du Delta du Chari. ORSTOM Fort-Lamy, 44 pp., mimeo.

Mann, M. J., 1962. Fish production and marketing from the Nigerian shores of Lake Chad (1960–61). Fed. Fish. Service Lagos, Nigeria, 50 pp., mimeo.

Monod., T., 1928. L'industrie des pêches au Cameroun. Paris Larose Ed., 504 pp.

Quensière, J., 1976. Influence de la sècheresse sur les pêcheries du delta du Chari (1971–73). Cah. ORSTOM Sér. Hydrobiol. 10: 3–18.

Stauch, A., 1977. Fish statistics in the Lake Chad basin during the drought (1969–76). Cah. ORSTOM Sér. Hydrobiol. 11: 201–215.

Vidy, G., 1982. Organisation et production de la pêche traditionnelle sur le Logomatia, drain naturel de la grande plaine inondée nord-camerounaise. Revue Hydrobiol. Trop., in press.

Welcomme, R. L., 1979. Fisheries ecology of floodplain rivers. Longman, 317 pp.

IV. Trophic relations

14. Trophic relations between the phytoplankton and the zooplankton

André Iltis and Lucien Saint-Jean

The study of trophic relations within planktonic populations in Lake Chad is based on gut content analyses of living individuals immediately after capture. They were conducted during the 'Normal Chad' period in March 1967, November 1968 and October 1970 (Gras et al. 1971).

Their interpretation is based on two indices:
— the percentage occurrence of a particular food, that is the ratio of the number of individuals containing it to the total number of individuals under consideration;
— the relative percentage occurrence, that is the ratio of the previous percentage to the entire percentage occurrence of the various foods divided by 100 (Table 1).

14.1 Diet of the main species

14.1.1 *Cladocera, Ostracods and Calanoids*

The first group was represented by eight species (*Diaphanosoma excisum, Moina micrura, Ceriodaphnia affinis, C. cornuta, Daphnia barbata, D. lumholtzi, D.*

Table 1 Percentage occurrence of the algal species in the gut contents of Ostracods, Cladocerans and Calanoids according to morphological and size characteristics. The numbers designating algal species are those mentioned in Fig. 1. The relative percentage of occurrence is indicated in brackets.

Algae	2.3.4.	5	6	7	8.9.10.11.12	13.14	15
Ostracods	8(2)	61(17)	15(4)	8(2)	154(42)	119(33)	0(0)
Cladocera	18(8)	43(10)	27(12)	5(2)	74(32)	58(25)	3(1)
Daphnia	33(10)	64(20)	37(12)	8(3)	100(31)	64(20)	4(1)
Bosmina	7(3)	36(13)	31(11)	9(4)	65(24)	118(44)	4(1)
Calanoids	19(16)	66(13)	62(13)	32(7)	170(34)	58(12)	23(5)

J.-P. Carmouze et al. (eds.) Lake Chad
© 1983, Dr W. Junk Publishers, The Hague/Boston/Lancaster
ISBN 978-94-009-7268-1

longispina, *Bosmina longirostris*) and the Ostracods by a single unidentified species. The calanoids were represented by two main species, *Tropodiaptomus incognitus* (most abundant during the observations), and *Thermodiaptomus galebi*. As a rule, observations were conducted only on adults. All these species were phytophagous or sestonophagous (feeding on algae, detritus, bacteria), no animal remains being found in the stomach contents, with the exception, however, of Calanoids where a few remains of Rotifers or crustaceans were observed. As the percentage occurrences of these prey were very low, their ingestion can be considered accidental, as the Calanoids depend primarily on particle filtration.

14.1.2 Cyclopoids

Three species were present in the lake: *Mesocyclops* cf. *leuckarti*, *Thermocyclops incisus circusi* and *Thermocyclops neglectus neglectus*.

The male and female adults of the first two species had a carnivorous diet; the small number of algae observed in a few individuals could either have been ingested occasionally, or been present in the guts of the captured prey.

In the copepodids of these two species, a gradual increase in the intake of animal matter was observed from the first stages to the adult stage, while the plant fraction decreased. The change in diet appeared to be gradual and did not occur at a particular stage. Gophen (1977) observed the same variation in diet of the individuals in Lake Kinneret. Therefore the diet of the first copepodid stages may be considered as mixed; that of copepodid stage 5 being almost entirely carnivorous, at least in females with a greater mean size.

Male and female *Thermocyclops neglectus neglectus* were herbivorous and carnivorous. Copepodids of this species were mainly herbivorous, the percentage occurrence of animal remains being very low: 7% for the copepodid stage 5 and 3% for the copepodid stage 4.

14.2 Composition of the gut contents

Cladocerans as a whole ingested unicellular or colonial algae ranging in size from 4 to 7 µm (coccoïd Chlorophyceae and isolated cells of *Oocystis*) up to about 30 µm (colonies or cenobes). The percentage occurrence of large algae (2, 3 and 4 in Fig. 1) was rather high for the whole group (18%) but it varied considerably especially in *B. longirostris* (7%) and the three species of *Daphnia* (33%) (Table 1).

If the consumption of filaments of *Anabaena* appears to be very low (percentage occurrence of 3), small filamentous Cyanophyceae were generally collected and absorbed by all the species of Cladocerans.

	O (26)	D (75)	C a(23)	C c(47)	M m(63)	D e(63)	B l (55)	T i (90)	T g (32)	T n(69)
1 GUT EMPTY										
2 MICROCYSTIS (G^des Col)										
3 PEDIASTRUM (FRAGMENTS)										
4 SPHAEROCYSTIS										
5 MICROCYSTIS (P^tes Col) ←20μ										
6 MELOSIRA										
7 OTHER DIATOMS										
8 SCENEDESMUS										
9 TETRAEDRON, CRUCIGENIA										
10 COSMARIUM, DESMIDIACÉES										
11 CHLOR.COCCOÏDES										
12 OOCYSTIS										
13 CEL.ANKYSTRODESM.										
14 CYANO.FILAMENTEUSES										
15 ANABAENA										
16 ANIMALS REMAINS										

Fig. 1 Occurrence of various planktonic algae in the gut contents of planktonic Crustaceans. The number of individuals of each species examined is given between parentheses. Each point corresponds to the presence of an alga or a group of algae found in the gut contents of an individual of the species concerned; C.c. = *Ceriodaphnia cornuta*; C.a. = *C. affinis*; D. = *Daphnia* (three species); D.e. = *Diaphanosoma excisum*; M.m. = *Moina micrura*; B.l. = *Bosmina longirostris*; O = ostracods; T.i. = *Tropodiaptomus incognitus*; T.g. = *Thermodiaptomus galebi*; T.n. = *Thermocyclops neglectus*.

The algae ingested by the Ostracods ranged from *Oocystis* (isolated cells) and coccoïd Chlorophyceae from 4 to 7 μm, to small colonies of *Microcystis* (from about 25 to 30 μm). Big *Microcystis* colonies (particularly *M. aeruginosa*), *Sphaerocystis* and portions of *Pediastrum* cenobes, rather abundant in the environment during the observations, were not found in the stomach contents. Within the filamentous algae, the small species were the only ones to be ingested; *Lyngbya*, small *Oscillatoria*, *Spirulina laxissima* and Ankistrodesmiform cells, etc ... *Anabaena* was not consumed and the *Melosira* found in the guts was in the form of cell portions, such as occurred in the environment.

The two species of Calanoids under study could be distinguished from the two previous orders by the increased ingestion of large algae whose percentage occurrence reached about 79%. *Tropodiaptomus incognitus* seemed to ingest filaments of *Anabaena flos-aquae* (the percentage occurrence of this algae reached 23% for all observations and 54% during the first series of analyses) but the stomach contents of *Thermodiaptomus galebi* collected at the same stations, included no heterocysts of *Anabaena*, but numerous unicellular algae, both colonial and filamentous.

485

Animal remains were found in 63% and 67% of the gut contents of adult *Mesocyclops* cf. *leuckarti* and *Thermocyclops incisus circusi* (Fig. 1 and Table 2); the highest predation is on the Cladocerans. In Copepodids with a mixed diet the range of algae found in the digestive tract was roughly the same as that of the filter-feeding species; thus Chlorophyceae, small filamentous and colonial Cyanophyceae and *Anabaena* were observed. Adults of *Thermocyclops neglectus* consumed all zooplanktonic organisms but the highest predation was on the Cladocerans. The percentage occurrence of animal remains was lower in males (12%) than in females (27%). The algae identified in the gut contents were the species that were usually ingested by Cladocerans and Calanoids. The consumption of *Anabaena* was active and undigested heterocysts were often observed in the foregut and the faecal pellets (Fig. 1). The copepodids belonging to this species were predominantly herbivorous: *Microcystis* and *Anabaena* dominated the stomach contents.

14.3 Conclusions

The examination of the gut contents of the microcrustaceans made it possible to specify their diets and to determine the animal or plant groups on which the greatest predation pressure was exerted.

Within the framework of the relations between the phytoplankton and zooplankton, it must first be pointed out that Rotifers (a score of species recorded by Pourriot (1968) in the southern basin) were not considered in this study. Their density was estimated at 44 individuals per liter in 1964–65 (Gras et al. 1967) and at 80.1^{-1} in the archipelago of the southern basin in February 1971. Density ranged from 9 to 14 individuals per liter in 1967–68 (Robinson 1971) and up to 39 (see zooplankton) in the northern basin. These Rotifers fed

Table 2 Percentage occurrence of animal remains, identified or not, in the three Cyclopoid species; N = number of individuals examined.

Species	♀		♂		C5		C4		C3		C1 + C2	
	N	%	N	%	N	%	N	%	N	%	N	%
Mesocyclops leuckarti	185	63	51	35	37	45	70	31	88	15	49	6
Thermocyclops incisus circusi	42	67	10	0	–	–		–		–		–
Thermocyclops neglectus	166	27	43	12	30	7	36	3	14	0	6	0

mainly on nannoplankton, small detritus and bacteria with the exception of a predatory species, *Asplanchna brightwelli*, which fed on smaller Rotifers, *Bosmina* and *Ceriodaphnia*; the abundance of this species remained rather low in the lake.

In terms of biomass, however, Rotifers were insignificant when compared to microcrustaceans, at least during the period of 'Normal Chad', so that the filter-feeding species of the last group represented almost the total herbivorous zooplankton of the lake. The same holds true for the next trophic level.

During the 'Normal Chad' period (1964–65, 1971), it can be estimated that the strictly carnivorous species (adults of *Mesocyclops* and *Thermocyclops incisus circusi*) represented from 2.5 to 4% of the total biomass; species which were predominantly carnivorous or herbivorous (Cyclopoids, Copepodids and adults of *Thermocyclops neglectus*) represent from 8% to 16% and the species which were strictly herbivorous-detrivorous made up the remaining 81 to 90%. Thus, approximately 5–8% of the zooplankton stock were secondary consumers, while 92–95% fed on phytoplankton.

The accuracy of these percentages is difficult to evaluate as the rather abundant adults and copepodids of *Thermocyclops neglectus* had a mixed diet and the animal fraction in the gut contents was variable and difficult to estimate.

At the end of the 'Normal Chad' period, in 1972–1973, the percentage of strictly carnivorous species (*Mesocyclops* cf. *leuckarti* and adult *Thermocyclops incisus circusi*) remained in the range from 1.4% to 3.8%, while the population of *Thermocyclops neglectus* with a mixed diet developed and predominated over the herbivorous species during the first half of 1973; therefore, it can be taken that the percentage of plant seston consumed decreased sharply.

After 1973, during 'Lesser Chad', zooplankton density decreased greatly, by about four fifths, and strictly carnivorous species as well as the Calanoids disappeared; the only remaining species were *Thermocyclops neglectus*, with a mixed diet, and the herbivorous Cladocerans.

Finally, it seems that part of the phytoplankton, algae or large colonies and *Anabaena* were not directly ingested by the zooplankton at any stage in the lake. This part of the phytoplankton could be utilized in the form of detritus or bacteria after degradation with the exception of *Anabaena flos aquae*. The latter which represented nearly the entire algal population in low waters seemed to be directly consumed by *Thermocyclops neglectus* (adults and last copepodid stages) and *Tropodiaptomus*, i.e., a little more than 30% of the zooplankton. Then it would be reintroduced into the trophic chain at the level of the zooplankton which fed on the nutritive film deposited on the sediment or on any other substrate.

References

Gophen, M., 1977. Food and feeding habits of *Mesocyclops leuckarti* (Claus) in Lake Kinneret (Israel). Freshwat. Biol. 7: 513–518.

Gras, R., Iltis, A. and Lévêque-Duwat, S., 1967. Le plancton du Bas-Chari et de la partie est du lac Tchad. Cah. ORSTOM Sér. Hydrobiol. 1: 25–96.

Gras, R., Iltis, A. and Saint-Jean, L., 1971. Biologie des crustacés du lac Tchad. II. Régime alimentaire des Entomostracés planctoniques. Cah. ORSTOM Sér. Hydrobiol. 5: 285–296.

Pourriot, R., 1968. Rotifères du lac Tchad. Bull. IFAN 30: 471–496.

Robinson, A. H. and Robinson, P. K., 1971. Seasonal distribution of zooplankton in the northern basin of Lake Chad. J. Zool. Lond. 163: 25–61.

15. Trophic relations of fishes in Lake Chad

Laurent Lauzanne

In a stable ecosystem, three main types of organisms can be distinguished: the producers, the consumers and the 'decomposers-transformers' (Dussart 1966).

The producers, which are autotrophic organisms, synthesize their own matter from mineral elements present in the environment. The energy used is usually solar (photosynthetic plants), but it can also be of chemical origin (chemosynthetic organisms, represented by some bacteria). The consumers use the organic matter produced by the autotrophic organisms to make their own biomass. All these consumers are connected by feeding relationships to form the predatory food chain.

The 'decomposers-transformers' degrade the organic matter of plants and dead animals to transform it into mineral salts. They are mostly heterotrophic bacteria and constitute the degradation food chain. The organic matter that is produced can then be used again by the autotrophs.

In fact the situation is more complex than this classical description of the food cycle (Elton 1927). In addition to the plant matter produced by photosynthesis, all the detritus from the predatory chain that is almost degraded and yet mineralized is available to the consumers. This detritus and the decomposer organisms, settle to the lake bottom and constitute the benthic organic cover. In the case of a deep lake, this organic cover is generally less important because mineralization will occur during settlement of the detritus. In shallow Lake Chad, the detritus settles rapidly and this organic cover is important. The detrivores consuming this cover, therefore, reintroduce organic matter into the food cycle which, without them, would have had to pass, as a whole through the group of the 'decomposers-transformers' to be reutilized by the ecosystem.

From these two original food sources of plant and detrital organic cover, we are therefore able to distinguish two food chains namely a grazing food chain and a detritus food chain (Fig. 1).

For each of these two chains, we will make the traditional distinction of trophic levels where the organisms from a level feed on organisms from the level immediately below it. We can distinguish fairly easily the first three levels where trophic relationships are relatively direct. The first level is composed of the

J.-P. Carmouze et al. (eds.) Lake Chad
© *1983, Dr W. Junk Publishers, The Hague/Boston/Lancaster*
ISBN 978-94-009-7268-1

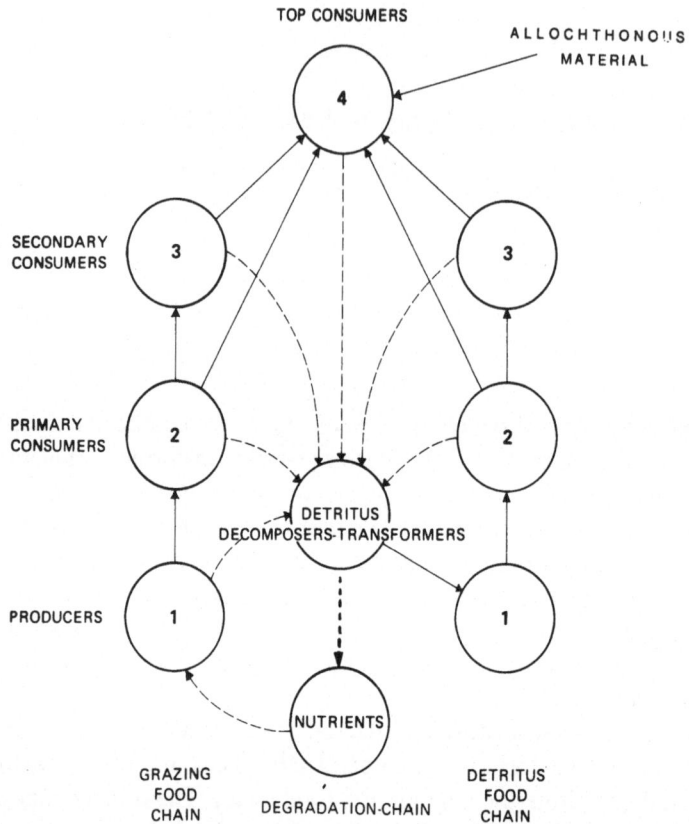

TOP CONSUMERS

ALLOCHTHONOUS
MATERIAL

SECONDARY
CONSUMERS

PRIMARY
CONSUMERS

DETRITUS
DECOMPOSERS-TRANSFORMERS

PRODUCERS

NUTRIENTS

GRAZING
FOOD
CHAIN

DEGRADATION-CHAIN

DETRITUS
FOOD
CHAIN

Fig. 1 Food cycle pattern in Lake Chad. Numbers correspond to trophic levels; arrows indicate the direction of energy transfers — continuous lines: by predation; dashed lines: by degradation (from Lauzanne 1976).

plants and detrital organic cover.* The second level includes the primary consumers which feed on plants and detritus. The third level is composed of the secondary consumers which feed mainly on the invertebrates of the zooplankton and benthos. A fourth level that is common to both chains gathers together the top consumers. These animals have complicated trophic relationships with the other three levels and even with organisms not belonging to the aquatic ecosystem (mainly terrestrial insects). In this case, we cannot refer to it as a food chain but rather as a food web (Cooper and Fuller 1945). In this report, we will deal with the study of trophic relationships through two approaches. First, a qualitative approach based mainly on the knowledge of the

* The first level ought to be strictly reserved for autotrophic organisms such as macrophytes, phytoplankton and some bacteria. The complex organic cover is however composed of a very high percentage of sedimented algae and fine plant debris, and thus placed in the first level.

diets of the fish and a quantitative approach describing the transfer of energy as organic matter moves from one trophic level to a higher one through predation.

15.1 Qualitative aspects

15.1.1 *The environment and available food*

The trophic relationships of the fishes in Lake Chad were studied mainly in its southeastern part which includes a zone of open water and an archipelago (Lauzanne 1976). The geographic, climatic and physico-chemical features of these two zones are described in detail at the beginning of this book and will not be repeated. Nevertheless, we should point out that the annual fluctuations in lake level follow the variations in the Shari flood after a time lag. Therefore, it is possible to distinguish a flood period in the lake from July to December and a period of low water from January to June. It will be shown later that this alternation of high and low water has an influence on the abundance of some prey.

In this ecosystem, nine major food types can be distinguished: phytoplankton, macrophytes, organic cover of the bottom, zooplankton, benthos, aquatic insects, shrimps, preyfishes and a food source from outside the aquatic ecosystem which is represented by terrestrial insects. These different types of food are described in varying detail in this volume. However, we will reconsider these different food classes by trying to define the features typical of the two main zones under study (archipelago and open water) and specifying the feeding preferences of each major group.

The phytoplankton

Generally, Cyanophyceae dominated the phytoplankton and always represented more than 90% of the total number of cells (Gras et al. 1967). The genera *Microcystis*, *Aphanocapsa* and *Anabaena* were the best represented. The seasonal variations in density were very low in the archipelago, while they were considerable in the open water. In the latter zone, phytoplankton density was very low from August to December due to a direct influence of the Shari flood.

The macrophytes

The open water did not contain macrophytes, while in the archipelago, each island was surrounded by a plant fringe several meters wide that was composed mainly of Cyperaceae and Graminaceae. Submerged aquatic plants, mainly *Potamogeton* and *Vallisneria* along with *Najas* and *Ceratophyllum* occurred in patches. The leaves of the aquatic plants and the seeds of various plants were utilized by several species of fish.

491

The organic bottom layer

This flaky-looking cover was composed of a detrital layer (fine plant debris, algae and planktonic crustacea which settled after death, faeces of the different organisms present in the overlying water and colloidal clay) and a live organic layer (bacteria, benthic diatoms, protozoa and Rotifera). it was very difficult to determine the importance of these different constituents, but the dominance of planktonic algae was shown through simple microscopical examination.

The zooplankton

In Lake Chad, the zooplankton was characterized by the predominance of crustacea such as the Copepoda and the Cladocera over the Rotifera, since the first grouping represented about 85% of the total number of individuals (Gras et al. 1967). The mean biomass in the archipelago was well above that of the open water and it was more stable over the year. The zooplankton in the open water, like the phytoplankton, underwent a sharp decrease from August to December, probably for the same reasons. The trophic structure of the zooplankton communities was similar in the different regions of Lake Chad (Gras et al. 1971) and most of the zooplankton was phytophagous since the predatory species represented only 6% of the total biomass.

The benthos

The benthic fauna in Lake Chad was composed mainly of molluscs, insect larvae and oligochaetes. Nematodes and Ostracoda were also present but they did not seem to make up a major part of the biomass (Dejoux et al. 1969). The true benthic species were of minor importance. The molluscs represented by seven main species and the Oligochaeta by three main species. The insect larvae were more diverse and were mainly chironomids.

The benthic molluscs were, either Gasteropoda that browsed on the organic cover of the bottom or Lamellibranchia that filtered the same cover. The Oligochaeta ingest the surface sediment from which they extract the organic matter (detritus, algae, bacteria). Generally, insects are detrivores, except the carnivorous Tanipodinae (Dejoux 1974). On the whole, it can be estimated that most of the benthic invertebrates were detrivores, and obtained their food from the organic layer on the bottom, the complexity of which has been emphasized.

The aquatic insects

The larvae of the chironomids, Ephemeroptera and Trichoptera were considered to be part of the benthos. Therefore, this section of the community consisted of the nymphs of Hemiptera, Chaoborus and Ephemeroptera adults of *Chaoborus* and imagos of chironomids and Trichoptera as well as Coleoptera.

The swimming insects which were only abundant in the submerged plants, were probably of little importance in high water.

Shrimps

During the period of 'Normal Chad', the shrimps (*Caridina africana* and *Macrobrachium niloticum* were very abundant in the archipelago and almost absent in the open water. The former was attached to the submerged plants while the latter was also found in offshore water. It appeared that the shrimps were much more abundant in low water than during the flood. *Caridina* feeds mainly on epiphytes and detritus (Fryer 1960) while *Macrobrachium* is clearly a detritus eater (Hopson 1972).

Prey fishes

On reaching a certain stage of growth, all the species of fish could be eaten by larger fish. However, certain species were consumed more regularly than others, probably because they were more abundant. The following list mentions the species most often caught by predators. It includes young fish belonging to large and small species (the latter indicated with an asterisk).

Characidae	*Alestes baremoze*
	Alestes dentex
	*Alestes dageti**
	*Micralestes acutidens**
Schilbeidae	*Eutropius niloticus*
Mormyridae	*Pollimyrus isidori**
	Petrocephalus bane
Citharinidae	*Distichodus rostratus*
Cyprinidae	*Labeo* sp.
	Barbus sp.*
Cichlidae	*Tilapia* sp.
	*Haplochromis bloyeti**
Mochocidae	*Brachysynodontis batensoda*

Among the small species, *Barbus* and *Haplochromis* only occurred among submerged plants, and were not found in open water. The young of large species such as *Labeo, Distichodus, Alestes baremoze* were abundant during the flood. This can be explained by the fact that most species in the Chad basin reproduce at the beginning of the flood.

The terrestrial insects

For certain fishes, terrestrial insects were of major importance as a food source, especially in the open water. These insects which live and feed on the vegetation of the islands and reed islands were carried away by the winds to fall into the water and drown. The insects consumed were Coleoptera, Hemiptera and especially Orthoptera which were often relatively large. The fall-out of terrestrial insects was especially important during flooding of the lake.

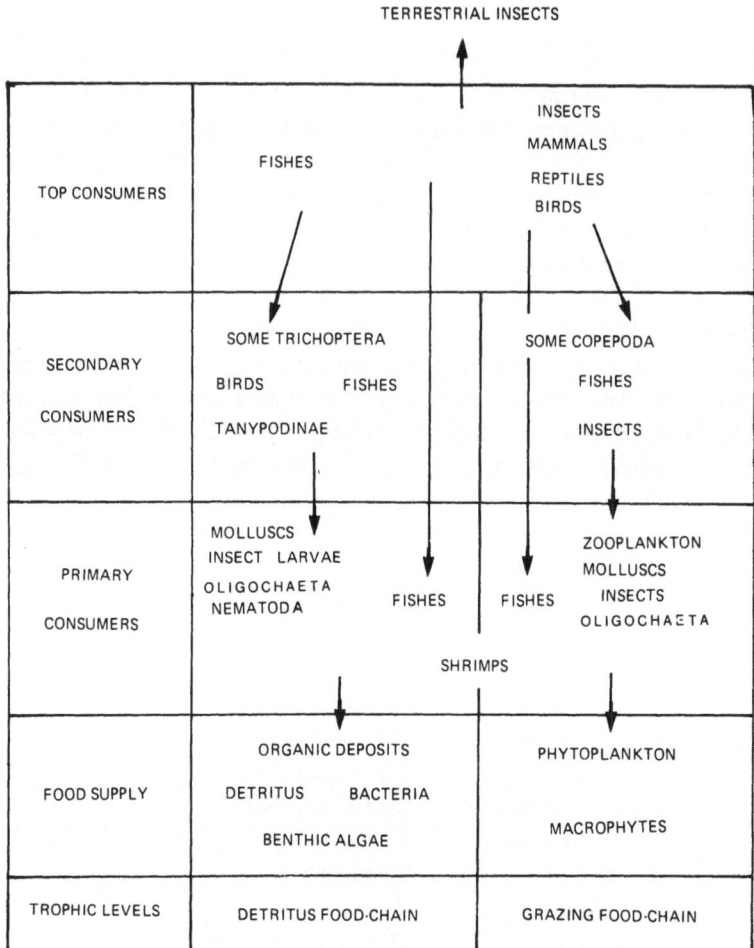

Fig. 2 Trophic relationships in Lake Chad; arrows indicate the direction of predation.

Based on the preceding observation on the feeding of the different groups of organisms, we have classified them according to their trophic levels (Fig. 2). Later, a discussion of their diets will show that the different species of fish were represented in all the consumer groups.

15.1.2 *Selection of species studied*

The choice of species was based on their importance in the population samples. In the archipelago, the results of catches by a large beach seine were used. This was relatively unselective since, theoretically, it took samples of all the species within a certain size range. In water where it was impossible to use this type of seine, we used as a base the catches from a set of 10 gill nets, in spite of doubts raised by their selectivity. We chose for study the species representing at least 1% of the total weight of the catches. They included 17 species in the archipelago and 17 in the open water (Table 1) of which 13 were common to the two biotopes. Thus, the total number of species was 21.

15.1.3 *The main types of diets – trophic levels*

The detailed results of the stomach content analyses of the species studied (Lauzanne 1976) were arranged according to ten food types. They include the nine types mentioned above, as well as fish debris such as large scales, vertebrae and spines. These results collect together the data without considering possible differences in diets related to the two hydrological seasons (flood and recession) and the two biotopes (archipelago and open water). For each species, the diets were determined by the percentage occurrence and the volumetric percentages

Table 1 Ponderal index of the main species in the Archipelago (Arch.) and the Open water (O.W.) expressed as a percentage of the biomass.

Species	Arch.	O.W.
Brachysynodontis batensoda (Mochocidae)	16.1	2.1
Sarotherodon galilaeus (Cichlidae)	12.5	–
Lates niloticus (Centropomidae)	11.5	12.4
Alestes baremoze (Characidae)	10.5	4.1
Alestes dentex (Characidae)	10.4	–
Hemisynodontis membranaceus (Mochocidae)	7.1	4.2
Hydrocynus forskalii (Characidae)	4.6	20.3
Hydrocynus brevis (Characidae)	3.4	4.3
Labeo senegalensis (Cyprinidae)	3.2	2.0
Schilbe uranoscopus (Schilbeidae)	2.7	14.1
Alestes macrolepidotus (Characidae)	1.8	–
Hyperopisus bebe (Mormyridae)	1.7	1.8
Eutropius niloticus (Schilbeidae)	1.7	10.8
Synodontis schall (Mochocidae)	1.6	3.3
Citharinus citharus (Citharinidae)	1.5	1.6
Heterotis niloticus (Osteoglossidae)	1.4	–
Bagrus bayad (Bagridae)	1.2	1.9
Distichodus rostratus (Citharinidae)	–	3.5
Labeo coubie (Cyprinidae)	–	2.9
Citharinus distichodoides (Citharinidae)	–	2.6
Synodontis clarias (Mochocidae)	–	1.8
Miscellaneous	7.1	6.3

(Hynes 1950) of each type of food and a food index, IA, which takes these two factors into account (Lauzanne 1975):

$$IA = \frac{\% \ OC \ X \ \% \ V}{100}$$

The results (Table 2) are illustrated by Fig. 3 where the species were grouped according to their food preferences and ranked among the major categories of consumers defined above.

15.1.3.1 *Dominant primary consumers.* This group of consumers included the phytoplanktophagous, the detritivorous and the macrophytophagous fishes. The phytoplanktophagous fishes were represented by a single species, *Sarotherodon galilaeus* and the detrivorous fishes by *Labeo senegalensis*, *Labeo coubie*, *Distichodus rostratus*, *Citharinus citharus* and *Citharinus distichodoides*. These five species fed on the organic bottom layer. In fact, the main difference between these two diets arose from differences in the feeding behaviour of the five species. *Sarotherodon galilaeus* is a filter-feeder which selects mainly algae and even certain types of algae (Lauzanne and Iltis 1975), while *Labeo*, *Citharinus* and *Distichodus* sample the whole bottom layer and even take a small amount of the underlying sediments. However, most of this cover was composed of sedimented algae, which led to the classification of these detrivorous fishes among the primary consumers. The dominant macrophyte-consuming fish was *Alestes macrolepidotus* (macrophytes forming 100% OC, 59% V) with a preference for the young leaves of *Potamogeton* and *Ceratophyllum*. However, insects, particularly terrestrial, were also important in the diet.

15.1.3.2 *Secondary consumers*

15.1.3.2.1 *Dominant zooplankton feeders.* This group was composed of the following four species: *Alestes baremoze*, *Hemisynodontis membranaceus*, *Brachysynodontis batensoda* and *Alestes dentex*. The first two species fed strictly on zooplankton, but the diets of the last two had secondary components. *Brachysynodontis* fed on insects, especially swimming larvae and nymphs. In addition to the zooplankton and the insects, *Alestes dentex* also consumed seeds (Graminaceae and Cyperaceae). The zooplankton taken by these four species was mainly crustacea (Copepoda and Cladocera), with Rotifera only of secondary importance.

15.1.3.2.2 *Dominant benthos feeders.* The four species representing this group were: *Synodontis clarias*, *Synodontis schall*, *Hyperopisus bebe* and *Heterotis niloticus*. These fishes fed mainly on benthic invertebrates present in the organic bottom cover (insect larvae, Ostracoda and molluscs). *Synodontis schall* was strictly benthophagous, while the other three species were less selective. *Synodontis clarias* fed almost entirely on benthos with the exception of a few terrestrial insects. *Hyperopisus bebe* also ingested seeds (*Hypomea*) and

496

Table 2 Percentages of occurrence and volume, food indices (IA = %C × %V/100) for the 21 species under study.

Species	Entire fishes			Fish detritus			Shrimps			Aquatic insects		
	%OC	%V	IA	%OC	%V	IA	%OC	%V	IA	%OC	%V	IA
Sarotherodon galilaeus												
Labeo senegalensis												
Labeo coubie												
Citharinus citharus												
Citharinus distichodoides												
Distichodus rostratus										83.8	11.4	9.5
Alestes macrolepidotus												
Alestes baremoze												
Hemisynodontis membranaceus										27.8	18.7	5.2
Alestes dentex										30	4.3	1.3
Brachysynodontis batensoda												
Synodontis schall												
Synodontis clarias												
Hyperopisus bebe												
Heterotis niloticus							18.9	10.7	2.0			
Lates niloticus	100	100	100									
Hydrocynus brevis	100	100	100									
Hydrocynus forskalii	86.1	74.5	64.1				24.0	25.3	6.1	0.2	0.1	
Bagrus bayad	57.0	68.3	38.9	34.0	9.4	3.2	44.0	7.1	3.1	29.0	1.4	0.4
Eutropius niloticus	36.0	30.0	10.8	31.8	16.0	5.1	2.5	0.8	0.1	13.8	1.2	0.2
Schilbe uranoscopus	52.5	67.0	35.2	18.2	13.5	2.4	24.2	9.6	2.3	17.2	0.6	0.1

497

Table 2 (continued).

Species	Terrestrial insects			Benthos			Zooplankton			Phytoplankton		
	%OC	%V	IA	%OC	%V	IA	%OC	%V	IA	%OC	%V	IA
Sarotherodon galilaeus										100	100	100
Labeo senegalensis												
Labeo coubie												
Citharinus citharus												
Citharinus distichodoides												
Distichodus rostratus												
Alestes macrolepidotus	78.4	30.0	23.5									
Alestes baremoze							100	100	100			
Hemisynodontis membranaceus							100	100	100			
Alestes dentex							94.4	68.0	64.2			
Brachysynodontis batensoda							100	95.7	95.7			
Synodontis schall	20.0	2.9	0.6	100	97.1	97.1						
Synodontis clarias	27.0	1.4	0.4	100	98.6	98.6						
Hyperopisus bebe				97.7	96.2	94.0						
Heterotis niloticus				100	61.1	61.1	32.4	11.8	3.8			
Lates niloticus												
Hydrocynus brevis												
Hydrocynus forskalii												
Bagrus bayad	16.0	7.1	1.1	20.0	6.7	1.3						
Eutropius niloticus	67.1	50.3	33.7	17.3	1.7	0.3						
Schilbe uranoscopus	20.2	6.9	1.11	38.4	2.4	0.9						

Table 2 (continued).

Species	Organic deposits			Macrophytes			Number of stomachs studied	Limits of standard lengths (mm)
	%OC	%V	IA	%OC	%V	IA		
Sarotherodon galilaeus							*	155–275
Labeo senegalensis	100	100	100				81	200–460
Labeo coubie	100	100	100				22	180–450
Citharinus citharus	100	100	100				55	160–500
Citharinus distichodoides	100	100	100				16	350–550
Distichodus rostratus	100	100	100				38	180–400
Alestes macrolepidotus				100	58.7	58.7	37	125–210
Alestes baremoze							*	150–265
Hemisynodontis membranaceus							118	250–340
Alestes dentex				18.9	13.3	2.5	90	145–265
Brachysynodontis batensoda							110	100–160
Synodontis schall							135	145–260
Synodontis clarias							37	180–240
Hyperopisus bebe				9.4	3.8	0.4	128	200–440
Heterotis niloticus				81.1	16.3	13.2	37	350–435
Lates niloticus							73	390–1310
Hydrocynus brevis							86	270–610
Hydrocynus forskalii							251	150–380
Bagrus bayad							100	160–435
Eutropius niloticus							283	110–235
Schilbe uranoscopus							99	180–235

* Some hundreds.

TROPHIC LEVELS		TERRESTRIAL INSECTS
4 TOP CONSUMERS	*LATES NILOTICUS* *HYDROCYNUS BREVIS* *HYDROCYNUS FORSKALII*	*BAGRUS BAYAD* ← *SCHILBE URANOSCOPUS* *EUTROPIUS NILOTICUS*
3 SECONDARY CONSUMERS	*SYNODONTIS SCHALL* *SYNODONTIS CLARIAS* *HYPEROPISUS BEBE* *HETEROTIS NILOTICUS*	*BRACHYSYNODONTIS BATENSODA* *HEMISYNODONTIS MEMBRANACEUS* *ALESTES DENTEX* *ALESTES BAREMOZE*
2 PRIMARY CONSUMERS	*CITHARINUS CITHARUS* *DISTICHODUS ROSTRATUS* *CITHARINUS DISTICHODOIDES* *LABEO COUBIE* *LABEO SENEGALENSIS*	*SAROTHERODON GALILAEUS* *ALESTES MACROLEPIDOTUS*
1 FOOD SUPPLY	ORGANIC DEPOSITS	PHYTOPLANKTON EPIPHYTES MACROPHYTES
TROPHIC LEVELS	DETRITUS FOOD-CHAIN	GRAZING FOOD-CHAIN

Fig. 3 The different kinds of consumers in Lake Chad.

Heterotis niloticus consumed shrimps and zooplankton in addition to benthos and seeds.

The main insect larvae consumed were Chironomids (Chironominae and Tanipodinae), Ephemeroptera (*Povilla adusta*) and Trichoptera (*Dipseudopsis* and *Ecnomus*).

Predation upon molluscs was mostly on undersized individuals. They were small species (young and adult *Gyraulus, Bulinus, Anisus, Segmentorbis, Gabbia, Pisidium, Eupera*), but also young immature individuals of larger species (*Bellamya, Cleopatra, Biomphalaria, Melania, Corbicula*).

15.1.3.3 *Top consumers.* This carnivorous group was composed of six species of which two were piscivorous only, while four of them had diets with more or less varied secondary components.

15.1.3.3.1 *Strictly piscivorous group. Lates niloticus* and *Hydrocynus brevis* are predators which feed only on living fishes. *Lates niloticus* can grow very large (maximum length observed: 132 cm for a weight of 78 kg), and has a stocky poorly streamlined shape.

Hydrocynus brevis does not grow so large (maximum length observed: 80 cm for a weight of 10 kg), but unlike *L. niloticus*, it is extremely streamlined and swims very fast. It is a tireless pursuer which does not give its prey many chances. Its jaws have formidable teeth enabling it to cut its prey into two parts with a single bite. It will even attack large fishes to take a bite from them as shown by Lewis (1974) in Lake Kainji.

15.1.3.3.2 *Less strictly piscivorous group.* The common characteristic of the four species in this group was that all of them consumed shrimps and aquatic insects in addition to fish. *Schilbe uranoscopus, Eutropius niloticus* and *Bagrus bayad* consumed not only whole fishes which were probably caught when alive but also a considerable amount of fish debris composed mainly of very large scales, large vertebrae and several bones such as spines and pectoral fins of *Synodontis*. It initially seemed as if this debris came from whole prey which were degraded by digestive juices, but it became obvious after investigation that the size of the predator was not great enough for ingestion of prey corresponding to the size of debris found. In these three species, the terrestrial insects played a significant role and were even very important for *Eutropius niloticus*. These three species with saprophagous tendencies were clearly different from *Hydrocynus forskalii* which fed mainly on live prey such as fish but also on large quantities of shrimps.

15.1.4 *Comparison of trophic relationships in the archipelago and in the open water*

As already observed, the trophic relationships of fish in trophic levels 2 and 3, that is the primary and secondary consumers, were rather direct. When, most of the food consumed by a given level comes from the next lower level it is part of a food chain. The trophic relationships of top consumers are much more complex as the food comes from all the trophic levels and even from food sources outside the aquatic ecosystem (terrestrial insects). Moreover, nutritional relationships can exist between the different constituents of this level, forming a food web. Lauzanne (1976) quoted figures which showed the trophic relationships of fishes in the archipelago and the open water where the inputs from each trophic level were indicated (as volumetric percentages) in the diet of each species.

15.1.4.1 *Primary and secondary consumers.* It was observed that trophic relationships were direct in the open water with the exception, however, of the inclusion of a small number of terrestrial insects in the diet of the secondary consumers of the detrital chain. In the archipelago (Fig. 4), the trophic relationships were less direct. For instance only 59% of the food of *Alestes macrolepidotus* (dominant primary consumer) consisted of macrophyte leaves. Some dominant secondary consumers obtained a certain amount of food from level 1 (macrophyte seeds).

15.1.4.2 *Top consumers.* The food web of the top consumers in the archipelago (Fig. 6) was complex and requires further explanation.

1. The secondary benthos feeders were not part of the food of the top consumers.

2. Of the primary consumers, *Sarotherodon* (phytoplankton feeder) and shrimps (detritus feeders) were important in the diet of *Lates niloticus* (75% *Sarotherodon galilaeus*) and *Hydrocynus forskalii* (56% shrimps). Fishes that ate detritus occurred in the diets of *L. niloticus*, *H. forskalii* and *H. brevis*.

3. The secondary consumers of the algal chain represented an important part of the diets. Zooplankton feeders were consumed by all the predators but were particularly important especially in the diets of *H. brevis* and *S. uranoscopus*. Fish eating periphytic zooplankton which were small species found among submerged plants (*Barbus*, *Haplochromis*) formed the diet only of the smaller predators (*E. niloticus*, *S. uranoscopus* and *H. forskalii*).

4. The input of aquatic insects, benthic invertebrates and fish debris could be of some significance for *Eutropius* and *Bagrus*.

5. The food supply from outside the aquatic ecosystem was composed of terrestrial insects and was only of major importance in the diet of *Eutropius*.

6. The nutritional relationships between the top consumers were fairly limited; *Eutropius* was only consumed by *H. brevis* (19%) and *Bagrus* (15%).

The food web of the top consumers in the open water was much simpler than that in the archipelago (Fig. 7).

1. As in the archipelago, predators did not feed on the secondary benthos feeders

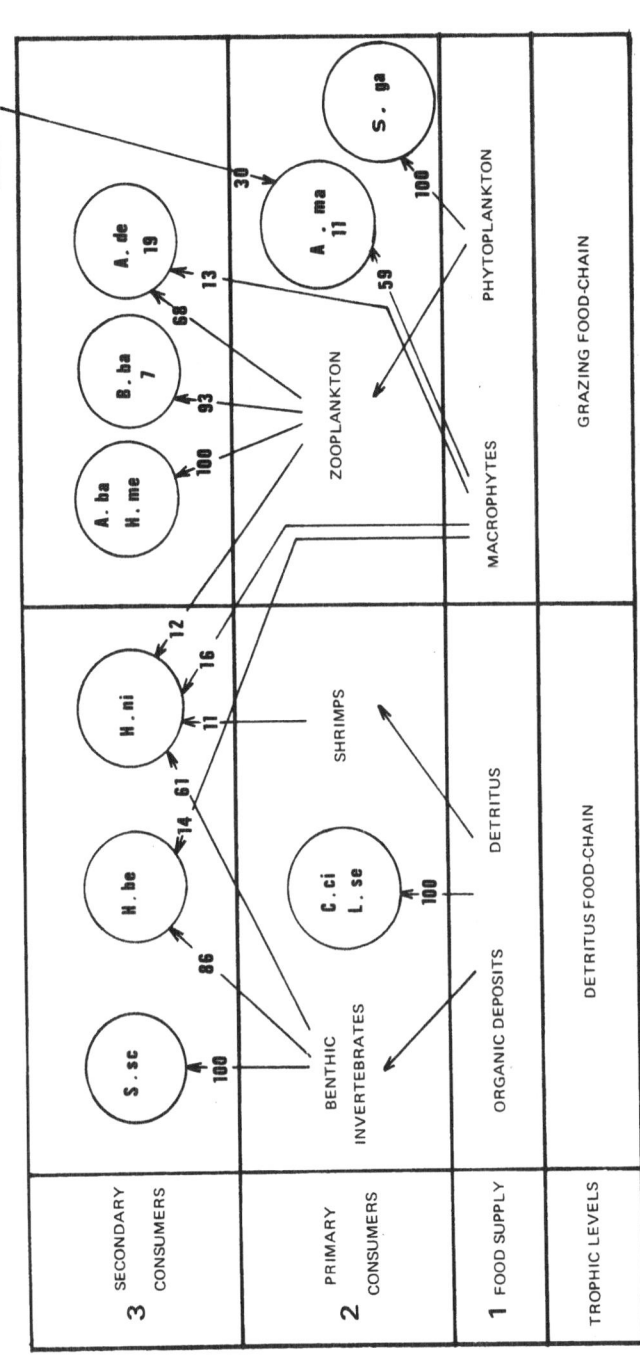

Fig. 4 Food webs of primary and secondary consumers in the Lake Chad archipelago. Arrows indicate the direction of transfers and the numbers correspond to the volumetric percentages of the input to the diets. Numbers in circles correspond to prey whose trophic position is unknown; S. .sc = *Synodontis schall*; H. be = *Hyperopisus bebe*; H. ni = *Heterotis niloticus*; A. ba = *Alestes baremoze*; H. me = *Hemisynodontis membranaceus*; B. ba = *Brachysynodontis batensoda*; A. de = *Alestes dentex*; A. ma = *Alestes macrolepidotus*; S. ga = *Sarotherodon galilaeus*; C. ci = *Citharinus citharus*; C. di = *Citharinus distichodoides*; L. se = *Labeo senegalensis*; L. co = *Labeo coubie*; D. ro = *Distichodus rostratus*; S. cl = *Synodontis clarias*.

503

504

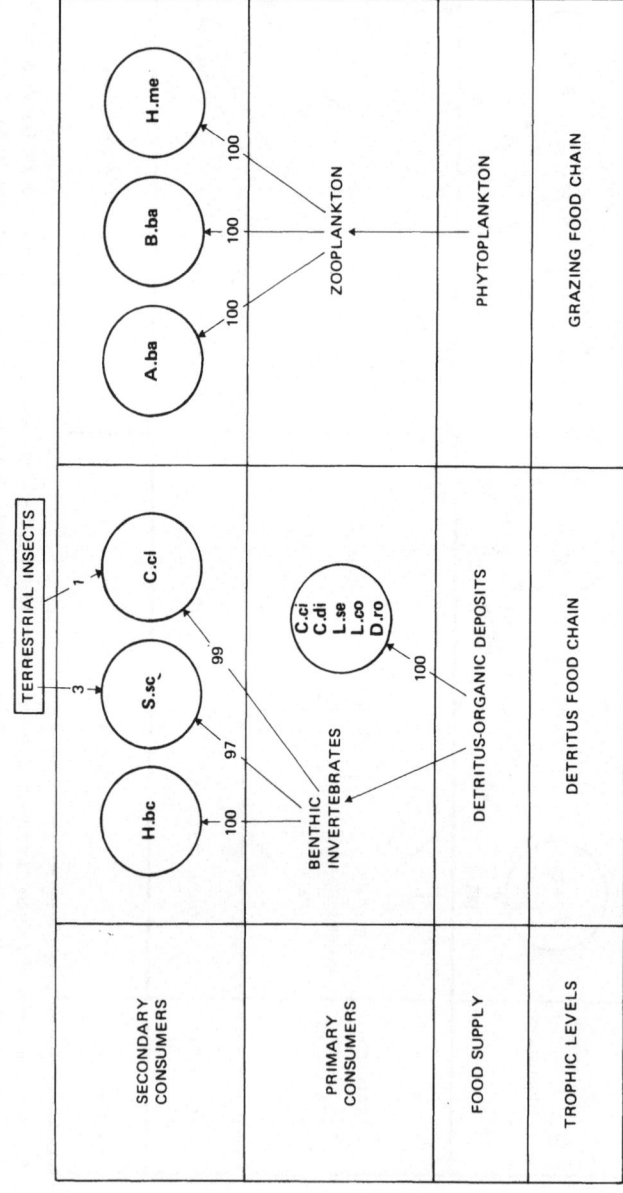

Fig. 5 Food webs of primary and secondary consumers in Lake Chad open water (see Fig. 4).

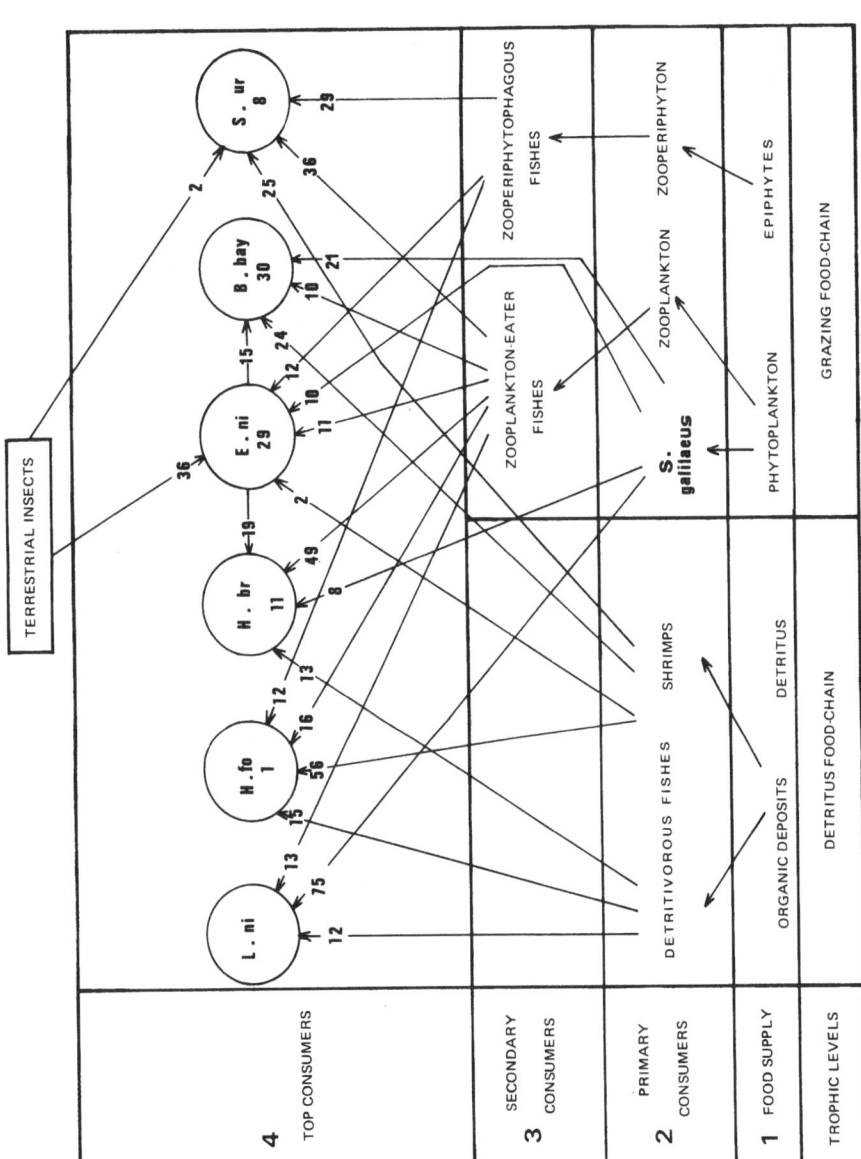

Fig. 6 Food webs of top consumers in the Lake Chad archipelago; L. n = *Lates niloticus*; H. fo = *Hydrocynus forskalii*; H. br = *Hydrocynus brevis*; E. ni = *Eutropius niloticus*; B. bay = *Bagrus bayad*; S. ur = *Schilbe uranoscopus*.

505

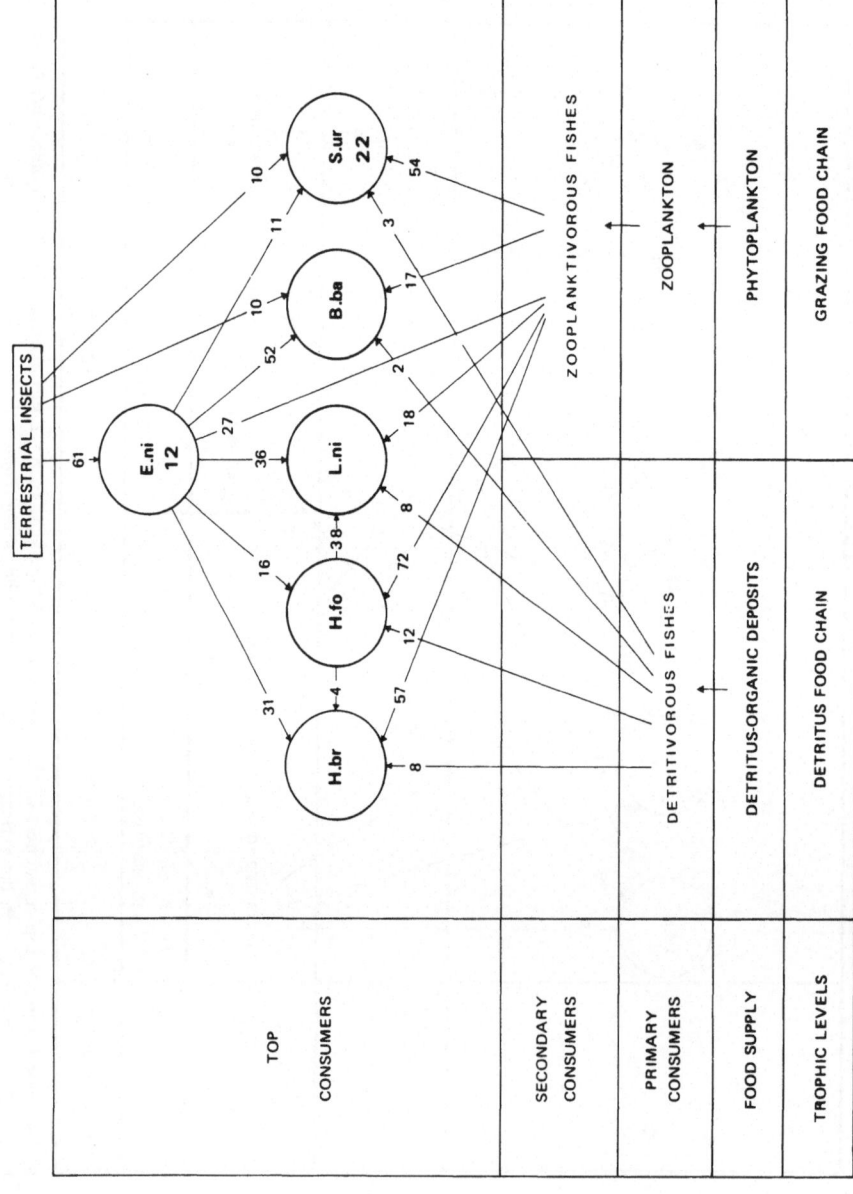

Fig. 7 Food webs of top consumers in Lake Chad open water (see Fig. 6).

2. The primary consumers were represented only by detritivores which constituted a small part of the diets.

3. The secondary consumers of the grazing food chain were important in the diets of all the predators.

4. The supplies of aquatic insects, benthos and fish debris were less important than in the archipelago.

5. The terrestrial insects played a secondary role in the feeding of *Bagrus* and *Schilbe* but were important for *Eutropius* (61%).

6. The food relationships between the predators were characterized by the role of *Hydrocynus forskalii* in the diet of *Lates* (38%), and especially by the considerable importance of *Eutropius* in the diets of the five species. In short, it can be estimated that the trophic relationships were much more diversified in the archipelago than in the open water. This was because of the absence of certain types of food (macrophytes, shrimps and periphytic zooplankton) in the open water.

15.1.5 *Comparison of the importance of the various consumer groups between the archipelago and the open water*

In the previous section, it was shown that within the four major group of consumers, the diets could differ considerably according to the biotope.

They were more varied in the archipelago where the types of food were more diversified. It was also observed that certain groups of organisms could become very important quantitatively in the diets, depending on the zone under study (the molluscs and terrestrial insects in the open water and shrimps in the archipelago). Nevertheless, these four groups were present in the two zones and it is interesting to compare their relative importance. From Fig. 8 (Lauzanne 1976) the following can be concluded:

1. The primary consumers, whether they were detrivorous or phytophagous, were of moderate importance in the archipelago (19%) as well as in the open water (13%).

2. The secondary consumers were present in small numbers in the open water (benthos, feeders, 7%; zooplankton feeders, 10%), while they dominated in the archipelago (benthos feeders, 5% and particularly zooplankton feeders: 44%).

3. The top consumers which were well represented in the archipelago (25%) played a major role in the open water (64%).

These two zones in the southeast of Lake Chad differed greatly in the abundance of the various consumer groups. In the archipelago, the zooplankton feeders dominated, while the open water was characterized by the abundance of the top consumers, due to the particular trophic relationships in each zone. On the one hand these consumers fed on zooplanktophagous fishes such as the two small species *Pollimyrus* and *Micralestes* which probably had

507

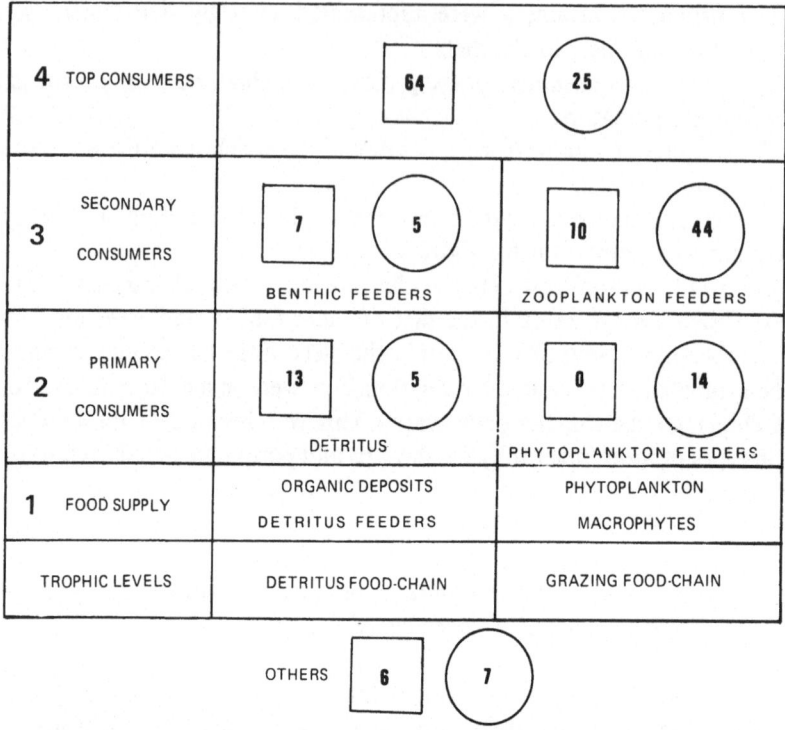

Fig. 8 Comparison of the importance of various consumer groups in the archipelago (circles) and the open water (squares). Numbers expressed as percentage of the biomass.

high production. On the other hand, they fed on the terrestrial insects which constituted the major food of *Eutropius niloticus* which was itself consumed in large quantities by all the other predators.

15.1.6 *Variations in diets and changes of trophic levels*

In Lake Chad, where food was abundant and varied, the adult fishes fed permanently on the same trophic level. However, their diets could undergo small variations depending upon the biotopes and the seasons. As already described, the diets of fishes in the open water were thus less diverse than those in the archipelago, mainly because there was a shortage of the food supply such as the shrimps and macrophytes in the first biotope while these were abundant in the second one. The influence of the hydrological seasons in Lake Chad upon the diets (Lauzanne 1976), was especially pronounced in the top consumers. So, during the flood, the predators consumed a greater number of young fishes of the large species than during the fall (Fig. 9b). This phenomenon was due to the

508

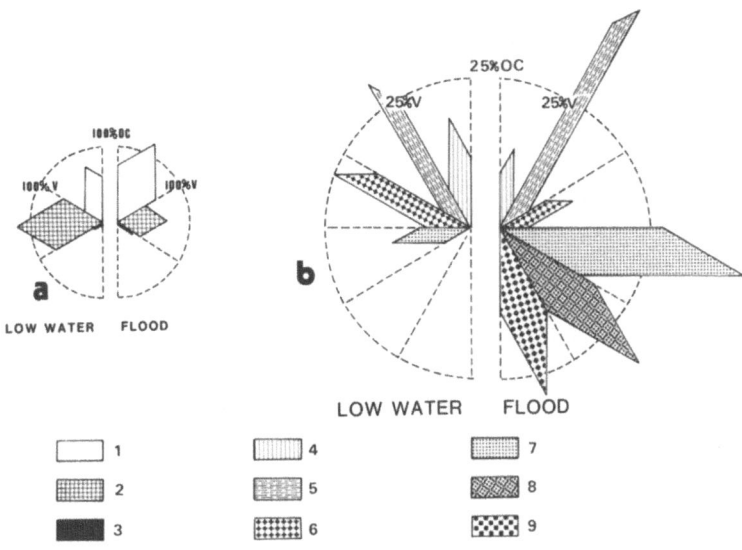

Fig. 9 Diet of *Hydrocynus forskalii* in the Lake Chad archipelago; (a) diet composition; (b) fishes; 1 = fish; 2 = shrimps; 3 = *Micronecta*; 4 = *Barbus callipterus*; 5 = *Micralestes acutidens*; 6 = *Haplochromis bloyeti*; 7 = *Alestes dentex* and *A. baremoze*; 8 = *Labeo* sp.; 9 = *Distichodus* sp. Each sector corresponds to a type of food. The percentage occurrence is put on the upper radius of the sector and the volumetric percentage on the lower one. The area of the parallelogram is proportional to the food index.

spawning of most species at the beginning of the flood and the abundance of young fish during the period of high water level. The terrestrial insects were also consumed in greater quantities during the flood which corresponded partly with the rainy season, and with the growth of herbaceous vegetation which was not the case during lake contraction. This phenomenon was well illustrated by *Eutropius niloticus* (Fig. 10). In the archipelago, the consumption of shrimps was greater during low water when these crustacea occurred in high densities, as shown by *Hydrocynus forskalii* in Fig. 9a.

It was found that the diets of the fishes changed profoundly during their life cycles. Most of the young fishes were initially zooplankton feeders before becoming adult. All the top consumers (level 4) underwent a secondary consumer period (level 3) as young fish. For instance, the young of *Hydrocynus forskalii* (Fig. 11a) were zooplanktophagous (level 3) and then had a temporary insectivorous period before consuming fishes (level 4). Some fishes changed their chain without changing their trophic level, as in the case of *Tetraodon fahaka*. (Fig. 11b). Adult fish fed on molluscs (level 3 – detritus food chain) but the young consumed zooplankton (level 3 – grazing food chain). The example of *Alestes baremoze* (Lauzanne 1973) provides a good summary of changes in diet during the life cycle. *Alestes baremoze* is a migratory fish which reaches

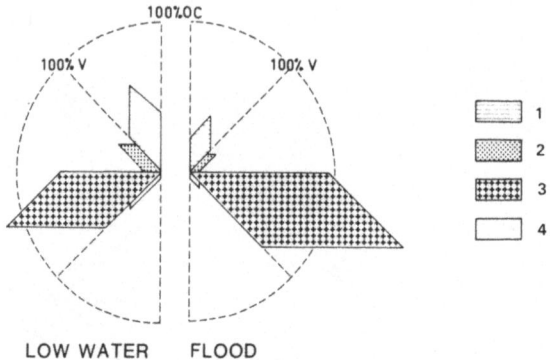

Fig. 10 Diet of *Eutropius niloticus* in open water. 1 = whole fish; 2 = fish debris; 3 = terrestrial insects; 4 = benthic invertebrates.

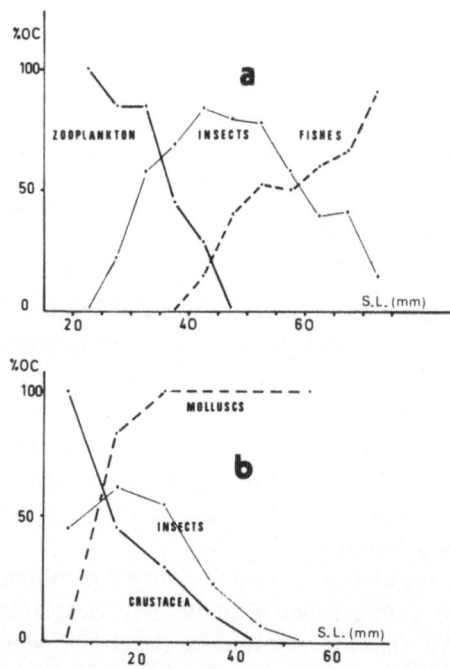

Fig. 11 Variation in diet of young *Hydrocynus forskalii* (a) and young *Tetraodon fahaka* (b).

adulthood in Lake Chad where it feeds strictly on zooplankton. During the dry season (low water), the brood fish moved up the Shari and the Logone Rivers where they found poor nutritional conditions and consumed rare Chironomids, terrestrial insects and crustacea. After spawning, the adults followed the rising

510

waters and entered the flood zones where they found an abundance of food composed mainly of leaves and seeds. The yearlings also entered the flooded plain where their food was composed of epiphytic organisms such as Copepoda, Cladocera, Ostracoda and Chironomidae. During their catadromous migration towards the lake, when there was a lowering of water, they continued to feed upon crustacea and insects. So, the young fishes reached the lake where they were able to grow rapidly by feeding on the zooplankton. Thus, this species changed its trophic level from the third level (zooplankton feeder) to the second level (feeding on leaves and seeds), over its migratory cycle.

Changes of trophic levels also occurred as a result of profound modifications of the biotopes. So, after the considerable drought of 1972–1973, the archipelago was divided by the lowering of water in the lake. There was a sharp decrease in the zooplankton and benthos due to drying up. In a study of diets of Mochocidae in 1974, Im (1977) observed that all the diets were then based upon fine plant detritus. During the period of 'Normal Chad', these fishes were either zooplankton or benthos feeders and therefore, ecological disturbances forced them from the third to the second trophic level.

15.2 Quantitative aspects

15.2.1 *General*

When organic matter is transferred from one trophic level to a higher level, it is accompanied by a loss of energy which depends mostly upon the precise energy efficiency of the consumer. Following the works of Ivlev (1939–1961) and Winberg (1956), Warren and Davis (1967) suggested an equation which considered the energy balance in fishes:

$$C = F + U + \Delta B + R$$

in which C is the amount of energy contained in the food consumed, F is the energy value of the excrements, U is the energy lost through urine and through the skin and gills, ΔB is the amount of energy corresponding to the increase in weight and R is the energy necessary for metabolism. The food efficiency can be characterized by several coefficients. The most commonly used one is the first order energy coefficient of growth (K1)

$$K1: \frac{\Delta B}{C}$$

This ratio which is calculated from the wet weight is called the conversion rate and the inverse ratio which is often used is called the food coefficient or trophic coefficient. This coefficient, K1, characterizes a gross efficiency, as, the amount of energy consumed is not used only to achieve growth. So, a second order energy coefficient of growth was proposed:

$$K\,2 = \frac{\Delta B}{C - (F + U)}$$

Some authors also use an assimilation index: AS:

$$AS = \frac{\Delta B + R}{C}$$

This coefficient accounts for the energy that is really used for growths and metabolism. These last two indices whose calculation is difficult were not considered because they must be studied in detail in the laboratory. From the field methods used, the conversion rate of food could be calculated from caloric equivalents (Lauzanne 1978) for three species, each characterizing a trophic level of consumers. The second trophic level was characterized by *Sarotherodon galilaeus*, a phytoplankton eater (Lauzanne 1978). The third level was characterized by the zooplanktivorous *Alestes baremoze* and the fourth level by the piscivorous *Lates niloticus* (Hamblyn 1966, Lauzanne 1977).

15.2.2 *Conversion rate and first order growth energy coefficient* (K1)

We will point out again that the food conversion rate is the ratio of the increase in weight of the fish during a given period of time to the weight of the food ingested during the same period. To evaluate the first parameter, it is necessary to know the growth curve of the fish under study. The second parameter (weight of food ingested) was obtained from a knowledge of the daily food intake which depends upon the weight of the fish and the water temperature. We will not dwell on the method used to estimate daily food intake (Lauzanne 1969, 1978), but only mention that is is based on a knowledge of the daily feeding periodicity and the rate of gastric evacuation. Conversion rates are listed in Table 3 for the three species under consideration. The piscivorous *Lates niloticus* had the highest rate and the poorest was for the phytoplankton eater *Sarotherodon galilaeus*, while the zooplankton eater, *Alestes baremoze*, had an intermediate value. This conversion rate (or its inverse, the food coefficient) which was calculated from the wet weights is interesting for the fish culturist since it provides information about the amount of food required to produce a certain amount of fish tissue. However, it could be misleading in connection with the energy efficiency of the predator. As a matter of fact prey and predators are far from being composed of equal proportions of water, mineral salts and organic matter, the only components used to obtain an energy value. The relationships between the various prey and predator constituents as well as the caloric equivalents of the organic matter were calculated (Lauzanne 1978). These relationships permitted the calculation of the conversion rate from

Table 3 Conversion rate (%) for the three species under study.

	S. galilaeus	L. niloticus	A. baremoze
Wet weight	3.1	22.4	8.8
Dry weight	5.5	24.9	34.7
Organic matter	11.5	26.4	39.2
Calories (K1)	18.9	27.3	44.8

dry matter to organic matter and finally to express it in terms of energy (Table 3). From this it was apparent that *Alestes* (zooplankton feeder) had the best energy efficiency, above that of *Lates* (piscivorous), while the K1 for *Sarotherodon* remained by far the lowest.

The differences between the various conversion rates for *Lates niloticus* were not very great. This phenomenon was due to the similar water and mineral salt contents of predator and prey so that the caloric equivalents of the organic matter were not very different for *Lates* and the fishes on which it fed. For contrary reasons, *Sarotherodon* and *Alestes* had conversion rates which were very different according to the calculation method.

Generally, results in the literature are not directly comparable with ours. Most of them are results of laboratory experiments where the conditions differ greatly from those existing in the natural environment. Nevertheless, these various results (Table 4) which must be prudently considered, suggest at least three important points:

1. For the same type of food, it is observed that K1 is always higher in warm-water fishes than in fishes living in temperate waters.

2. For the same thermic preference of fishes (warm waters, temperate waters), it seems that the lowest efficiency is obtained in phytoplankton eaters and particularly plant feeders. In this last class, *Ctenopharyngodon* is a particularly good example since this species uses only 2% of the energy consumed for its growth and moreover, it assimilates only 13% of the energy ingested, while 81% of the latter is lost in the form of waste products (Fisher 1970). Although it is very long, the digestive system of the phytophagous fishes does not seem to be as well adapted to the assimilation of plant matter, especially cellulose as that of some insects and mammals.

3. For carnivorous fishes the highest efficiency is obtained in fishes that eat crustacea such as zooplankton, *Gammarus* and shrimps and benthic inverte-brates, while the lowest efficiency is obtained in the ichthyophagous fishes. This last remark is contrary to the general opinion that efficiencies increase as we go up the trophic chain. In fact, it seems that two groups must be considered, the herbivores with low efficiency and the carnivores with higher efficiency. The energy efficiency of the detritivorous fishes on which we have no information, ought logically to be intermediate, since their food is composed of more or less

Table 4 Values of K1 for the different fish species. The warm-water species are underlined.

Species	Food	T°C	K1 (%)	Authors
Tilapia mossambica	phytoplankton	25	22.2	Mironova (1974–75)
Sarotherodon galilaeus	phytoplankton	26	18.9	present study
Ctenopharyngodon idella	macrophytes	23	1.9	Fisher (1970)
Alestes baremoze	zooplankton	26	44.8	present study
Perca fluviatilis	*Gammarus*	14	20.4	Solomon, Brafied (1972)
Salmo trutta[a]	*Gammarus*	?	25.1	Surber (1935)
Salmo trutta	*Gammarus*	?	33.1	Pentelow (1939)
Salmo trutta	*Gammarus*	?	42.5	Schaeperclaus (1933)
Ophiocephalus striatus	*Metapeneus*	28	26.0–51.1	Pandian (1967)
Pseudopleuronectes americanus	*Nereis*	10	23.5–20.6 ⎱	Chesney, Estevez (1976)
		20	23.9–19.7 ⎰	
Limanda yokohamae	?	?	15.8–21.8	Hatanaka et al. (1956)
Stizostedion vitreum vitreum	amphipodes	20	14.3 ⎱	
	crayfish	16	12.7 ⎬	Kelso (1972)
	fish	12	13.9 ⎰	
Esox lucius	fish	?	14.9 ⎱	
Stizostedion luciopera	fish	?	15.1 ⎬	Backiel (1971)
Silurus glanis	fish	?	13.9 ⎰	
Lates niloticus	fish	26	27.3	present study

[a] The results for *Salmo trutta* were given as fresh weight. They were converted to obtain K1 by taking 845 cal/g for *Gammarus* (Mann 1965) and 1400 cal/g for *S. strutta* which is a mean value given for Salmonidae (Cummins and Wuycheck 1971).

degraded plants and animals. Within these groups, the value of K1 can vary greatly (Table 4). It does not seem that these differences can be explained only through variations in energy values of the various foods. Their protein, lipid and glucose contents are likely to be important along with the varying ability of fishes to assimilate these constituents.

15.2.3 *Energy transfer along food chains*

In the first part of this chapter, we described the qualitative relationships between the different trophic levels. These results provided an example of the energy transfer in the food pyramid. We will consider two hypothetical cases. In the first one, *Lates niloticus* feeds only on *Sarotherodon galilaeus* and in the second one on *Alestes baremoze*. In the first case, the food chain is composed of three links (3 trophic levels) and of four links in the second one (Fig. 12). The amount of energy accumulated by the zooplankton (considered here only as phytophagous) was estimated by using the results of Petipa et al. (1973) where

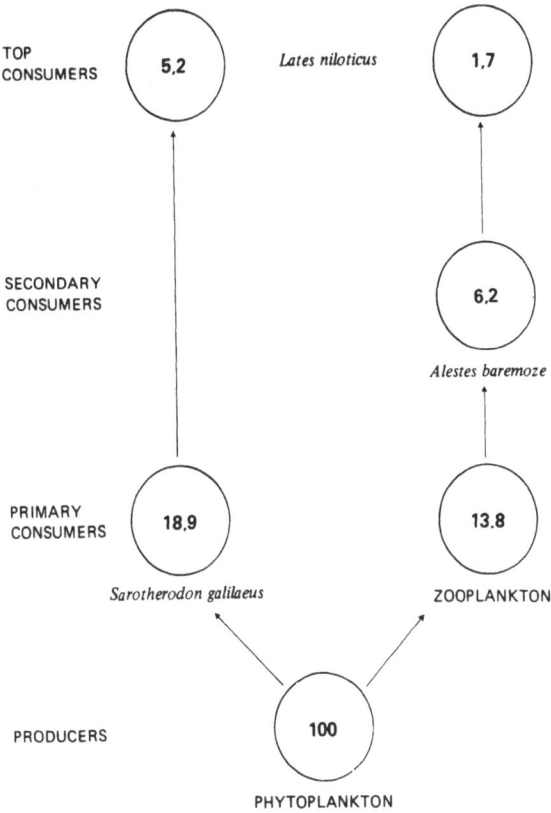

TROPHIC LEVELS

Fig. 12 Energy accumulated by *Lates niloticus* (in calories) according to the food chain used, starting from 100 calories for phytoplankton (from Lauzanne 1977).

K1 is equal to 13.8%. We observe that of 100 calories supplied by the algae (level 1), *Lates niloticus* accumulates 1.7 of them with the chain composed of two intermediate links (zooplankton and *Alestes baremoze*) and 5.2 of them, that is three times more, with a chain composed of a single intermediate link (*Sarotherodon galilaeus*). This example shows that the longer the food chain, the greater the loss in energy. The most efficient hypothetical cycle would be one where the algal production is consumed by a phytoplanktivore such as *S. galilaeus*, and the waste products partly transformed by a detritivore.

15.3 Conclusions

During the period of 'Normal Chad', the main species of the archipelago and the open water in the southeastern part of the lake could be classified into major consumer groups according to the trophic levels. The first level was composed of the original food sources such as algae and detritus and the second one of primary consumers such as phytoplankton feeders, macrophyte feeders and detritivores. The third level consisted of the benthos and zooplankton feeders and the fourth one of the top consumers especially piscivores.

The diets could undergo variations according to the biotopes, the hydrological seasons, the age of the fish or changes in the environment. These modifications could be modest if the fish remained in the same trophic level. They could also be very pronounced and then, the fish changed its trophic level. The importance of the different groups of consumers which was evaluated in the southeast of the lake was considerably different in the archipelago and the open water. The archipelago was characterized mainly by the abundance of the planktivores, especially zooplanktivores which represented 44% of the fish biomass. The open water was largely characterized by the top consumers which represented 64% of the fish biomass. The dominance of the planktivores in the archipelago resulted doubtless from the high planktonic biomass which was stable throughout the year. In the open water, the top consumers consisted of six species of which five were mainly piscivores and one, *Eutropius niloticus*, consumed mainly terrestrial insects. This last species was eaten in great quantities by the other five which also ate many small zooplanktivores such as *Micralestes* and *Pollimyrus*. Therefore, it seemed that this high biomass depended indirectly on terrestrial insects and the abundance of the small zooplanktivores which probably had a high production.

The food supply seemed to be rather well utilized. However, the phytoplanktivores such as *Sarotherodon* which were important in the archipelago were absent from the open water almost certainly for reasons of reproduction. Submerged macrophytes of the archipelago would certainly support denser population of grazers. The zooplankton was particularly well used by species of commercial importance (*Alestes* and *Synodontis*), but also by many young fishes belonging to the large species and by small prey species (*Micralestes, Pollimyrus*). The benthophages which did not represent a significant biomass consumed mainly insect larvae, especially Chironomidae, and Ostracoda and molluscs. Worms (Oligochaeta and Nematodes) which had, however, a high biomass seemed to be ignored by most of the benthic feeders. The first order energy coefficient of growth (K1) which was determined for three species, each representing a trophic level, was lowest for the phytophagous fishes and highest for the zooplanktivores, while the piscivores had intermediate efficiencies. Similarly, we showed that the energy lost when organic matter moved to an upper level was considerable. These two remarks led to the conclusion that the total energy efficiency of the fish community in the archipelago which was composed mainly of zooplantivorous fish must be well above that of the community in the open water which was composed mainly of top carnivores.

References

Backiel, T., 1971. Production and food consumption of predatory fish in the Vistula river. J. Fish. Biol. 3: 369–405.

Chesney, E. J. and Estevez, J. I., 1976. Energetics of winter flounder (*Pseudopleuronectes americanus*) fed the polychaete, *Nereis virens*, under experimental conditions. Trans. am. Fish. Soc. 105 (5): 592–595.

Cooper, G. P. and Fuller, J. L., 1945. A biological survey of Moosehead lake and Haymock lake, Maine. Maine Dept. Inland Fish. Game, Fish. Surv. Rept. 6: 1–160.

Cummins, K. W. and Wuycheck, J. C., 1971. Caloric equivalents for investigations in ecological energetics. Mitt. int. Ver. Limnol. 18, 158 pp.

Dejoux, C., 1974. Synécologie des Chironomides du lac Tchad (Diptères Nematocères) Trav. Doc. ORSTOM, No. 56, 161 pp.

Dejoux, C., Lauzanne, L. and Lévêque, C., 1969. Evolution qualitative et quantitative de la faune benthique dans la partie est du lac Tchad. Cah. ORSTOM. Sér. Hydrobiol. 3: 3–58.

Dussart, B., 1966. Limnologie. Gauthier-Villars, Paris, 677 pp.

Elton, C., 1927. Animal Ecology. MacMillan Co., New York, 107 pp.

Fischer, Z., 1970. The elements of energy balance in grass carp (*Ctenopharyngodon idella* Val.) Part I. Pol. Arch. Hydrobiol. 17, 4: 421–434.

Fryer, G., 1960. The feeding mechanism of some Atyid prawns of the genus *Caridina* Trans. Soc. Edinb. 64: 217–244.

Gras, R. Iltis, A. and Lévêque-Duwat, S., 1967. Le plancton du Bas Chari et de la partie est du lac Tchad. Cah. ORSTOM. Sér. Hydrobiol. 1: 25–96.

Gras, R., Iltis, A. and Saint-Jean, L., 1971. Biologie des crustacés du lac Tchad. II. Régime alimentaire des Entomostracés planctoniques. Cah. ORSTOM. Sér. Hydrobiol. 5: 285–286.

Hamblyn, E. L., 1966. The food and feeding habits of Nile perch *Lates niloticus* ((Linné) (Pisces: Centropomidae). Rev. Zool. Bot. Afr. 74: 1–28.

Hatanaka, M., Kosaka, M. and Sato, Y., 1956. Growth and food consumption in plaice. Part I. *Limanda yokohamae* (Günther). Tohoku J. Agric. Res. 7: 151–162.

Hopson, A. Y., 1972. A study of the Nile perch in Lake Chad. Overseas Res. Pupl. London, 19, 93 pp.

Hynes, H. B. N., 1950. The food of freshwater stickbacks (*Gasterosteus aculeatus* and *Pygosteus pungitius*) with a review of methods used in studies of the food of fishes. J. anim. Ecol. 19: 36–58.

Im, B. H., 1977. Etude de l'alimentation de quelques espèces de *Synodontis* (Poissons, Mochocidae) du Tchad. Thèse de spécialité, Toulouse, 150 pp.

Ivlev, V. S., 1939. The energy balance of the growing larva of *Silurus glanis*. Dolk. Akad. Nauk. S.S.S.R. new ser. 25: 87–89.

Ivlev, V. S., 1961. Experimental ecology of the feeding of fishes. Yale Univ. Press, New Haven, Conn., 302 pp.

Kelso, J. R. M., 1972. Conversion, maintenance and assimilation for walleye, *Stizostedion vitreum vitreum*, as affected by size, diet and temperature J. Fish. Bd. Can. 29: 1181–92.

Lauzanne, L., 1969. Etude quantitative de nutrition des *Alestes baremoze* (Pisc. Charac.). Cah. ORSTOM Sér. Hydrobiol. 3: 15–27.

Lauzanne, L., 1970. La sélection des proies chez *Alestes baremoze* (Pisc., Charac.). Cah. ORSTOM Sér. Hydrobiol. 3: 15–27.

Lauzanne, L., 1972. Régimes alimentaires des principales espèces de poissons de l'archipel oriental du lac Tchad. Verh. int. Ver. Limnol. 18: 636–646.

Lauzanne, L., 1973. Etude qualitative de la nutrition des *Alestes baremoze* (Pisces, Characidae). Cah. ORSTOM Sér. Hydrobiol. 7: 3–15.

Lauzanne, L., 1975a. La sélection des proies chez trois poissons malacophages du lac Tchad. Cah. ORSTOM Sér. Hydrobiol., 9: 3–7.

517

Lauzanne, L., 1975b. Régimes alimentaires d'*Hydrocyon forskalii* (Pisces, Characidae) dans le lac Tchad et ses tributaires. Cah. ORSTOM Sér. Hydrobiol. 9: 105–121.

Lauzanne, L., 1976. Régimes alimentaires et relations trophiques des poissons du lac Tchad. Cah. ORSTOM Sér. Hydrobiol. 10: 267–310.

Lauzanne, L., 1977. Aspects qualitatifs et quantitatifs de l'alimentation des poissons du Tchad. Thèse d'Etat, Paris, 284 pp.

Lauzanne, L., 1978a. Croissance de *Sarotherodon galilaeus* (Pisces, Cichlidae) dans le lac Tchad. Cybium S.3, 3: 5–14.

Lauzanne, L., 1978b. Equivalents caloriques de quelques poissons et de leur nourriture. Cah. ORSTOM. Sér. Hydrobiol. 12: 89–92.

Lauzanne, L., 1978c. Etude quantitative de l'alimentation de *Sarotherodon galilaeus*. Cah. ORSTOM Sér. Hydrobiol. 12: 71–81.

Lauzanne, L. and Iltis, A., 1975. La sélection de la nourriture chez *Tilapia galilaea* (Pisces, Cichlidae) du lac Tchad. Cah. ORSTOM Sér. Hydrobiol. 9: 193–199.

Mann, K. H., 1965. Energy transformation by a population of fish in the river Thames. J. anim. Ecol. 34: 253–275.

Mironova, N. V., 1974. The energy balance of *Tilapia mossambica*. J. Ichtyol. 14: 431–438.

Pandian, T. J., 1967. Food intake, absorption and conversion in the fish *Ophiocephalus striatus*. Helgoländer Wiss. Meeresunters 15: 637–647.

Pentelow, F. T. K., 1939. The relation between growth and food consumption in the brown trout (*Salmo trutta*). J. exp. Biol. 16: 446–473.

Petipa, T. S., Pavlova, E. V. and Mironov, G. N., 1973. The food web structure, utilization and transport of energy by trophic levels in the planktonic communities, 142–167 In J. H. Steele, (ed.), Marine Food Chains, Oliver and Boyd, Edinburgh. Reprint by Otto Koeltz Antiquariat, Keenigstein, 552 pp.

Schaeperclaus, W., 1933. Text book of pond culture. Translated by E. Hund, U.S. Fish and Wildlife Service fish leaflet 311, 260 pp.

Solomon, D. Y. and Brafield, A. E., 1972. The energetics of feeding, metabolism and growth of perch (*Perca fluviatilis L.*) J. anim. Ecol. 41: 699–718.

Surber, E. W., 1935. 65th annual meeting. Trans. am. Fish. Soc.

Warren, C. E. and Davis, G. E., 1967. Laboratory studies on the feeding, bioenergetics and growth of fish, 175–214, In S. D. Gerking, (ed.), The Biological Basis of Freshwater Fish Production. Blackwell Scientific Publications. Oxford, 495 pp.

Winberg, G. G.–956. Rate of metabolism and food requirements of fishes. Traduit du russe par: J. Fish. Bd. Can. trans. ser. 194 (1960), 253 pp.

16. The impact of birds on the lacustrine ecosystem

Claude Dejoux

Lake Chad is situated on one of the major pathways of migratory birds between Europe and the tropical countries of Central and Eastern Africa. In most cases it is only a halting spot for destinations further south and the migrants only use the lake as a resting place, not intensively utilizing its food resources. However, the morphology of the lake basin varies with changes in water level and the relationship between the lake and its bird fauna can be profoundly changed, so that while the periods of high water provide birds with a resting place only, the periods of low water provide the birds with vast mud flats or large sandy beaches with abundant food, for which the birds stop.

Apart from these periods when the migratory birds are particularly abundant there is a mainly sedentary bird fauna which makes regular and more or less intensive use of the lake ecosystem.

According to oral tradition, there were periods when the lake was visited by numerous birds nesting in the islands! If these data are accurate, there must be a period of heavy predation in the aquatic ecosystem corresponding to major nesting, if only to feed the young. However, during the observations made between 1965 and 1973, we did not notice any considerable increase in the lacustrine fauna, although the lowering of the lake water contributed to the establishment of a population of Limicolae resulting from the development of muddy areas. The only large aggregations of birds observed were temporary and did not correspond to a nesting.

It was possible to determine the diets of numerous bird species observed in Lake Chad through the analysis of more than one thousand stomach contents. They can be divided into four main groups according to their use of the aquatic biotopes.

16.1 Birds closely dependent upon the aquatic environment

There are two distinct groups of birds: the pelicans and shags, and the river eagles. The first two are mainly ichthyophagous and inhabit privileged zones

J.-P. Carmouze et al. (eds.) Lake Chad
© *1983, Dr W. Junk Publishers, The Hague/Boston/Lancaster*
ISBN 978-94-009-7268-1

such as the Shari delta. Here, food is abundant, and, in certain periods, they make use of the fish aggregations that migrate upstream for spawning. Shags are found along reed belts where they catch fish in submerged vegetation and use papyrus as a resting place. They are also found in large quantities in the arms of the drying up lake where the ichthyofauna is confined. The river eagles have an ichthyophagous diet and although dead fishes represent their main food, they will occasionally feed on other dead animals.

During flight many birds use the environment in relation to the aggregations of encountered prey. Although the scissor-bills are present mainly on the large rivers, they do not avoid the delta zone or the southeastern aquatic vegetation mats. They feed mainly on small fishes captured by scooping through the water with their wide open bills.

Caspian and gull-billed terns are mainly ichthyophagous, while the whiskered and white-winged black terns have a mixed diet of fish and insects. The large emergence of Chironomids or May-flies often leads to considerable aggregations of white-winged black terns which occasionally feed on aquatic vegetation.

The occurrence of red-necked little bitterns and king-fishers is related to the presence of resting places. Although they are very abundant on the river banks, they also occupy the reed belts of the islands belonging to the lake archipelagoes. Red-necked little bitterns feed almost exclusively on small fishes, the big aquatic insects being ingested only accidentally.

Finally, two groups of birds occupy two large biotopes; the submerged vegetation and the mud flats.

— Aquatic mats with *Potamogeton* are occupied by the largest density of birds. The lily trotter and the lesser lily trotter are found on the surface leaves and utilize insects from the vegetation as well as the insect larvae living near the surface (Odonata and Hemiptera). In certain periods, this submerged vegetation is also occupied by Anatidae, especially the common teals, the pygmy teals, the hottentot teals and the cape widgeons. The shovellers are much less numerous but they are also found in these biotopes. Anatidae use a part of the aquatic mats up to 10 cm in thickness in addition to the emergent zone. Common and hottentot teals are mainly insect feeders, while the other two species feed on grains and vegetation and also feed on a small amount of insects.

— Mud flats are certainly the most utilized biotopes in Lake Chad, within the limits mentioned at the beginning of the chapter. They represent the favourite habitat for waders: African great white herons, spoonbills, black-winged stilts, open-bills. The last two species feed mainly on molluscs, while the first two capture mainly fishes and insects.

While the large waders most often live isolated or in small groups, the small limicolae are sometimes found in large aggregations. The redshanks, the marsh sandpipers or the wood sandpipers should be mentioned here as well as various small sandpipers which feed on the epifauna of the wet mud flats or the fauna of the shallow flooded zones. In the latter case, they feed mainly on chironomids, small Hemiptera and Coleoptera: Dytiscidae.

16.2 Birds with a mixed diet

About 50% of the mixed diet consisted of aquatic organisms and 50% of it was terrestrial.

At first, we find a group of Passeriformes which catch prey during flight and stay fairly close to the reed belts that are used as resting places. Similarly, the sand martins and the wire-tailed swallows, the big and small cane-warblers, the Senegal fire-finch, the beautiful long-tailed sunbird and the West African prinia chase prey above the open water. Close to the papyrus or *Phragmites* reeds, the various bee-eaters intensively feed on insects and accordingly their diet contains elements from the aquatic environment only (Odonata, Chironomids) or is of terrestrial origin (Acrididae, Diptera and Carabidae).

Some birds found in marshy zones or along the islands lined with aquatic mats, use either the insects, the invertebrate fauna or the small aquatic vertebrates. They include the black crake, the African green-backed heron, the African little grebe, the pygmy kingfisher and the African moorhens. Finally, along beaches or in flood zones, there are birds with a mixed diet that feed on the dominant prey organisms present. They include among others a few Anatidae such as the spurwinged goose, the pintail, the white-faced duck or certain limicolae whose dominant species are the little stint and the ruff and reeve.

16.3 Birds using the aquatic environments as a supplementary feeding ground

This group include a collection of birds which, apart from the garganey are not dependent upon the aquatic environment but make slight and occasional use of it from time to time.

This collection is very dissimilar and includes small passeriformes such as the lesser white-throats and grey backed camaroptera, the rufous grass-warbler, the Niger black headed weaver as well as birds of prey such as the West-African black kite, the spotted and milky eagle owls. Finally, along the banks or on the beaches the Egyptian plovers, the dusky redshank and the West African pratincole use equally the terrestrial or aquatic insect fauna.

16.4 Birds occasionally using the environment

It is hardly possible to speak of exploitation in this case as the predation by certain species is so low and occasional. We will merely give a list of birds which can move near the aquatic environment and which can take prey from these biotopes without particularly searching for them. Their impact can be considered as insignificant. Among these species are the little egret, the West African hadada, the comb goose, the Senegal thick-knee and the plovers, the nightjars, the little African swift, the wood-warbler, the redpate grass-warbler, the slender-billed weaver ...

They are usually entomophagous birds which occasionally ingest aquatic insects after their emergence.

16.5 Conclusion

Considering its size and variety, the aquatic environment of Chad in general and Lake Chad in particular is not greatly used by the birds. Out of 183 taxa studied, only 84 contained a significant quantity of aquatic organisms in their stomach and hardly twenty of them can be considered as closely related to the freshwater environment. Apart from the open waters, the zones used are usually very localized as shown in Fig. 1 and the coastal environments are the only ones to be visited regularly.

Of course, the birds do not escape the general and sometimes favourable 'glut of prey', and predation on the environment is therefore exclusive and intensive. So, we could observe considerable aggregations of terns and swallows that were very localized at the time of large emergence of chironomids in the Archipelago or considerable aggregations of ichthyophagous birds during traditional fisheries. Finally, the combined presence of important bird aggregations and a very dense aquatic fauna leads to a considerable but temporary use of the environment as sometimes observed in the developing polders of the arms of the Lake Chad archipelago. It was not uncommon then to find 10 to 15 000 Chironomid larvae in the stomach and crop of a cape widgeon. When it is known that several thousands of such birds occupied this kind of environment in certain periods, it is not necessary to dwell on the importance of their impact.

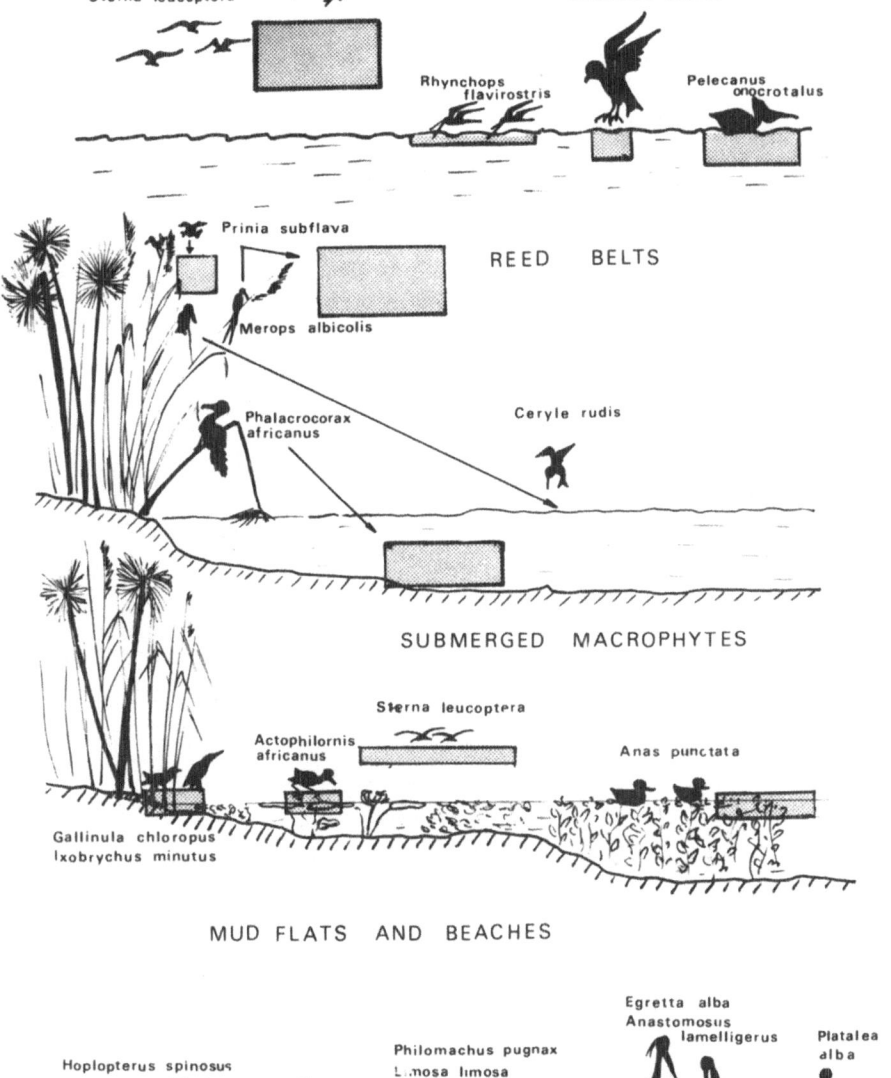

Fig. 1 Zonation of the major sites of the main bird species which depend upon Lake Chad.

523

Table 1 Different uses of the aquatic environment by the birds; list of the common names and their scientific equivalents.

A. Birds closely dependent upon the aquatic environment

Pelecanus onocrolatus L.	Rosy pelican
Phalacrocorax a. africanus (Gmelin)	Long-tailed shag
Egretta alba melanorhynchos (Wagler)	African great white heron
Ixobrychus m. minutus (L.)	Red-necked little bittern
Anastomus l. lamelligerus Temminck	African open-bill
Platalea alba Scopoli	African spoonbill
Anas c. crecca L.	Common teal
Anas capensis Gmelin	Cape widgeon
Anas clypeata L.	Shoveller
Anas punctata Burchell	Hottentot teal
Nettapus auritus (Boddaert)	Pygmy goose
Haliaetus vocifer clamans (Daudin)	River eagle
Actophilornis africana (Gmelin)	Lily trotter
Microparra capensis (Smith)	Lesser lily trotter
Limosa l. limosa (L.)	Black-tailed godwit
Tringa glareola L.	Wood sandpiper
Tringa stagnatilis (Bechstein)	Marsh sandpiper
Tringa totanus (L.)	Common redshank
Himantopus h. himantopus (L.)	Black-winged stilt
Rhynchops flavirostris Vieillot	Scissor-bill
Sterna n. nilotica Gmelin	Gull-billed tern
Sterna leucoptera Temminck	White-winged black tern
Sterna h. hybrida Pallas	Whiskered tern
Sterna tschegrava Lepechin	Caspian tern
Ceryle r. rudis (L.)	Pied kingfisher
Alcedo cristata Pallas	Malachite kingfisher

B. Birds with mixed diet

Podiceps ruficollis capensis Salvadori	African little grebe
Butorides striatus atricapillus (Afzelius)	African green-backed heron
Dendrocygna viduata (L.)	White-faced duck
Plectropterus g. gambensis (L.)	Spur-winged goose
Anas a. acuta L.	Pintail
Gallinula c. chloropus (L.)	African moorhen
Limnochorax flavirostra (Swainson)	Black crake
Philomachus pugnax (L.)	Ruff and reeve
Calidris minuta (Leisler)	Little stint
Ceyx p. picta (Boddaert)	Pygmy kingfisher
Merops albicollis Vieillot	White-throated bee-eater
Merops p. pusillus Muller	Least bee-eater

Table 1 (continued).

B. Birds with mixed diet

Merops b. bulocki Vieillot	Red-throated bee-eater
Merops superciliosus	
chrysocercus	
(Cabanis & Heine)	Blue-cheeked bee-eater
Riparia r. riparia (L.)	Sand martin
Hirundo s. smithii Leach	Wire-tailed swallow
Acrocephalus rufescens chadensis	
(Alexander)	Rufous cane-warbler
Acrocephalus gracilirostris neglectus	
(Alexander)	
Prinia subflava pallescens Madarasz	West african prinia
Nectarinia p. pulchella (L.)	Beautiful long-tailed
	sunbird
Lagonosticta s. senegala (L.)	Senegal fire-finch

C. Birds using the environment as a supplement

Tringa erythropus (Pallas)	Dusky redshank
Pluvianus aegyptius (L.)	Egyptian plover
Glareola p. pratincola (L.)	West african pratincole
Bubo africanus cinerascens Guerin	Spotted eagle-owl
Bubo lacteus (Temminck)	Milky eagle-owl
Sylvia c. curruca (L.)	Lesser whitethroat
Cisticola galactotes Lynes	Rufous grass-warbler
Camaroptera brachyura brevicaudata	
(Cretzschmar)	Grey-backed camaroptera
Ploceus melanocephalus capitalis	
(Latham)	Niger black headed weaver

D. Birds using the environment occasionally)

Egretta g. garzetta (L.)	Little egret
Bostrychia hagedash (Reichenow)	West african hadada
Sarkidiornis m. melanota (Pennant)	Comb goose
Burhinus senegalensis (Swainson)	Senegal thick-knee
Vanellus spp.	Plovers
Caprimulgus eximius Temminck	Zinder golden nightjar
Caprimulgus c. climacurus	
Vieillot	Long-tailed nightjar
Apus a. affinis (Gray)	Little african swift
Phylloscopus sibilatrix (Bechstein)	Wood-warbler
Cisticola r. ruficeps (Cretzschmar)	Redpate grass-warbler
Ploceus l. luteolus (Lichtenstein)	Slender-billed weaver

525

V

17. The lacustrine ecosystem during the 'Normal Chad' period and the drying phase

Jean-Pierre Carmouze, Jean-René Durand and Christian Lévêque

Between 1964 and 1978, multidisciplinary research was carried out on Lake Chad, a shallow endhorheic lake which is particularly sensitive to variations in water supply. From 1964 to 1971, its area did not change much, in spite of a significant lowering of water level. It was during this period called 'Normal Chad' that most of the quantitative studies were carried out. Here we will draw up a general balance of physical, chemical and biological characteristics of the environment in an attempt to understand its functioning, especially the hydrochemistry and the balance of energy and matter.

Due to decreased floods, the area of the lake was considerably reduced after 1972, involving at first the separation of the two basins, then the partition of the south basin and the drying up of the north basin in 1975. Over the course of this period, there was considerable transformation of the lacustrine environment and communities and we will try to draw up a general scheme of these changes.

It is obvious that considerable gaps exist in our knowledge of the functioning of the ecosystem. This is especially true for all bacterial activity (decomposition, mineralization, production, and role in the trophic systems) which almost certainly play a major role but cannot be quantified. However, the diversity of data collected and especially the length of the period of observations make the studies on Lake Chad almost unique for tropical Africa. The only ecosystem investigated in a similar way was Lake George, Uganda, studied between 1966 and 1972, as part of the International Biological Programme thus allowing a comparison of two shallow lakes, one equatorial and one tropical. It would also have been interesting to make an analogous comparison for the drought period but the study on Lake Chilwa, which had a similar dry phase did not give sufficient quantitative results for that period.

17.1 'Normal Chad': aspects of its functioning

17.1.1 *The environment*

17.1.1.1 *The recent quaternary lakes and the present topography.* The Chad basin extends over 250 000 km² between 5° and 25° N and 7° and 25° E. The

J.-P. Carmouze et al. (eds.) Lake Chad
© *1983, Dr W. Junk Publishers, The Hague/Boston/Lancaster*
ISBN 978-94-009-7268-1

lacustrine environments of this area have been considerably modified over the recent quaternary, during which three main periods can be distinguished within the sensitivity of C_{14} measurements:

— between about 40 000 and 18 000 years BP the lake basin was occupied by much more numerous lakes than now, in which Ostracod shells and diatoms were deposited. Some eolian layers in these deposits indicate that the lakes were subjected to temporary drying periods;

— between 18 000 and 13 000 years BP the lakes disappeared and the region was altered considerably by wind. The southern limit of the desert zone moved 500 km towards the equator;

— from 13 000 years BP the depressions of the basin were occupied by several lakes that were maintained for more than ten thousand years. However the existence of eolian sands and of layers with dessication cracks between the lacustrine series prove that these lakes also encountered several periods of drought.

The present endorheic lake occupies a 25 000 km² basin between 12° and 14° 20′ N and 13° and 15° 20′ E. During the 'Normal Chad' period (water level higher than 280 m) the area of water varied between 15 000 and 21 500 km².

The lake is divided into two basins of fairly equal area and of shallow mean depth: 3 to 5 m for the north basin; 0.75 to 2.75 m for the south basin. Each of these basins is subdivided into three main regions that correspond to three habitat types: the open water; the reed islands which are islands of vegetation (*Cyperus papyrus* and *Phragmites*) colonizing shallows; the archipelagoes whose islands correspond to emergent crests of an immersed erg. The two basins are connected by a vast zone of reed islands called the 'Great Barrier'.

17.1.1.2 *The climate.* Situated in the intertropical convergence zone, the lake is subjected to a Sahelian type of climate. The dry winds from the northeast (harmattan) blow from October to April, while the humid winds from the southwest occur between May and September. The winds blow especially between 06 and 12 a.m. and the speed is on average greater than 5m/s for 6 hours per day.

The rainy season begins in May–June and ends in October with 50% of the precipitation occurring in August. The mean annual rainfall increases from the north to the south of the lake, from 150 to 250 mm. Annual variability is high.

The mean annual air temperature is 28°C while the monthly average varies from 29° to 32°C between March and October and from 22° to 24°C between December and February. The minimum diurnal differences occur in August (7° to 9°C) and maxima in January (16° to 17°C):

On the edge of the lake, at Bol, the humidity of the air reaches a maximum in August (72 to 81%) and a minimum in February–March (23 to 31%). The evaporation is considerable, between 2.05 and 2.20 m/year.

The mean annual total incident radiation is 2310×10^4 joules $m^{-2} day^{-1}$

with a minimum of 2140×10^4 in January and a maximum of 2580×10^4 in March. The seasonal variations are thus low. The mean insolation is about 280 hours a month.

17.1.1.3 *The fluvial system and the hydrological regulation of the lake.* From the hydrological and biological point of view (fishes in particular), the lacustrine environment essentially depends on the fluvial system. The hydrographic system consists mainly of the Shari and the Logone which flows into the Shari near N'Djamena, 250 km from the lake. These two rivers have a tropical regime characterized by an annual flood (September to December) and a well-marked period of low water (from March to June). Due to these seasonal variations 40 to 50% of the flow occur during the four months of the flood. At this time some of the water leave the river bed to flood vast flat plains. The floodplain of North Cameroon (or Yaéré) (5000 km^2) is fed by the overflowing of the left bank of the Logone in September–October and the waters recede from December to February, drained by the El Beid which flows northwards into the south of the lake. The overflows of the right bank of the Logone give rise to the 'Grand Courant' which contributes to the flooding of a large plain between the Logone and the Shari.

The fluvial water (between 20 and 50×10^9 m^3 with a mean of 40×10^9 m^3) represents 86–87% of the total supply to the lake; the rest comes from rain. These supplies are counterbalanced by losses of 92% due to evaporation and 8% by seepage losses. The volume of the lake is maintained at a low level between 40 and 85×10^9 m^3. The annual turnover of the lake water is very important with an average of 2/3 being renewed each year. Because of this rapid turnover time the lake does not efficiently absorb the irregularities of annual supplies.

Water movements are under the influence of the wind regime and fluvial water supplies. At the end of the low water in June, the monsoon winds favour a general displacement of the southern waters towards the north and movement begins in the north basin at the level of the northeastern part of the 'Great Barrier'. In August this is accentuated by a push from the waters of the Shari flood which then begins. The lake receives 50% of its fluvial supplies in October and November, when the harmattan begins, displacing the overflowing zone towards the south of the 'Great Barrier'. At the maximum of the flood the waters then penetrate into the north basin along the southeast coast and drive back all the residual water in the direction of the Northeastern Archipelago and Reed Islands. The flood waters also begin to invade the reed islands and the archipelago in the south basin. From the end of January there is no longer a flow of water into the north basin and the last supplies of the Shari that spread out in the south basin involve a general displacement of water towards the periphery. From April to June, there is no notable water circulation.

17.1.1.4 *The physical characteristics of the waters.* The mean annual temperature ranges between 25.5° and 27.5°C. During the cold season (December to February) the water temperature varies from 17° to 21°C whereas in the warm season (May–June) it is generally over 30°C. The average diurnal difference is about 2°C. A thermal stratification can develop over the day during rare periods when the wind does not blow. Thus Lake Chad is a tropical polymictic lake.

The oxygen distribution is homogenous over the water column in the early morning when the concentration approached saturation. There is an increase of surface oxygen tension over the course of the day (120% saturation) and a diurnal stratification appears during calm periods.

Transparency varies according to the season, the region of the lake and the water level. At the 282 m level, in the open waters of the south basin, it is maximal in December–January (1 m) and minimal in August (20 cm) when the Shari flood waters arrive loaded with silt. However, in the archipelago of the south basin the seasonal variations are less marked. In the entire north basin the transparencies range between 60 and 90 cm.

17.1.1.5 *The chemical characteristics of the waters and salinity regulation.* Lake Chad waters are fresh, but their salinity changes, according to the regions considered. The salinity of the Shari water that enters the lake is about 60–65 mg l^{-1} (or about 50 μS cm^{-1}). The relative importance of the different elements is 48.4% HCO_3, 6.3% Ca, 2.95% Mg, 4.6% Na, 2.86% K and 34.8% SiO_4H_4. Salinity increases with distance from the delta: the open waters of the south basin are 1.2 to 1.5 times more concentrated than those of the Shari while those of the archipelago in the north basin are 10 to 20 times higher. On average, the waters of the north basin (625 mg l^{-1}) are four times more saline than those of the south basin, which are themselves two and a half times more saline than those of the Shari. The chemical composition changes with the concentration: HCO_3 and Na become predominant at the expense of SiO_4H_4, Ca and Mg and their relative importance in the north basin is 61.4% HCO_3, 7% Ca, 3.8% Mg; 9.9% Na, 4.5% K and 5.25% SiO_4H_4. Seasonal variations are more accentuated in the regions close to the Shari delta than in the archipelago.

The pH of the river water varied from 7 to 8; in the lake, the pH does not exceed 8 in the south basin, and 8.9 in the north basin.

These results show that Lake Chad is not a basin of high salt concentration as might be assumed, *a priori*, from both its lack of surface outlet and the fact that it is subjected to strong evaporation. With an average conductivity of 450 μS, it falls in the middle of the list of major African lakes classified according to increasing conductivity (Talling and Talling 1965; Symoens 1968). It is on an average more saline than Lake Victoria, George, Malawi and less than Lakes Tanganyika, Albert, Kivu and Turkana.

Moreover, the salinity of the water during the 'Normal Chad' period changes only slightly from one year to another. So, between the severe 1972 recession and the large flood of 1957, the average salinity decreased by 30%. By comparison, the salinity of Lake Chilwa can triple from year to year (Kalk 1979).

These apparent contradictions result from several factors which combine to maintain the dissolved salt stocks proportionally close to those of the water volume.

(a) The low salinity of the water is maintained by three factors:
— the low salinity of the river water (50–60 mg l^{-1}) is half the mean salinity of the lake water;
— climato-geographical regulation results from a combination of water transfer factors (river supplies, rain, evaporation, seepage) to give an 11-fold concentration of the river water which is not very high for a closed lake in an arid zone. This is because seepage losses are relatively important (7.5% of the total annual water losses);
— considerable biogeochemical sedimentation of SiO_4H_4, Ca, Na, HCo_3/CO_3, and, to a lesser degree, K^+ occurs in the lake itself. This geochemical precipitation comes from the neoformation of smectites and the precipitation of calcite. The clayey neoformations are favoured by a relatively high concentration of dissolved silica from the river waters (20 mg l^{-1}) and by some supplies of solids rich in iron and aluminium hydroxides. SiO_4H_4, Ca and Mg are sedimented. The precipitation of calcite is favoured by the fact that the anions are only represented by HCO_3^- and CO_3^{--}; saturation is thus quickly reached and calcite forms in some environments of relatively low salinity (700–900 mg l^{-1}). Biochemical sedimentation, mainly due to molluscs and macrophytes, mainly involves Ca, HCO_3, Mg and K.

All these processes of chemical sedimentation lead to a 45% decrease from the value that the salinity would reach by climato-geographical regulation only. Given that the different elements are sedimented to different degrees, the chemical composition of the water evolves towards sodium-bicarbonate facies (cf. Chapter 4).

(b) The attenuation of temporal fluctuations in salinity

At the time of the flood, the volume of the lake increases more quickly than the salt stocks in all the salt concentration basins. This causes a decrease of the water while during its recession the inverse happens. Usually these effects are poorly dampened in a shallow lake. In Lake Chad, however, salinity fluctuations are attenuated by two principal mechanisms. The first results from the morphology of the south basin as the erg bordering it from the southeast to the southwest favours the isolation of ponds which rapidly dry up when the water subside. This results in important losses of salty waters in the marginal zones. Thus at recession, the lowering of salt stocks is accelerated and the increase of salinity is diminished in proportion. The increase of salinity during the fall in water level from 1967 to 1972 was attenuated by 18%, due to this mechanism.

In contrast, during a flood, some of the salts deposited during the preceding dry phase are partially redissolved. The difference between the increase in salt stocks and that of water volume is reduced and thus the decrease of the water's salinity is attenuated.

The second mechanism concerns the rate of biogeochemical sedimentation. It tends to increase during flood subsidence and to decrease during a flood. In both cases this mechanism attenuates the difference between the change of the lake volume and the salinity. The increase in salinity was reduced by 15% during the contraction of 1967–1972 due to this mechanism.

The combination of these two mechanisms dampens variations in salinity very efficiently:
— during the contraction of 1967–1972, the average salinity of the waters was 67% of the value which would have been obtained without these two mechanisms;
— during the expansion period from 1945 to 1957, the salinity only reached 50% of the value expected without these mechanisms.

It should finally be noted that 60% of the salts deposited in the marginal zones are dissolved when water returns. In other words, each flood-recession cycle involves a definite loss of dissolved salts. This appears to be a supplementary means of elimination of salts supplied by the rivers which contributes to the maintenance of a low mean salinity.

17.1.1.6 *Sedimentology*. Four main types of sediment are found:
— mud, the most abundant sediment in the lake is a structureless material generally present in a fine and homogenous suspension. The water content is 2.5 to 5 times the dry weight which includes a significant organic fraction (10 to 15%) with a mean C and N content of 20 and 8% respectively. Peat is frequently found at the edge of vegetation fringes where decomposing organic matter is very abundant;
— clay can be in a soft and structureless, a more or less structured, or in a granular form. Its water content varies between 40 and 130% of dry weight and its organic fraction does not exceed 5%. This sediment is well represented in the open waters and to a lesser extent in the archipelago where it occupies an average of 30% of the bottom surface;
— the pseudo-sands are granules, from 0.2 to 0.3 mm mean diameter, composed of goethite and nontronite. They are especially abundant in the open waters adjacent to the Shari delta.
— part of the sand comes from the flooded erg of the north (0.25 mm size, quartz) and another part from fluvial supplies (micaceous sand of small size: 0.16 mm). The latter are mainly localized in the coastal zone adjacent to the delta.

17.1.2 *The organisms*

17.1.2.1 *The phytoplankton.* More than 1000 species and taxa were identified in Lake Chad where the algal flora was dominated qualitatively by the Desmids and the Diatoms whereas the Cyanophycea played a very important role quantitatively (30 to 50% of algal biovolume in 1971–1972).

During the 'Normal Chad' period four regions were distinguished which were characterized by the dominance of species or groups of species:

— the open waters of the north where the Desmids (*Closterium aciculare*) were predominant for the major part of the year; *Pediastrum, Botryococcus* and *Microcystis* were abundant while the diatom *Melosira granulata* were absent;

— the northeastern archipelago where Cyanophycea (*Anabaena* and *Microcystis*) were dominant but *Closterium, Pediastrum* and *Botryococcus* were still abundant. In the open waters of the south and the southeast, which are under the influence of the Shari flood, the diatoms *Melosira granulata* and *Surirella muelleri* make up the bulk of the phytoplankton;

— the eastern and southeastern archipelago where the Cyanophycea, *Microcystis* and *Anabaena* were abundant and occasionally *Surirella, Pediastrum* and *Melosira*.

The highest densities were observed in the archipelagoes (1.4 to 2 $\mu l \, l^{-1}$) and they were much lower in the open waters of the south and the southeast (0.03 to 0.22 $\mu l \, l^{-1}$). The total biomass was estimated at 40 800 tons fresh weight for the lake in 1971, or 6200 tons dry weight if a conversion factor of 15% is used. From the primary production measurements made over several years in the eastern archipelago, a model was established to estimate the hourly production around midday (Lemoalle 1973). The mean gross daily production in this region was 4.2 g O_2 m^{-2} day^{-1} or 550 g $C\,m^{-2}$ $year^{-1}$. It was a little lower in the southern open waters and a little higher in the north basin.

17.1.2.2 *The macrophytes.* Twelve plant associations were really important in Lake Chad where the aquatic plants occupied about 2400 km^2 during the period of 'Normal Chad'. *Vossia cuspidata* which was very abundant in the Shari delta as well as *Cyperus papyrus* in the south basin disappeared progressively towards the north, as a function of salinity, whereas *Typha australis* appeared. *Phragmites australis* was well represented everywhere and there was also much submerged vegetation containing *Potamogeton, Ceratophyllum, Vallisneria, Utricularia* and *Nymphea*. The biomass of the aerial parts was estimated, in dry weight at 31 t ha^{-1} for *Phragmites*, 28 t ha^{-1} for *Cyperus papyrus*, 17.5 t ha^{-1} for *Vossia* and 15.5 t ha^{-1} for *Typha*. The total biomass for the entire lake was calculated at 7.2×10^6 t for the aerial parts of these four species and at 13×10^6 t (also dry weight) for the roots.

The annual production of macrophytes was not studied, but it was certainly important and probably roughly similar to the biomass of the aerial parts.

17.1.2.3 *The zooplankton*. The population of planktonic crustacea in Lake Chad was fairly rich as 8 species of Cladocera and 4 species of Copepods were well represented. In comparison only four species were abundant in Lake Chilwa (Kalk 1979), Lake Turkana (Ferguson 1975) and Lake Naivasha (Litterick et al. 1974) and two in Lake George (Burgis 1974). There were also several Rotifer species (Pourriot 1968) but they only represented a small percentage of the zooplankton biomass.

There is a seasonal fluctuation in the density of organisms which is strongly influenced by the Shari flood, in the open waters of the south basin (Gras et al. 1967). In the zones that are well protected from the influence of the flood, the minimum density is also observed during the cold season and a maximum in April–May.

There were no notable variations in the zooplankton population structure over the lake as a whole. Nevertheless, three major ecological zones can be distinguished during the period or 'Normal Chad' based on the density of organisms and the seasonal cycle of abundance. The open waters of the south and the southeast were the poorest (95 ind. 1^{-1}, 110 g m^{-3}) and the quantitative variations there, were very high over the course of the year. In the archipelago and the reed islands of the southern basin, the seasonal variations were very reduced and the density was higher (318 ind. 1^{-1}; 333 mg m^{-3}). Finally, in the northern basin where the seasonal variations were also low, the density was intermediate (224 ind. 1^{-1}, 216 mg m^{-3}).

The total biomass of zooplankton crustacea was estimated at 12 200 tons, dry weight, during 1971 or an average of 0.81 g m^{-2}. This biomass was unevenly distributed: 0.9 g m^{-2} in the northern basin which had 60% of the stock, 0.7 g m^{-2} in the eastern archipelago and 0.25 g m^{-2} in the eastern and southeastern open waters.

The duration of development of the embryonic stages, determined *in vitro* ranged between 1 and 1.5 days at 30°C, and 1.4 and 2 days at 25°C, for both the Cladocera and the Copepoda. The post embryonic development was more rapid in Cladocera. Juvenile development varied between 1.2 and 2.3 days at 30°C in *Moina micrura, Diaphanosma excisum* and *Ceriodaphnia cornuta* and between 2.9 and 4.4 days in *Bosmina longirostris* and the three species of *Daphnia*. In the Copepods, juvenile development at 30°C lasted about 6 days in the Cyclopoids, 11 days in *Thermodiaptomus galebi* and 18 to 25 days in *Tropodiaptomus incognitus*.

The annual P/B ratio of the major species was determined over the course of a year of observations in the eastern archipelago. On average it was 23 for the Calanoids, 63 for the Cyclopoids and 113 for the Cladocera (191 for *Moina*, 151 for *Diaphanosoma*, 151 for *Ceriodaphnia* and 72 to 76 for *Daphnia* and *Bosmina*). The annual production was estimated at about 860 000 tons in 1971 (47 g m^{-2} dry weight, or 260 kcal m^{-2}).

17.1.2.4 *The zoobenthos.* The benthic fauna was essentially composed of three groups of invertebrates: oligochaetes and molluscs, represented by a small number of species, and many species of insect larvae.

The distribution of the oligochaetes and the molluscs depended especially on the nature of the sediment although the chemical composition of the water could have had an influence on the distribution of some species. The insects appear to be less sensitive to these ecological factors but they had, like the oligochaetes, a seasonal cycle of abundance with a maximum during the cold season and a minimum at the end of the warm season.

The benthic biomass expressed as dry weight (shell not included for the molluscs) was estimated at 71 000 tons for the entire lake in 1970, or 3.7 g m^{-2} on the average. Molluscs represented the bulk of the biomass, over 90% or 3.3 g m^{-2} on the average. They were particularly represented by three species of Prosobranchs: *Melania tuberculata* numerically dominant in the northern basin, *Cleopatra bulimoides*, abundant in the southern basin and *Bellamya unicolor* which represented 40% of the benthic biomass. The molluscs were particularly abundant in the open waters and the archipelago of the northern basin as well as near the 'Great Barrier' where the biomass was between 4.8 and 7.2 g m^{-2}. They were rare in the northern and the eastern part of the lake where the biomass was respectively 0.02 and 1.1 g m^{-2}.

The Oligochaetes were essentially represented by the Tubificids (*Aulodrilus remex* and *Euilodrilus* sp.) in the north basin, whereas the Alluroididae (*Alluroides tanganyikae*) were dominant in the open waters of the south and the southeast. Their average biomass of 0.3 g m^{-2} generally varied between 0.1 and 0.2 g m^{-2} except in the open waters of the north where they reached 0.8 g m^{-2}. Finally, the insects were relatively poorly represented in the benthic biomass with a mean of 0.1 g m^{-2}.

Only the production of benthic molluscs has been studied over the course of a seasonal cycle in different biotopes and different regions of the lake (Lévêque 1973). The mean annual P/$\bar{\text{B}}$ ratio was estimated at 5.8 for *Bellamya unicolor*; 4.4 for *Melania tuberculata*; 2.6 for *Cleopatra bulimoides* and *Corbicula africana*; 2 for *Caelatura aegyptiaca*. The annual production was 279 000 tons of dry organic matter in 1967–1970 or 14.5 g m^{-2} year or 58 Kcal m^{-2} year on average. The Prosobranchs and more particularly *Bellamya* were responsible for the major part of this production. The shell production was about 100 g m^{-2} year.

The biomass of worms and benthic insects was low but the corresponding production was probably quite significant. In fact, biological cycles in the tropics are short and the more short-lived the organisms, the higher their P/B ratios (Lévêque et àl. 1976). In these conditions, a production of about 5 g m^{-2} year for these two groups is a likely value and the total benthic production was probably close to 20 g m^{-2} year or 85 Kcal m^{-2} year^{-1}.

17.1.2.5 *The fish*. Among the one hundred and forty species of fish identified by Blache (1964) in the Chad basin, one hundred and twenty were observed in Lake Chad and the lower reaches of its tributaries. Most of these species also occur in the basins of the Nile and/or the Niger. There is no endemism with the exception of *Alestes dageti* which is found only in Lake Chad.

The lacustrine species are represented in the fluvial system but many fluvial species are not found in the lake. The fluvial system is actually the permanent environment from which the recolonization of the lacustrine environment starts following the periodic drying of the latter.

The distribution of these species in the lake depended on both the distance from the fluvial system and on the type of habitat (open water, archipelago, reed islands ...). There were fewer species in the north basin than in the south basin, several of these disappeared north of the line from Baga Sola to Malamfatori probably because of the increase in salinity of the water. This was particularly so in the case of the Mormyridae and *Schilbe uranoscopus*.

The archipelago zones of the south basin were characterized by the abundance of *Alestes baremoze* and *A. dentex*, Cichlidae and *Heterotis niloticus*. These species were rare or absent in the open waters where some small planktivorous fish serving as food for the predators (*Hydrocynus, Lates*, etc. ...) are found. The largest number of species were found in the southern border of the lake and the delta region because of the proximity of the fluvial system. In particular, the presence of *Ichtyoborus besse, Siluranodon auritus* and *Polypterus senegalus* was noted as they are fluvial species which were absent from the rest of the lake.

In the rivers as in the lake, the seasonal variations in catches related to species abundance proved the importance of migration, which were of two main types: longitudinal spawning migrations and lateral migrations linked mostly in searching for food. Taking into account their possible migratory behaviour and knowledge of their distribution, it was possible to differentiate several species groups from those whose distribution was widespread, to those which were only found in particular environments or circumstances. Four species appeared to be really ubiquitous: *Lates niloticus, Labeo senegalensis, Distichodus rostratus*, and *Hemisynodontis membranaceus*. Some other apparently ubiquitous species are only found in permanent environments: *Micralestes acutidens, Hydrocynus forskalii, Eutropius niloticus. Alestes dageti* only frequented the lake whereas some species, such as *Alestes nurse, Ichtyborus besse*, were caught only in the rivers and their temporary annexes. Several species were restricted to the flooded zones and to their more permanent surrounding: *Clarias* sp., *Synodontis nigrita, Brienomyrus niger, Ctenopoma* spp. The extreme and well-known example of the dipnoid, *Protopterus annectens*, should also be mentioned. It was very common in all the zones of temporary flooding along with *Nothobranchus*, whose very short cycle includes the laying of long-lived eggs, which was generally caught in isolated temporary ponds.

The relative abundance estimated using gill nets showed that the mean catches over the course of an annual cycle were much lower in the south basin than in the north basin. The open waters of the south basin were also much poorer than the archipelago where some direct biomass estimates were made. The use of ichthyotoxins gave mean biomass figures of about 500 kg ha^{-1} with a range of 110 to 820 kg ha^{-1}.

The beach seine fishing gave much lower values of about 90 to 100 kg ha^{-1}. This was essentially due to differences between the environments sampled as the coastal coves were richer than the open waters of the archipelago, and also to the beach seining itself which must have involved a certain underestimation of biomasses.

Some similar sampling was also carried out in the rivers where clear differences were again found for similar reasons. The average biomasses obtained with beach seines in the Shari upstream of the confluence with the Logone, was near 120 kg ha^{-1} whereas by poison fishing in a secondary arm of the river in the same region the mean figure for biomass was about 2400 kg ha^{-1}.

The characteristically diverse traditional fishing techniques, the dominance of river fishing over lake fishing and the very moderate level of exploitation before 1960, were totally modified after 1963. This was due to the introduction of nylon gill nets which resulted in a decrease in the importance of traditional techniques, a very rapid increase of fishing effort, and an increase of fishing in the lake. Most of the catches (80 to 90%) were made with large mesh nets (80 to 120 mm). Before the contraction of the lake, the fishing was most developed in the north basin and concentrated on *Lates niloticus, Heterotis niloticus, Citharinus* spp., *Distichodus rostratus, Labeo* spp., *Gymnarchus niloticus* and *Hemisynodontis membranaceus*.

The catch per unit effort decreased considerably from about 15 kg 100 m^{-2} night^{-1} in 1963 to 1.5 in 1967–1969 and 1.0 in 1971 whereas the total catches increased from 30 000 tons in 1963, to 46 800 tons in 1970, and 86 300 tons in 1971. Fishing effort thus multiplied by about 40 between 1963 and 1971.

The stock of *Alestes baremoze*, a fluvio-lacustrine migratory species was studied most intensively (Durand 1978). Total catches of this species were estimated at 6360 tons in 1969, 8280 tons in 1970 and 4970 tons in 1971. Exploitation was almost exclusively fluvial during this period with lacustrine catches not exceeding 5%. In a 'Greater Lake Chad' the optimal mean catches could reach 9000 to 11 000 tons depending on fluvial or fluvio-lacustrine exploitation.

Some active traditional fishing still occurred during recession of the river at the outlets of floodplains, the most important being that on the El Beid which connects the North Cameroon floodplains to Lake Chad, where about 1200 tons of fish, almost exclusively a few months old, were caught. This corresponds to a yield of 4 kg ha^{-1} for all the flooded zones.

The total catches before the drying period were estimated at 60 000 tons in

1969, 65 000 tons in 1970 and 115 000 in 1971 for the fluvio-lacustrine environment. Although the lake had then been exploited more than previously, there was apparently no overexploitation. In 1972, the fishing rose to 165 000 tons or 104 kg ha^{-1} $year^{-1}$. Some similar values (100 to 130 kg ha^{-1} $year^{-1}$) were found from 1974 to 1977 in the south basin which by then represented the only lacustrine environment.

From these observations, the mean sustainable yield could be about 90 kg ha^{-1} or 180 000 tons from a large Lake Chad with an area of 20 000 km^2.

17.1.2.6 Trophic relations

Planktonic Crustacea. The principal Cladoceran species fed on unicellular or colonial planktonic algae whose size is between 4 μ and 30 μ. However they did not usually consume the Cyanophycean, *Anabaena flos-aquae* which represented a major part of the phytoplankton. This species was however important in the diet of *Tropodiaptomus incognitus* and of *Thermocyclops neglectus* but irregularly, depending on the abundance and the state of the colonies.

Fishes. From the study of their stomach contents (Lauzanne 1976), the principal fish species could be classified into major consumer groups according to their trophic level. If the first level was represented by primary food sources (algae and detritus), three other levels could be recognized.

The second was formed by the primary consumers, includes the phytoplankton filter-feeders (*Sarotherodon galilaeus*), detritivores which fed on the bottom organic film made up mostly of sedimental algae (*Labeo senegalensis, L. coubie, Distichodus rostratus, Citharinus citharus* and *C. distichodoides*) and phytophages which consume higher plants (*Alestes macrolepidotus*).

The third level includes the secondary consumers among which were the strict zooplanktivores (*Alestes baremoze* and *Hemisynodontis membranaceus*), predominantly zooplanktophagous species (*Brachysynodontis batensoda* and *Alestes dentex*). Several benthophagous species also belonged to this category which exclusively or essentially fed on benthic invertebrates: *Synodontis schall, S. clarias, Hyperopisus bebe* and *Heterotis niloticus*.

The fourth level is made up of top consumers; some were strictly ichthyophagous (*Lates niloticus, Hydrocynus brevis*), whereas others had a more varied diet of fish, shrimp or insect larvae (*Schilbe uranoscopus, Eutropius niloticus, Bagrus bayad, Hydrocynus forskalii*).

Food was abundant and varied in Lake Chad. The adult fish belonged to a well-determined trophic level but their diet was subject to slight modifications according to habitat and season. The benthophagous species of the open waters of the south basin seemed to have a simplified diet in comparison to those of the archipelago due to the absence or scarcity of food sources such as shrimps, insect larvae or macrophytes in the open waters. Conversely the benthic feeders of the open waters consumed molluscs in greater abundance than those of the archipelago where molluscs were not so common.

Some changes of trophic level were observed during the period of growth. Thus most of the terminal consumers (level 4) were initially secondary consumers (level 3) during their young stage.

The trophic relations between fishes of level two and three were relatively direct and most of the food consumed by the fish of a given level came from the immediately lower level, forming a food chain. The problem was more complex for the less-strictly ichthyophagous terminal consumers whose food came from all trophic levels and even from some sources outside the aquatic ecosystem (terrestrial insects).

The importance of different consumer groups was variable according to the region of the lake. In the eastern archipelago mostly planktophagous and especially zooplanktophagous species were found, representing 44% of the ichthyomass. In the eastern open waters the terminal consumers dominated (64% of the biomass) due to the decreased importance of terrestrial insects and the abundance of small zooplankton feeders such as *Micralestes acutidens* and *Pollimyrus isidori* which appeared to have a high production. The benthophagous species do not represent a significant biomass and did not appear to play a role in the nutrition of terminal consumers in the archipelago or in the open waters.

The conversion rate of food has been determined quantitatively for three species of fish belonging to three different trophic levels. In fresh weight or energetic equivalents these rates are respectively 3.1 and 18.9 for *Sarotherodon galilaeus* (phytoplanktophagous), 8.8 and 44.8 for *Alestes baremoze* (zooplanktophagous), and 26.4 and 27.3 for *Lates niloticus* (ichthyophagous).

17.1.3 *Matter and energy budgets*

From the various results given above many aspects of transfer and yield can be estimated for the Lake Chad ecosystem. These data will then be compared to those obtained in some other aquatic ecosystems, both tropical and temperate.

17.1.3.1 *Lake Chad.* It has long been argued whether the body of water corresponding to Lake Chad is really a lake and some authors consider that its relatively low volume, uniformly flat morphology and its endorheic character make it nothing more than an area inundated by river waters. Daget (1967) considered Lake Chad more similar to floodplains of tropical rivers than to true lakes. The absence of an endemic fauna could be an argument in favour of this hypothesis. However, this absence results from the frequent drying of the lake and its recolonization which is carried out by some organisms sheltered in the lower reaches of rivers and the few permanent waters of the south basin.

In view of recent research this hypothesis must now be modified because the main characteristic of these zones is a very marked seasonal rhythm of flooding and drying, and also because some clearly lacustrine characters appear in the north basin of Lake Chad during the period called 'Normal Chad'. This north basin which is deeper and partially sheltered from the Shari flood, represents a much more stable environment than the south basin which can be considered as a transitional environment where the fluvial influences impose marked seasonal rhythms.

This distinction between the south and north basin is also evident in their populations since the two basins are ecologically different. In fact the zonation is more complex because it varied according to the taxonomic group considered, and their ecological requirements. However, taking into account the lacustrine habitats (open waters and archipelagoes), it is possible to consider only four major, relatively homogenous, areas on the basis of biomass and production. The corresponding data expressed in kg per hectare for biomasses and in kg ha^{-1} year^{-1} for production are presented in Table 1.

The phytoplankton biomasses were particularly low in the open waters of the south basin and about 70 times more important in the archipelagoes of the north basin; the archipelagoes of the south basin and the open waters of the north basin are at an intermediate level. Primary production is only known through the values for the southeastern archipelago; thus 550 g C m^{-2} year^{-1}

Table 1 Biomass (kg ha^{-1}, dry weight) and productions (kg ha^{-1} year^{-1}) in the four main natural regions of Lake Chad.

		South basin		North basin	
		Open water	Archipelago	Open water	Archipelago
Phytoplankton					
	B̄	0.09	4.13	3.84	7.28
Gross P.P.			18 000		
Macrophytes					
	B̄	←————————— 11 000 —————————→			
	P	←————————— ? —————————→			
Zooplankton					
Copepods		2.5	6.9	8.9	
Cladocera	B̄				
Copepods		159	438	732	
Cladocera	P				
Benthos					
Molluscs	B̄	25.8	10.6	64.2	33.6
Worms	B̄	2.0	0.8	8.0	1.8
Insects	B̄	0.1	0.6	2.1	1.8
Molluscs	P	77	30	353	136

corresponds to an approximate value of 1800 kg ha^{-1} (net production) of phytoplankton. No regional values are available for the macrophytes but their average biomass for the entire lake is 11 000 kg ha^{-1}.

The same relative poverty of zooplankton was found for the open waters of the south basin but less marked than for the phytoplankton. The north basin as a whole is the richest environment. The high turnover rates (from 60 to 80 on the average) lead to fairly high production.

The results obtained for the principal benthic groups were more heterogenous. The open waters of the south basin were very poor in insects whereas in the entire north basin, the biomasses were 20 times greater. Molluscs and worms had equally high biomasses and production in the north basin, but in each of the two basins the biomasses of open waters were twice those of the archipelagoes. The production could only be calculated for the molluscs whose average turnover rate varies between 2 and 6.

The mean values obtained in the four main lacustrine zones were used to calculate some average values for the entire lake, taking into account the relative areas of the zones (Table 2). The energy equivalents have been calculated by taking as mean calorific values (cal g^{-1} dry weight), 3000 for the phytoplankton, 4300 for the macrophytes, 5600 for the zooplankton and 1400 (cal g^{-1} fresh weight) for the fish.

The production of macrophytes has not been measured, but Thomson (1975)

Table 2 Lake Chad: biomasses and energy budgets.

		Mean annual biomass and productions (kg ha^{-1}, dry weight)	Energy equivalents (Kcal m^{-2})	Total biomass and productions (tons, dry weight)
Incident energy			201.5×10^4	
Phytoplankton	B	3.4	1	6200
	P	(1800)[a]	(540)[a]	3.2×10^6
Macrophytes	B or P	(11 000)	(4730)[b]	(20×10^6)
Zooplankton	B	6.8	3.8[c]	12 200
	P	474	265	860 000
Benthos	B	37	15.3	71 000
	P	(180)	(90)	(350 000)
Fish	B or P	(250 f.w.)	(37.5)[d]	(450 000 f.w.)

[a] Gross P.P. = 550 g C m^{-2} yr^{-1}; net P.P. is assumed to be 10% of gross P.P.; 1 g C = 3.3 d. w; 3000 cal g^{-1} d.w.
[b] 4300 cal g^{-1} d.w.
[c] 5600 cal g^{-1} d.w.
[d] 1500 cal g^{-1} f.w.
N.B.: fresh weight is used for fish.

estimated the production of *Papyrus* (stems and roots) at about 100 tons of dry matter per hectare and per year in some Uganda swamps. If these results were extrapolated to other macrophytes (*Phragmites* in particular), they may represent a net production similar to the biomass, which is considerable. This net production is equivalent to the gross production of phytoplankton.

Moreover, it is probable that the production of phytoperiphyton which has not been measured, is also very important as is generally the case in shallow lakes (Wetzel 1979). The total primary productivity from all sources is thus very high in Lake Chad.*

If it is supposed that the turnover rate of the worms and the insects, which only represent 10% of the biomass, is close to that of molluscs, the benthic production is estimated to be about 2.5 times lower than the zooplankton production, although the benthic biomass is five times higher than the biomass of planktonic crustaceans. This is due to the fact that the planktonic organisms have much shorter generation times than the benthic organisms and thus a much higher rate of production.

We do not have an exact estimate of the biomass of fishes, but we saw (in Chapter 13) that the mean sustained yield must be about 100 kg ha^{-1} year^{-1} and it can reasonably be supposed that the biomass must be between 200 and 300 kg ha^{-1}.**

The only data available on fish production concern the common zooplankto-phagous species *Alestes baremoze* for which the P/B ratio is close to 1 with a mean longevity of about 5 years. By supposing that this value is reasonable the biological production would be approximately equal to the biomass of the fish or about 450 000 tons per year for the entire, the energetic equivalent of which is 37.5 kcal m^{-2} year^{-1}.

The incoming solar energy measured at N'Djamena is 201.5×10^4 Kcal m^{-2} year^{-1} and the total transfer of energy in the lake would therefore be as follows: gross phytoplankton production represented 0.25% of the solar energy and the macrophyte production would be of the same order of magnitude. The zooplankton production corresponded to only 0.013% and benthic production to 0.0045%. Finally the production of all the fish can be calculated at 0.002% of the total incident energy.

The relative importance of plant and animal production can also be compared. By supposing that the entire primary production (phytoplankton, macrophytes and phytoperiphyton) is about 6000 Kcal m^{-2} year^{-1}, secondary production represents 6.5% and the fish production 0.6% of this value.

---

* Bacterial activity, has not been studied but it must probably be very important in Lake Chad, given the abundance of plant detritus and decomposition. We do not know the role of these organisms in the remineralization of nutritive elements.
** Some close links exist between rivers and the lake for several species of fish which make seasonal spawning migrations. The values used here are for the entire fisheries (lake and lower reaches of rivers) but the lacustrine part is by far more predominant.

Finally, the ichthyological production represents about 10% of the secondary production.

We will see later a comparison of these values with results obtained from other lakes. It is nevertheless evident that Lake Chad is a rich ecosystem primarily due to certain factors favourable to a high productivity.

The shallow depth initially allows the development of a vast littoral zone colonized by several emergent or submerged macrophytes. In Lake Chilwa, Howard-Williams and Lenton (1975) showed the fundamental role played by this littoral zone in the functioning and regulation of shallow lakes. It provides some specific habitats for plants and animals, contributes greatly to autotrophic production and constitutes an important source of food and detritus for the deeper zones. The shallow depth also allows good oxygenation of the bottom water favouring the development of fairly dense benthic populations and the aerobic decomposition of organic detritus. Conversely the stirring of water by wind disturbs the sediment below shallow water and fine particles are resuspended decreasing transparency and thus the euphotic layer. In addition, disturbance of the sediment is unfavourable to the maintenance of a rich benthic fauna.

The high solar energy input in the tropics and its lack of seasonal variation compared with higher latitudes are also surely favourable to autotrophic production. But the role of salinity regulation must also be noted. Its effects are such that the spatio-temporal variations of salinity and ionic composition do not induce fundamental modifications of the communities. It is, however, impossible to know if a much higher salinity would have only negative effects. It is probable that, as the type of changes noted in the north basin of the lake would suggest, as long as concentrations were not too high, there would be a decrease in population diversity by selection of well-adapted species, and an increase in biomass and production of those remaining.

In conclusion, let us recall that one of the reasons for expecting a high productivity should be *a priori*, the stability of the lacustrine ecosystem and this depends on its hydrological regulation. This is not, however, the case in Lake Chad as the normal lacustrine volume is only twice the average yearly fluvial supplies which is highly variable.

17.1.3.2 *Comparison with other lacustrine ecosystems*

(a) *Lake George*. Among the African tropical lakes, Lake George is the only one which has been studied in detail during the I.B.P.. The main results are given in Table 3 (after Burgis 1978).

Both Lake Chad and Lake George are shallow lakes with very low mean depth. Due to their location global incident radiation is high and of the same order of magnitude although slightly lower at Lake George, because of greater cloudiness. However they differ in several ways. Firstly Lake Chad is of the endorheic type and its area 80 times greater than that of Lake George which has

Table 3 Comparison of characteristics and productivity of some tropical (Lake Chad and Lake George) and temperate lakes.

Lakes	Latitude	Altitude a.s.l. (m)	area (km²)	Z (m)	Solar energy Kcal m^{-2} yr^{-1} (×10⁴)	Gross P.P. Kcal m^{-2} yr^{-1}	Zooplankton B Kcal m^{-2} yr^{-1}	Zooplankton P Kcal m^{-2} yr^{-1}	Zoobenthos B Kcal m^{-2} yr^{-1}	Zoobenthos P Kcal m^{-2} yr^{-1}	
George	0	913	250	2.4	172	19 710	5.2	200	3.3		Burgis (1978)
Chad	12–14	282	20 000	4.0	201.5	5040	3.8	265	15.3	90	this study
Findley	47		0.1	7.8		25	1	5	3	7	Wissmar and
Mirror	43		0.15	5.8		380	3	21	9	30	
Marion	49	300	0.1	2.4		50	2			34	Wetzel (1978)
Lawrence	42		0.05	12.0		430					
Wingra	43	97	1.4	2.4		4300	20	230	2	22	
Kiev (reservoir)	50	103	992	4.0		3590	8.1	206	44.1	156.6	Gak (1972)
Baikal	51–55	455	31 500	730	105.6	875	9.4	88.6	5.5	13.0	Moskalenko (1972)
Naroch	54		80	9.0		1975	4.0	75	0.9	4.0	Winberg et al. (1972)
Myastro	54		13.1	5.4		2260	10.1	161	2.7	12.9	
Batorin	54		6.2	3.0		2329	10.0	192			
Mikolajskie	54		4.6	11.0		4140		430		120	
Taltowisko	54		3.3	14.0		4370		400		200	Kajak et al (1972)
Sniardwy	54		109.7	5.9		3300		110		50	
Flosek	54		0.04	3.0		2200		470		0.5	
Warniak	54		0.4	1.2		1600		—		140	
Esrom	56		17.3	12.3		2440		100		103	Jonasson (1972, 1977, 1979)
Loch Leven	56	107	13.3	3.9	93	6000 to 9000		100	95	386	Morgan (1974)
Rybinsk (reservoir)	59	102	4300	5.4		650	1.6	78	3.3	12.8	Winberg (1972)
Krasnoe	66	66	9	6.6		1570	5	111	4.6	19.4	
Pääjärvi	61	78	13.4	14				55.0		19.2	Sarvala (1978)
Øvre Heimdalsvatn	61	1090	0.8	4.7	94.9	(110)		11.0		12.2	Larsson (1978)
Red Lake	64	277	9.1	6.6		1093	5.7	98.3	4.8	20.4	Andronikova (1972)
Myvatn	65		37	2	79	1180		40		160	Jonasson (1979)
Krivoe	66		0.5	12.0		160	1.2	12.5	1.7	2.2	Alimov et al. (1972) and
Krugloe	65		0.1	2.1		55	0.8	10	2.7	5.2	Winberg (1972)

an outlet. But the most important is the climatic difference as the seasonal variations of different climatic factors are very muted in Lake George due to its location under the equator (Burgis et al. 1973). This results in an exceptionally constant physico-chemical environment for an aquatic ecosystem (Greenwood 1976; Burgis 1978), which is evidently not the case in Lake Chad where considerable differences are observed from one season to another such as the variation in water temperature from 18°C during the cold season to 32°C during the warm season. Populations were relatively stable over the course of the year in Lake George whereas some marked seasonal variations were observed in Lake Chad.

One of the major characteristics of Lake George was the importance of phytoplankton in the biomasses and in the trophic cycles of aquatic organisms. Composed mainly of Cyanophacea (80%), the phytoplankton represented an average biomass of 300 Kcal m^{-2} and they made up 98% of the planktonic biomass and 90% of the total biomass exluding the macrophytes. In Lake Chad the phytoplankton biomass was considerably lower (0.34 g m^{-2} in dry weight or about 1 Kcal m^{-2}) and only represented about 30% in dry weight or 20% in calories of the plankton biomass. This low proportion of phytoplankton in the plankton biomass is remarkable. The difference between the algal biomass (from 1 to 300) was not reflected in the values of primary production since the ratio of gross primary production between Lake George and Lake Chad was only about 4. Thus, it may be deduced that algal population cycles were much more rapid in Lake Chad and that turnover rates eighty times higher on an average, reduced the tremendous gap between the two lakes biomasses.

The zooplankton biomass was of the same order of magnitude in the two lakes: 3.8 Kcal m^{-2} in Lake Chad, 5.2 in Lake George (of which 3.7 is for planktonic Crustacea and 1.5 for *Chaoborus*). But only one Copepod species (*Thermocyclops hyalinus*) is dominant in the latter whereas seven species of Cladocera and Copepods are well represented in the zooplankton populations of Lake Chad.

The benthos of Lake George was poor (about 3 Kcal m^{-2}) probably due to the very high fluidity of the mud which was the most common sediment and which was not a good substrate for the establishment and maintenance of benthic species especially molluscs. The latter represent by contrast most of the benthic biomass of Lake Chad estimated at an average of 15.3 Kcal m^{-2}.

The predominance of phytoplankton in the total biomass of Lake George was reflected in the composition of fish populations. They were dominated by some herbivorous species that are able to use the Cyanophycea. Thus two species (*Sarotherodon niloticus* and *Haplochromis nigripinnis*) which represented 60% of the ichthyomass fed directly on the phytoplankton. However, in Lake Chad secondary (mostly zooplanktophagous and benthophagous) or terminal (ichthyophagous) consumers were dominant in the biomass.

Thus there were some fundamental differences between the two lakes although they both are tropical and shallow. In Lake George, the environmental conditions as well as the populations and the productivity of the ecosystem were stable over the course of the year. The phytoplankton biomass was high but the production relatively low and the animal biomass was dominated by a few essentially herbivorous species. The trophic cycles are thus short, but it appears that the efficiencies are low (Burgis 1978).

In Lake Chad, however, the ecological factors showed some annual and yearly variations that were sometimes considerable. The phytoplankton biomass was fairly low but the production relatively high. The fauna was more diverse and several species were well represented in the herbivorous zooplankton. The trophic chains to fish were generally longer with a high proportion of secondary consumers and ichthyophagous species. These trophic structures explain that although it is very productive, Lake Chad has average fish yields lower than Lake George: 100 to 120 kg ha^{-1} year^{-1} instead of 140 kg ha^{-1} year^{-1} (Burgis and Dunn 1978).

(b) *Middle and high latitude lakes.* We compared the observations made on Lake Chad to some data collected in other lakes and reservoirs belonging to various regions of the earth, mostly located in temperate zones (Table 3).

In 1973, Brylinski and Mann compiled data gathered from 55 lakes and reservoirs for the International Biological Program. The analysis of these results show that the amount of solar energy available appears to have a greater influence on production, on a global scale, than the amount of available nutrients. The latitude, which integrates the duration of light, the temperature of the air, the duration of the growing season etc. appears to explain a large part of the statistical variability. On the other hand, when comparing the same latitudes, the quantity of available nutrients has a very considerable influence.

In reality the most productive lakes must be those for which the vegetative period is the longest, thus allowing the development and maintenance of a diversified population of primary producers. This duration of vegetative period increases from arctic zones where it is reduced to some months, towards the tropics. The ultimate state is exemplified by Lake George located on the equator where the environmental conditions are relatively constant throughout the year. Lake Chad is at 14°N, and there is a well-marked seasonal cycle with a cold season from November to February during which a notable slackening is observed.

The influence of latitude between the extremes of arctic and tropical lakes is clearly seen in the data in Table 3, but some very diverse values are found for temperate lakes. Thus Loch Leven which receives half of the incident energy, has a gross primary phytoplankton productivity approaching that of Lake Chad. Some Polish lakes also have a gross primary productivity approaching that of Lake Chad (Kajak et al. 1972).

The zooplankton biomass was not particularly high in Lake Chad (6.8 kg ha^{-1}) in dry weight or 3.8 Kcal m^{-2}) but the production was fairly high although it stayed lower than that observed in some temperate Polish lakes. The $P/\bar{B}$ ratios calculated (annual mean of 65 for the zooplankton) are, however, much higher than those which have been observed until now and have been reviewed by Waters (1977). These high $P/\bar{B}$ values are probably partially related to the fact that the growing period and thus the production, last throughout the year in Lake Chad. The mean benthic biomass of 37 kg ha^{-1} is fairly high but not exceptional since some very high values have been observed in several temperate lakes (Larsson 1973; Jonasson 1972). In most cases, however, the high biomasses are due to some Chironomid larvae, whereas, in Lake Chad, the molluscs are largely dominant. The benthic production estimated at 180 kg ha^{-1} year^{-1} of dry weight (90 Kcal m^{-2} year^{-1}) is also among the mean values of several temperate lakes having higher production (Table 3, and Waters 1977). However, as for the zooplankton, the $P/\bar{B}$ ratios of Lake Chad molluscs are fairly high and it is probably the same for the insects and the Chironomids, in particular, whose life cycles are often shorter than one month.

Terminal production cannot be estimated through the average fishing yield of 100 to 120 kg ha^{-1} but it is an index of a high productivity. This is all the more true as, unlike Lake George, it is not dominated by short food chains and phytoplankton-feeding species. The zooplanktophagous and ichthyophagous species have a particularly important place. The high terminal production could be explained by the average or high productivity at all trophic levels and by a major diversification of terminal consumers, which use most of the available trophic resources.

In conclusion, if Lake Chad does not appear as an exceptionally rich lake in each of its major biological components, it appears that it is an example of a system where most of the classic trophic levels (phytoplankton, macrophytes, zooplankton, benthos and fish) are very well represented. The organic degradation of plants and animals and the rapid recycling of nutrients is probably one of the main explanations of the richness of this lacustrine ecosystem. Due to the diversity of trophic pathways and species composition it shows a remarkable plasticity which allows it to adapt to environmental changes and to maintain a high productivity.

17.2 Modification of Lake Chad during a period of drought (1972–1978)

In all the preceding chapters, we have only considered two major periods offering the greatest contrast by calling one 'Normal Chad', from 1965 to 1972 and the other a drying state or 'Lesser Chad', from 1972–1973.

The end of this last period cannot be ascertained because it continues in 1981, and may still continue for several years. Comparisons between the two periods explain the changes in the ecosystem but we should not forget that it did not remain stable and unvarying between 1965 and 1972. The water level progres-

547

sively dropped more than two meters, involving a decrease to nearly half of the original volume. This evolution, though gradual had consequences for the populations and we must take it into account in order to understand the evolution of the ecosystem during this period of drought.

17.2.1 *Hydrology and environment*

17.2.1.1 *The hydrological evolution.* The habitats of the lake were completely transformed at the time of the exceptionally low floods of the Shari in 1972/1973 (17×10^9 m^3 against 40×10^9 m^3 on the average). In April 1973 the 'Great Barrier' dried up and from this period the south and north basins had completely different hydrological regimes.

The drying of the whole lake worsened with the Shari flood of 1973–1974 which was nearly as low as that of the preceding year (18.4×10^9 m^3). However, it did allow reflooding of a large part of the south basin which had been divided in April–May 1973 by the drying out of the southeastern reed islands. Nevertheless this region was dried out again in May 1974. As the Shari waters hardly crossed the 'Great Barrier', the north basin continued to dry and from July 1973 to July 1974, the water level fell by 0.90 m, leading to the appearance of shallows in the open waters of the North.

During 1974–1975, the Shari flood was much higher (30.5×10^9 m^3) though still far from the median flood volume. The height of the water level at Bol, by the end of 1974, was close to that registered at the end of 1971.

Following this period, the south basin had a 'new average situation' close to that of 'Normal Chad', that is, without a hydrological break between the open waters and the archipelago at low water period. However, this new situation differed from the preceding one by the more marked seasonal fluctuations in water level and the presence of more abundant vegetation. Unlike the south basin, the north basin was no longer fed normally. The development of a thick plant cover at the 'Great Barrier' during the very low water period of 1974 was an obstacle to the passage of flood waters. The inundated areas were very rapidly reduced to no more than a few ponds in the center of the basin by May 1975, but these ponds dried up completely by October.

The 1975–76 flood which was close to the median flood with 36.6×10^9 m^3 caused a clear rise in water level in the south basin and the variations of the water level at Bol were close to those registered in 1971–72. In spite of this, the vegetation of the 'Great Barrier' again restricted penetration of flood waters into the north basin and the latter was only partially reflooded in the south but dried rapidly before the flood receded. During 1976–77 and 1977–78 the Shari floods were low, sufficient to maintain the new hydrological situation of the south basin, but insufficient for a permanent reflooding of the north basin, even in the south.

On the whole, two periods can be distinguished over the course of this severe drought. The first began at the end of 1972 during a flood that was insufficient to feed the north basin and whose most spectacular consequence was the separation of the two basins in May 1973. It ended with the complete drying of the north basin at the end of 1975. The following period corresponded to a 'Lesser Chad' limited to the south basin and receiving a fluvial water supply sufficient to maintain it in a situation close to that characteristic of 'Normal Chad'. This is the present situation.

17.2.1.2 *Evolution of the physical and chemical characteristics of the water.* A slow evolution of the abiotic characteristics of the environment was accelerated, from 1973. Three regions of the lake were isolated and evolved differently: the open waters of the south basin, the region of the southeast archipelago and the north basin.

(a) The drying period

The open waters of the south were the least modified because they receive the Shari waters. However from the 1973 low water to the flood at the end of 1973 this environment had considerable seasonal modifications in that its area and its volume almost doubled (2500 km^2 $- 1200$ km^2 and 7.5×10^9 m$^3 - 4.5 \times 10^9$m^3). The mean annual salinity increased only slightly (10%) but the seasonal differences were more pronounced than during the period of 'Normal Chad' in that salinity varied from 45 to 110 mg l^{-1} against 45 to 70 mg l^{-1} during 'Normal Chad'. A similar seasonal variation was registered in 1974 but with less amplitude.

From April–May 1973 the archipelago of the south basin was isolated from the open waters. Deprived of supplies the habitat was rapidly subdivided and some ponds isolated. The salinity of the water increased rapidly reaching values that were four times higher (530 mg l^{-1}) five months later. Moreover, the development of vegetation modified the relative composition of this salinity by reducing the percentage of potassium and dissolved silica; pH values stayed between 6.9 and 8.2. In some bays and channels that were well sheltered from the wind by the presence of macrophytes newly developed during the decrease in water level, a stratification was established as the water warmed up in February–March and it was maintained until the arrival of the flood in September. During this period, these areas behaved as warm monomictic lakes. Under these conditions oxygen levels fell and during the cold season the oxygen content at the surface was 12% of saturation during the day, but was more variable during the warm season (between 5 and 60%). The hypolimnion generally stayed anoxic. In 1974–75 similar seasonal variations were registered.

The north basin was cut off from the south basin in April–May 1973 and became progressively dried out. During this period which was complete in

September–October 1975, the transparency of the water rapidly decreased to reach very low Secchi disc values of about 10 cm; its oxygen content was very variable and marked by some frequent periods of anoxia; and the salinity tripled from mid-1973 to the beginning of 1974, going from 1000 to 3000 mg l^{-1} (or from 1200 to 3700 μS cm^{-1}). This is a low value for waters on the way to drying up and it can partly be explained by an increase in chemical sedimentation (the salinity would be about 4000 mg l^{-1} without it) and probably by communication with the open water and the diluting effect of the underlying aquifer. The waters became very alkaline, and the pH increased to 9.2. The 1974–75 flood fed only a very small area of the north basin next to the 'Great Barrier', and transformed it into a marsh. As the flood waters spread the transparency of the water rose to 30–60 cm, the anoxia temporarily disappeared and the salinity again fell to a normal value. A few months later this region dried up once more as the drying process was repeated.

(b) The 'Lesser Lake Chad'

During the 1974–75 flood the open water of the south basin returned to normal conditions. The hydrochemical characteristics again became close to those of 'Normal Chad' with the mean salinity equal to 70 mg l^{-1} (against 60 mg l^{-1} for 'Normal Chad'). On the other hand, the seasonal variation was more pronounced with the maximum salinity in June equal to 110 mg l^{-1} against 75 mg l^{-1} during a normal period and the minimum salinity in September equal to 43 mg l^{-1} against 47 mg l^{-1}.

In the archipelago of the south basin the characteristics of the water were also close to those of the 'Normal Chad' period but with more severe seasonal fluctuations mostly due to the vegetation which remained abundant.

17.2.2 *The populations*

The analysis of changes in the animal and plant populations was very uneven. During the first period (1972–75) the information gathered was relatively complete since it includes some observations on the phytoplankton, macrophytes and fish for the entire drying period and for the beginning of this period on molluscs, insects and zooplankton. On the other hand, the beginning of a 'Lesser Lake Chad' can be described only through some partial data on the macrophytes and some fairly rich data on the fish populations and stocks.

17.2.2.1 *The populations during the drying period (1972–1975).* The drying period extends from 1972 to mid-1975 for the north basin of the lake and led to its total drying up. It ceased towards the end of 1974 in the south basin which was by then reduced to 1500 km^2 of open water and some areas of water in the archipelago, when the river flooded in November 1974. The collection of data was particularly difficult during the drying phase since the low water level and the development of plants often made communications impossible.

(a) The macrophytes

The evolution of the vegetation was characterized by a general increase in the south basin and showed various degrees of development depending on the species. The most spectacular development was that of *Aeschynomene* (*A. elaphroxylon* very dominant). The 'Ambatch' were true forests which invaded zones on the way to drying; prairies of *Vossia cuspidata*, formerly fairly localized, where found in all regions of open water and reed islands of the south basin. In contrast, *Typha australis* decreased in area.

In the north basin, the vegetation could not have a comparable development because of the very rapid retreat of the water. But plant formations that were formerly present returned and were accentuated by the grazing of herds. Only some small plants of *Aeschynomene elaphroxylon* and some stands of *Typha australis* were noted.

(b) The phytoplankton

In the Southeastern Open Water the plankton evolved towards a fluvial type increasingly marked by a high index of diversity, higher densities and some considerable differences between low water and flood periods. A considerable development of Euglenoids was noted at low water. The phytoplankton composition remained the same (with the exception of *Synedra berolinensis*, a diatom absent, or rare until then) and this region was, as a whole, the least modified during the drying period.

In the archipelago of the south basin, isolation led to some areas of water that were cut off or only temporarily linked to each other. The phytoplankton biomasses were, in general, much higher than before (7 to 8 mg l^{-1}) and decreased rapidly with the arrival of the flood water, filtering through the vegetation barrier of reed islands. This group of ponds became a marsh in which the Euglenophytes predominated while the Cyanophytes were less prevalent.

In the north basin, as in other regions of the south basin, the period preceding its isolation showed a clear increase in algal biomasses. Soon after it became isolated, biovolume increased considerably: 179 μl l^{-1} in November 1974 corresponding to 1658 mg of chlorophyll *a* in the central basin. After dilution by some temporary water supplies, the amount of chlorophyll reached 3600 mg m^{-3} before the completion of drying. Some species characteristic of saline lakes of Kanem (Iltis 1974) appeared at the end of 1974 and in 1975, and the population was almost entirely composed of Chlorophycea and Diatoms with the latter predominating. At the end of this period the remaining water bodies in the north basin became more like saline ponds with a marked concentration of dissolved salts (3–5 g l^{-1}) and a high pH (9.1 to 9.5).

(c) The benthos

The observations on the benthic molluscs were mostly made before the drying period, from 1967 to 1972, in the south basin of the lake. In the archipelago, there was a general decrease of the number of abundant species. First, *Melania tuberculata* became dominant everywhere except on the mud substrate but the molluscs became generally scarce throughout the archipelago after 1972. In the open water the data obtained in 1972 showed a clear decrease in density of molluscs on the blue clay whereas the population of the pseudo-sand had not changed since 1968.

The aquatic insect populations also showed some evidence of pronounced evolution. In the Southeastern Archipelago the dominant species changed and above all there was a clear decrease in the number of species collected between 1965 and 1974. The species that were abundant and characteristic of the northern lake in 1970 (*Cryptochironomus stilifer* and *Tanytarsus nigrocinctus*) became dominant in 1972 and 1974 in the archipelago of the south basins and in 1974 *T. nigrocinctus* became equally abundant in the Shari delta.

It is likely that the main factor controlling the benthic populations is the partial resuspension of superficial layer of loose sediments. This resuspension is itself linked to the decrease in water level since it makes the water close to the substrate more and more sensitive to disturbance, thus creating environmental conditions that are unfavourable to the development of molluscs (very loose substrates). This effect had probably occurred early in the slow process of lake level decrease between 1967 and 1970. This hypothesis is corroborated by the contrasting stability of populations on the pseudo-sand substrates which include very few fine particles. This explanation is probably also valid for the oligochaetes and the insects, with the latter having the additional advantage of considerable powers of dispersion (either active or passive) which must also be a factor in the very rapid evolution of their populations.

(d) The zooplankton

The evolution of the zooplankton populations can be followed mostly in the southeastern archipelago. Until 1972, their general characteristics stayed the same with the total biomass about 300 mg m^{-3} and a diverse population dominated by copepods. In 1972, the first signs of change were noted with a decrease in density of *Bosmina* and a considerable increase of *Thermodiaptomus* and of littoral Cladocera. In 1973 the changes were still more noticeable with the reduction in Cladocera to *Moina* and *Diaphanosoma* only, increasing dominance of *Thermocyclops neglectus* and spectacular increase of rotifers whose densities reached 2500 to 4000 individuals per liter against 50 to 100 between 1965 and 1971. Their demographic structures also changed with a relative increase in adult biomass in spite of which the total biomass was not modified and stayed at about 300 mg m^{-3} until August 1973. The effects of the

552

isolation of the archipelago were were then felt and biomass decreased rapidly with the disappearance of Diaptomids.

The period 1974–75 in the Southeastern Archipelago was characterized by low biomasses (55 mg m^{-3} on average), the disappearance of *Mesocyclops leuckarti* and lower abundance than expected of littoral forms. The two species of Cladocera maintained densities that they had during the high water period. The main reason for this general decrease in density was probably a more or less permanent deficit of oxygen, causing increased mortalities and/or lower reproduction.

In the open water of the south basin, during 1973 and 1974, the population seemed to retain its previous general characteristics, both of densities, which stayed in the same range, and of population diversity since seven species were well represented. Due to lack of data we cannot characterize the changes in the north basin, but the population was well diversified (9 species) and abundant (500 mg m^{-3}) at the beginning of the period. Later changes must have had some analogy with those of the Southeastern Archipelago in its isolated phase, but it must have been more strongly modified by the much higher conductivities prevailing in the north than in the south.

The factors influencing zooplankton abundance can be reduced to two: on the one hand the abundance and availability of phytoplankton for the phytoplanktophagous microcrustaceans and on the other hand the importance of the zooplanktophagous predators. The abundance of algae never seems to be a limiting factor, except seasonally in the impoverished waters, close to the Shari delta, because no correlation was found between the variation in abundance of phyto- and zooplankton. The regulation of densities therefore, must essentially be a function of the presence of abundant predators and of the important pelagic food webs. The long-term stability of densities would then be explained by their independence from predation pressure as long as the conditions for zooplankton nutrition remain favourable.

(e) The fishes

During the drying phase, three distinct processes of change can be described from experimental data. In the north basin the changes at the end of 1974 favoured three groups of species for different reasons: *Sarotherodon* suffered considerable mortality but still proliferated due to its high reproductive capacity. The survival of *S. aureus* is the most obvious as it is most resistant to lack of oxygen. The second group includes *Polypterus* species (mainly *P. senegalus*) which can supplement branchial with aerial respiration; whether they spawned or not is unknown. The third group included the Mochocidae whose importance increased although they did not reproduce, due to the concentration of the water and their better resistance to the worsening environmental conditions. In the final phase, the *Synodontis* disappeared and in the marsh the *Sarotherodon* and *Polypterus* populations were joined by *Brienomyrus niger* (which must have an accessory mode of aerial respiration, Bénech and Lek 1980) and especially *Clarias*, which dominates the biomass.

From the beginning of 1973 to the end of 1974, the Southeastern Archipelago was completely isolated and the periodic water supplies it received did not carry any fishes, since the flood waters had to cross a dense barrier vegetation of several kilometers wide. The deep water pelagic species, mainly predators and zooplanktophagous species, disappeared first. Two types of severe mass mortality occurred due to phenomena leading to hypoxic conditions: first the storms occurring during the rainy season brought about a rapid resuspension of reducing compounds; secondly the arrival of the flood waters which were very poor in oxygen and relatively acid (enrichment of dissolved free CO_2) after passing through the vegetation barrier.

Only the species with a particular resistance to hypoxia remained until the end of 1979 after the flood. These were the species possessing the possibility of aerial respiration (*Polypterus, Clarias, Brienomyrus*); they also included species with tolerance of low oxygen concentrations, such as *Distichodus rostratus, Brachsynodontis batensoda, Sarotherodon niloticus* that were checked experimentally by Bénech and Lek (1980). In total only seven species were found at the beginning of 1975.

The zone in contact with the rivers: the open water of the south basin were characterized by their permanent connection with the Shari and in 1973–74 by the appearance of a new feature of large vegetation belts which developed. Thus the rivers appeared to be a refuge for those stocks of the south basin which suffered from the changes in environmental conditions.

The observations in the delta region during 1972–73, are much more instructive however, because they reflect the evolution of the fluvio-lacustrine exchanges at the beginning of the drying period. In 1972 and 1973 there was a diminution in the catches, signifying that their migration behaviour was already disturbed and that the northern basin also acted as a temporary refuge. This quantitative decrease was accompanied by changes in the composition of the populations from which *Alestes baremoze* and *A. dentex* disappeared and previously abundant species, such as *Distichodus rostratus, Citharinus* spp. etc., were replaced by species such as *Synodontis clarias* and *S. frontosus*, etc. . These changes showed the relative weakness of the south basin stocks which were then the only ones to sustain the delta fisheries.

The changes in the aquatic environments during the drying period were accompanied by a parallel change in the total fisheries catches in the Lake Chad region. This phase corresponded to the exhaustion of stocks, especially in the north basin (cf. Fig. 3, Chapter 13). It was followed by a considerable increase in the total catches which tripled between 1970 and 1976, reaching a maximum figure of 200 000 tons.

In conclusion, the decline of the water level during the drying phase, acted on the fish populations by reducing the water volume and thus by increasing natural and fishing mortality. There was also a considerable variation in dissolved oxygen content due to various causes such as the decomposition of higher plants and the resuspension of reducing compounds. These phenomena promoted the dominance of marshy species adapted especially to hypoxia and able to survive in unstable environments.

17.2.2.2 *The 'Lesser Lake Chad' (1975–1978)*. The drying phase was at its height during August–September, 1974. Half of the north basin was then dry while the south basin was greatly reduced after the very low second flood of 1973. The total water area hardly exceeded 6000 km² (cf. Fig. 20, Chapter 13); With the fluvial flood of 1974, the south basin was reflooded and has since remained the only permanent lacustrine environment.

In the north basin the drying was accentuated and resulted in a complete drying in October, 1975. Since then there were periodic resurgences of water during which half the southern part of the north basin was transformed into a vast marsh.

Some observations have been made during this period, on higher plants but the main point here also concerns the fishes. Data collection was very difficult in all the marsh zones, where the shallow water and the development of vegetation hampered normal movement.

(a) The macrophytes

In 1974 areas of the south basin which have been dry for one or two years were reflooded and there followed seasonal variations in water level similar to those of the lake, but with a bigger amplitude. The evolution of the vegetation can be summarized as follows:

— total disappearance of certain species such as *Ipomea aquatica, Lemma perpusilla* and two of the three species of *Aeschynomene: A. afraspera*, and *A. pfundii*;

— considerable regression of other species such as *Cyperus papyrus* and *Typha australis*;

— stabilization of the areas covered by the Ambadjs (*Aeschynomene elaphroxylon*);

— the massive appearance of *Pistia stratiotes, Nymphea lotus* and the development of *Vossia cuspidata*.

At the end of 1976 two species dominate the aquatic vegetation of the south basin: *Aeschynomene elaphroxylon* and *Vossia cuspidata*.

In the north basin, existing local vegetation disappeared during the drying and the young shoots were heavily grazed. Two species were still common just before the drying: *Typha australis* and *Aeschynomene elaphroxylon* (heavily grazed small plants). These two species were found after 1975 in the south and *Typha* in the depressions that functioned as temporary ponds.

(b) The fishes

The gross changes in the vegetation of the north basin were followed by aerial observations, but this was hardly possible for fish populations. One can only say that these marshy zones are not similar to the classical flooded zones, because the populations there are very specific. They are dominated by *Clarias* and secondarily by *Protopterus*, a well-adapted species, as well as individuals of various other species, depending on the years and the quantity of excess water from the south basin. From the fishery statistics of 1976 and 1977 it seems that the productivity of these temporary environments was very high.

The fish populations of the permanent environments (south basin and rivers) between 1976 and 1978 are much better known. The species richness of the south basin increased after a minimum following the flood of 1974, without reaching the 1973 level: 20 species at the end of 1975 instead of 34. The seasonal water level variations were lower than during the drying phase, although still greater than those observed at the establishment of 'Normal Chad'. The higher mean levels prevented the resuspension of the sediment fraction during storms, eliminating one of the causes of oxygen shortage. However, the other cause of anoxia remained with the arrival of the hypoxic flood waters. There was however still a periodic reappearance of species sensitive to the shortage of oxygen, that must have come with the rising water from the nearby open water through the partially degraded vegetation. These recolonizations were noted for *Schilbe, Siluranodon* ... and some pelagic species such as *Alestes baremoze* and *A. dentex*.

The establishment of a new fish community with increasing diversity (the Shannon index went from 0.92 in 1975 to 1.89 in 1977) was seen in the open water of the south basin between 1975 and 1977. This new lake population was analyzed from data covering a complete annual cycle (1976–1977) obtained on the delta. From this, three trends appear:
— species abundant before the drying whose stocks did not recover or did so very slowly, in the context of 'Lesser Chad': e.g. *Citharinus* spp., *A. baremoze* and *A. dentex*;
— initially rare species which became abundant such as *Schilbe uranoscopus, Distichodus brevipinnis, Gymnarchus niloticus, Brienomyrus niger*;
— species that were as abundant before as after the drying whether through rapid recovery of their stocks, e.g. *Synodontis schall, Hydrocynus forskalii* or *Eutropius niloticus*, or not having had any clear decline, e.g. *Brachysynodontis batensoda*.

These modifications of population composition and the relative importance of the principal species depended on complex interspecific competition. Here, fundamental biological characters such as reproductive behaviour, fecundity, feeding adaptability as well as the size of the stock to survive the drying etc., played a major role. The seasonal fisheries for juveniles in the El Beid, followed from 1974 to 1978, showed the variability of recruitment after 1973 with an apparently anarchic succession of abundant species such as *Sarotherodon* niloticus in 1974–75, *Marcusenius cyprinoides* in 1975–76, etc. However the species diversity increased slowly and the relative abundance seems to have stabilized progressively.

During the establishment phase of the 'Lesser Chad', fishing activities were kept at a high level and the total catches remained above 100 000 tons from 1975 to 1977. The composition of the catches changed noticeably but the yield per hectare remained high from 100 to 120 kg ha^{-1}. This indicates that the new ecosystem stabilized around 'Lesser Chad' also included conditions favourable to considerable terminal production such as normal floods causing inundation of the plains and favourable to the survival of young fry, and a permanent lacustrine environment with diverse habitats.

17.2.3 *Conclusions*

A period of high waters in 1963 was followed by one of relative stability. From 1972 the level of Lake Chad lowered rapidly, as a result of the drought and the north basin dried up in 1975. After this drying period a period called 'Lesser Chad' occurred during which only the south basin remained under water.

The drying period included some rapid changes in habitats and populations. Three major zones could be distinguished which had different fates: the north basin which dried, the eastern archipelago which underwent an initial drying before being reflooded, and the region close to the Shari delta in relation to the fluvial system, which showed some fluvio-lacustrine characteristics throughout. A considerable development of macrophytes and the appearance of marshy conditions was however observed in all these regions from time to time.

The instability of environmental conditions which characterized the drying phase generally included the disappearance of several species from the populations. However, some others developed, especially those which were well adapted to the new environmental conditions, such as marsh fish species and algal species of meso-carbonated waters. Hence due to its floral and faunal richness the ecosystem rapidly recovered and found new ways of functioning. This enables overall productivity to remain fairly high including the yields of fish.

The period of 'Lesser Chad' was characterized by a relative stability. The new lake was not however a homologue of the old one because it was clearly more limited (40% of the area) than the 'Normal Chad'. The seasonal variations were, moreover, much more marked due to the smaller volume of water and the influence of the fluvial system was therefore increased. Thus the lake was effectively a simple expansion of the fluvial system. During this period, some of the water brought in by the rivers, temporarily invade the southern part of the north basin where an abundant marsh fauna developed.

Unfortunately it was not possible to study the disturbances in the populations since the drying period in detail, but we have gathered the most spectacular aspects. This kind of ecosystem where the ecological conditions are subject to

557

considerable variations are of particular interest because they comprise experimental natural environments which could be used for modelling attempts. Moreover, such environments are fairly abundant not only, in Sudanian Africa (southern zone of Sudan, central delta of the Niger, etc.) but also south of the equator (Lake Chilwa, delta of the Okavango, for example). Some of these are actually in danger, especially due to management projects, and knowledge of their functioning is necessary to assist all attempts at protection.

References

Alimov, A. F., Boullion, V. V., Finogenova, N. P., Ivanova, M. B., Kuzmitskaya, N. K., Nikulina, V. N., Ozeretskovskaya, N. G. and Zharova, T. V., 1972. Biological productivity of lakes Krivoe and Krugloe. In: Z. Kajak and A. Hillbricht-Ilkowska (eds.), Productivity Problems of Freshwaters, pp. 39–56, Warsaw and Krakow, Polish Scientific Publishers.

Andronikova, I. N., Drabkova, V. G., Kuzmenko, K. N., Michailova, N. F., and Stravinskaya, E. A., 1972. Biological productivity of the main communities of the Red Lake. In: Z. Kajak and A· Hillbricht-Ilkowska (eds.), Productivity Problems of Freshwaters, pp. 57–71, Warsaw and Krakow, Polish Scientific Publishers.

Bénech, V. and Lek, S., 1981. Résistance à l'hypoxie et observations écologiques pour seize espèces de poissons du Tchad. Rev. Hydrobiol. trop. 14: 153–168.

Blache, J., 1964. Les poissons du bassin du Tchad et du bassin adjacent du Mayo-Kebi. Mém. ORSTOM No. 4, 483 pp.

Brylinsky, M. and Mann, K. H., 1973. An analysis of factors governing productivity in lakes and reservoirs. Limnol. Oceanogr. 18: 1–14.

Burgis, M. J., 1974. Revised estimates for the biomass and production of zooplankton in Lake George, Uganda. Freshwat. Biol. 4: 535–541.

Burgis, M. J., 1978. Case studies of lake ecosystems at different latitude: the tropics. The Lake George ecosystem. Verh. int. Ver. Limnol. 20: 1139–1152.

Burgis, M. J., Darlington, J. E. P. C., Dunn, I. G., Ganf, G. G., Gwahaba, J. J. and McGowan, L. M., 1973. The biomass and distribution of organisms in Lake George, Uganda. Proc. r. Soc., B, 184: 271–278.

Burgis, M. and Dunn, I. G., 1978. Production in three contrasting ecosystems. In S.D. Gerking (ed.), Ecology of Freshwater Fish Production, pp. 137–158, Blackwell Scientific Publications.

Daget, J., 1967. Introduction à l'étude hydrobiologique du lac Tchad. C. R. Somm. Séanc. Soc. Biogéogr. Paris, 380: 6–10.

Durand, J. R., 1978. Biologie et dynamique des populations d'Alestes baremoze du lac Tchad. Trav. Doc. ORSTOM, No. 98, 332 pp.

Ferguson, A. J. D., 1975. Invertebrate production in Lake Turkana (Rudolf); Symposium on the Hydrobiology and Fisheries of Lake Turkana, 25–29th May 1975, 28 pp., mimeo.

Gak, D. Z., Gurvich, V. V., Korelyakova, I. L., Kostikova, L. E., Konstantinova, N. A., Olivari, G. A., Priimachenko, A. D., Tseeb, Y. Y., Vladimirova, K. S. and Zimbalevskaya, L. N., 1972. Productivity of aquatic organisms communities of different trophic levels in Kiev Reservoir. In Z. Kajak and A. Hillbricht-Ilkowska (eds.), Productivity Problems of Freshwaters, pp. 447–455, Warsaw and Krakow, Polish Scientific Publishers.

Gras, R., Iltis, A. and Lévêque-Duwat, S., 1967. Le plancton du Bas-Chari et de la partie Est du lac Tchad. Cah. ORSTOM Sér. Hydrobiol. 1: 25–96.

Greenwood, P. H., 1976. Lake George, Uganda. Phil. Trans. r. Soc. Lond. (B) 274: 375–391.

Howard Williams, C. and Lenton, G., 1975. The role of the littoral zone in the functioning of a shallow lake ecosystem. Freshwat. Biol. 5: 445–459.

Iltis, A., 1974. Phytoplankton des eaux natronées du Kanem (Tchad); VIII-Classification des milieux étudiés et espèces caractéristiques. Cah. ORSTOM Sér. Hydrobiol. 8: 81–91.

Jonasson, P. M., 1972. Ecology and Production of the Profundal Benthos in Relation to Phytoplankton in Lake Esrom. Oikos, suppl. 14: 1–148.

Jonasson, P. M., 1977. Lake Esrom Research 1867–1977. Folia limnol. Scand. 17: 67–89.

Jonasson, P. M., 1979. The Lake Myvatn ecosystem, Iceland. Oikos 32: 289–305.

Kajak, Z., Hillbricht-Ilkowska, A. and Pieczynska, E., 1972. The production processes in several Polish Lakes. In: Z. Kajak and A. Hillbricht-Ilkowska (eds.), Productivity Problems of Freshwaters, pp. 129–147, Warsaw and Krakow, Polish Scientific Publishers.

Kalk, M., 1979. Zooplankton in Lake Chilwa: adaptations to changes. In M. Kalk, A. J., McLachlan and C. Howard-Williams (eds.), Lake Chilwa, studies of change in a tropical ecosystem, pp. 125–141, Monographiae Biologicae, Vol. 35, Dr. W. Junk Publishers, The Hague.

Kalk, M. and Schulten-Senden, C. M., 1977. Zooplankton in an endorheic lake (Lake Chilwa, Malawi) during drying and recovery phases. J. limnol. Soc. sth. Afr. 3: 1–7.

Larsson, G. L., 1973. A limnological study of a high mountain lake in Mount Rainier Park, Washington State, USA. Arch. Hydrobiol. 72: 10–48.

Larsson, P., Brittain, J. E., Lien, L., Lillehammer, A. and Tangen, K., 1978. The lake ecosystem of Øvre Heimdalsvatn. Holarct. Ecol. 1: 304–320.

Lauzanne, L., 1976. Régimes alimentaires et relations trophiques des poissons du lac Tchad. Cah. ORSTOM Sér. Hydrobiol. 10: 267–310.

Lemoalle, J., 1973. L'énergie lumineuse et l'activité photosynthétique du phytoplancton dans le lac Tchad. Cah. ORSTOM Sér. Hydrobiol. 7: 95–116.

Lévêque, C., 1973a. Dynamique des peuplements, biologie et estimation de la production des mollusques benthiques du lac Tchad. Cah. ORSTOM Sér. Hydrobiol. 7: 117–147.

Lévêque, C., 1973b. Bilans énergétiques des populations naturelles de mollusques benthiques du lac Tchad. Cah. ORSTOM Sér. Hydrobiol. 7: 151–165.

Lévêque, C., Durand, J. R. and Ecoutin, J. M., 1977. Relations entre le rapport P/B et la longévité des organismes. Cah. ORSTOM Sér. Hydrobiol. 11: 17–32.

Litterick, M. R., Gaudet, J. J., Kalff, J. and Melack, J. M., 1979. The limnology of an African lake: Lake Naivasha, Kenya. Paper presented to the Workshop on African Limnology, Nairobi, December 1979, 73 pp., mimeo.

Morgan, N. C., 1974. Historical background to the International Biological Programme project at Loch Leven, Kinross. Proc. R. Soc. Edinb. (B) 74: 45–55.

Morgan, N. C. and McClusky, D. S., 1974. A summary of the Loch Leven IBP results in relation to lake management and future research. Proc. r. Soc. Edinb., 74: 407–416.

Moskalenko, B. K. and Votinsev, K. K., 1972. Biological productivity and balance of organic substance and energy in Lake Baikal. In: Z. Kajak and A. Hillbricht-Ilkowska (eds.), Productivity Problems of Freshwaters, pp. 207–226, Warsaw and Krakow, Polish Scientific Publishers.

Pourriot, R., 1968. Rotifères du lac Tchad. Bull. IFAN, A, 52: 535–543.

Rigler, F. H., 1978. Limnology in the high Artic: a case study of Char Lake. Verh. int. Ver. Limnol. 20: 127–140.

Sarvala, J., 1978. An ecological energy budget of the Lake Paajarvi. Lammi Notes 1: 12–16.

Symoens, J. J., 1968. La minéralisation des eaux naturelles. Exploration Hydrobiologique du bassin du lac Bangweolo et du Luapula, Vol. 2, fasc. 1, 199 pp.

Talling, J. F. and Talling, I. B., 1965. The chemical composition of African lake waters. Int. Rev. Ges. Hydrobiol. 50: 241–263.

Thompson, K., 1975. Productivity of *Cyperus papyrus*. In Cooper, (ed.), Photosynthesis and Productivity in Different Environments, Cambridge University Press.

Thompson, K., 1976. The primary productivity of African wetlands with particular reference to the Okavango Delta. Proc. Symp. Okavango Delta: 67–79, Botswana, Gaborone.

Waters, T. F., 1977. Secondary production in inland waters. Advances in Ecological Research 10: 91–165.

Wetzel, R. G., 1979. The role of littoral zone and detritus in lake metabolism. Arch. Hydrobiol. Beih. 13: 145–161.

Winberg, G. G., 1972. Some interim results of Soviet IBP investigations on lakes. In Z. Kajak and A. Hillbricht-Ilkowska (eds.), Productivity Problems in Freshwaters pp. 363–381, Warsaw and Krakow, Polish Scientific Publishers.

Winberg, G. G., Babitsky, V. A., Gavrilov, S. A., Gladky, G. V., Zakharenkov, I. S., Kovalevskaya, R. Z., Mikheeva, T. M., Nevyadomskaya, P. S., Ostapenya, A. P., Petrovich, P. G., Potaenko, J. S. and Yakushko, O. F., 1972. Biological productivity of different types of lakes. In Z. Kajak and A. Hillbricht-Ilkowska (eds.), Productivity Problems of Freshwaters, pp. 383–404, Warsaw and Krakow, Polish Scientific Publishers.

Wissmar, R. C. and Wetzel, R. G., 1978. Analysis of five north American lake ecosystems. VI-Consumer community structure and production. Verh. int. Ver. Limnol. 20: 587–597.

Systematic index

J.-P. Carmouze et al. (eds.) Lake Chad
© *1983, Dr W. Junk Publishers, The Hague/Boston/Lancaster*
ISBN 978-94-009-7268-1

568

General index

J.-P. Carmouze et al. (eds.) Lake Chad
© *1983, Dr W. Junk Publishers, The Hague/Boston/Lancaster*
ISBN 978-94-009-7268-1